# Managing Forest Ecosystems

Volume 44

The aim of the book series Managing Forest Ecosystems is to present state-of-the-art research results relating to the practice of forest management. Contributions are solicited from prominent authors. Each reference book, monograph or proceedings volume will be focused to deal with a specific context. Typical issues of the series are: resource assessment techniques, evaluating sustainability for even-aged and uneven-aged forests, multi-objective management, predicting forest development, optimizing forest management, biodiversity management and monitoring, risk assessment and economic analysis.

Seppo Kellomäki

# Forest Management for Timber Production and Climate Change Mitigation

## Linking Dynamics of Carbon Cycle in Ecosystem Management

Springer

Seppo Kellomäki
School of Forest Sciences
University of Eastern Finland
Joensuu, Finland

ISSN 1568-1319 ISSN 2352-3956 (electronic)
Managing Forest Ecosystems
ISBN 978-3-031-71574-7 ISBN 978-3-031-71575-4 (eBook)
https://doi.org/10.1007/978-3-031-71575-4



This Springer imprint is published by the registered company Springer Nature Switzerland AG
The registered company address is: Gewerbestrasse 11, 6330 Cham, Switzerland

# Preface

Forested areas account for 4 billion hectares of land globally. This is about 30% of the total land area and contains a large part of the biomass in terrestrial ecosystems. Global forest cover is divided into three different forest biomes, namely tropical, temperate, and boreal forests extending from the Equator toward the Poles. Forest biomes are primarily distinguished by the thermal conditions they experience. This makes the global forest cover sensitive to global climatic warming and likely changes in precipitation. The effects of climatic warming are large at high latitudes in the Northern Hemisphere, where coniferous forest dominates. Regardless of the forest biomes, the management and harvest of forest resources are local, but their impacts on ecological systems through wood-based products and services are global. In this context, forestry produces tangible benefits (goods) like timber, and intangible benefits (services) like the absorption of atmospheric carbon to mitigate climatic warming.

In forestry, the production of goods and services is based on the management of forest structure toward a particular function. Forest management aims at controlling the genetic properties of tree populations, changing the properties of the habitats occupied by these populations, or both, to maintain or increase the production of goods and services. The physiological and ecological performance of tree populations (and populations of other species) is optimized to create ecosystem structures that help to produce the specified goods and services desired, like timber and carbon. The demand for wood-based products may rise in future through the increased use of wood in the construction and manufacturing industries. This implies that more carbon dioxide is stored in wood-based materials outside forests, thus mitigating warming by slowing down carbon dioxide emissions to the atmosphere.

The sustainable use of forests has been a major principle in forestry since the early 1950s but is still a challenge. In the context of forestry, sustainable management and use refers to the use of forests and forested land to maintain the biodiversity, productivity, and regeneration capacity of forest ecosystems. The aim is to meet the ecological, economic, and social functions of forests without damaging other ecosystems. International environmental and forest policy, as outlined in the Rio

Convention and the Kyoto Protocol, emphasizes the increasing role of forests in responding to the need to reduce the net carbon dioxide atmospheric emissions and conserve biodiversity. In this respect, there is a clear need to make forests capable of adapting to climate change, sustainably produce timber and other goods and services, sequester carbon, and conserve biodiversity. Forests and wood-based products provide a large carbon sink, thus removing carbon dioxide from the atmosphere in a cost-efficient way, as convincingly shown in the reports of the Intergovernmental Panel on Climate Change.

The Kyoto Protocol addresses forest management from two perspectives. Firstly, there is an increasing interest in developing improved forest management strategies that enhance carbon sequestration in forests (link to Article 3.4 of the Kyoto Protocol). Secondly, improper management techniques may result in a net release of carbon from the forest ecosystem, including from soil carbon stocks. In short, the main problem faced in the management of carbon resources in forest ecosystems is how to maintain and enhance the capacity to sequester and store atmospheric carbon in trees and forest ecosystems while managing forest resources for different purposes. Multifunctional forestry emphasizes the need for a balanced approach toward satisfying varying needs in a sustainable manner. The sequestration of carbon in forestry emphasizes the need for integrated forest production or multipurpose forestry to avoid potential problems and conflicts. Forest-based carbon sequestration reduces atmospheric carbon dioxide levels, thus mitigating climatic warming.

The carbon absorbed and stored in a boreal forest is affected by land use and land-use changes, management activities, disturbances, the use of harvested wood and biomass, and other uses of boreal forest resources. Carbon is stored in living trees, standing dead trees, downed wood, the forest understory, and soils. It can also be transferred among different carbon pools and to the atmosphere. Over time, forest landscapes change through succession driven by the structural and functional dynamics of separate tree stands and sites. In this succession, one population/community of plants and other organisms is gradually replaced by another population/community, with increasing complexity. The consequent changes in the primary production of trees and other plants further affect the herbivores and organisms of higher trophic levels that occupy these sites. In forest succession, living and non-living ecosystem components form the natural capital. This capital can be linked to carbon cycle dynamics through management to provide ecosystem goods and services for direct and indirect uses.

This book is written under the main heading: "Management of boreal forests for producing timber and climate change mitigating." Its specific aim is to link the carbon cycle dynamics in ecosystems with forest management to integrate timber production with carbon retention in ecosystems, thus mitigating climatic warming. The focus of the book is on forests and forestry in the boreal zone, but the discussion also utilizes findings for selected areas in the temperate zone. The book addresses carbon dynamics in boreal forest ecosystems in relation to the growth and development of forests under management, with impacts on the carbon uptake, emissions, and stocks in ecosystem structures. It discusses the potential of forest-based carbon for mitigating climatic warming based on the carbon retained in ecosystems and wood-based products outside ecosystems.

This book cites many studies on the likely effects of forest management operations and strategies on carbon dynamics in the harvest of timber and biomass for different purposes. Global-scale meta-analyses have been important for the creation of generalizable, model-based simulations over larger areas. These simulations are valuable for understanding the links between forest management and the structure and function of carbon in forest ecosystems. These links provide important information on how to control carbon fluxes and stocks to influence the interaction between forest ecosystems and the atmosphere toward mitigating climatic warming. In forestry, this information is valuable for developing and using forest management and harvesting methods that enhance carbon uptake and accumulation in forest ecosystems, which is necessary to mitigate climatic warming.

This book has 14 chapters, which are described briefly as follows: Chap. 1 addresses the main properties of global forests, while Chap. 2 focuses on boreal forests and their contribution to global terrestrial carbon storage. Chapters 3 and 4 address the carbon dynamics in natural and managed forests in the short- and long-term, with links to forest succession. This provides the background for Chap. 5, in which these carbon dynamics are related to the growth, development, and mortality of trees, and which addresses the impacts of management operations on carbon stocks. Thereafter, Chap. 6 further addresses the management operations and strategies for mitigating climate change based on carbon fluxes and stocks in the ecosystem. In this context, Chap. 7 analyzes the impacts of management on soil carbon, with a specific focus on reforestation. In Chaps. 8, 9, and 10, reforestation, pre-commercial management, and commercial management are discussed regarding the successional dynamics of managed forests, with further discussion in Chap. 11 on how forest fertilization affects the carbon sequestration and stocks in boreal forest ecosystems. Finally, Chaps. 12, 13, and 14 address the effects of forest management on climate benefits through sequestrating and retaining carbon in managed boreal forest ecosystems.

I acknowledge the support of my colleagues when compiling this book. Especially, the roles of Prof. Heli Peltola, Mr. Harri Strandman, and Mr. Veli-Pekka Ikonen have been decisive in different phases in preparing the book. I further acknowledge the School of Forest Sciences, University of Eastern Finland, for the facilities and support provided to this book project.

Joensuu, Finland
April 22, 2024

Seppo Kellomäki

# Disclaimer

The back cover text was generated using Augmented Insights, an internal Springer Nature tool built using Artificial Intelligence (generative pre-trained transformer approach) and underwent subsequent human editing to ensure accuracy.

# Contents

# About the Author

**Seppo Kellomäki** The research career of Prof. Seppo Kellomäki has spanned from the early 1970s to today. His research activities deal mainly with the structure and functioning of boreal forests and sustainable management of boreal forests for varying ecosystem goods and services. His early studies focused on the multiple uses of forest and management of Scots pine forest for high-quality timber. However, his main contribution to forest science deals with climate change impacts on the dynamics of the boreal forest ecosystem in conditions where forests and forest-based industry are important for the welfare of society. In this context, his focus has been on experimental studies on how the elevating temperature, atmospheric carbon dioxide, and changes in precipitation likely affect the physiology and growth of boreal tree species. These studies have provided a solid base for modelling. This has made it possible to understand how climate change likely affects the long-term dynamics of the boreal forest ecosystem and the future of forest management. The main questions have been how management should be used and developed to avoid detrimental effects of successional dynamics and management of boreal forest ecosystem for different ecosystem services. Many of his research papers further address the problems of how forest management and forests should be adapted to climate warming in the boreal conditions, and how management might be used to mitigate warming.

# Chapter 1
# Forests and Global Change

**Contents**




**Abstract** This book outlines forest management approaches for integrating timber production and carbon retention in ecosystems to mitigate climatic warming. Carbon dynamics in boreal forest ecosystems are addressed in relation to the growth and development of trees under management, with impacts on carbon uptake, emissions, and storage in ecosystem structures. Furthermore, carbon dynamics affect the properties of forest ecosystems with high variability across different habitats, thus maintaining and increasing the species and genetic richness in natural and managed forests. There is a need to integrate different management objects to satisfy various needs, such as timber production, climatic warming mitigation, and forest-based biodiversity conservation, thus linking forest management and harvest with ecosystem carbon dynamics.


## 1.1 Forests and Global Change

### *1.1.1 Forests: Resources for Many Purposes*

Forests are a major renewable and multifunctional natural resource and make up the largest terrestrial ecosystems in the world. The concept of multiple-use forestry, or forest production, has widely been used to emphasize (i) how forests can be managed for different purposes; (ii) how various products and services are valued; and (iii)




S. Kellomäki, *Forest Management for Timber Production and Climate Change Mitigation*, Managing Forest Ecosystems 44,
https://doi.org/10.1007/978-3-031-71575-4_1

how management should be organized in forestry. Forest production involves the direct and indirect use of the forest environment, using materials and resources available in the forest ecosystems to manufacture different goods and provide diverse services. In forestry, management and harvest operations are local, whereas the impacts of forest-based products and services on ecological systems are global.

Forestry produces tangible benefits like timber, energy biomass, and habitat for wildlife. Forests also provide intangible benefits like scenic beauty and the absorption of atmospheric carbon to mitigate climatic warming. Forest-based production is commonly integrated, i.e., the same forested land may be used for timber production and game management, or for producing other environmental benefits such as the conservation of biodiversity. Therefore, forests have ecological, social, economic, and cultural functions for people. These main dimensions are the basis of sustainable forest management, which aims at providing future generations with the same goods and services to which the current generation has access.

Globally, forested areas comprise four billion hectares of land. This is about 30% of the total land area and contains a large part of the biomass in terrestrial ecosystems. Forests are present in tropical, temperate, and boreal biomes or formations from the Equator toward the Poles. These forest formations are distinguished by temperature and water conditions. Consequently, climatic warming has large effects on global forest cover, especially at high latitudes in the Northern Hemisphere, with the dominance of boreal forests. They retain large amounts of carbon, especially in soils, with important contributions toward (i) mitigating climatic warming, and (ii) providing tangible and intangible products and services. In this regard, the structure of forest ecosystems is controlled in management to make the ecosystem function to produce desired goods and services. Proper management of the genetic properties of tree populations, the properties of sites/habitats occupied by these populations, or both, maintains or increases production of these goods and services sustainably. In management and harvest, the physiological and ecological performance of tree populations (and populations of other species) is directed towards creating ecosystem structures that are needed to produce the specified goods and services Maintaining the multiple roles of boreal forests is key for the sustainable management of boreal forests (Kellomäki 2009).

### *1.1.2 Global Forest Loss, Carbon Dioxide Emissions, and Climate Change*

Global deforestation is an important cause of the increasing carbon emissions driving climate change. Forest loss dates back thousands of years, but the enhancing forest loss is related to the increasing growth of human population and increasing demand for agricultural land and forest-based products. Currently, deforestation is low in the boreal climatic domains but high in the tropical climatic domain, with the result that forest area is still reducing globally. The Global Forest Resources

Assessment (FRA), coordinated by the Food and Agriculture Organization of the United Nations (FAO), found that the world's forest area decreased from 31.6% to 30.6% of global land area between 1990 and 2015, but that the rate of loss has slowed in recent years (FAO 2016). All forests have an important role in the conservation of biodiversity. However, deforestation of tropical forests has been of major concern, as the remaining primary tropical forests are estimated to contain 50–70% of all terrestrial plant and animal species in the world. Tropical deforestation and degradation of all types of forests are among the main causes of global biodiversity loss.

In addition to being major causes of global biodiversity loss, deforestation and related land-use changes are estimated to contribute approximately 20–25% of global total carbon emissions at present. Deforestation is thus the common denominator for biodiversity loss and climate change, as emphasized in the United Nations Framework Convention on Climate Change (FCCC), signed at the United Nations Conference on Environment and Development (UNCED) in Rio de Janeiro in 1992. The FCCC established the legal basis for a system of international cooperation for the reduction of greenhouse gas emissions to mitigate climate change. The detailed provisions and obligations were concluded in the Kyoto Protocol in 1997. The Kyoto Protocol came into force in 2005. Thereafter, its content was specified in a series of the Conferences of the Parties (COP) following the increasing knowledge on the likely changes in climate and global societal development.

### *1.1.3 Links of Climate Change to Forested Land Use*

Forest-based carbon sequestration decreases atmospheric carbon dioxide ($CO_2$) levels, thus mitigating climatic warming. The production of wood-based products may be increased further by increasing the use of wood in the construction and manufacturing industries. This implies sequestering carbon in wood-based materials outside forests for climatic warming mitigation, thus slowing down carbon dioxide emissions to the atmosphere. Forests and wood-based products provide a large carbon sink, thus removing carbon dioxide from the atmosphere in a cost-efficient way, as convincingly shown in the reports of the Intergovernmental Panel on Climate Change (IPCC) (e.g., IPCC 2014).

The IPCC is an intergovernmental body of the United Nations (UN) that was established in 1988 by the World Meteorological Organization (WMO) and the United Nations Environment Programme (UNEP). It is open to all the members of the WMO and UN. The IPCC provides objective, science-based assessments on (i) how the global climate is likely to change; (ii) how to respond to the changes to mitigate climatic warming; and (iii) how to adapt to changing environmental conditions. In short, the IPCC assesses the likely impacts of climate change on natural, political, and economic systems and on the risks to nature and humans, and develops appropriate responses for mitigating and adapting to climate change. Although the IPCC does not carry out original research, nor monitor climate or related phenomena, its assessments are based on published scientific literature, including peer-reviewed and non-peer-reviewed literature.

The IPCC reports to the United Nations Framework Convention on Climate Change (UNFCCC), which is committed to "stabilizing greenhouse gas concentrations in the atmosphere at a level that would prevent dangerous anthropogenic (human-induced) interference with the climate system." The IPCC estimates that land-use change, including conversion of forests into agricultural land, contributes a net increase of $1.6 \pm 0.8$ Gt carbon per year to the atmosphere. The Kyoto reduction targets imply that signatory countries must account for changes in carbon stocks in land use, land-use change, and forestry (LULUCF). The LULUCF findings are important for forest-rich countries that are committed to following the rules of the Kyoto Protocol. The national Panels monitor and assess the implementation of the LULUCF suggestions for mitigating climate change in the context of forestry. Wherever the management of timber production, carbon sequestration, and conservation of biodiversity are among the main forest-based services.

### *1.1.4 Carbon Dioxide and Other Atmospheric Greenhouse Gases*

The balance between $CO_2$ uptake and emission is disturbed by the burning of fossil fuels and the decay of biomass and organic materials based on net primary production. Carbon emissions are further related to land use changes, including the conversion of forested land into agricultural land and pastures (deforestation), and the building of urban development infrastructure such as houses, roads, and sport fields. The devastation of forests further reduces $CO_2$ uptake and enhances the decay and oxidation carbon compounds in soil organic matter (SOM). $CO_2$ emissions are an indicator for secondary production through the decay of detritus into $CO_2$ in aerobic conditions, and into methane ($NH_4$) and nitrous oxide ($N_2O$) in anaerobic conditions, like in peatlands (Fig. 1.1).

Atmospheric $CO_2$ increased from 280 ppm (parts per million) in the year 1850 to 410 ppm by the year 2020. The increase in atmospheric $CO_2$ has originated largely (80%) from the combustion of fossil fuels (coal, oil) and the production of cement, while the rest (20%) has originated from land use. The lifetime of $CO_2$ in the atmosphere is 30–95 years (e.g., Sathre et al. 2013) (Fig. 1.1). The lifetime of other greenhouse gases is longer (e.g., 120 years for $N_2O$) than that of $CO_2$. The lifetime indicates the time over which equilibrium in the given $CO_2$ (or other greenhouse gases, GHGs) concentration is achieved after a sudden increase/decrease in concentration (Fig. 1.1, insert). Molecules of GHGs are removed from the atmosphere in deposition and adsorption on the surfaces of land and water. On the other hand, molecules of $CO_2$ are absorbed in photosynthesis and bound in biomass.

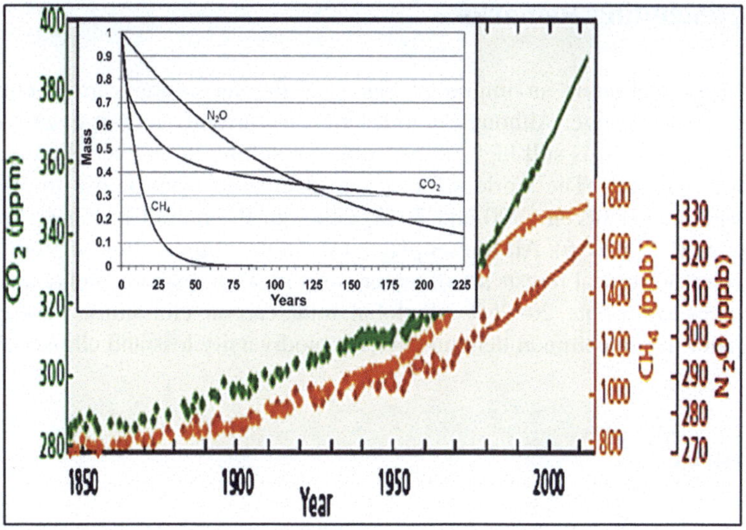


**Fig. 1.1** Concentration of greenhouse gases in the atmosphere from 1850 to 2010, with green indicating carbon dioxide ($CO_2$), orange indicating methane ($NH_4$), and red indicating nitrous oxide ($N_2O$). The values are based on ice cores (dots) and direct atmospheric measurements (lines) (IPCC 2014, Fig. SPM 1, p. 3). Permission of IPCC. Insert: Decay of pulse emission of $CO_2$, $N_2O$, and $CH_4$ in the atmosphere in natural processes (Sathre et al. 2013). Courtesy of Springer Nature

## 1.2 Scope of This Book

This book will outline forest management approaches for integrating timber production and carbon retention in ecosystems to mitigate climatic warming. The focus of the book is on forests and forestry in the boreal zone, but the discussion is extended to cover selected parts of temperate zone wherever appropriate to scale local findings into a larger context. Carbon dynamics in boreal forest ecosystems are discussed in relation to growth and development under management, with impacts on carbon uptake, emissions, and storage in ecosystem structures. To a smaller extent, the potential of forest-based carbon to mitigate climatic warming based on the carbon retained in ecosystems and wood-based products outside ecosystems is discussed. Furthermore, carbon dynamics affect the properties of forest ecosystems with high variability across different habitats, thus maintaining and increasing the species and genetic richness in natural and managed forests. In multifunctional forestry, there is a need to integrate different management objects to satisfy various needs, like the production of timber, mitigation of climatic warming, and conservation of forest-based biodiversity.

## 1.3 Concluding Remarks

Global deforestation is an important cause of the increasing carbon emissions driving climate change. Although deforestation is currently low or even reducing in boreal domains, it is still high in the tropics, resulting in an overall reduction in global forested area. The world's forest area decreased from 31.6% to 30.6% of global land area between 1990 and 2015, although the rate of loss has slowed in recent years (FAO 2016). All forests play a role in the conservation of biodiversity, but the roles of tropical forests are of major concern. Deforestation and related land-use changes contribute 20–25% of global total carbon emissions. Furthermore, deforestation is the common denominator for biodiversity loss and climate change.

## References

FAO (2016) Global Forest Resources Assessment 2015: How are the world's forests changing? 2nd edn. Food and Agriculture Organization of the United Nations, Rome

IPCC (2014) Climate change 2014: Synthesis report. Contribution of Working Groups I, II and III to the fifth assessment report of the intergovernmental panel on climate change [Core Writing Team, R.K. Pachauri, and L.A. Meyer (eds.)]. IPCC, Geneva, Switzerland, 151 pp

Kellomäki S (ed) (2009) Forest resources and sustainable management, 2nd edn. Paper Engineers' Association/Paperi ja Puu Oy, Jyväskylä

Sathre R, Gustavsson L, Haus S (2013) Time dynamics and radiative forcing of forest bioenergy systems. In: Kellomäki S, Kilpeläinen A, Alam A (eds) Forest bioenergy production. Springer Science + Business Media, New York, pp 185–206

# Part I
# Boreal Forests and Carbon Dynamics in Forest Ecosystems

# Chapter 2
# Global Forests, with a Focus on Boreal Forests

## Contents




**Abstract** Globally, forests cover about 30% of the world's land area but contain nearly 80% of the total terrestrial biomass. In humid conditions, annual dry matter production increases with increasing temperature from less than 2 Mg ha$^{-1}$ yr.$^{-1}$ in the boreal zone to more than 20 Mg ha$^{-1}$ yr.$^{-1}$ in the tropical zone. On a global scale, the total amount of carbon in forest ecosystems is 758 Gt. Boreal forests extend over Canada, China, Finland, Japan, Norway, Russia, Sweden, and the United States, where there are still large areas of unmanaged forest across the high-latitude regions. The mean carbon stock in boreal forests is 60–100 Mg C ha$^{-1}$, i.e., 60–70% in the belowground biomass (roots and dead biomass), and the rest in the aboveground in living biomass of trees, shrubs, and ground vegetation. The mean carbon residence time in boreal forest is 100 years, but the decay rate varies depending on the temperature and moisture conditions.





S. Kellomäki, *Forest Management for Timber Production and Climate Change Mitigation*, Managing Forest Ecosystems 44,
https://doi.org/10.1007/978-3-031-71575-4_2

## 2.1 Carbon Sequestration and Carbon Stocks in Global Forests

### 2.1.1 *Forest Area and Carbon Stocks in Different Climate Domains*

Forests cover about 30% of the world's land area (Table 2.1) but contain nearly 80% of the total terrestrial biomass. Over 40% of forests are in the tropics, one-quarter in the temperate zone, and one-third in the boreal zone. Deserts and semi-deserts cover about 30% of the world's land area, but desert vegetation represents less than 2% of the total terrestrial biomass. These numbers are based on the IPCC Special Report on Land Use, Land-use Change, and Forestry, published in 2000 (IPCC 2000, Table 1, p. 4). These values may deviate from those of the FAO, which are used later, but the magnitude is the same in both cases. In particular, variability in the area of the climatic domains is responsible for differences between estimates published in different studies. Globally, the annual harvest of industrial round wood from boreal forests is 500–600 million $m^3$ (Fischlin et al. 2009), which is 35–40% of the global round wood harvest.

**Table 2.1** Distribution of the world's land area and carbon stocks in vegetation and soils (to a depth of 1 meter in soils)

| | | Carbon stock (Gt C) | | |
|---|---|---|---|---|
| Climatic domain | Area ($10^9$ ha) | Vegetation | Soil | Total |
| Forests | | | | |
| • Tropical forests | 1.76 (42) | 212 (59) | 216 (27) | 428 (27) |
| • Temperate forests | 1.04 (25) | 59 (16) | 100 (13) | 159 (14) |
| • Boreal forests | 1.37 (33) | 88 (25) | 471 (60) | 559 (49) |
| Total | 4.17 (100) | 359 (100) | 787 (100) | 1146 (100) |
| Other land types | | | | |
| • Tropical savannahs | 2.25 | 66 | 264 | 330 |
| • Temperate grasslands | 1.25 | 9 | 295 | 304 |
| • Deserts and semideserts | 4.55 | 8 | 191 | 199 |
| • Tundra | 0.95 | 6 | 121 | 127 |
| • Wetlands | 0.35 | 15 | 225 | 240 |
| • Croplands | 1.60 | 3 | 128 | 131 |
| Total | 10.95 | 107 | 1224 | 1331 |
| Grand total | 15.12 | 466 | 2011 | 2477 |

The numbers in parentheses indicate the percentage (%) of area and carbon stock in vegetation and soil per climatic domain relative to the total global values (IPCC 2000, Table 1, p. 4)

### 2.1.2 *Growth, Biomass, and Carbon Stocks in Different Climate Domains*

The productivity of vegetation in different climatic zones depends on the variability of temperature and precipitation. In humid conditions, the annual dry matter production increases with increasing temperature from less than 2 Mg ha$^{-1}$ yr.$^{-1}$ in the boreal zone to more than 20 Mg ha$^{-1}$ yr.$^{-1}$ in the tropical zone (Aber and Melillo 1991). Similarly, the biomass per unit area increases with increasingly favorable temperature conditions from about 120 Mg ha$^{-1}$ in boreal forests to 400 Mg ha$^{-1}$ in tropical rainforests. The amount of soil organic matter is higher in the boreal zone, at 400–1000 Mg ha$^{-1}$, than in the tropics, at 250–300 Mg ha$^{-1}$. This reflects the faster decay of dead plant material in warm and humid conditions. On the other hand, a scarce supply of water substantially reduces the productivity and biomass per unit area even in warm conditions. The difference in productivity between humid and semiarid regions is not as large as the difference in biomass per unit area. In these conditions, the role of herbs and grasses, which undergo rapid decay of biomass, is more important than that of woody species, in which biomass decays slowly (Aber and Melillo 1991).

Since the early 1990s, global forest resources assessment (FRA) has been carried out by the Food and Agriculture Organization of the United Nations (FAO). According to FRA 2015, the worldwide carbon stock in trees is 296 Gt, including 234 Gt carbon in aboveground biomass and 62 Gt in belowground biomass. Forests contain a further 41 Gt carbon in dead trees, 23 Gt in litter, and 398 Gt carbon in soil. Globally, the total amount of carbon in forest ecosystems is 758 Gt. The total carbon in the above- and belowground biomass of trees is nearly half of the total carbon in forests. However, carbon in the biomass and soil varies substantially between climatic domains (Table 2.1). The carbon density of biomass is highest in South America and Western and Central Africa. In these areas, the mean carbon stock is 120 Mg ha$^{-1}$, which is more than one and a half times larger than the global mean (74 Mg ha$^{-1}$). Along with the reduction of forest cover, the carbon in forest biomass has decreased 442 million Mg per year since 1990, which is equal to 1.6 Gt of carbon dioxide. This reduction in forest carbon is mainly related to the conversion of forest land to pastures and agriculture land and the degradation of forested land.

**Box 2.1 FAO concepts used in inventories of global forest and forest resources**

Boreal forest is controlled by several sovereign countries, making it difficult to obtain precise statistics on boreal forest resources. The concept of "forest" is vague, making it difficult to estimate forest resources. In the context of the Food and Agriculture Organization of the United Nations (FAO), land containing trees and other woody plants are classified as "forest" if an

(continued)

**Box 2.1** (continued)
area ≥ 0.5 hectares is occupied by trees taller than 5 m, with total canopy cover ≥10% possible to reach in situ. In the global forest inventories, the FAO further uses the concepts "other wooded land" and "other land." "Other wooded land" is land ≥0.5 hectares, with trees higher than 5 m and a canopy cover of 5–10%, or trees that can reach these thresholds in situ, or with a combined cover of trees and shrubs ≥10%, excluding land under agricultural or urban land use. Land that does not meet the criteria for "forest" or "other wooded land" is "other land." This is land spanning more than 0.5 hectares, with a canopy cover ≥10% of trees able to reach a height of 5 m at maturity (FAO 2006, 2016).

### 2.1.3 *Changes in Global Forest Cover*

Land use and changes in land cover have a large effect on carbon sequestration, or uptake and stocks, in the global carbon cycle. The stocks of carbon bound in soil exceed that in biomass and the atmosphere. Using the IPCC climatic regions, carbon stocks are largest at higher latitudes in the Northern Hemisphere (Table 2.1), including in boreal forests and forest tundra, the latter representing permafrost. In these areas, the carbon stock in biomass is small compared to that in the tropics, where the carbon in biomass substantially exceeds that in soil. However, carbon loss from forest ecosystems has been large in Africa, South America, and South and Southeast Asia (Fig. 2.1). At the same time, the stock of forest carbon has increased in Europe, East Asia, and North America. Over the period 1990–2015, the net losses of forest carbon decreased from 0.5 Gt per year in the 1990s to 0.3 Gt per year from 2010 to 2015. This reduction has been especially pronounced in South and Central America and in Asia but is slowing with the slowing loss of forest cover. According

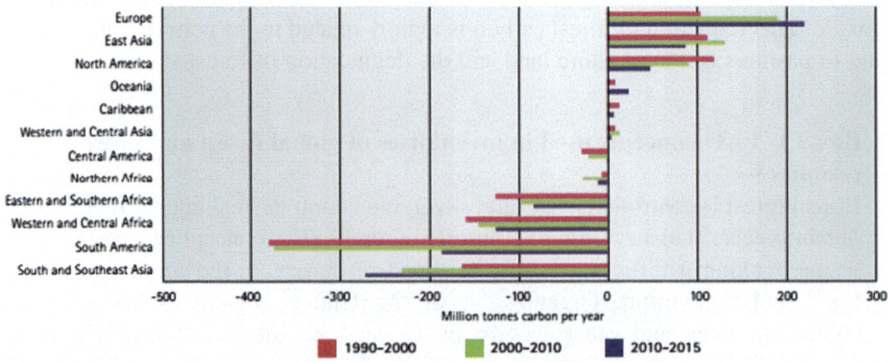


**Fig. 2.1** Gain and loss of carbon in forest biomass in selected global subregions in the periods 1990–2000, 2000–2010, and 2010–2015 (FAO 2016). Courtesy of FAO

to FRA 2015, "Brazil alone reported that the annual loss of carbon in above- and below-ground biomass was reduced from 240 million tonnes of carbon per year in the 1990s to about 80 million tonnes per year for 2010–2015" (FAO 2016). The FAO expects that the current global trend of decreasing carbon stocks is likely to continue, but losses might further be reduced.

## 2.2 Main Properties and Ecosystem Services of Boreal Forests

### *2.2.1 Bioclimatic Limits and Geographical Distribution*

Boreal forests are located around the Northern Hemisphere, adjacent to the tundra, and only patches of boreal forests are located further south at high altitudes (Burton et al. 2010). Although forests at higher elevations in Kazakhstan, Mongolia, China, and Japan have a boreal character, the total area of boreal forest in these countries is small. Boreal forests account for 30% of the global forest area, including peatlands (bogs, mires) and permafrost soils, of which half is covered by forest (DeLuca and Boisvenue 2012). Boreal landscapes are mixtures of closed forests and open woodlands, which split by bogs, mires, lakes, and rivers into separate patches of varying size. The boreal landscape is thus characterized by a great variability of habitats. The forests themselves are a mosaic of successional and subclimax plant communities, which are susceptible to varying disturbances (e.g., fire, wind, and snow).

Boreal forest (forests and woodlands) is circumpolar at high latitudes (45–70°), extending over Canada, China, Finland, Japan, Norway, Russia, Sweden, and the United States, and comprising about 30% of the global forest area (Fig. 2.2). Boreal forests are characterized by the dominance of coniferous species, but these are mixed with deciduous species on fertile sites (Bonan and Shugart 1989; IBFRA 2020). The northern boreal forests is limited by treeless tundra, whereas the southern boreal forest gradually changes to temperate deciduous forest and even to grasslands/steppes in the most continental parts of North America and Eurasia (Sirois 1992). The limits of the treeless tundra are undefined, especially in Alaska, Canada, and Siberia. In these regions, large areas between the treeless tundra and the boreal forest are forest tundra, where the soil temperature regularly remains below 0 °C in summer. This is due to the cold winters and thin snow cover that allow deep soil frost (Walter 1979). Forest tundra is especially vulnerable to climate warming, which is causing boreal forest to move northwards. The southern boreal forest is also vulnerable to a climate, with its replacement by temperate forest or even grasslands/steppes (IPCC 2007a, b).

Boreal forests fall between the summer and winter limits of the arctic air mass (Gordon et al. 1989), in areas characterized by annual mean temperatures of −5 °C to +5 °C, although this may fall to −10 °C in the continental parts (Fig. 2.2). The northern limit of forest cover is set by the 10 °C July isotherm, whereas the southern

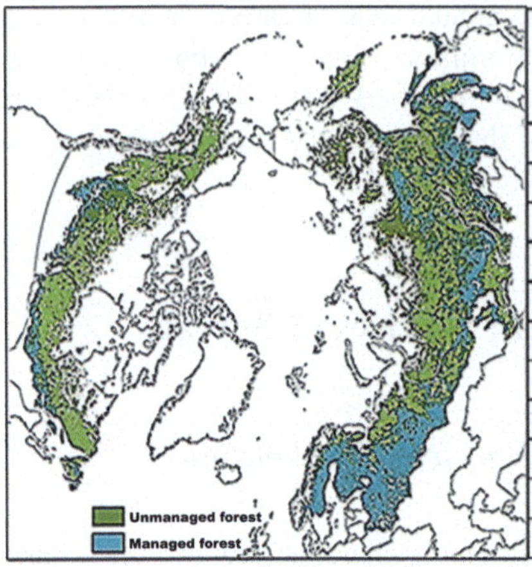


| Country | Boreal land area, km² | %, of total land area in boreal | Area in boreal, mill. ha | %, of total forest in boreal | Forest area, % of boreal land area |
|---|---|---|---|---|---|
| China | 140 300 | 18 | na | na | na |
| Canada | 54 446 170 | 56 | 224.0 | 72 | 41 |
| Finland | 293 321 | 97 | 19.5 | 98 | 66 |
| Iceland | 1 340 | 1.3 | 0.03 | 100 | 23 |
| Kazakhstan | 15 000 | ~1 | na | na | na |
| Mongolia | 63 000 | 4 | ~1.2 | 12 | 20 |
| Norway | 270 752 | 84 | 5.1 | 77 | 1 |
| Russia | 12 484 890 | 73 | 673.0 | 85 | 54 |
| Sweden | 306 260 | 74 | 18.0 | 75 | 55 |
| USA/Alaska | 656 600 | 41 | 17.0 | 33 | 26 |
| World | 19 678 000 | | 957.0 | | 49 |

**Fig. 2.2** Distribution of global boreal forests. Left: Boreal forests around the northern hemisphere divided into unmanaged and managed forests (IBFRA 2020). Courtesy of IBFRA. Right: Key statistics on boreal forests around the world (Burton et al. 2010). Courtesy of IUFRO

limit is set by the 18 °C July isotherm. In this range, the climate is characterized by short summers; the number of days with a daily mean temperature ≥ 10 °C is 30–120 per year. In the most climatically extreme areas bordering the tundra, the soil may be frozen throughout the year (permafrost). Winters are severe, with a mean daily temperature < 0 °C for five to seven months. The within-year temperature variability is also large, from −50 °C to +40 °C, and the variability of the mean daily summer temperature is 15–20 °C. The mean maximum temperature of the warmest summer month is more than 10 °C, and the duration of summer is no longer than four months.

In boreal forests, the mean annual precipitation is 200–750 mm, excluding most maritime areas (e.g., western parts of the North American boreal forest), where the annual precipitation may exceed 1000 mm. In boreal forests, the climate is humid, and precipitation exceeds evaporation (e.g., in the Nordic countries). In the most continental areas of boreal forest (e.g., in Siberia), the climate is subhumid or semiarid due to limited precipitation. Under humid conditions, the most common soil formation is podsolization due to the acidic soil solution produced under a coniferous canopy. In many areas of boreal forest (e.g., in Nordic countries), soils are still young due to continental glaciation, which disappeared only 10,000–15,000 years ago.

### 2.2.2 *Tree Species Composition*

In boreal forests, most tree genera occur throughout the boreal zone, and they are characterized by a low diversity of species. Gymnosperm species such as pines (*Pinus*), spruces (*Picea*), firs (*Abies*), larches (*Larix*), junipers (*Juniperus*), thujas

(*Thuja*), and hemlocks (*Tsuga*) are common, while the most common deciduous species include aspens (*Populus*), birches (*Betula*), willows (*Salix*), and alders (*Alnus*) (Table 2.2). Most boreal tree genera represent transcontinental distributions across Eurasia or North America (Nikolov and Helmisaari 1992). The number of coniferous species is largest in North America and in the southern parts of East Asia. The number of tree species is particularly small in the north-western areas of Eurasia, where Scots pine (*Pinus sylvestris*) and Norway spruce (*Picea abies*) dominate the forested landscapes. Large parts of boreal forests are still natural (unmanaged) across the high-latitude regions in Canada, Russia, and the United States.

### 2.2.3 Forest Resources and Industrial Stem Wood Production

Globally, industrial round wood harvest (timber) and biomass, carbon sequestration, and biodiversity are the main services provided by boreal forests (e.g., Fischlin et al. 2009). This implies that more than 33% of lumber and 25% of paper on the export market originate from boreal regions. Furthermore, boreal forests regulate climate through the exchange of energy and water. They also function as a large reservoir of biogenic carbon, on a level comparable to that of tropical forests, thus helping to mitigate climatic warming. Furthermore, boreal forests provide ecosystem services, such as facilitating hunting, berry and fungi, foraging, leisure, and spiritual activities, and providing medicines and economic opportunities to local, regional, and global populations.

Wulder et al. (2007) estimated the total area of boreal forest and woodland to be 1960 million ha, including 1105 million ha of proper forests. Based on Kuusela (1990) and global forestry statistics (FAO 2006, 2016), global boreal forest area is 1470–1640 million ha, in North America (Canada, Alaska), the Nordic countries (Finland, Sweden, Norway, Iceland), and Russia. Of this, two thirds of all boreal forest is in Eurasia, mostly in Russia (69% of the total global boreal forest area). Globally, mean stem wood stock is about 120 $m^3$ $ha^{-1}$, but this varies from region to region (Table 2.3). In the Nordic countries, for example, the mean wood stock is about 90 $m^3$ $ha^{-1}$. These forests have been intensively harvested and managed for timber production for decades, and even for hundreds of years. In Alaska, the mean wood stock is about 150 $m^3$ $ha^{-1}$, including large areas of old-growth forest left outside harvest and management. Regardless of boreal region, the mean annual growth (0.8–3.3 $m^3$ $ha^{-1}$ $yr.^{-1}$) exceeded the mean annual cutting (0.06–1.7 $m^3$ $ha^{-1}$ $yr.^{-1}$) in the late 1970s. The mean annual cutting was also lower than mean annual growth by the early 1980s (Kuusela 1990). For boreal forests, growth implies an increase of stocking by 1300 million $m^3$ $yr.^{-1}$. This is approximately 30% of global forest growth. Globally, the annual industrial round wood harvest from boreal forests is 500–600 million $m^3$, providing 35–40% of the global round wood harvest. The boreal forests in northern Europe or Fennoscandia (including Norway, Sweden, Finland, and north-western Russia) provide about 40% of the timber used in Europe (Fischlin et al. 2009).

**Table 2.2** Main coniferous and broad-leaved boreal tree species (Kuusela 1990)

| Coniferous/North America | | Coniferous/Eurasia | |
|---|---|---|---|
| *Pinus contorta* Dougl. | Lodgepole pine | *Pinus sylvestris* L. | Scots pine |
| *Pinus banksiana* Lamb. | Jack pine | *Pinus cembra* L. var. *sibirica* | Siberian cembra |
| *Pinus resinosa* Ait. | Red pine | *Pinus pumila* Rig. | Japanese stone pine |
| *Pinus strobus* L. | Eastern white pine | *Larix sibirica* Ledeb. | Siberian larch |
| *Pinus rigida* Mill. | Pitch pine | *Larix gmelinii* (Rup.) Litv. | Dahurian larch |
| *Larix laricina* (Du Roi) K. | Tamarack | *Picea abies* (L.) Karst. | Norway spruce |
| *Picea mariana* (Mill.) B.S.P. | Black spruce | *Picea abies* var. *obovata* Ledeb. | Siberian spruce |
| *Picea glauca* (Moench) Voss | White spruce | *Picea jezoensis* (Sieb. & Zucc.) Carr | Yeddo spruce |
| *Picea rubens* Sarg. | Red spruce | *Picea glehnii* (Fr. Schmidt) Mast. | Sakhalin spruce |
| *Tsuga canadensis* (L.) Carr. | Eastern hemlock | *Abies sibirica* Ledeb. | Siberian fir |
| *Thuja occidentalis* (L.) | Eastern white cedar | *Abies nephrolepis* Maxim. | Khingan fir |
| *Abies balsamea* (L.) Mill. | Balsam fir | *Abies sachalinensis* Mast. | Sakhalin fir |
| | | *Abies gracilis* Komar. | Kamchatka fir |
| Broad-leaved/North America | | Bread-leaved/Eurasia | |
| *Alnus rugosa* (Du Roi) Spreng. | Hazel alder | *Alnus incana* L. | Grey alder |
| *Alnus incana* L. Moench | Speckled alder | *Alnus hirsute* Tures. | Manchurian alder |
| *Betula neoalaskana* Sarg. | Alaskan birch | *Alnus kamschatica* Call. Kam. | Creeping alder |
| *Betula papyrifera* Marsh. | Paper birch | *Betula pendula* Roth. | White birch |
| *Betula occidentalis* Hook. | Water birch | *Betula pubescens* Ehrh. | Downy birch |
| *Betula alleghaniensis* Britton | Yellow birch | *Betula platyphylla* Sukacz. | Asian white birch |
| *Populus balsamifera* L. | Balsam poplar | *Betula ermanii* Cham. | Erman's birch |
| *Populus tremuloides* Michx. | Trembling aspen | *Betula davurica* Patt. | Dahurian birch |
| *Populus deltoides* Bartr. | Eastern cottonwood | *Betula japonica* Sieb. | Japanese white birch |
| *Populus grandidentata* Michx. | Largetooth aspen | *Betula costata* Trautv. | Yellow birch |
| | | *Populus tremula* L. | Aspen |
| | | *Populus suaveolens* Fisch. ex Poit. & A. Vilm. | Mongolian poplar |
| | | *Populus sieboldii* Miq. | Siebold aspen |

**Table 2.3** Forest resources in the boreal regions and their share of global totals (Kuusela 1990; FAO 2006, 2016)

| Region | Total values | | | | Mean values | | |
|---|---|---|---|---|---|---|---|
| | Forest area (million ha) | Growing stock (million $m^3$) | Annual growth, 1000 $m^3$ | Annual removal, 1000 $m^3$ | Stock ($m^3$ $ha^{-1}$) | Annual growth ($m^3$ $ha^{-1}$ $yr.^{-1}$) | Removal ($m^3$ $ha^{-1}$ $yr.^{-1}$) |
| Alaska | 48 | 1300 | 3600 | 3000 | 160 | 0.8 | 0.06 |
| Canada | 436 | 23,000 | 356,000 | 152,000 | 110 | 1.7 | 0.3 |
| Nordic countries | 60 | 4500 | 157,000 | 102,000 | 90 | 3.3 | 1.7 |
| Russia | 930 | 67,000 | 750,000 | 357,000 | 130 | 1.4 | 0.4 |
| Total | 1474 | 96,000 | 1,270,000 | 614,000 | 120 | 1.6 | 0.4 |

Stock, growth, and removal refer to stem wood

Large parts (50%) of boreal forests are still outside commercial use and regular management (Olsson 2011). In Canada, for example, 80% of the boreal forest is still untouched by settlements or roads (Ruckstuhl et al. 2008). This is not true for the Nordic countries, where most of the forested land (locally up to 90%) is used and managed for timber and biomass. Even the northern boreal forests are used in forestry under conditions where climatic warming will be pronounced by the end of this century. In this context, there are still many open questions concerning how managed boreal forests can be adapted to climate change (e.g., Innes et al. 2009), including through tree species choice and planting practices, tending of seedling stands and proper tree species mixtures, thinning practices, and the source/sink dynamics in carbon sequestration for climate change mitigation (Kellomäki et al. 2005, 2008).

### 2.2.4 *Carbon Sequestration and Carbon Stocks*

Keith et al. (2009) estimated that the mean carbon stock in boreal forests is 60–100 Mg C $ha^{-1}$, comprising 60–70% in the belowground biomass (roots and dead biomass), and the rest in the aboveground living biomass. They estimated that the mean carbon residence time in boreal forests is 100 years, although the decay rate varies depending on the temperature and moisture conditions (Raich and Schlesinger 1992). In this regard, Fischlin et al. (2007) estimated that boreal forests worldwide contain 703 Pg ($10^{15}$ g) carbon (trees, soils), which is 13% of the carbon in forests globally. Most carbon (70–90%) is in soils that are cold, facilitating only slow decay and the consequent development of deep organic soils. This is especially the case in peatlands, where waterlogged conditions further reduce decay. Globally, boreal peatlands contain 400–500 Pg of carbon, which is one third of the world's organic soil carbon (e.g., Gorham 1991; Apps et al. 1993) or the equivalent of 40 ppm in atmospheric $CO_2$ concentration (Moore et al. 1998).

In boreal forests, the mean carbon budgets include net sinks in the range of 0.5–2.5 Mg C $ha^{-1}$ $yr.^{-1}$ (Shvidenko and Nilsson 2003), while global boreal forests are a carbon sink of 0.50 ± 0.08 Pg C $yr.^{-1}$ (Pan et al. 2011). The estimates are still uncertain, as shown by Kurz and Apps (1999) for Canadian forests. They found that the forests represented a carbon sink (0.075 Gt C $yr.^{-1}$) from 1920 to 1970, which was changed to a carbon source of 0.050 Gt C $yr.^{-1}$ in 1994, when changes due to disturbances (fire, harvest) were included in the estimates. The exclusion of management effects may also bias the estimates of the carbon sink/source relations. Intensive management may increase carbon uptake more than it increases emissions, whereas less intensive management may create the opposite situation (Shepashenko et al. 1998). Factors such as nitrogen deposition, atmospheric $CO_2$ fertilization, and climatic warming, with a consequent increase in growth, have not been included in the carbon sink/source analyses. These factors may increase carbon uptake and thus offset losses of carbon due to disturbances and cuttings (Schimel et al. 2001). The recent global estimates of carbon stocks and fluxes in boreal forests indicate that the total *carbon* store is from 367.3 to 1716 Pg ($10^{15}$ g), with the mid-point (1095 Pg) being between 1.3 and 3.8 times larger than in any previous estimates (Bradshaw and Warkentin 2015).

In the Northern Hemisphere, peatlands cover 250–350 million ha of land in the boreal zone (e.g., Apps et al. 1993; Strack 2008). Peatland is waterlogged land in which the net primary production exceeds the decay of organic matter, leading to the accumulation of incompletely decayed organic matter in the soil. Peatlands cover 18–28% of boreal land area, the percentage being higher in the maritime than in the continental parts (Kauppi et al. 1997). Globally, boreal peatlands are a major sink for carbon dioxide ($CO_2$), and a major source of atmospheric methane ($CH_4$). The carbon bound in boreal peat deposits (419 Pg) exceed the amount of carbon bound in boreal upland forest soil (199 Pg), plant detritus (32 Pg), and plant biomass (64 Pg) (Apps et al. 1993). Overall nitrous oxide ($N_2O$) emissions from natural peatlands are low, but they may be increased when using peat soils in agriculture (Maljanen et al. 2001). In general, the uptake and losses of carbon in/from peatlands are related to the changes in the balance between the net exchange of $CO_2$, emission of $CH_4$, and losses of dissolved carbon controlled by the groundwater level and water flowing through mire ecosystems.

### 2.2.5 Biodiversity

Boreal forests are characterized by large numbers of individuals of few tree species with a wide ecological amplitude. This contrasts with tropical forests, which sustain a small number of individuals of many species with a narrow ecological amplitude. Genetic diversity in any species is in part the result of the opportunity for gene recombination. Genotypic variability represents adaptations to the specific conditions of local environments, suggesting a high degree of local adaptation within the boreal domain. In the continental parts of the boreal forests, fire controls the natural dynamics of the forests and consequently influences biodiversity. Some species are adapted to using the resources provided by standing and lying burnt trees in various stages of decay.

The share of mature forest is still high in many parts of the boreal forest, providing habitats for species dependent on specific conditions of old-growth forests. Ruckstuhl et al. (2008) estimates that some 25% of the total boreal forest area is still fully natural, not involved in harvest and other commercial use of forested land. They note further that the boreal forest is "one of the few biomes in which the large predator–prey systems are still widely operational, invasive species are few by proportion, and large-scale natural disturbances such as wildfire and epidemic insects still occur extensively at their natural frequencies and patterns." They estimate further that boreal forests provide habitats for 20,300 species, with a circumpolar distribution of similar taxa. However, the number of species per unit area is low at northern latitudes, and the total species richness in boreal forest is greater than that in the tundra but less than that in temperate forests. The differences in the number of species between tundra, boreal forests, and temperate forests are related to the productivity of land across the latitudinal temperature gradient across the vegetation zones (e.g., Waide et al. 1999; Hyvärinen et al. 2019). The boreal forests frequently give way to mires and small lakes, leading to a mosaic structure of forest and wetland, which provides a large variety of habitats, thus increasing the species richness in boreal landscapes.

Although the number of species in managed boreal forests is low, these species represent an important part of the global biological diversity. This holds good for many countries in the boreal domain, such as Finland, where more than 90% of the forested area is managed for timber production. The estimated total number of species is about 50,000, out of which about 43,000 are known. The reason for the low total number of species is the short time that has elapsed since the last glaciation (10,000 years ago), with the consequence that migration is still occurring. These species and the subsequent biodiversity include a large proportion of natives of the eastern taiga (e.g., flying squirrel, Ural owl, Siberian jay). Most of these taiga species are associated with spruce forests. On the other hand, a high proportion of the forest species in Finland (20–25%) are dependent on dead wood (800 coleopterans, 1000 dipterans, 1000 fungi, 200 lichens, etc.). Many of these species are specialized in living on recently burnt wood, while a high proportion lives in peatland forests or on mires (Hyvärinen et al. 2019).

The dynamics of boreal forest ecosystems have historically been driven by disturbances such as fire, storms, and insect population outbreaks. Fire controls biodiversity by providing standing and lying dead wood in different phases of decay. This sustains resources and sites for a set of species typical of early post-fire communities that are distinct from later successional species. These include species from a range of groups, including birds, beetles, spiders, and vascular and non-vascular plants. The success of many species is very closely linked to the regular fire cycle. If regular fires are absent, many species can build large populations only in situations with reduced species richness. Such effects can be observed in many managed boreal forests in Nordic countries. Moreover, where effective firefighting

**Table 2.4** Taxonomic breakdown (%) of threatened species in boreal forests (Bradshaw et al. 2009)

| Taxon | % of species in IUCN Red List in 2008 |
|---|---|
| Fungi and lichens | 5 |
| Plants | 5 |
| • Ferns | – |
| • Mosses | – |
| • Vascular plants | – |
| • Trees | – |
| Butterflies and moths | – |
| Birds | 50 |
| Mammals | 35 |
| Other vertebrates | 5 |

has made fire events rare, species that depend on fire-modified habitats are now threatened. In Finland, 14 species associated with burnt forest land, mostly beetles (Coleoptera) and bugs (Hemiptera), are threatened (Hyvärinen et al. 2019). Similarly, harvest and management increase fragmentation and reduce species richness. Bradshaw et al. (2009) estimated that 348 boreal species are threatened, and included these in the IUCN (International Union for the Conservation of Nature) Red List (Table 2.4). Across all taxa, the share of threatened species is highest in northern Europe. Bradshaw et al. (2009) emphasized the urgent need to preserve existing boreal forest and restore degraded areas to avoid a reduction in biodiversity and carbon sink, which the boreal forest provides in protection of the global environment.

## 2.3 Structure and Functioning of Managed Boreal Forests, with Reference to Finland

### *2.3.1 Boreal-Zone Tree Species Composition in Finland*

In Finland, forests extend across the boreal zone from hemiboreal forests through to southern boreal forests (60° N), to the northern boreal forests bordering the tundra in the north (70° N) (Fig. 2.3). The tree species composition is mainly characterized by Scots pine (*Pinus sylvestris*), Norway spruce (*Picea abies*), and birch (*Betula pendula* and *Betula pubescens*), while the abundance of other species is small. The stocking of Scots pine increases systematically from 30–40% in the south to 70–80% in the north. In contrast, the stocking of Norway spruce decreases from 35–40% in the southern and middle boreal forests up to 63° N, to 10–15% of total stocking further to the north. The stocking of birch is 10–20% of the total stocking throughout the whole country. Over the whole of Finland, the share of Scots pine is 50% while the shares of Norway spruce and birches are 30% and 20%. The share of other

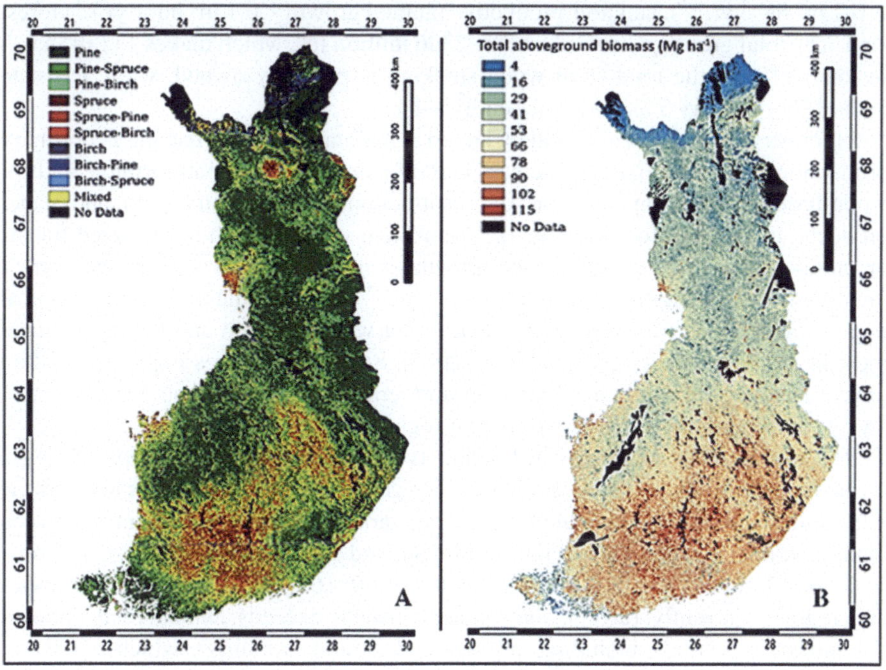


**Fig. 2.3** Distribution of tree species and stocking of stem wood over Finland (Lukeš et al. 2014). Permission of Elsevier. (**a**) Tree species composition and mixture based on stem wood stocking of trees. (**b**) Total aboveground biomass in stem wood stocking of trees

species (including aspen, alders, mountain ash, oak etc.) is only marginal even in the southernmost parts of the boreal zone, bordering the hemiboreal zone (Luke 2021).

## *2.3.2 Forest Resources and Growth*

Boreal forests are generally unmanaged, with the exception of forests in the Nordic countries (Finland, Sweden, and Norway). Over the last couple of centuries, these forests have been used for the large-scale production of different goods and services. In the past, cultivation shifts and tar distillation profoundly modified the structure and functioning of these forests. The impacts of past land use were local, when compared to the industrial harvest of timber for large-scale manufacturing (e.g., for the saw wood, pulp, and paper industries), which was established in the early 1800s. In the future, the importance of boreal forests will increase with the growing importance of bio-based materials and energy, which can provide substitutes for some fossil-based materials and energy.

In Finland, the total land area for forestry is 26 million ha, including productive (annual growth $\geq$1 $m^3$ $ha^{-1}$ $yr.^{-1}$ of stem wood), low-productivity (annual growth of

0.1–1 $m^3\ ha^{-1}\ yr.^{-1}$), and non-productive (annual growth $<0.1\ m^3\ ha^{-1}\ yr.^{-1}$) forest land. The total growing stock of trees is 2306 million $m^3$, which makes 111 $m^3\ ha^{-1}$. In total, 71% of the total stem wood stocking is privately owned, while the state owns 18%, industry 7%, and other entities 4%.

In forestry, management is widely aimed at maintaining or increasing net primary production. This is utilized to increase carbon storage from the atmosphere and for wood-based production, such as of construction materials and pulp and paper products. In Finland, for example, the mean stem wood growth in managed forests on productive forest land (stem wood growth $\geq 1\ m^3\ ha^{-1}\ yr.^{-1}$) is 3.2 $m^3\ ha^{-1}\ yr.^{-1}$ in northern boreal regions, and 6.6 $m^3\ ha^{-1}\ yr.^{-1}$ in the middle and southern boreal regions. This holds for even-aged management when medium and higher management intensity is used (Kellomäki et al. 2005). Such growth rates imply that carbon uptake is about 1.8 Mg C $ha^{-1}\ yr.^{-1}$ in northern boreal regions and 3.9 Mg C $ha^{-1}\ yr.^{-1}$ in the middle and southern boreal regions.

The total growth of forests in Finland is 104 million $m^3\ yr.^{-1}$, representing a mean annual growth of 4.6 $m^3\ ha^{-1}\ yr.^{-1}$. In general, growth decreases towards the north following the declining temperature but varies throughout the country between regions in relation to the variability of climate and soil fertility (Fig. 2.4). The time scale used in forestry is long, with a rotation of 60–160 years depending on species and region. Currently (2021), total annual growth exceeds annual total removal (71 million $m^3\ yr.^{-1}$), including cuttings and natural mortality), which is 68% of the total growth (Luke 2021).

### 2.3.3 Rotation-Based Forestry Management and Production Cycles

Growth and removal in tree populations/communities change through a chronosequence over time. The following concepts are widely used to describe the growth dynamics in growth and yield studies under annual time resolution (Fig. 2.5). Under even-aged management, the management cycle or rotation is related to the growth of trees from reforestation to the final cut (clear-cut or cut for natural regeneration). The optimal rotation length indicates the time from regeneration to the final cutting that generates the maximum economic yield from timber. The calculation of this period is specific to each stand and to the economic and sustainability goals of the harvest. In this regard, the optimal rotation length is one that maximizes the net present value (*NPV*) considering the net cash inflow (*PVR*) of timber value minus the outflow (*PVC*) costs of management and harvest ($R_t = PVR_t - PVC_t$):

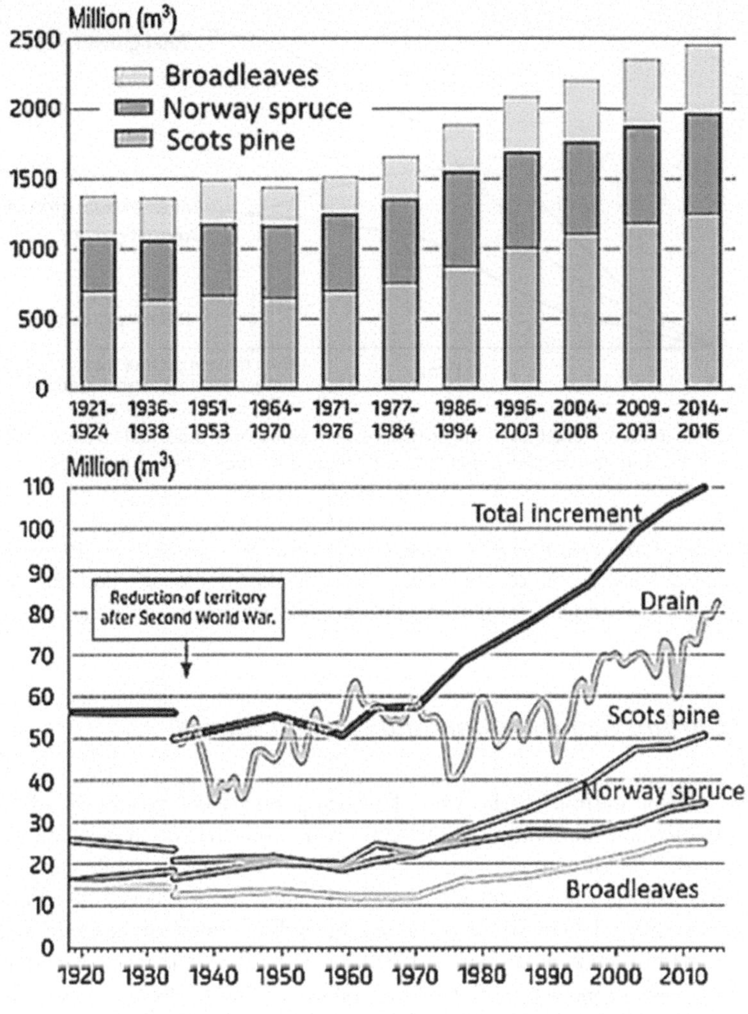


**Fig. 2.4** Forest resources in Finland from the 1920s to 2015s in Finland (Luke 2021). Courtesy of the Natural Resources Institute Finland. Upper: Total resources of major tree species over the period from 1920 to 2015. Lower: Total annual growth of stem wood stock of major tree species with the cutting drain over the period from 1920 to 2015

$$\mathrm{NPV} = \sum_{t=0}^{n} \frac{R_t}{(1+p)^t} \tag{2.1}$$

where $t$ is time and $p$ the discount rate.

In biological terms, the optimum rotation length is the point where the slope of the mean annual growth is equal to zero (Fig. 2.5) at the intersection of the mean annual growth and the current annual growth. This indicates the gradual decline of annual

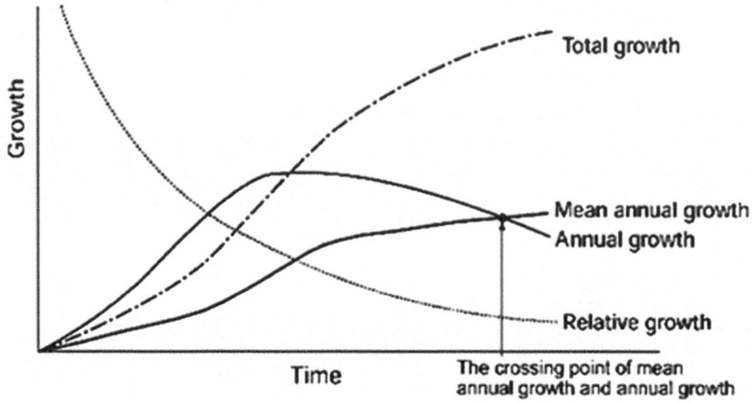


**Fig. 2.5** Schematic representation of the growth of stem wood over time, and an explanation of the main concepts used in growth and yield studies. Compiled by the author of this book

growth, with a reduction in the increase of the value of timber likely to be obtained for substantial elongation of rotation. These further increases carbon stocks in maturing trees but reduces the net present value expected in future harvests.

### *2.3.4 Carbon Sequestration and Carbon Stocks*

In boreal forests, carbon stocks vary depending on (i) the properties of the site; (ii) the structure and functioning of the ecosystem; and (iii) how forests are managed and harvested. The variability of carbon stocks in trees is large, while carbon stocks in the soil are stable even under intensive soil management. This is shown also by Liski and Westman (1997a, b), who studied the carbon density in upland forest sites, located in the temperature sum range of 800–1300 d.d. (between the latitudes 66–60° N). The study area represented the southern parts of the northern boreal forests and the whole of the middle and southern boreal forests in Finland (Fig. 2.6).

According to Liski and Westman (1997a, b), forest-based carbon increased from north to south, in tandem with increasing growth and litter fall. In upland soils, the carbon density was 2.2 kg C $m^{-2}$ in the organic soil layer at the most fertile sites (*Oxalis-Myrtillus* site type, OMT) and other sites with higher fertility. At the medium-fertile and medium-poor sites (*Mytillus* and *Vaccinium* site types, MT, and VT, respectively), carbon density was 1.9 kg C $m^{-2}$, and at the poor sites (*Calluna* site type, CT, and even poorer), 1.2 kg C $m^{-2}$. The organic layer contained 28% of total carbon in the soil, while the mineral soil to a depth of one meter contained 68%, and below that 4%. The amount of carbon in the soil was more than double that in trees. Increasing temperature increased the carbon density 0.266 kg C $m^{-2}$ per increase of 100 d.d. in the temperature sum from the north to the south, and the site fertility explained 36% and 70% of the variability of carbon in the organic and

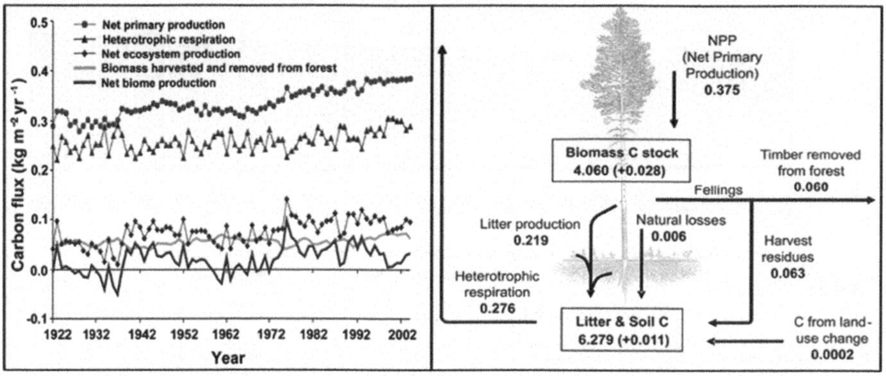


**Fig. 2.6** Annual carbon fluxes over Finland in the period 1922–2002 (left), and mean forest-based carbon budget over Finland in the 1990s (right) (Liski et al. 2006). Fluxes in the left panel are in kg C $m^{-2}$ $yr.^{-1}$, and carbon mass in the right panel is kg C $m^{-2}$. Open Access. Permission of EDP Sciences

mineral soil layers. Based on the National Forest Inventory data, Liski and Westman (1997b) estimated that the total carbon storage in the upland sites throughout Finland falls in the range of 1109–1315 Tg (Fig. 2.6).

Forest management and harvest affect carbon density and the total amount of carbon in forest ecosystem. Based on the Forest National Inventory (FNI) data, Liski et al. (2006) found that over an 80-year period, net primary production (NPP) increased from 0.29 to 0.39 C kg $m^{-2}$ $yr.^{-1}$ (35% increase) and heterotrophic respiration (Rh, decay of soil organic matter) increased from 0.25 to 0.29 C kg $m^{-2}$ $yr.^{-1}$ (16% increase) (Fig. 2.6). At the same time, the net ecosystem production (NEP) (or net ecosystem exchange, NEE) increased from 0.04 to 0.10 C kg $m^{-2}$ $yr.^{-1}$ (150% increase). The net biome production (NBP) (carbon in NEE plus carbon in harvests) fell below zero in the 1920s and 1930s, and occasionally in the 1960s. Thereafter, the net biome production remained positive, indicating that harvest remained smaller than growth. In the 1990s, the mean net ecosystem production was 0.38 kg C $m^{-2}$ $yr.^{-1}$, but 0.28 kg C $m^{-2}$ $yr.^{-1}$ (70% of NEE) was lost in heterotrophic respiration. At the same time, over half of the net ecosystem production (0.060 kg C $m^{-2}$ $yr.^{-1}$) was harvested (0.039 kg C $m^{-2}$ $yr.^{-1}$), but 72% (0.028 kg C $m^{-2}$ $yr.^{-1}$) of the net biome production accumulated in the tree biomass and the rest (0.011 kg C $m^{-2}$ $yr.^{-1}$) in soil. In addition to upland sites, peatland ecosystems also sequester carbon.

In Finland, the total land area for forestry is 26 million ha, mainly across the boreal zone. In the whole territory, the amount of stored carbon dioxide ($CO_2$) is 1300 million Mg in the upland soils and 5500 million Mg in the organic soils in peatlands, while the carbon stock in woody biomass is 700 million Mg (Fig. 2.7). Currently, the amount of $CO_2$ bound in woody biomass is increasing because growth exceeds cuttings (drain). At the same time, the carbon stock in soil tends to increase because of increasing production of litter related to growth and cuttings. In 2010, for

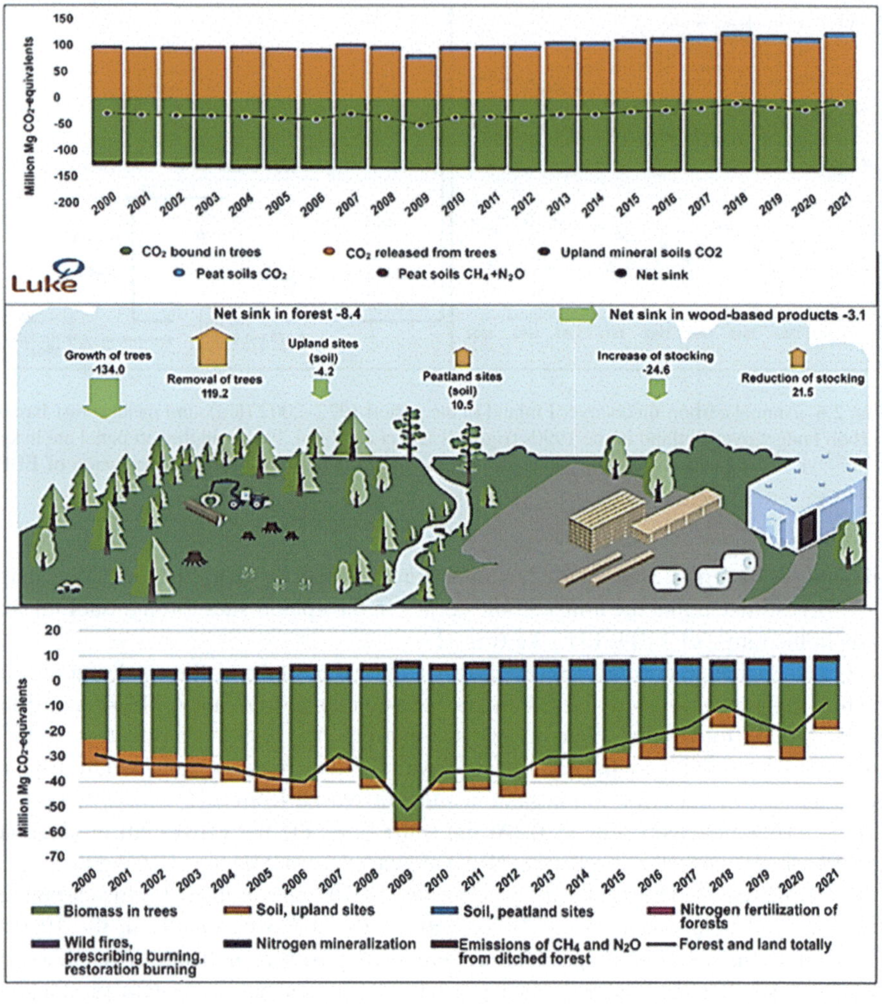


**Fig. 2.7** Sequestration of greenhouse gases and changes in greenhouse gas stocks in forests, peatlands, and bound in wood-based products in Finland in 2000–2021 (Luke 2022). Courtesy of the Natural Resources Institute Finland. Upper: Sequestration of greenhouse gases in woody biomass, dead organic matter in mineral soils, and organic soils in peatland, expressed in million Mg of $CO_2$ equivalents. Middle: Changes in carbon stock emissions of greenhouse gases in forests and peatlands related to disturbances and restoration of drained peatlands, and of drained peatlands and drainage of peatlands for forestry in million Mg of $CO_2$ equivalents. Lower: Greenhouse gas balance for forests and wood-based products in million Mg of $CO_2$ equivalents

example, forests sequestered 32.8 million Mg carbon in $CO_2$ equivalents. Trees and mineral soils were sinks, while peat soils were a source of $CO_2$. The total emissions of major greenhouse gases were 74.6 million Mg in $CO_2$ equivalents. In land use and land use changes (LULUCF), including in forestry, 44% of the total emissions were sequestrated in forest and mire ecosystems and in wood-based products (Luke 2021).

Management and harvest intensities affect carbon dynamics in upland and peatland sites, with impacts on carbon uptake and losses controlling the carbon sources and sinks. This is most evident at the site and stand scale, but the links between management and harvest are evident even at larger scales. The most recent inventory of greenhouse gases, in 2021, shows that carbon sequestration in forest ecosystems (trees and soil) was 8.3 million Mg of $CO_2$ equivalents of greenhouse gases (Fig. 2.7). In forests, the sequestration was less than that in 2020, when 12.2 million Mg of $CO_2$ equivalents were stored. This reduction is related to increased harvests, with removal being 11% larger than in the previous year. Concurrently, a decrease in sequestration occurred in upland sites, and emissions rose in drained peatlands. Together, this slight reduction of sequestration in upland sites and slight increase of emissions in peatland sites contributed to the lowest value of net sink over the period from 1990 to 2021 (Luke 2021).

## 2.4 Concluding Remarks

Boreal forests fall between the summer and winter limits of the arctic air mass. This makes them especially vulnerable to climatic warming, which is likely to modify tree species composition and enhance the growth and development of forests. On an annual basis, climatic warming in this century is likely to be 3–5 °C, which may trigger heat stress and more frequent drought episodes, leading to widespread forest death due to outbreaks of insects and other pests, and increasing fire episodes (e.g., Olsson 2011). Proper management is therefore necessary to adapt boreal forests to climate change and mitigate the harmful effects climatic warming, which is likely to have on their growth and development of boreal forests. The conversion of intact boreal forests into managed ones may reduce carbon stocks in boreal forests, while proper management may provide many opportunities to control the forest ecosystem/atmosphere interaction to mitigate climate change. Boreal peatlands are also principal factors in potential climate change mitigation. Globally, boreal peatlands are a major sink for carbon dioxide ($CO_2$), and a major source of atmospheric methane ($CH_4$). In general, the uptake and losses of carbon in/from peatlands are related to the changes in the balance between the net exchange of $CO_2$, emission of $CH_4$, and losses of dissolved carbon controlled by the ground water level and water flowing through peatland ecosystems.

## References

Aber JD, Melillo JM (1991) Terrestrial ecosystems. Saunders College Publishing, Philadelphia

Apps MJ, Kurz WA, Luxmore RJ, Nilsson LO, Sedro RA, Schmidt R et al (1993) Boreal forests and tundra. Water Air Soil Pollut 70:39–53

Bonan GB, Shugart HH (1989) Environmental factors and ecological processes in boreal forests. Annu Rev Ecol Syst 20:1–28

Bradshaw CJA, Warkentin IG (2015) Global estimates of boreal forest carbon stocks and flux. Glob Planet Chang 28:24–30

Bradshaw CJ, Warkentin IG, Sodhi NS (2009) Urgent preservation of boreal carbon stocks and biodiversity. Trends Ecol Evol 24(10):541–548

Burton PJ, Bergeron Y, Bogdanski BEC, Juday GP, Kuuluvainen K, McAfee BJ, Ogden A, Teplyakov VK, Alfaro RI, Francis DA, Gauthier S, Hantula J (2010) Sustainability of boreal forests and forestry in a changing environment. In: Mery G, Katila P, Galloway G, Alfaro RI, Kanninen M, Lobovikov M, Varjo J (eds) Forests and society – responding to global drivers of change. IUFRO World Series, Vienna, pp 249–282

DeLuca TH, Boisvenue C (2012) Boreal forest soil carbon: distribution, function, and modelling. Forestry 85(2):161–184

FAO (2006) Global Forest Resources Assessment 2005. Progress towards sustainable forest management. FAO Forestry Papers 147:1–320

FAO (2016) Global Forest Resources Assessment 2015. How are the world's forests changing? 2nd edn. Food and Agriculture Organization of the United Nations, Rome

Fischlin A, Midgley GF, Prince JT, Leemans R, Gopal B, Turley C, Rounsevell M et al (2007) Ecosystems, their properties, goods, and services. In: Parry ML, Canziani OF, Putikof JP, van der Linden JP, Hanson CE (eds) Climate change 2007: impacts, adaptation, and vulnerability. Contribution of Working Group II to the Fourth Assessment Report of the Intergovernmental Panel on Climate Change (IPCC). Cambridge University Press, Cambridge

Fischlin A, Ayres M, Karnosky D, Kellomäki S, Louman S, Ong C, Plattner G et al (2009) Future environmental impacts and vulnerabilities. In: Seppälä R, Buck A, Katila P (eds) Adaptation of forests and people to climate change – a global assessment report. IUFRO World Series, Vienna, pp 53–100

Gordon B, Bonan GG, Shugart HH (1989) Environmental factors and ecological processes in boreal forests. Annu Rev Ecol Syst 20:1–28

Gorham E (1991) Northern peatlands: role in the carbon cycle and probable responses to climatic warming. Ecol Appl 1:182–195

Hyvärinen E, Juslén A, Kemppainen E, Uddström A, Liukko U-M (toim.) (2019) Suomen lajien uhanalaisuus – Punainen kirja 2019. Ympäristöministeriö & Suomen ympäristökeskus, Helsinki

IBFRA/International Boreal Forest Research Association (2020) About boreal forests. http://ibfra.org/about-boreal-forests/. Accessed 2 Aug 2021

Innes J, Joyce LA, Kellomäki S, Louman B, Ogden A, Parrotta J, Thompson I (2009) Management for adaptation. In: Seppälä R, Buck A, Katila P (eds) Adaptation of forests and people to climate change. A global assessment reports. IUFRO World Series, Vienna, pp 135–186

IPCC (2000) Special report: land use, land-use change, and forestry (summary for policymakers). Intergovernmental Panel on Climate Change

IPCC (2007a) Climate change 2007: the physical science basis. Contribution of working group I to the fourth assessment. In: Report of the intergovernmental panel on climate change. Cambridge University Press, Cambridge

IPCC (2007b) Climate change 2007: contribution of working group II to the fourth assessment report of the intergovernmental panel on climate change, 2007. Cambridge University Press, Cambridge

Kauppi P, Posch M, Hänninen P, Henttonen H, Ihalainen A, Lappalainen E et al (1997) Carbon reservoirs in peatlands and forests in the boreal region of Finland. Silva Fennica 31(1):13–25

Keith H, Mackey BG, Lindenmayer DB (2009) Re-evaluation of forest biomass carbon stocks and lessons from the world's most carbon-dense forests. PNAS 106(28):11635–11640

Kellomäki S, Strandman H, Nuutinen T, Peltola H, Korhonen KT, Väisänen H (2005) Adaptation of forest ecosystems, forests, and forestry to climate change. Finnish Environment Institute, FinAdapt Working Paper 4:1–50

Kellomäki S, Peltola H, Nuutinen T, Korhonen KT, Strandman H (2008) Sensitivity of managed boreal forests in Finland to climate change, with implications for adaptive management. Philos Trans R Soc B 363:2341–2351

Kurz WA, Apps MJ (1999) A 70-year retrospective analysis of carbon fluxes in the Canadian forest sector. Ecol Appl 9(2):526–547

Kuusela K (1990) The dynamics of boreal coniferous forests. Gummerus Kirjapaino Oy, Jyväskylä

Liski J, Westman CJ (1997a) Carbon storage in forest soil of Finland. 1. Effect of thermoclimate. Biogeochemistry 36:239–260

Liski J, Westman CJ (1997b) Carbon storage in forest soil of Finland. 2. Size and regional patterns. Biogeochemistry 36:261–274

Liski J, Lehtonen A, Palosuo T, Peltoniemi M, Eggers T, Muukkonen P, Mäkipää R (2006) Carbon accumulation in Finland's forests 1922–2004 – an estimate obtained by combination of forest inventory data with modelling of biomass, litter, and soil. Ann For Sci 63(7):687–697

Luke (2021) Finnish Statistical Yearbook of Forestry (2021). Luonnonvarakeskus (Luke), Helsinki

Luke (2022) Suomen LULUCF-sektorin 2021–2025 velvoitteen toteutuminen. Report to Ministry of Agriculture and Forestry

Lukeš P, Rautiainen M, Manninen T, Stenberg P, Mõttus M (2014) Geographical gradients in boreal forest albedo and structure in Finland. Remote Sens Environ 152:526–535

Maljanen M, Hytönen J, Martikainen PJ (2001) Fluxes of $N_2O$, $CH_4$ and $CO_2$ on afforested boreal agricultural soils. Plant Soil 231:113–121

Moore TR, Roulet NT, Waddington JM (1998) Uncertainty in predicting the effect of climatic change on the carbon cycling of Canadian peatlands. Clim Chang 40:229–245

Nikolov N, Helmisaari H (1992) Silvics of the circumpolar forest tree species. In: Shugart HHR, Leemans R, Bonan GB (eds) A system analysis of the global boreal forest. Cambridge University Press, New York

Olsson R (2011) To manage or protect. Air Pollut Climate 26:1–67

Pan Y, Birdsey RA, Fang J, Houghton R, Kauppi PE, Kurz WA et al (2011) A large and persistent carbon sink in the world's forests. Science 333:988–993

Raich JW, Schlesinger WH (1992) The global carbon dioxide flux in soil respiration and its relationship to vegetation and climate. Tellus 44B:81–99

Ruckstuhl KE, Johnson EA, Miyanishi K (2008) Introduction. The boreal forest and global change. Philos Trans R Soc B 363:2245–2249

Schimel DS, House IJ, Hibbard KA, Bousquet P, Cialis P, Peylin P et al (2001) Recent patterns and mechanisms of carbon exchange by terrestrial ecosystems. Nature 414:169–173

Shepashenko A, Shvidenko AD, Nilsson S (1998) Phytomass (live biomass) and carbon of Siberian forests. Biomass Bioenergy 14(1):21–31

Shvidenko AD, Nilsson S (2003) A synthesis of the impact of Russian forests on the global carbon budget for 1961–1998. Tellus 55B:391–415

Sirois L (1992) The transition between boreal forest and tundra. In: Shugart HH, Leemans R, Bonan GB (eds) A system analysis of the global boreal forest. Cambridge University Press, Cambridge

Strack M (ed) (2008) Peatlands and climate change. International Peat Society, Saarijärven Offset Oy, Saarijärvi

Waide RB, Willig MR, Steiner CF, Mittelbach G, Gough L, Dodson SI et al (1999) The relationship between productivity and species richness. Annu Rev Ecol Syst 30:257–300

Walter H (1979) Vegetation of the earth and ecological system of the geo-biosphere, 2nd edn. Springer, New York

Wulder MA, Campbell S, White JC, Flannigan M, Campbell ID (2007) National circumstances in the international circumboreal community. The Forestry Chronical 83(4):539–556

# Chapter 3
# Dynamics of Carbon in Boreal Forest Ecosystems

## Contents




**Abstract** In forest ecosystems, gross primary production (*GPP*) is the fundamental supporting service under ecosystem management. Energy uptake maintains ecosystem structure and functioning on different trophic levels, including for first-order consumers (herbivores) and second- and higher-order consumers (carnivores) in the hierarchy of energy flow through the ecosystem. In boreal conditions, the scarce availability of nitrogen is among the main factors limiting gross primary production. Carbon uptake is further controlled by temperature, air humidity, atmospheric $CO_2$, and the availability of soil water. *GPP* releases oxygen into the atmosphere and binds atmospheric carbon in biomass. Carbon remains in the ecosystem for some time but is lost to the atmosphere in respiration and oxidative processes.





S. Kellomäki, *Forest Management for Timber Production and Climate Change Mitigation*, Managing Forest Ecosystems 44,
https://doi.org/10.1007/978-3-031-71575-4_3

## 3.1 Carbon in Forests and Forest Ecosystems

### *3.1.1 Interactions Between Atmospheric Carbon and the Biosphere*

In the lower atmosphere (the troposphere, i.e., up to 15 km above the Earth's surface), the current content of carbon dioxide ($CO_2$, in parts per million, ppm) is about 410 ppm in the summertime and 420 ppm in the wintertime in the upper latitudes of the Northern Hemisphere. Atmospheric carbon is a major component of the global carbon cycle, comprising approximately 720 Gt of carbon (IPCC 2007). Atmospheric $CO_2$ represents the difference between carbon emissions from and uptake into geological and ecological processes on Earth.

The exchange of atmospheric carbon between water and land systems implies that the atmosphere is both a sink and a source of carbon. Concurrently, the global increase in atmospheric $CO_2$, temperature, and the availability of nitrogen and water are affecting the current forest sinks and sources of carbon in boreal regions (e.g., Hyvönen et al. 2007). Climate change has a multitude of impacts on the functioning and structure of forest ecosystems. This is very evident in the boreal forests, located at higher latitudes in the Northern Hemisphere (45–70° N). The impacts of climate change on the structure and functioning of forests will depend on (i) the current climatic conditions; (ii) the rate of future climatic change; and (iii) the adaptive capacity of trees and other plants to grow and regenerate under climatic variability, in terms of both absolute and relative changes in climatic conditions.

The main function of ecosystems involves the interception and movement of energy, water, and nutrients. Movement of energy represents a trophic hierarchy through primary production to secondary production, thus taking up and further cycling elements (nutrients) in the environment for further uptake in primary production (Fig. 3.1). Movement of energy is based on carbon compounds, comprising sugars, proteins, cellulose, and lignin, produced by green plants or autotrophs in primary production. These compounds are used by heterotrophs in secondary production (e.g., Aber and Melillo 1991; Chertov et al. 2001). Dead material from autotrophs and heterotrophs is further used as an energy source by saprotrophs, which feed and decompose dead organic matter originating from plants, animals, and other decomposers. Organic matter in litter and humus mixes with soil particles to form the soil organic matter (SOM). Carbon in plant communities and SOM form the main carbon stocks in terrestrial ecosystems (Hyvönen et al. 2007).

### *3.1.2 Carbon Cycle, with Flows and Stocks in Forest Ecosystems*

Forest ecosystems contain biological and nonbiological components that interact, affecting regeneration, growth, and carbon uptake in plant communities (trees and

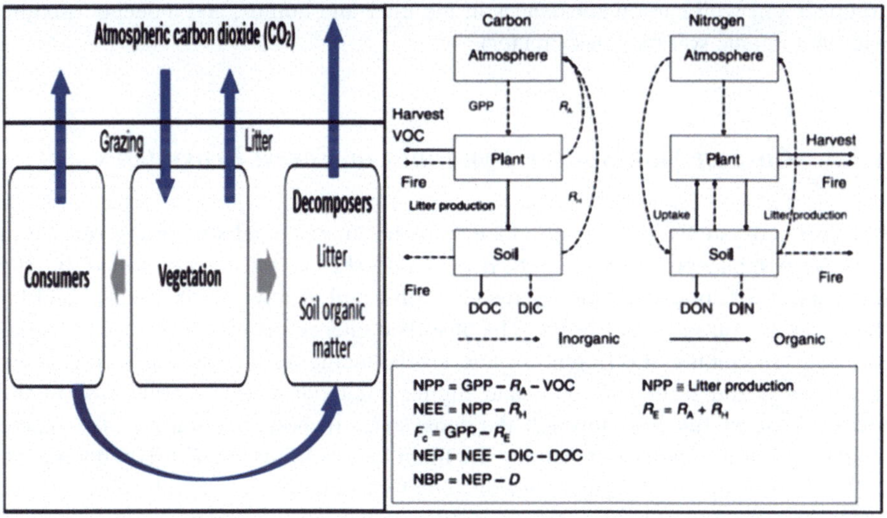


**Fig. 3.1** Interactions between atmospheric carbon and the biosphere, with overviews of the carbon and nitrogen dynamics in forest ecosystems under harvest- and fire-induced disturbance of carbon fluxes and stocks. The overview equations are for carbon uptake and losses and carbon dynamics at the ecosystem, landscape, and biome scale. Left: Schematic presentation of carbon flows in forest ecosystems showing forest–atmosphere interactions related to carbon uptake by vegetation, consumption via grazing, through to carbon loss via decomposition of litter and soil organic matter. Right: Carbon and nitrogen dynamics in forest ecosystems in gross primary production (*GPP*) with uptake of carbon through photosynthesis and carbon losses in autotrophic respiration ($R_A$) and volatile organic compounds (*VOC*) related to net primary production (*NPP*). Additional losses of dissolved inorganic (*DIC*) and dissolved organic carbon (*DOC*) occur through the soil profile into rivers and into the atmosphere, thus reducing net ecosystem production (*NEP*). At the scale of the forest and forest landscape, losses caused by disturbances (*D*) such as fire, herbivory, and management activities further reduce the amount of carbon stored in forest ecosystems. The resulting net imbalance of forest ecosystem carbon can be interpreted as the net biome productivity (*NBP*). The nitrogen cycle parallels the carbon cycle but has an additional flux through the uptake of nitrogen (*N*) from the soil (Hyvönen et al. 2007). Free Article. Courtesy of Wiley

other vegetation). Furthermore, the decay of dead material produced in plant communities involves the emission of carbon. In boreal forests, the vegetation comprises a mixture of different plant populations broadly divided into the horizontal layers: a tree layer, an understory layer of shrubs, and a layer of ground vegetation including dwarf shrubs, herbs, grasses, mosses, and lichens. Trees play a dominant role in boreal upland sites, affecting the amount and chemical properties of SOM, with large impacts on carbon sequestration in the forest ecosystem. This is most evident on fertile sites, where the litterfall from trees exceeds that from other layers of vegetation and forms the main source of SOM in boreal forests. Litter from different vegetation layers carries nutrients to the soil detritus, which decays and releases nutrients into soil solution for uptake in trees and other vegetations. In nutrient cycling, erosion and leaching of nutrients occur along with the decay of organic matter in the soil profile and outside of the ecosystem. The decay of litter produces an upper soil with carbon substances in the litter and humus, and a deeper mixture with nonorganic soil particles (matter).

### 3.1.3 Carbon Sink/Source Dynamics in Forest Ecosystems

In forest ecosystems, primary production is the fundamental supporting service in ecosystem management (Fig. 3.2; Box 3.1). *GPP* is mainly controlled by the interception of solar radiation by trees, shrubs, and ground cover species, and by the cycles of nutrients and water. The absorbed energy maintains ecosystem structure and functioning at different trophic levels, including for first-order consumers (herbivores) and further second- and higher-order consumers (carnivores) in the hierarchy of energy flow through the ecosystem. In boreal conditions, the scarce availability of nitrogen (*N*) is among the main factors limiting *GPP* (Hyvönen et al. 2007). Carbon uptake is further controlled by temperature, air humidity, atmospheric $CO_2$, and the availability of soil water. *GPP* releases oxygen into the atmosphere and binds atmospheric carbon in biomass. Carbon remains in the ecosystem for some time but is lost in respiration and oxidative processes. From a geologic perspective, carbon accumulated in sediments returns to the atmosphere.

In the context of a forest ecosystem, plants like trees in stands are carbon sinks whenever the carbon uptake in the gross photosynthesis (gross photosynthetic production) exceeds the losses of carbon in the total ecosystem respiration

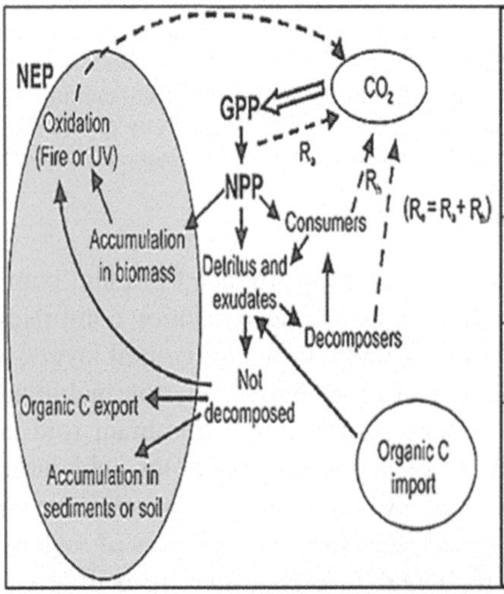


| | |
|---|---|
| • Gross primary production (GPP, g $m^{-2}yr^{-1}$)<br>• Autotrophic respiration (RA, g $m^{-2}yr^{-1}$) for maintenance ($R_m$) and growth ($R_g$) | • Gross carbon fixation<br>• Loss of carbon in respiration,<br>• RA= $R_m$ + Rg |
| • Gross primary production (NPP, g $m^{-2}yr^{-1}$) | • Gross primary production minus autotrophic respiration,<br>• NPP=GPP-RA |
| • Heterotrophic respiration (RH, g $m^{-2}yr^{-1}$) | • Loss of carbon through decay and secondary producers |
| • Net ecosystem production (NEP) or net ecosystem exchange (NEE , g $m^{-2}yr^{-1}$)<br>• Net biome production (NBP, g $m^{-2}yr^{-1}$) | • Gross primary production minus total respiration, NEE= GPP-RA-RH =NPP-RH<br>• Net biome production, NBP =NEE plus carbon lost in disturbances |

**Fig. 3.2** Carbon sink/source dynamics in forest ecosystems (Lovett et al. 2006). Permission of Springer Nature. Left: Overview of carbon stocks and flows through an ecosystem showing the key processes and stocks in ecosystem carbon dynamics. Right: Main concepts used in describing the carbon cycle and carbon stocks in the context of interactions between the atmosphere and biosphere

($R_E = R_A + R_H$) and volatile organic compounds (Fig. 3.2). At the seasonal scale, a boreal tree stand is a sink in the daytime in summer but a source during summer nights. A stand is also a source in winter during both the day and night. The sink strength of boreal tree stands varies depending on the developmental phase in the life cycle of the trees and on the management cycle of forest stands. The sink strength of trees follows their growth rate. Through a chronosequence of the tree stand, sink strength is highest in young and pole-phase trees, with a gradual decline as the trees approach full maturity in the old-growth phase. Concurrently, the carbon stock in trees increases. In old climax stands of trees, carbon uptake and loss balance each other in the absence of disturbance. However, there is evidence that even old-growth tree stands may sequester carbon in ecosystem stocks, especially through the growth of ground cover species.

## 3.2 Carbon Exchange Between Forest Ecosystems and Atmosphere

### *3.2.1 Carbon Sources and Sinks, Measurements, and Stocks*

On the global scale, the biosphere takes carbon from the atmosphere in primary production and emits it to the atmosphere in autotrophic respiration, driving and maintaining the regeneration and growth functions in the biosphere. In this context, the difference between *GPP* and autotrophic respiration ($R_A$) yields net primary production (*NPP*) (Fig. 3.2). *NPP* is the remaining share of *GPP* after carbon is lost in autotrophic respiration, for example in trees, in the growth and maintenance of the foliage, branches, stem, coarse roots, fine roots, and mycorrhizae. Carbon is also used in the synthesis of defence chemicals. In boreal upland sites, the mean *NPP* is 50–870 g C $m^{-2}$ $yr.^{-1}$, the values being larger in deciduous than in coniferous forests (Gower et al. 2001). On the other hand, the share of production belowground is greater in coniferous (36% of the total primary production) than in deciduous (19%) forests.

The carbon potentially available for longer-term storage refers to net ecosystem production (*NEP*). This can be calculated in two ways: $NEP = NPP - R_H$ or $NEP = GPP - R_E$, where $R_E$ is the ecosystem respiration. Furthermore, net ecosystem exchange (*NEE*) measures the net exchange of carbon between an ecosystem and the atmosphere per unit of ground area and time, i.e., the difference between *GPP* and the carbon lost in autotrophic and heterotrophic respiration. Net biome production (*NBP*) refers to the uptake of carbon through photosynthesis and the loss of carbon through disturbances, e.g., fire or timber harvest (Hamilton et al. 2002). The main concepts of carbon dynamics in forest ecosystems, including carbon sources and sinks, measurements, and stocks, are defined in Box 3.1.

**Box 3.1 Carbon sources and sinks, measurements, and stores**

Basic concepts

- Carbon store (storage, stock) refers to the quantity of carbon bound in given systems, such as growing trees and forest soil. Over time, the amount of carbon in storage changes due to carbon being bound in or removed from the system.
- Carbon sink refers to a system absorbing more carbon from the atmosphere than it emits, with a consequent increase of carbon in storage. Thus, the net ecosystem exchange or net ecosystem production ($NEE = NEP = NPP - R_H$) are positive: $NEE$ and $NEP > 0$.
- Carbon sequestration refers to the carbon removed from the atmosphere and stored temporarily in a sink.
- Carbon source refers to a system that emits more carbon into the atmosphere than it absorbs in sinks, with a reduction of carbon in storage. $NEE$ and $NEP$ ($NEE = NEP = NPP - R_H$) are negative: $NEE$ and $NEP < 0$.
- Carbon balance refers to the net sum of carbon absorbed into and emitted from the system over a defined period. The carbon balance varies over time, and it is zero at a given moment or in each period when the carbon absorption is equal to the carbon emission.

Measurements and units

- Absorption and emission of carbon dioxide ($CO_2$) are widely given in terms of carbon (C).
- 1 kg C is equal to 3.7 kg $CO_2$.
- 1 kg $CO_2$ is equal to 0.27 kg C.
- 1 $m^3$ of wood contains 0.2 Mg C.
- 1.55 kg of $CO_2$ is needed to produce 1 kg of wood.
- 1 $m^3$ of wood retains 1 Mg of $CO_2$, producing 0.7 Mg of oxygen.
- $CO_2$ equivalent refers to the combined effect of several greenhouse gases like $CO_2$ and methane ($CH_4$), the latter being 23 times more effective than $CO_2$ in trapping heat in the lower atmosphere. This implies that 1 kg of methane has the same warming effect as 23 kg of $CO_2$. Consequently, 1 kg of carbon and 1 kg of methane implies: 1 + 23 = 24 kg $CO_2$ equivalents.
- On larger scales, including globally, carbon stores are expressed in Mt. (megatons) or Gt (gigatons): 1 Mt. is one million ($10^6$) tons; 1 Gt is 1 billion ($10^9$) tons (i.e., 1 Gt is 1000 Mt).

### 3.2.2 Monitoring Carbon Exchange Between the Atmosphere and Forest Ecosystems

In forests, *NEE* indicates the balance between the carbon taken up into and emitted from the ecosystem. The *NEE* measured in daylight ($NEE_{c,day}$, g C m$^{-2}$ ground s$^{-1}$) is the gross canopy photosynthesis ($GPP_c$, g C m$^{-2}$ ground s$^{-1}$) minus the sum of the respiration in the canopy foliage ($R_{c,l}$, g C m$^{-2}$ ground s$^{-1}$), the respiration of sapwood in the stem and branches ($R_{c,w}$, g C m$^{-2}$ ground s$^{-1}$), and the carbon efflux from soil. Carbon efflux from soil includes all sources of autotrophic (e.g., coarse, and fine roots) and heterotrophic respiration in the forest floor ($R_s$, g C m$^{-2}$ ground s$^{-1}$) (e.g., decay of litter and humus, respiration of roots) (Wang et al. 2004):

$$NEE_{c,day} = GPP_c - (R_{c,l} + R_{c,w} + R_s). \tag{3.1}$$

The net ecosystem exchange at night ($NEE_{c,\ night}$, g C m$^{-2}$ ground s$^{-1}$) is:

$$NEE_{c,night} = R_{l,d} + R_{c,w} + R_s \tag{3.2}$$

where $R_{l,d}$ (g C m$^{-2}$ ground s$^{-1}$) is the dark respiration in the foliage.

*NEE* is commonly measured using eddy covariance technology (Box 3.2). If the flow velocity is high enough, the air flow in a forest canopy is turbulent with eddies (Fig. 3.3). Eddies carry $CO_2$ (and $H_2O$) back and forth between the atmosphere and canopy, whereas slower air flow is laminar with no/few eddies and only a small exchange of $CO_2$ (and $H_2O$) between canopy and atmosphere. The exchange of $CO_2$ occurs in the boundary layer of the leaves/canopy through the $CO_2$ uptake in photosynthesis and $H_2O$ losses in transpiration. The $CO_2$ content in eddies and three-dimensional wind velocity are used to calculate the exchange rate of $CO_2$. In eddy covariance measurements, the uptake (photosynthesis) is indicated by a negative flux into the ecosystem from the atmosphere, while emission (respiration) is a

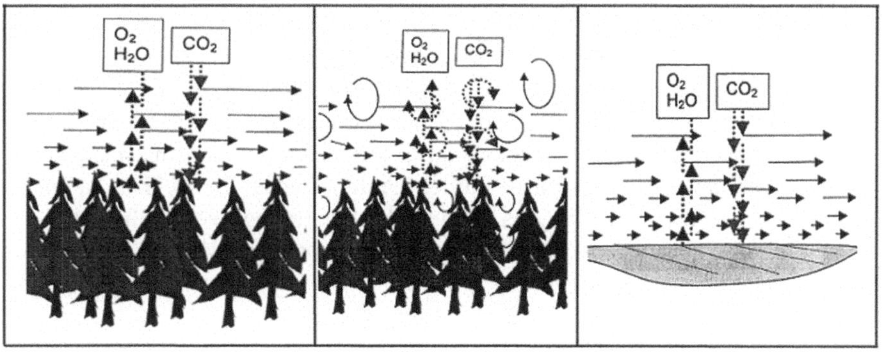


**Fig. 3.3** Schematic presentation of the $CO_2$ and $H_2O$ exchange in a forest ecosystem when air flow through the canopy is laminar (left), and turbulent (middle). The exchange of $CO_2$ and $H_2O$ takes place in the boundary layer of leaves (right). Compiled by author of this book

positive flux from the ecosystem to the atmosphere. In daytime, both fluxes occur, whereas only respiration occurs at nighttime.

**Box 3.2 Atmosphere–forest ecosystem exchange of carbon in eddy fluxes**

In eddy covariance measurements, photosynthesis causes a negative flux into the ecosystem from the atmosphere, while respiration produces a positive flux out from the ecosystem to the atmosphere. During the daytime, $NEE_{day}$ [g C $m^{-2}$ ground $day^{-1}$] is equal to the *GPP* in the canopy minus the carbon lost in autotrophic respiration in leaves ($R_l$, g C $m^{-2}$ ground $day^{-1}$), branches, sapwood, and roots [$R_{b,\ w}$, g C $m^{-2}$ ground $day^{-1}$] and in heterotrophic respiration in soil [$R_s$, g C $m^{-2}$ ground $day^{-1}$]. Furthermore, autotrophic respiration in foliage may be divided into photorespiration ($R_{l.\ p}$) and dark respiration ($R_{l.\ d}$) (Ge et al. 2011):

$$NEE_{\mathrm{day}} = GPP – (R_l + R_{b,w} + R_s) = GPP – (R_{l.p} + R_{l.d} + R_{b,w} + R_s) \tag{3.3}$$

Net ecosystem exchange at night ($NEE_{night}$, g C $m^{-2}$ ground $day^{-1}$) includes only autotrophic losses in dark and heterotrophic respiration: $NEE_{night} = R_{l.\ d} + R_{b,\ w} + R_s$. On a daily scale, the net $CO_2$ exchange between the forest ecosystem and the atmosphere is the sum of eddy fluxes ($F_e$, g C $m^{-2}$ ground $day^{-1}$) from the canopy monitored by the eddy covariance instrument, and the flux representing the storage of $CO_2$ ($F_s$, g C $m^{-2}$ ground $day^{-1}$) in the air layer below the canopy (Aubinet et al. 2000): $NEE = F_e + F_s$, with the flux from the canopy ($F_e$) referring to the mean covariance between the vertical wind velocity ($w'$) and the fluctuations in the $CO_2$ density ($c'$). This also includes a correction of the measured $CO_2$ flux for changes in air density (Aubinet et al. 2000; Ge et al. 2011):

$$F_e = \left(\frac{\overline{pT_i}}{\overline{p_i}\,\overline{T_a}}\right)\left[\overline{w'c'_i} + \frac{m_a}{m_v}\left(\overline{c'}/\overline{\rho_a}\right)\overline{w'\rho'_{vi}}\right] \tag{3.4}$$

where $p$ is air pressure (Pa), $T$ is temperature (°K), and $\rho_{vi}$ and $\rho_a$ are the mean density of water vapor and dry air, respectively. Furthermore, $m_a/m_v$ is the ratio between the molecular mass of dry air and that of water vapor. The bars over factors refer to the time averages, and primes to the instantaneous fluctuations around the mean quantities (Aubinet et al. 2000). The carbon storage below the canopy ($F_s$) is obtained by integrating over the height the rate of temporal change in the $CO_2$ concentration profile below the canopy:

$$F_s = \frac{v_a \Delta c_r}{m} \tag{3.5}$$

where $v_a$ is the volume ($m^3$) of air column below the eddy instrument height, $\Delta c_r$ is the change in $CO_2$ density per unit time at the canopy source height, and $m$ is the molar volume of $CO_2$ (Aubinet et al. 2000).

In a boreal Scots pine forest (63° N, 145 m a.s.l), the seasonal patterns of daily *NEE* values were similar throughout the 2000–2002 monitoring period (Wang et al. 2004; Zha et al. 2004, Fig. 3.4). In winter, the values were slightly positive (carbon source) until the end of April, and were negative thereafter, following the release of winter dormancy and increase of photosynthetic capacity from early May to August. In the growing season, the maximum daytime values were – 4.8, –5.4, and – 6.6 g C $m^{-2}$ $d^{-1}$ in 2000, 2001, and 2002. The mean daytime values were – 2.42 ± 0.83, –2.68 ± 0.97, and – 2.88 ± 0.93 g C $m^{-2}$ $d^{-1}$. In the growing season, maximum nighttime values were 1.8, 2.8, and 2.9 g C $m^{-2}$ $d^{-1}$, and the mean values were 0.98 ± 0.23, 1.24 ± 0.32, and 1.63 ± 0.28 g C $m^{-2}$ $d^{-1}$. From September to October, carbon uptake declined until the trees fell into winter dormancy. The site became a

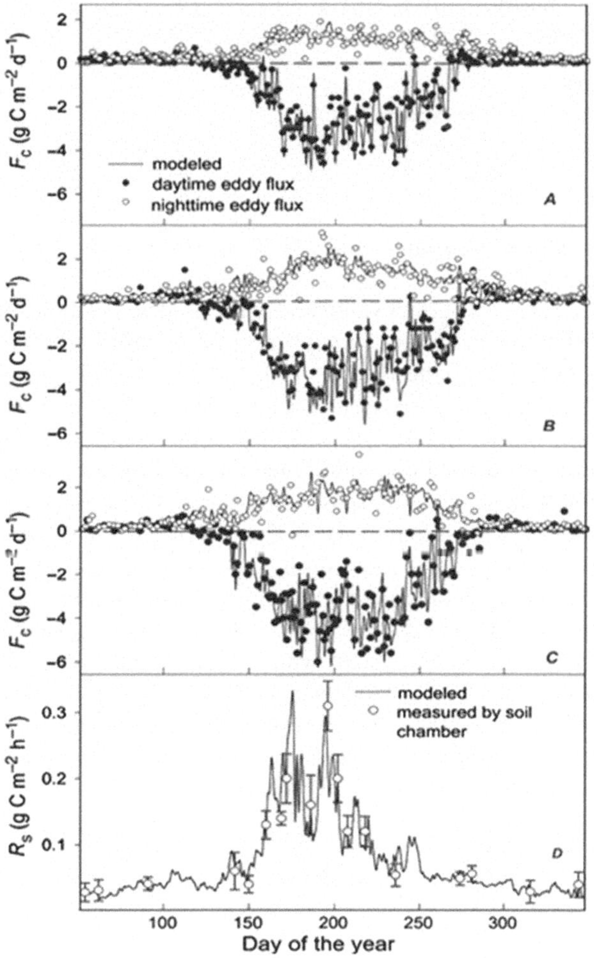


**Fig. 3.4** Seasonal and annual variability in daily net ecosystem exchange (**a**–**c**) and respiration in the forest floor (**d**) in a Scots pine (*Pinus sylvestris*) stand in middle-boreal conditions (63° N, 145 m a.s.l) in 2000–2002 (Wang et al. 2004). The points signify measurements, and the line indicates modeled values based on meteorological measurements. Permission of Oxford University Press

**Table 3.1** Modeled gross carbon uptake, ecosystem respiration, and net ecosystem exchange [g C $m^{-2}$ $yr.^{-1}$] in a boreal Scots pine (*Pinus sylvestris*) stand (62° N), with the contribution from diverse sources compared to the values based on a model utilizing meteorological measurements (Wang et al. 2004)

| Parameter | 2000 | | 2001 | | 2002 | |
|---|---|---|---|---|---|---|
| | Growing season | Non-growing season | Growing season | Non-growing season | Growing season | Non-growing season |
| Gross carbon uptake | −718 | −12 | −808 | −9 | −980 | −11 |
| Ecosystem respiration | 501 | 64 | 572 | 57 | 686 | 54 |
| Share of ecosystem respiration (%) for | | | | | | |
| • Needles | 48 | 17 | 33 | 18 | 28 | 18 |
| • Branches + stem | 10 | 4 | 9 | 5 | 11 | 4 |
| • Forest floor | 42 | 79 | 58 | 77 | 61 | 78 |
| Net ecosystem exchange (*NEE*) | | | | | | |
| • Modeled | −217 | 52 | −236 | 48 | −294 | 43 |
| • Measured | −194 | 63 | −261 | 51 | −316 | 58 |

small carbon source again (0.28 ± 0.09 g C $m^{-2}$ $d^{-1}$) when soil temperature fell below 5 °C.

Table 3.1 further summarizes the modeled carbon uptake through canopy photosynthesis and losses through ecosystem respiration presented in Fig. 3.4. Annual gross photosynthesis was −730, −817, and − 991 g C $m^{-2}$ $yr.^{-1}$ in 2000, 2001, and 2002. Similarly, ecosystem respiration was 565, 629, and 640 g C $m^{-2}$ $yr.^{-1}$, which accounted for 77%, 77%, and 65% of the annual gross photosynthesis. Carbon efflux from the forest floor contributed the most to ecosystem respiration: the share of the forest floor was 46–62%, while the share of foliage was 27–44%, and of wood (branches plus stem) 9–11%. The contribution of different components of ecosystem respiration varied more during the growing season than during the non-growing season, when the carbon efflux from the forest floor accounted for more than 77% of the total ecosystem respiration. Finally, the measured annual net carbon uptake (*NEE*) in the ecosystem was −131, −210, and −258 g C $m^{-2}$ $yr.^{-1}$ in 2000, 2001, and 2002. Modeling yielded values that were close to those measured (Wang et al. 2004). The measurements represented a peak *NEE* common for boreal tree stands at ages varying from 10 to 60 years, declining thereafter along with the maturation of trees (e.g., Hyvönen et al. (2007).

## 3.3 Seasonality of Carbon Sink/Source Dynamics in Boreal Conditions

*NEE* varies throughout the day and over seasons. In this regard, Ge et al. (2011) reported a time series of eddy covariance measurements over the period 1999–2008 in a boreal Scots pine forest located in eastern Finland (63° N, 145 m a.s.l). The site was of low fertility (*Vaccinium* site type) and was occupied by a tree stand. The mean density of the stand was 1175 trees $ha^{-1}$, with a mean height of 12 m, and a mean diameter of 11 cm. The soil at the site was a sandy loam with a water-holding capacity of 35 mm in the top 50 cm of soil. The climate at the site was characterized by cold winters with thick snow cover and a short growing season (140–175 days). Mean annual precipitation in the area for the period of 1961–2000 was around 700 mm, of which about 38% was of snow. The monthly mean temperature was –10.4 °C in January and 15.8 °C in July. Measurements were used as a reference to simulate *NEE* scaling from the photosynthetic biochemistry per unit leaf area to canopy photosynthesis per unit ground area. The sun/shade sub-model was used to consider the daily change in the fraction of sunlit ($LAI_{sun}$) and shaded leaf index ($LAI_{sh}$) in the canopy and the corresponding difference in photosynthetic rate and canopy stomatal conductance ($g_{cs}$).

Figure 3.5 shows the daily data points of the eddy covariance measurements and modelled values, which were calculated based on meteorological measurements using a process-based model outlined in Box 3.3 (Ge et al. 2011). Through most of the wintertime, the daily values of *NEE* were slightly positive (carbon source) until the end of April, representing the time of release of winter dormancy and increasing photosynthesis, i.e., the site became a carbon sink (negative values of eddy covariance measurements). In May, the carbon uptake increased rapidly, and through June to August the site continued to be a carbon sink, with the carbon uptake exceeding the carbon emissions. Thereafter, the carbon uptake declined through September to October until trees fell into winter dormancy and the site again became a small carbon source. The model adequately simulated the annual course of the daily values of carbon flux as measured by the eddy covariance method (Ge et al. 2011).

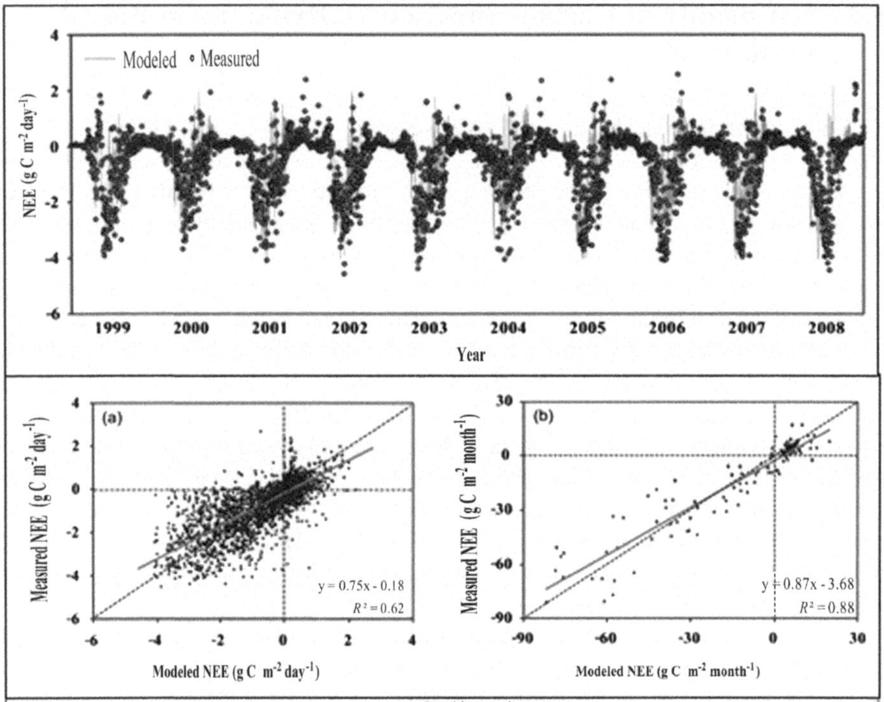


| Year | Growing season | | Outside growing season | | Whole year | |
|---|---|---|---|---|---|---|
| | Modeled | Measured | Modeled | Measured | Modeled | Measured |
| 1999 | -205.7 | -213.0 | 39.8 | 26.7 | -165.8 | -186.3 |
| 2000 | -189.0 | -195.0 | 59.2 | 34.0 | -129.8 | -161.0 |
| 2001 | -239.8 | -237.0 | 59.3 | 55.0 | -180.5 | -182.0 |
| 2002 | -246.7 | -250.4 | 43.9 | 36.7 | -202.8 | -213.7 |
| 2003 | -255.3 | -267.7 | 60.6 | 36.2 | -194.7 | -231.5 |
| 2004 | -189.8 | -206.8 | 34.3 | 48.6 | -155.4 | -158.1 |
| 2005 | -255.4 | -254.4 | 64.5 | 40.0 | -190.9 | -214.4 |
| 2006 | -252.4 | -240.9 | 65.9 | 50.2 | -186.5 | -190.7 |
| 2007 | -223.9 | -219.8 | 61.1 | 49.3 | -162.8 | -170.4 |
| 2008 | -263.1 | -257.0 | 68.8 | 67.2 | -194.3 | -189.8 |

**Fig. 3.5** Seasonality of carbon sink/source dynamics in a boreal Scots pine (*Pinus sylvestris*) stand (63° N, 145 m a.s.l) over a longer period based on eddy covariance measurements and model computations of net ecosystem $CO_2$ exchange (*NEE*, g C $m^{-2}$) over growing season (May–September), outside growing season (October–April), and whole year (Ge et al. 2011). Permission of Elsevier. Upper: Annual and daily variation of ecosystem net carbon flux (*NEE*) in the boreal Scots pine stand in the monitoring site in 1999–2008. Points are measured fluxes and solid lines are modeled fluxes. Middle: Relationship between measured and modelled ecosystem net carbon flux (*NEE*) on the daily (Left) and monthly (Right) basis. Lower: Measured and modelled amount of net ecosystem exchange (*NEE*) over growing season, outside growing season, and whole year

**Box 3.3 Overview for scaling up photosynthetic biochemistry per unit leaf area to canopy photosynthesis per unit ground area**

In calculations, net radiation absorbed in the canopy ($R_{nc}$) is divided into sunlit ($R_{nc.sun}$) and shaded ($R_{nc.sh}$) fractions of foliage. Thus, the net canopy photosynthesis ($A_c$) includes photosynthesis both in sunlit ($A_{c.sun}$) and shaded ($A_{c.sh}$) foliage (Ge et al. 2011):

$$A_c = A_{c.sun} + A_{c.sh} = \int_0^{LAI} f(R_{nc.sun}) f(T_c) f(c_a) f(N_L) f(LAI_{sun}) f(g_{s.sun}) dLAI + \int_0^{LAI} f(R_{nc.sh}) f(T_c) f(c_a) f(N_L) f(LAI_{sh}) f(g_{s.sh}) dLAI \quad (3.6)$$

$$R_{nc} = R_{nc.sun} + R_{nc.sh} \quad (3.7)$$

$$g_{cs} = \frac{LAI_{sun}}{LAI} \int_0^{LAI} g_{s.sun} + \frac{LAI_{sh}}{LAI} \int_0^{LAI} g_{s.sh} \quad (3.8)$$

where the functions $f(T_c)$, $f(C_a)$, $f(NL)$ and $f(g_{s.sun})$, and $f(g_{s.sh})$ refer to the effects of the air temperature, atmospheric $CO_2$, nitrogen content in foliage, and stomatal conductance in the sun and shade on the $CO_2$ uptake rate. Stomatal conductance (both on sunlit and shaded leaves) is controlled by radiation ($R_{nc}$), vapor pressure deficit ($D_a$), concentration of $CO_2$ ($C_a$), canopy temperature ($T_c$), soil moisture ($W_{soil}$), and soil temperature ($T_s$): $g_s = gs.\max f(- R_{nc}) f(D_a) f(c_a) f(T_c) f(W_{soil}) f(T_s)$ where $g_{s.max}$ is the maximum value of stomatal conductance. Dark respiration in foliage ($R_{l.\ d}$) is controlled by canopy temperature and the nitrogen content in foliage. Heterotrophic respiration in the forest floor is a function of soil temperature ($T_s$) and the volumetric water content ($\theta_s$) of soil:

$$R_{l.d} = R_{l.d(20)} f_1(T_c) \int_0^{LAI} f(N_L) f(LAI) dLAI \quad (3.9)$$

$$R_{b.w} = M_b R_{b(20)} + M_w R_{w(20)}) f_2(T_w) \quad (3.10)$$

$$R_s = R_{s(20)} \left(1 - e^{\theta_s c_\theta}\right) f_2(T_s) \quad (3.11)$$

where $R_{l.\ d(20)}$, $R_{b(20)}$, $R_{w\ (20)}$, and $R_{s(20)}$ are the rates of autotrophic respiration in the foliage, branches, sapwood, and the rate of heterotrophic respiration in

(continued)

**Box 3.3** (continued)
the forest floor at a temperature of 20 °C. Furthermore, $T_w$ refers to the temperature in sapwood. The functions $f(N_L)$, $f_1(T_c)$, and $f(LAI)$ represent the nitrogen content in foliage, temperature in the canopy, and the fraction of leaf area. The factors $M_b$ and $M_w$ (kg $m^{-2}$ ground) are the dry biomass of living branches and sapwood above the ground.

During the growing season (May–September), the amount of modeled *NEE* values varied from −189 to −263 g C $m^{-2}$, and from −195 to −268 g C $m^{-2}$ for eddy covariance measurements (Fig. 3.5). Outside the growing season, the simulated values varied from 34 to 69 g C $m^{-2}$, whereas the measured values fell in the range of 27–67 g C $m^{-2}$. Annually, the modeled values of *NEE* ranged from −129 to −203 g C·$m^{-2}$ and the measured values from −161 to −232 g C $m^{-2}$ for the period 1999–2008. During the growing season, the daily values of *NEE* were dependent on the air temperature ($T_c$), photosynthetically active radiation (*PAR*), and vapor pressure deficit ($D_a$). The modeled trend in *NEE* variation was congruent with the response of the net photosynthesis of Scots pine to corresponding environmental variables under laboratory conditions (Wang et al. 1996; Wang 1996).

## 3.4 From Source to Sink with Balancing Carbon Emissions and Uptake

### *3.4.1 Fluxes of Carbon Dioxide Over Boreal Clear-Cut Area and Forest-Covered Site*

Under even-aged forestry, management, and harvest affect climate warming through disturbances in the carbon dynamics of ecosystem from reforestation though the initial young phase to full maturity in the old phase. These successional processes are related to management-induced change in the structure and functioning of the ecosystem, with concurrent changes in sink/source interaction changes. The concurrent retention of carbon in an ecosystem is related to the production of litter, the carbon bound in biomass, and decomposing litter and organic matter in soil. After the final cut (clear-cutting, or cutting for natural regeneration), such sites may be carbon sources over several years before shifting to carbon sinks as reforestation progresses. This period covers the seedling phase, with links to precommercial management of seedling stands affecting the rate of the shift from a carbon source to a carbon sink.

The eddy covariance measurements of Rannik et al. (2002) show the effect of the growth and development of seedlings of Scots pine (*Pinus sylvestris*) and recovering ground vegetation on carbon dioxide ($CO_2$) and water vapor fluxes over forest cover and forest clearings (Fig. 3.6). The first site with forest cover was established in 1962, thus being 38 years old in the pole phase during the measurements. The height

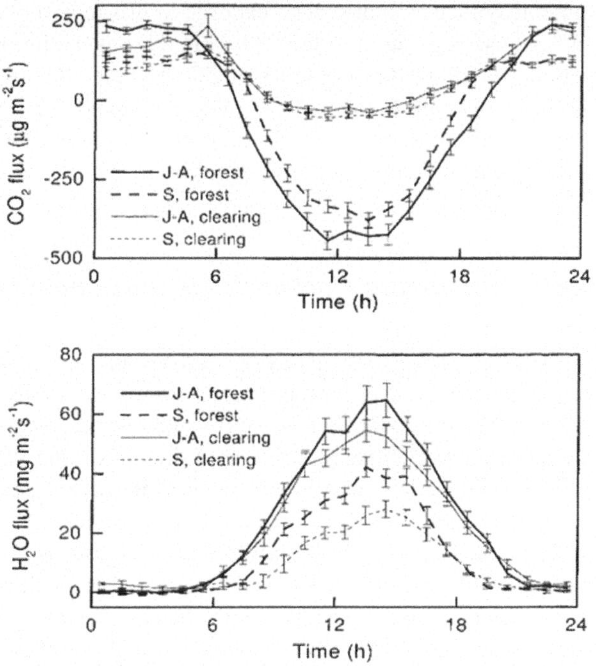


**Fig. 3.6** Mean diurnal course of $CO_2$ (Upper) and $H_2O$ (Lower) fluxes in July–August (J–A) and September (S) measured for a forest site (61°N) with forest cover containing a precommercial stand of Scots pine (*Pinus sylvestris*) and for a clearing on a site occupied by small seedlings of Scots pine The error interval denoted is ±1 S.E. (Rannik et al. 2002). Permission of Elsevier

of the dominant trees was 14 m in a pure stand of Scots pine. The stand density was 2500 trees ha$^{-1}$, with a biomass of 47 Mg ha$^{-1}$ and all-sided needle area of 7 m$^2$ m$^{-2}$. The ground vegetation included heather (*Calluna vulgaris*), cowberry (*Vaccinium vitis-idaea*), and blueberry (*Vaccinium myrtillus*). The dominating moss species was wind-blown moss (*Dicranum undulatum*). The annual mean temperature at the site was 3 °C, with annual mean precipitation of 700 mm. The soil was a coarse silty glacial till of low fertility (*Vaccinium* site type, VT).

The clearing site was clear cut 5 years before measurement (Fig. 3.6). Prior to clear-cutting, the site was covered by 90-year-old managed Scots pine forest on a site of low fertility (*Vaccinium* site type, VT). The soil was sandy and stony, representing an esker formation. After clear-cutting, harvesting residues were left on the site and the soil was scarified. One-third of the ground surface was represented by bare mineral soil, one-third was mostly bare humus and logging debris from the scarification, and one-third was intact. The ground vegetation covered about 40% of the area, mainly occupied by cowberry (*Vaccinium vitis-idea*), blueberry (*Vaccinium myrtillus*), and mosses (*Pleurozium schreberi* and *Dicranum* spp.). Based on the natural seeding, there was a sparse stand of small Scots pine seedlings (mean height < 20 cm) on the site.

According to Rannik et al. (2002), the nighttime respiration was of the same magnitude at both sites (Fig. 3.6). In the daytime, the forest was a strong sink, while at the clearing the carbon output and input were almost in balance. This implied that the $CO_2$ taken up by the ground vegetation balanced that released from the soil. In the forest-covered site, the $CO_2$ exchange was related to respiration from the soil and to photosynthesis in the forest canopy. In this regard, the contribution of understory and ground vegetation to the $CO_2$ balance was negligible. During the measurement period, the forest was a net sink of $CO_2$ and the clearing a $CO_2$ source. In the forest, the daily average uptake of $CO_2$ was −2.4 and −1.7 g $m^{-2}$ per day in the July–August and September periods, respectively, while the mean release of $CO_2$ in the clearing was 4.0 and 2.5 g $m^{-2}$ per day.

### *3.4.2 Balancing Carbon Emissions and Uptake over Time*

Grelle et al. (2023) also analyzed ecosystem carbon fluxes from eddy covariance measurements in five young stands on mesic sandy soils in southern Sweden (57–60° N) (Fig. 3.7). For all study sites, the climate was humid and continental, the mean annual temperature being 5.5–6.3 °C, and the mean annual precipitation 527–766 mm. The young forest stands with different productivity were in southern (four sites) and in south-central (one site) Sweden.

The measurements dealt with stands previously wind-felled or harvested in clear-cuttings. The measurements were concerned with three stands of Norway spruce (*Picea abies*), one of Scots pine (*Pinus sylvestris*), and one of larch (*Larix* × *eurolepis*). One of the spruce stands had the stumps harvested, while one was fertilized, and one had no specific treatment. The measurements were used for estimating (i) the carbon lost through emission of $CO_2$ over the regeneration phase after clear cutting; (ii) the link when a stand shifts from emitting to absorbing carbon dioxide; and (iii) the rate of recovery of the carbon balance in five differently disturbed and reforested clear-felled areas. The recovery of the carbon balance after clear cutting was studied in relation to site productivity, removal of coarse woody debris, and nitrogen fertilization (Fig. 3.7).

According to Grelle et al. (2023), the annual carbon fluxes returned from positive (sources) to negative (sinks) in 8–13 years after disturbance depending on the site productivity and management. The impact of productivity was especially pronounced when comparing the sites in southern and northern Sweden, which showed decreasing productivity northwards. The return took place in a period equal to 15% of the rotation length at these sites. This implied that the stands reached a neutral carbon balance under conventional reforestation in the period equal to 30% of the rotation period in these sites. Grelle et al. (2023) further found carbon emissions being the lowest, and recovery time shortest, in a stand where the stumps and logging residues had been harvested. In this stand, the total carbon gains since disturbance equaled the total losses after 11 years. Overall, Grelle et al. (2023) summarized that the managed forest in the southern boreal zone of Sweden is a carbon source during a small part of the rotation period. The length of the source period can be shortened by increasing the productivity of forest sites and/or reducing woody debris left on the site. Consequently, seedling establishment, stand productivity, and biomass removal in pre-commercial management are of importance in effecting the recovery of the carbon balance in commercial forestry in boreal forests.

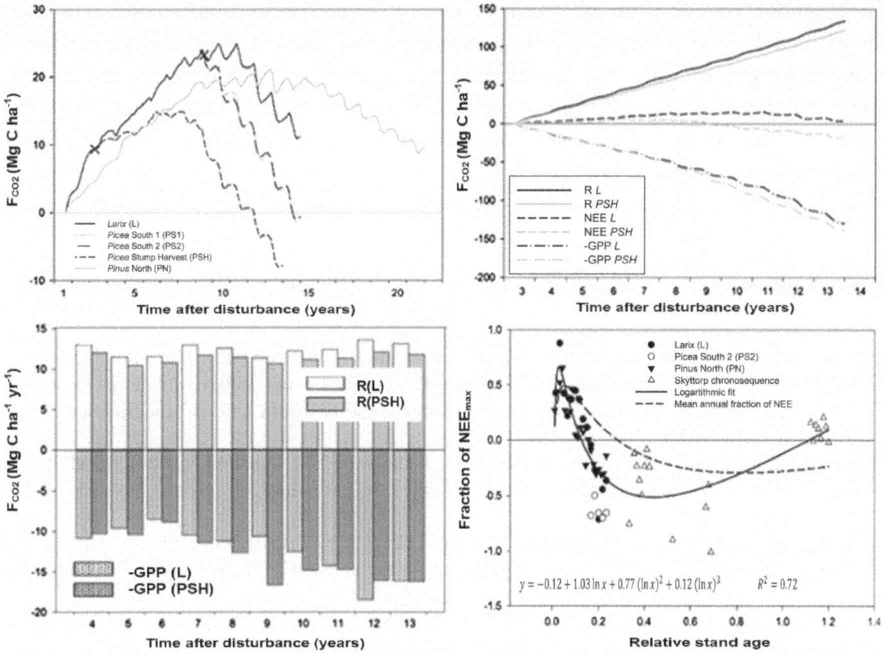


**Fig. 3.7** Recovery of the carbon balance in young forests across southern Sweden (Grelle et al. 2023). Permission of Elsevier. Legend for all the subfigures concerning tree species with treatment and location for figure panels: (i): PN is a *Pinus sylvestris* stand; (ii): PS1 and PS2 are *Picea abies* stands; (iii): L is a *Larix* × *eurolepis* stand; (iv): PSH is a *Picea abies* stand with stump harvest, and PN stands for *Pinus* sylvestris in the North. Left/Upper: Cumulative carbon dioxide fluxes by mass for reforestation after disturbance. The x-axis denotes the years after disturbance. Measurements for *Picea abies* (PS1, PS2, and PSH) started some years after the disturbance (starting points marked by crosses), and similarity with *Larix* × *eurolepis* (L) was assumed during the first, missing years. Right/Upper: Cumulative gross carbon dioxide fluxes in mass units of carbon of the larch site (L) and the stump-harvested site (PSH). *R* = ecosystem respiration, *NEE* = net ecosystem exchange, *GPP* = gross primary productivity. Left/Lower: Annual sums of gross ecosystem carbon dioxide fluxes from sites for PSH and L as a function of time since disturbance. Right/Lower: Normalized relationship between relative stand age and net ecosystem exchange (*NEE*). The equation is valid for the solid line. The stand age is normalized with rotation period and *NEE* is normalized with the maximum *NEE* during the rotation. The intersection of the solid and the dashed line denotes the point where harvest yields maximum mean annual *NEE*

## 3.5 Concluding Remarks

The findings of Rannik et al. (2002) and Grelle et al. (2023) deal with single stands that form chronosequences in the regularly managed forest landscape over separate stand-based production cycles (rotations). At the single-stand scale, the carbon balance fluctuates from carbon uptake to loss at final felling, inducing shifting spatial fluctuation in the carbon stock in forest landscapes depending on the geographical scale. The regular, appropriate management and allocation of harvest throughout the forest landscape allow maintenance of carbon sequestration and stocks at a level determined by the productivity, management, and harvest. Management-induced changes in the structure and functioning of ecosystems, including harvest cycles, affect the potential sequestration of carbon in tree stands and in the soil occupied by trees. This affects the carbon sequestration in the whole forest ecosystem following stand-based successional processes.

## References

Aber JD, Melillo JM (1991) Terrestrial ecosystems. Saunders College Publishing, Philadelphia

Aubinet M, Grelle A, Ibrom A, Rannik Ü, Moncrieff T, Foken T et al (2000) Estimates of the annual net carbon and water exchange of European forests: the EUROFLUX 612 methodology. Adv Ecol Res 30:113–175

Chertov OG, Komarov AS, Nadporozhskaya M, Bykhovets SS, Zudin SL (2001) ROMUL – a model of forest soil organic matter dynamics as a substantial tool for forest ecosystem modelling. Ecol Model 138:289–308

Ge Z-M, Kellomäki S, Zhou X, Wang K-Y, Peltola H (2011) Evaluation of carbon exchange in a boreal coniferous stand over a 10-year period: an integrated analysis based on ecosystem model simulations and eddy covariance measurements. Agric For Meteorol 151:191–203

Gower ST, Krankina O, Olson RJ, Apps RM, Linder S, Wang C (2001) Net primary production and carbon allocation patterns of boreal forest ecosystems. Ecol Appl 11(5):1395–1411

Grelle A, Hedwall P-O, Strömgren M, Håkansson C, Bergh J (2023) From source to sink – recovery of the carbon balance in young forests. Agric For Meteorol 330:109290

Hamilton JG, DeLucia EH, George K, Naida SL, Finzi AC, Schlesinger WH (2002) Forest carbon balance under elevated $CO_2$. Oecologia 131:250–260

Hyvönen R, Ågren GI, Linder S, Persson T, Cotrufo MF, Ekblad A, Freeman M, Grelle A, Janssens IA, Jarvis PG, Kellomäki S, Lindroth A, Loustau D, Lundmark T, Norby RJ, Oren R, Pilegaard K, Ryan MG, Sigurdsson BD, Strömgren M, van Oijen M, Wallin G (2007) The likely impact of elevated [$CO_2$], nitrogen deposition, increased temperature, and management on carbon sequestration in temperate and boreal forest ecosystems: a literature review. New Phytol 173:463–480

IPCC (2007) Climate change 2007: impacts, adaptation, and vulnerability. In: Working Group II contribution to the Fourth Assessment Report of the Intergovernmental Panel on Climate Change. Cambridge University Press, Cambridge

Lovett GM, Cole JJ, Pace ML (2006) Is net ecosystem production equal to ecosystem carbon accumulation? Ecosystems 9:152–155

Rannik Ü, Altimir N, Raittila J, Suni T, Hussein AGT, Hölttä T, Lassila H, Latokartano M, Lauri A, Natsheh A, Petäjä T, Sorjamaa R, Ylä-Mella H, Keronen P, Berninger F, Vesala T, Hari P, Kulmala M (2002) Fluxes of carbon dioxide and water vapour over Scots pine forest and clearing. Agric For Meteorol 111:187–202

Wang K-Y (1996) Effects of long-term $CO_2$ and temperature elevation on gas exchange of Scots pine. PhD thesis,. University of Eastern Finland

Wang K-Y, Kellomäki S, Laitinen K (1996) Acclimation of photosynthetic parameters in Scots pine after three years of exposure to elevated temperature and $CO_2$. Agric For Meteorol 82:195–217

Wang K-Y, Kellomäki S, Zha T, Peltola H (2004) Component carbon fluxes and their contribution to ecosystem carbon exchange in a pine forest: an assessment based on eddy covariance measurements and an integrated model. Tree Physiol 24:19–34

Zha T, Kellomäki S, Wang K-Y, Rouvinen I (2004) Carbon sequestration and ecosystem respiration for 4 years in a Scots pine forest. Glob Chang Biol 10:1492–1503

# Part II
# Carbon Dynamics in Boreal Forest Landscapes

# Chapter 4
# Carbon in the Shifting Mosaics of Forest Landscapes

## Contents



 
S. Kellomäki, *Forest Management for Timber Production and Climate Change Mitigation*, Managing Forest Ecosystems 44,
https://doi.org/10.1007/978-3-031-71575-4_4

**Abstract** In forest ecosystems, succession refers to the gradual replacement of one population/community of trees and various ground vegetation species by another population/community in a process that inclines toward increasing complexity of habitats and diversity of organisms. At the landscape scale, the separate forest stands (or patches, or pieces of landscape mosaic) shift from one developmental phase to another. During the shift, separate mosaic pieces show structural differentiation based on the life cycle of trees. Natural disturbance and comparable forestry management measures tend to reduce pre-disturbance carbon stocks (legacy carbon) from the previous production cycle. Management and harvest operations affect carbon stocks in trees and ground vegetation, as well as carbon in soil litter and detritus. Concurrently, soil carbon increases in litterfall and is lost through decay and respiration in soil litter and detritus. In reforested clear-cut areas, carbon loss decreases until the uptake of carbon rises again in the recovering ground vegetation and growth of seedlings, thus compensating the carbon lost from soil after clear cutting and contributing to the recovery of the ecosystem carbon balance.

## 4.1 Succession in Forest Ecosystems

### *4.1.1 Succession, Growth, and Development of Forests*

In boreal forest landscapes, natural disturbance and forest management/harvest change the carbon dynamics in ecosystems at varying spatial and temporal scales. White and Pickett (1985) define disturbance to be "any relatively discrete event in time that disrupts ecosystem, community, or the physical environment." In this regard, carbon stocks and fluxes are affected by both natural and human-induced disturbances, with effects on the terrestrial carbon cycle in ecosystems and on biogeochemical interactions between ecosystems and the atmosphere. At spatial and temporal scales, carbon dynamics are linked to the succession of populations/communities of trees and other vascular (shrubs, herbs, grasses) and non-vascular (mosses, lichens) plants depending on the properties of the forest landscape (Fig. 4.1).

Disturbance disrupts the structure and functioning of ecosystems, communities, or populations, with changes in the availability of resources and substrate, or physical environment (light, temperature, soil moisture, and nutrients) (e.g., White and Pickett 1985). Disturbance occurs at variable spatial and temporal scales related to, for example, attacks of fungi and insects, and to exposure to wildfires, storms, snow load, and so forth, thus increasing the mortality of trees, ground vegetation, and other organisms. Death of a tree or trees forms gaps (openings) in the canopy of communities/populations of variable size, from the crown area of single trees to forest landscapes representing many separate trees and stands (Fig. 4.1).

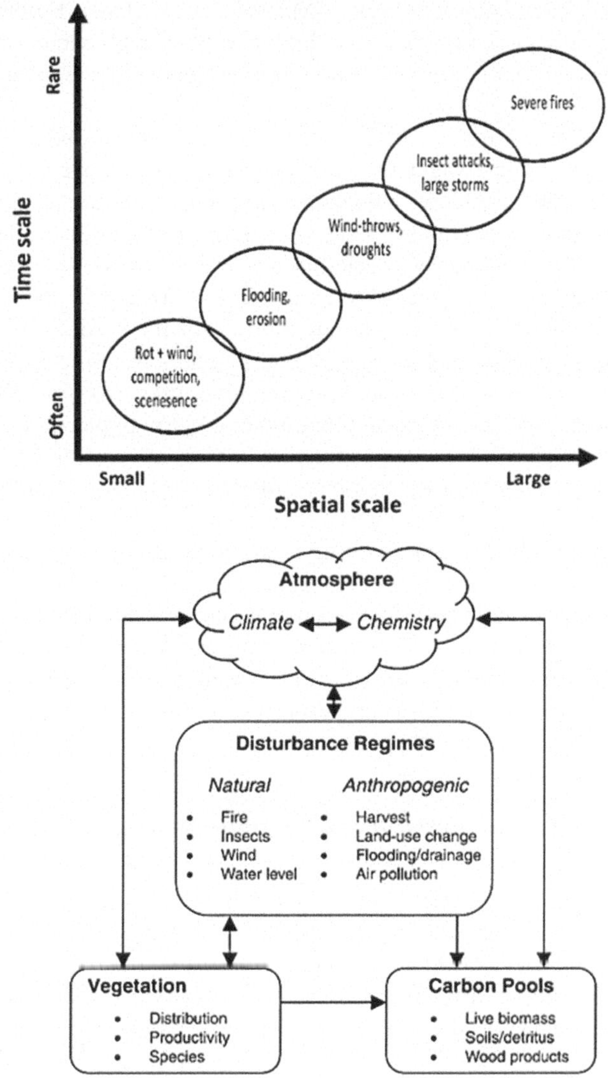


**Fig. 4.1** Outline of the spatial and temporal scales of selected disturbance events impacting boreal forests and, consequently, secondary succession. Upper: Disturbance events occurring at a small spatial and temporal scales include the windfall of individual trees, competition for growth resources between individual trees, and local fungal and insect attacks that kill individual trees or small groups of trees. Large-scale and temporal disturbance events include storms and wildfires, which destroy forest and tend to have long-lasting effects on forest ecosystems at the landscape scale. Small- and large-scale disturbance impact both natural and managed forests (Framstadt et al. 2013, pp. 27–30). Open Access. Courtesy of Nordic Council of Ministers. Lower: Outlines of the effects of disturbance on the dynamics of terrestrial carbon, and feedback between the atmosphere and terrestrial carbon through various boreal forest pools and pools outside (Bhatti et al. 2003, pp. 799–855). Open Access. Courtesy of NRC Research Press

### *4.1.2 Succession Related to Disturbance Dynamics*

In the context of forest ecosystems, succession refers to the gradual replacement of one population/community of trees and various ground vegetation species by another population/community inclining toward increasing complexity of habitats and diversity of organisms. Successional dynamics are driven by disturbance, which controls (i) the structure and properties of a site and the plant populations/communities occupying the site; and (ii) establishment of new communities/populations growing and developing toward maturity and stability in the prevailing environmental conditions (Muscolo et al. 2014). Furthermore, succession affects herbivores and organism of higher trophic levels occupying the sites. Here, the focus is on the primary producers (trees, ground vegetation) providing energy for organisms at higher trophic levels.

Succession is divided into primary and secondary ones. Primary succession occurs when vegetation occupies sites that never contained vegetation previously, such as vegetation colonizing land uplifted at the shores of lakes or from the sea (Fig. 4.2). Secondary succession refers to the recovery of vegetation at sites previously occupied by vegetation that was destroyed, for example, by wildfire or land management. Secondary succession is driven by population processes, including regeneration, growth, and mortality, which control the dynamics of populations and communities of trees and ground vegetation. In the cycle of secondary succession, the recovery of tree stands/communities proceeds from disturbance and regeneration and consequent growth toward the old-growth phase. This forms a chronosequence through different developmental phases. Succession of a chronosequence involves the growth and development of trees from the seedling phase through pole and middle-age phases to the old-growth phase, where full maturity is reached toward the end of the life span specific to different tree species. The phase-by-phase development of single patches or separate patches of trees in different developmental phases takes up carbon through growth but loses it through respiration in growth processes and in the decay of litter, dead trees and detritus in soil. In chronosequences, many previous ecological interactions (ecological legacy) are present in managed forests, where clear-cut areas are reforested through planting/seeding after cutting.

### *4.1.3 Succession, Management, and Ecosystem Services*

In forest succession, living and non-living components of forest ecosystems form and maintain the natural capital, which provides ecosystem goods and services obtainable for direct and indirect use (Fig. 4.3). Goods refer to the use/non-use of the output from the ecosystem, and services to the processes producing goods (Bateman et al. 2011; Sing et al. 2015). Ecosystem goods and services are thus a result of the structure and functioning of ecosystems. They are related to successional phases at the tree, patch, and forest landscape scales under no management, or under the management needed for producing specific goods and services in forestry. In this process, succession of tree/plant communities gradually transforms habitats from open to closed ones in a chain involving the initial tree mortality, through to the growth of seedlings and, eventually, mature trees.

Each phase of a chronosequence is successive, but the process reverts to previous phases following subsequent disturbance (Fig. 4.3). Forest communities may be characterized by floristic and/or physiognomic properties representing the overstory (upper and lower tree layers of the canopy) and the understory (lower layers of shrubs, herbs, mosses, and lichens). Tree seedlings may belong to the understory (e.g., shrub layer and the ground cover, including grass, herb, and moss layers)

**Fig. 4.2** Total carbon stores and net change in total carbon, and total primary growth and change in primary and secondary succession in forest ecosystems over time (Harmon 2002). Permission of Oxford University Press. Upper: Net primary production (solid line) and the biomass change (dotted line) as a function of time since disturbance. Middle/ Upper: Change in net primary production (solid line) (dotted line) and biomass change (solid line) as function of rotation length. Middle/Lower: Total carbon stocks for secondary (solid line) and primary (dotted line) as a function of time since disturbance. Lower: Net change in carbon stocks for secondary succession (solid line) and primary succession (dotted line) as a function of time since disturbance

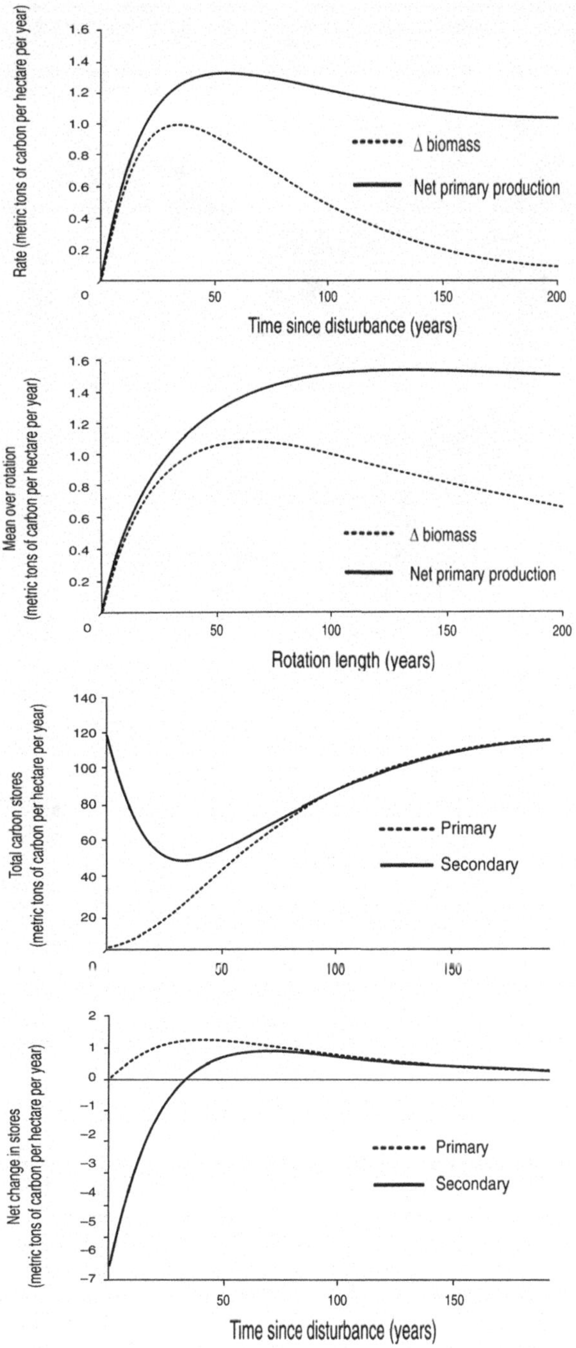

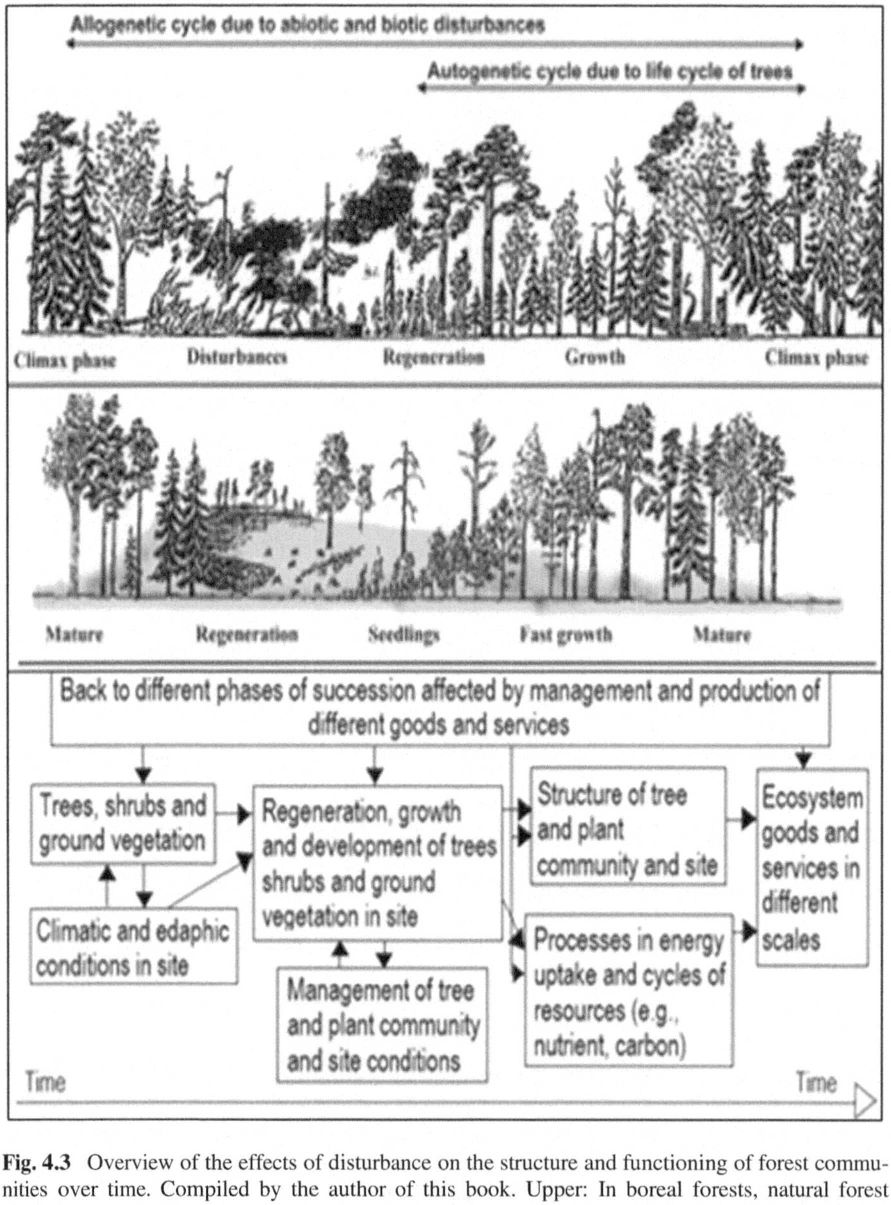


**Fig. 4.3** Overview of the effects of disturbance on the structure and functioning of forest communities over time. Compiled by the author of this book. Upper: In boreal forests, natural forest succession is a combination of autogenic and allogenic successions. *Autogenic* succession is driven by minor disturbance events such as the fall of individual trees in locally high winds. In *allogenic* succession, the dynamics of forest communities are driven by major disturbance such as large-scale forest damage due to high wind. This landscape structure is related to the mosaic-like variability in environment throughout the landscape. Middle: Outline of successional dynamics of trees in managed boreal forest ecosystems following the regenerative cutting of mature trees. Regeneration starts with natural seeding like from parent trees after seed-tree cutting. This is followed by the growth of seedlings and trees of intermediate age under pre-commercial and commercial management toward maturity in the chronosequence. Below: Processes driving succession initiated in an opening created by disturbance from reforestation to maturity of trees and vegetation in patches, which have a potential to produce ecosystem goods and services in different spatial and temporal scales

below the dominating tree canopy. Over time, growing tree seedlings form the overstory, constituting the upper canopy when trees achieve dominating height over other plant layers (Oliver and Larson 1990, pp. 140–162).

In natural succession, the structure and functioning of forest ecosystems change gradually over time (Fig. 4.3). Succession is controlled by abiotic (wind, wildfire, snow) and biotic (mortality of trees due to limited space, attack by insects and other pests) disturbance. This modifies the structure and the consequent functioning of the ecosystem from moment to moment, causing reversions in the developmental phase of the ecosystem. Dying trees provide space for regeneration, with subsequent increases in growth and the accumulation of carbon in biomass in trees and other plants. Tree populations and communities of varying age grow and develop from populations of seedlings to populations of mature trees, with wide variations in their prevailing function.

Management and harvest produce a forest landscape whose dynamics follow the overall dynamics of natural forest landscapes in chronosequences at different developmental phases (Fig. 4.3). Under harvest and management, the chronosequences are related to timber and biomass production of goods with a limited production cycle. The production cycle is based on the ecological and economic limits, with social acceptance. Under even-aged management, the production cycle involves the rotation from reforestation to the final cut. Under uneven-aged management, the production cycle involves an interval between selective cuttings with no specific regeneration phase, unlike even-aged management. In natural chronosequences with no specific time span, the succession proceeds related to the life span of different tree species, with close links to natural disturbance.

## 4.2 Gap Dynamics, from Trees and Patches to Landscapes

### *4.2.1 Spatial Patterns of Gaps at Varying Temporal Scales*

In boreal forest landscapes, the regeneration, growth, and development of trees occur in patches of land if even-aged management is used. A patch refers to an area of forest land occupied by a given tree population/community. Patches are the smallest features used for classification of forest land/landscapes for ecology, forestry, and mapping. Trees in patches form stands of trees, which refer to a population/community of trees sufficiently uniform in composition, and spatial arrangement. Thus, a tree stand refers to an aggregation of trees occupying a specific area (patch) being distinguishable from the tree stands in adjoining areas. Furthermore, landscape patterns consist of separate habitats, including patches and tree stands. A habitat is any discrete area with a definite shape and spatial configuration used by a species for breeding or obtaining resources.

At the landscape scale, the separate forest stands, or patches, or pieces of landscape mosaic shift from one developmental phase to another (Box 4.1). During the shift, separate pieces show structural differentiation related to the life cycle of trees. An episode of high wind, for example, turns the development of single or several stands, back to an earlier successional phase. The risk of such shifts increases when trees grow taller with likely uprooting of trees or breaking of stems for excess wind force. At the same time, the risk and intensity of fire may increase due to the increasing accumulation of litter and other flammable material on sites. The structures and functioning of stands/patches over an entire region vary over time following the successional dynamics of each stand or patch of trees from post-disturbance reorganization toward stability. The mass of trees (living or dead) throughout a region oscillates temporally and spatially following the succession of the patch mosaic of sites and tree stands (shifting-mosaic in steady state) (e.g., Bormann and Likens 1979, pp. 164–176).

Patch dynamics is an approach emphasizing spatial patterning at a variety of scales, including that of individual trees, populations of trees, communities of trees, and forest landscapes (e.g., Pickett et al. 1987; Pickett and Rogers 1997). The patch dynamics of individual trees are driven by the scale of the canopy gap created by the dying tree (e.g., a radius of 10 m, with the crown area up to 100 $m^2$ in mature trees) that allows space for the regeneration and growth of offspring. At the population and community scale, patch dynamics concern several trees in each space, with a gap area extending over the radius covering the length of two to three trees in the space (e.g., a radius of up to 20 m, with a crown area up to 1500 $m^2$), representing also clearing of land for the regeneration and growth of offspring. Landscape patch dynamics extend over several populations and communities, allowing space for regeneration and growth of offspring over areas previously occupied by several populations and communities (Box 4.1).

**Box 4.1 Reforestation, growth, and development of trees in gaps**

Gaps make resources available and change the microclimate based on the size of the gap, through increasing radiation, temperature, and moisture on the soil surface. In addition, gaps increase nutrients in the soil due to the increasing infiltration of water, decay of soil organic matter, and release of nutrients. At the individual-tree or small-patch scale, regeneration may occur through seeding from trees surrounding the gaps or from the buds and roots of destroyed trees in the gap. Establishment of seedlings depends on the properties of the seedbed, which determines the number of seeds germinated, as well as on herbivores and pests that destroy seeds and small seedlings. Shade-tolerant Norway spruce (*Picea abies*) is better adapted to small canopy gaps than the shade-intolerant Scots pine (*Pinus sylvestris*) and birches (*Betula* sp.), which are better adapted to large gaps.

In general, the shade-intolerance and shade-tolerance of plant species, especially those of trees, drive succession in boreal forests. The dominant species in early phases (immediately after disturbance) are widely called pioneer species, and those in the late phase (long after disturbance) are called

(continued)

**Box 4.1** (continued)
climax species. Typically, pioneer species produce large quantities of wind-dispersed seeds and/or regenerate from buds in a widely spread root system. Once pioneer species have occupied sites, their regeneration declines, allowing climax species to become dominant and even replace pioneer species. Shade-tolerant species tend to form climax or self-perturbating populations/communities. When a subsequent disturbance occurs, pioneer species again occupy the site in a new successional cycle. In the boreal forests, shade-intolerant birches are adapted to exploit large gaps after fire or wind disturbances, whereas shade-tolerant Norway spruce establishes itself in small gaps below the canopy of shade-intolerant species in the absence of disturbance. A canopy opening (gap) area $> 0.04$ ha is required for the successful regeneration of many shade-intolerant trees, especially deciduous ones (e.g., Busing 1994; Bolton and D'Amato 2011).

### *4.2.2 Carbon Stock Changes Related to Disturbance*

The impacts of disturbance on the dynamics of forest ecosystems may be divided into those affecting at the patch or stand scale, or at the landscape scale. Patch-scale disturbance involves single trees or small groups of trees downed by high winds or snow load. Small-scale disturbance also includes management operations like the harvesting of trees thinned for spacing in tree stands. Similarly, thinning to a wide spacing for natural regeneration (e.g., shelterwood and seed-tree methods), thus maintaining the vigor of trees or improving reforestation for the next production cycle. At the landscape scale, natural disturbance like high winds may destroy trees in several patches. This is typical also for wildfires, which reduce the carbon stored in trees and in the litter and soil detritus. Clear-cutting over several adjacent patches may be comparable to a large-scale disturbance, which tends to reduce ecosystem carbon at the landscape scale.

Regardless of the scale of disturbance, the impacts of disturbance on the dynamics of ecosystem carbon stocks over time can be visualized as in Fig. 4.4. Natural disturbance and comparable management measures tend to reduce the initial stock of legacy carbon from the previous production cycle. Management and harvest operations also affect the carbon stocks in trees and ground vegetation, as well as that in the litter and soil detritus, with a subsequent reduction of carbon uptake and autotrophic respiration. This increases heterotrophic respiration and loss of carbon from the soil litter and detritus. Rapid and immediate decay leads to a brief period in which the ecosystem is a carbon source, followed by an extended period of carbon uptake. A delayed peak in microbial respiration ($R_H$), caused by delayed snag falls and decay of coarse woody debris in the soil, thus delaying the onset of sustained net carbon uptake. In clear-cut areas, at the ecosystem level, the carbon decreases until the uptake of carbon rises with the recovering growth of ground vegetation and tree

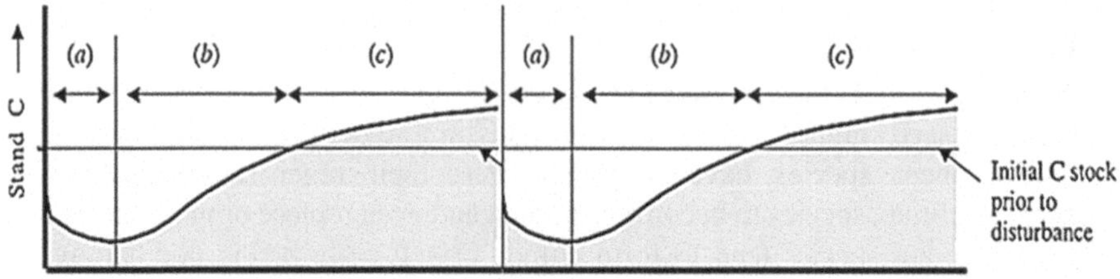


**Fig. 4.4** Stand-level carbon dynamics in boreal forests after two consecutive disturbance events (clear-cutting). During the first phase after disturbance (**a**), the carbon losses in decomposition and respiration exceed the carbon uptake (i.e., the stand is a carbon source). During the next phase (**b**), carbon uptake exceeds carbon losses (i.e., the stand is a net carbon sink), but the stand has not yet regained the carbon level prior to disturbance. During the final phase (**c**), the stand is a net sink, and carbon stocks exceed pre-disturbance (pre-clear-cutting) levels (Bhatti et al. 2003, pp. 799–855). Open Access. Courtesy of NRC Research Press

seedlings, thus balancing the carbon lost from litter and soil detritus since disturbance. The pre-disturbance carbon stocks in the patch are re-established, implying that the growing and developing tree stand has recaptured the carbon released in the disturbance event (Bhatti et al. 2003).

### *4.2.3 Impacts of Disturbance on the Carbon Cycle in Forest Ecosystems*

The effects of disturbance-induced changes in forest structure and functioning on ecosystem carbon dynamics are related to tree species composition, size, and density in populations/communities as well as to the properties of tree cover (leaf area, canopy coverage), ground vegetation, coarse wood litter, and soil detritus (Fig. 4.5). Disturbance transfers carbon from one pool to another, such as from the stems of standing living trees to dead coarse woody debris. On the other hand, carbon transfer also occurs from living biomass in trees and ground vegetation and from soil detritus to the atmosphere through wildfires and decomposition processes. In addition, disturbance changes microclimatic conditions and the physical and chemical properties of soil, thus producing spatial and temporal variability in forest structure, successional phases, and carbon cycle trajectories through chronosequences. Liu et al. (2011) further summarize how the legacy impacts of disturbance affect the carbon cycle via mortality, mass transfer, and succession, with changes in heterotrophic respiration, forest succession, and soil/climate interaction (Fig. 4.5). The mortality of trees, such as in wildfires, transfers carbon from the ecosystem directly to the atmosphere, or releases carbon from standing dead trees and downed dead trees exposed to decay.

Carbon in pools of living organs is transferred to the carbon pools of dead organs, as in leaves to litter, trees to snags or logs, living coarse roots to coarse woody debris,

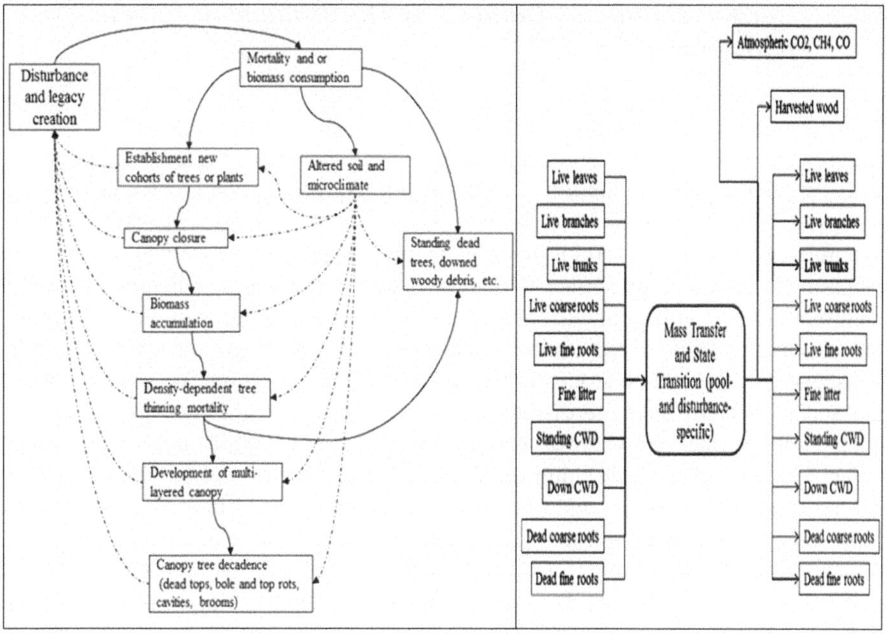


**Fig. 4.5** Carbon dynamics in boreal forest ecosystems, with disturbance-induced changes in forest structure and functioning (Liu et al. 2011). Open Access. Courtesy of Henday Scientific Journals. Left: Impacts of disturbance on structural development and creation of disturbance legacies in forest ecosystems. Disturbance legacy refers to the influence of the structure and functioning of an ecosystem prior to disturbance on post-disturbance ecosystem structure and functioning. The influence of legacy on ecosystem structure and functioning persists over a given period depending on the properties of landscape mosaics. Right: Framework outlining the mortality and carbon transfer among carbon pools in a forest ecosystem following disturbance, which are pool- and disturbance-specific. Coarse woody debris (CWD) refers to the stems of dead trees blown down by high winds. In this context, the rate of mass transfer indicates the intensity of disturbance, including the mortality and the mass of carbon transferred between pools. Harvest wood refers to harvest of stem-wood only, or whole-tree harvest, with effects on the carbon pools in different tree organs, thus further affecting greenhouse gas emissions from the forest ecosystem (Liu et al. 2011)

living fine roots to dead fine root debris, and so forth (Fig. 4.5). Hence, disturbance drives carbon transfer among pools and subsequent impacts on the carbon cycle. According to Liu et al. (2011), "the carbon transferred from any living carbon pool to its dead carbon equivalent is generally calculated as the fraction that dies (carbon transfer coefficient) multiplied by the pre-disturbance living carbon pool." To this end, Liu et al. (2011) quantified the effect of disturbance by assigning coefficients to describe fractions of the carbon and nutrient pools (Fig. 4.5), representing (i) those that are consumed and emitted to the atmosphere; (ii) those that die and fall to the ground; (iii) those that die and remain standing; and (iv) those that are harvested and transported out of the forest site. Therefore, mass transfer rates indicate the intensity and frequency of disturbance, including the reduction of living carbon pools and the mass transferred among pools.

### *4.2.4 Carbon Dynamics Induced by Disturbance in Forest Ecosystems*

Figure 4.6 shows, in theory, how a stand-replacing event (e.g., clear-cutting) changes the carbon stocks and fluxes in a forest ecosystem. Following the disturbance, gross primary production (*GPP*), net primary production (*NPP*), net ecosystem production (*NEP*), and net biome production (*NBP)* decrease substantially. According to Liu et al. (2011), disturbance may switch the forest ecosystem from a carbon sink to a carbon source. This post-disturbance change indicates a decrease in carbon uptake, with a concurrent increase in respiration. In a span of 20–30 years, the carbon uptake and respiration balance each other to pre-disturbance levels. Consequently, the carbon stocks in living biomass (trees, ground vegetation), litter, and soil recover to their pre-disturbance levels. However, the recovery of carbon stocks implies the recovery of carbon losses in autotrophic and heterotrophic respiration, as well as in logging removal. The recovery of carbon losses tends to stabilize earlier than the growth of ecosystem carbon stocks. Liu et al. (2011) found that regenerating temperate forest ecosystems remain a source of carbon for at least 14 years after clear-cutting before switching to a carbon sink, as indicated by the net ecosystem carbon exchange. Similarly, Amiro et al. (2010) found the switch from source to sink to occur in a span of 10–20 years from the stand-replacing disturbance event when eddy covariance measurements were used to analyze the carbon dynamics after reforestation.

## 4.3 Shifting Mosaics of Tree Stands, with Effects on Carbon Distribution in Forest Landscapes

### *4.3.1 Succession of Forest from Patches to Landscapes over Time*

The successional dynamics of boreal forests are mainly based on the responses of the regeneration, growth, and mortality of trees to prevailing environmental conditions in interaction with autogenic and allogenic disturbance. In natural boreal forests, the location, size, and shape of given patches are related to disturbance induced by events like strong wind, snow load, wildfire, attack by insects and fungi, and mortality of trees due to interspecific competition for resources and to life span. The long-term growth and development of tree stands on given sites form chronosequences associated with given disturbance events. These chronosequences are either in synchrony with those of other patches or not, depending on the timing of natural disturbance or management/harvest operations in each patch. This disturbance induces variable endogenous sequences of developmental phases of tree stands, from reforestation to full maturity depending on the size and severity of the disturbance.

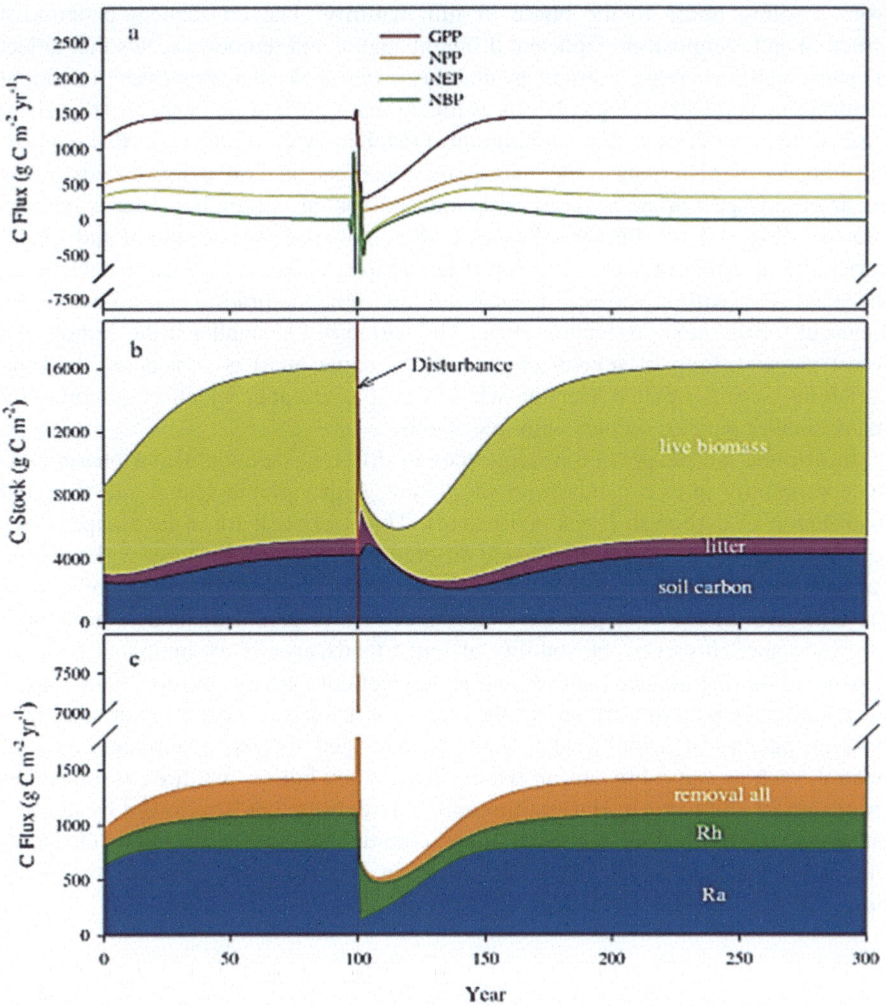


**Fig. 4.6** Theoretical carbon stocks and fluxes in a forest ecosystem following a stand-replacing disturbance event (Liu et al. 2011). Open Access. Courtesy of Hindawi Scientific Journals. Upper: Changes in gross primary productivity (*GPP*), net primary productivity (*NPP*), net ecosystem productivity (*NEP*), and net biome productivity (*NBP*) following a disturbance. Middle: Changes in three major carbon pools as affected by disturbance. The ecosystem loses a significant amount of carbon during and after the disturbance event, but it will have net carbon gain after about 20 years due to forest regrowth. Lower: Changes of ecosystem autotrophic respiration ($R_a$), heterotrophic respiration ($R_h$), and removals (e.g., logging, fire removal, insect consumption, erosion, etc.). $R_a = GPP - NPP$; $R_h = NPP - NEP$; removal all $= NEP - NBP$

According to White et al. (2000) and Jentsch et al. (2002), the final location and shape of a given patch are determined by random processes, but the pattern of subsequent chronosequence (structure and composition) is non-random proceeding from seedling phase to the phase of full maturity. The consequent patterns of structure and composition represent different spatial and temporal scales that reflect (i) site conditions (e.g., nutrient availability); (ii) ecological processes, including disturbance; and (iii) ecological units as functional groups (a set of species that share similar characteristics within a community) (McKinley et al. 2011). In this context, the impacts of disturbance and management/harvest on boreal forest ecosystems occur on an appropriate scale of temporal and spatial variability. This is demonstrated in Fig. 4.7 for the variability of carbon storage across temporal and spatial variability at different scales in a forest landscape. Under even-aged management, the stand-scale carbon varies from clear-cutting to the maturation of trees before the clear-cut for the next production cycle. The variability is smaller if the carbon in a forest parcel (group of several patches in the same area) is considered, and the variability is still smaller over the whole forest landscape, which is compiled by many smaller patches of sites with tree stands.

In a forest landscape, chronosequences in different developmental phases produce variability in tree stands from site to site, with variable spatial and temporal distribution of carbon in forest ecosystems. Thus, a boreal forest landscape represents a shifting mosaic of tree stands in different developmental phases with variable carbon stocks in the trees and soil. In a large area, this variability is stable relative to the long-term disturbance patterns and induced successional dynamics throughout the landscape. However, the stability of forest landscapes is dynamic and is scale-dependent on disturbance patterns and ecological complexity. In forest landscapes, the disturbance pattern summarizes the effects of disturbance events over space and time on patches at a multi-patch scale. In managed forests, disturbance may be natural, such as caused by strong winds, heavy snowfall, or wildfires, or related to management and harvest. Harvest through clear-cutting that is followed by reforestation operations, and pre-commercial and commercial cuttings, modifies the structure of tree stands and soil, with effects on the growth and development of tree stands on the sites exposed to these activities.

### *4.3.2 Carbon Dynamics in Natural Boreal Forest Chronosequences*

The successional dynamics of boreal forests are based on the responses of regeneration, growth, and mortality to prevailing environmental conditions and disturbance patterns. In this regard, Shorohova et al. (2009) used a chronosequence to define the endogenous patterns of growth and development of Norway spruce in the southern boreal zone in Karelia, Russia (Fig. 4.8). Stem wood stocking had accumulated for 140 years since regeneration at the study site. The stocking declined until the regeneration of the second post-disturbance cohort, thus representing the establishment, growth, and mortality of a cohort with a life span of about 240 years. Shorohova et al. (2009) further found that the stocking in consecutive cohorts tended

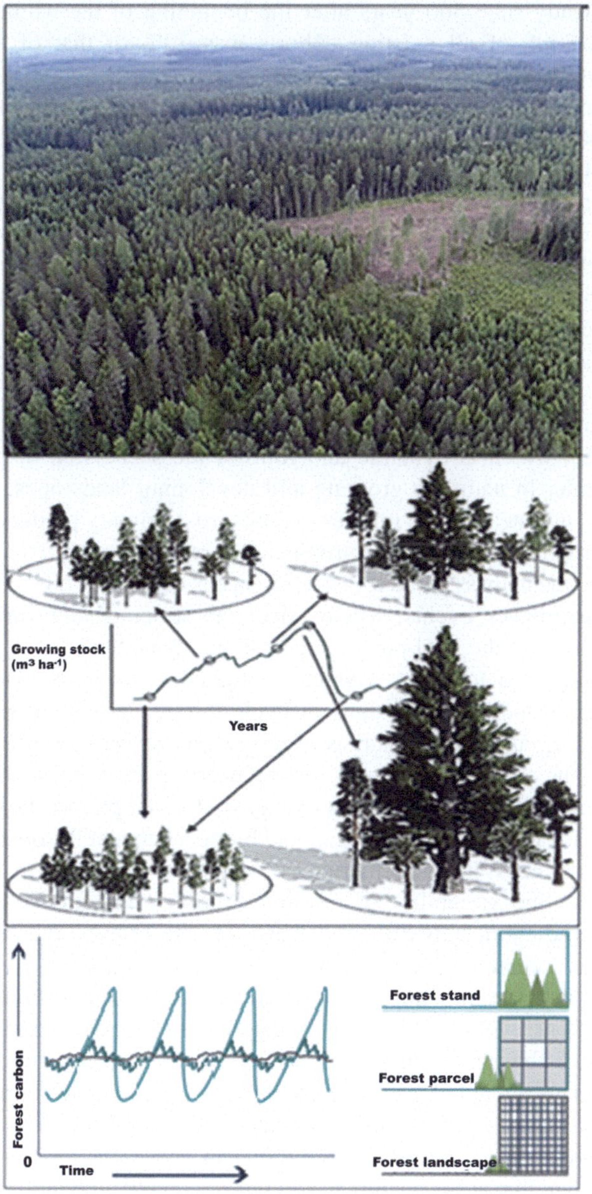


**Fig. 4.7** *Patterns* emerging at various scales in time and space, ranging from populations through to communities and ecosystems. Upper: An aerial photograph of a managed boreal forest area with tree stands in different developmental phases. Courtesy of Metsähallitus Finland. Middle: Stand-scale variability in the growth and development of a tree stand in different phases of production cycles, from small seedlings to mature trees and to the next production cycle, over the life span of trees in a chronosequence of boreal Norway spruce forest. Compiled by the author of this book. Lower: Forest carbon fluctuation through several rotations over time at varying spatial scales under even-aged management: Upper (i) for a single stand with wide temporal variability of carbon over time; Middle (ii) for a forest parcel containing separate stands with one at once under management with temporal variability of carbon over time through the whole parcel including a single stand

to achieve a steady state 560 years after the beginning of the whole successional cycle. The stocking of all existing cohorts was 70% of that of the first post-disturbance cohort. Concurrently, the density of trees per cohort oscillated with the regeneration and death of trees, with impacts on stem wood growth. In the first post-disturbance cohort, the growth maximum was achieved at the age of 80 years since regeneration. The growth collapsed when 220 years had passed following regeneration but increased again with cycling in stem wood stocking. The long-term growth stabilized around zero, representing the growth equal to mortality as observed in steady state development (Kazimirov 1971; Shorohova et al. 2009).

Over any forest landscape, the structure and functioning of ecosystems vary over time. Carbon dynamics in each patch follow the growth and development of trees from disturbance with regeneration, through reorganization, to the long-term stability of carbon stocks in trees and soil over time (Bormann and Likens 1979, pp. 151–181). Carbon in trees and soil oscillate around the long-term balanced mean (Fig. 4.9), which is dynamic and follows the succession of patches in the landscape mosaic. In naturally growing and developing landscapes, the long-term shifting mosaic in a steady state implies that the gross primary production equals the losses in autotrophic ($R_A$) and heterotrophic ($R_H$) respiration: $GPP = R_A + R_H$. In a steady state, the mean mass in the forest landscape is dependent on the frequency and intensity of disturbance events, which affect the structure and variability of the landscape mosaic and the total rate of successional dynamics.

The dynamics of a forest ecosystem are linked to its mass dynamics, which control several feedbacks in the nutrient cycle through growth/litterfall/mortality/detritus/nutrient uptake. The dynamics of a forest ecosystem are therefore biologically related to the amount of biomass (living, dead) in the site (Fig. 4.9). The mass in the ecosystem buffers the impacts of energy (radiation, gravity, heat, wind, water flow) affecting the energetic, hydrologic, and biogeochemical factors controlling the functioning of the ecosystem. Furthermore, the amount and structure of mass in the ecosystem affect the variability of resources through several feedbacks via the availability of radiation, nutrients, and water for the growth and regeneration of trees and ground vegetation.

Over time, the accumulation of mass in the ecosystem restores the within-ecosystem nutrient cycle to one resembling that before clear-cutting (Fig. 4.9). This implies that the phases in which primary production and the availability of water and nutrients are in balance. The accumulation of mass declines or even stops in the transient phase, when the tree population/community achieves full maturity (old-growth forest, climax forest). In the transient phase, the total amount of mass may be larger than that at full maturity, when the accumulation and loss of mass are balanced. The decline of mass is also related to the gradual death of trees belonging to the initial tree cohorts and replaced by new tree cohorts. In general, this creates

**Fig. 4.7** (continued) under management; and Lower (iii) for a forest landscape containing several stands with one at once under management with lowest variability of carbon through the parcel over time (McKinley et al. 2011). Permission of Wiley

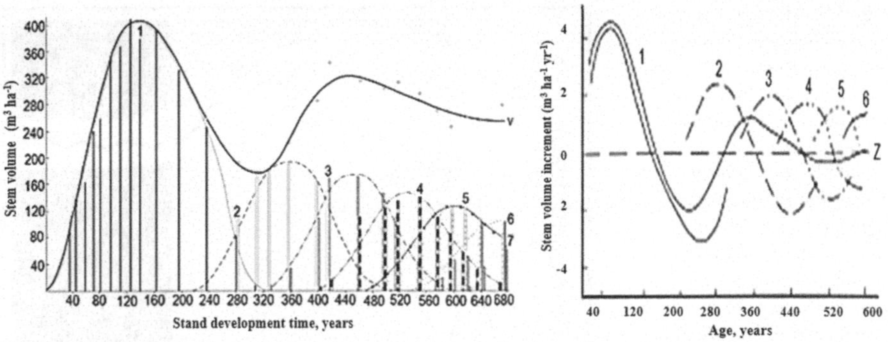


**Fig. 4.8** Stem wood stocking in Norway spruce (*Picea abies*) chronosequences over the growth and developmental phases of trees over time in boreal forests (Kazimirov 1971; Shorohova et al. 2009). Open Access. Courtesy of EDP Open. The numbers 1–7 represent the cohorts; the circles, empirical data from sample plots; and the solid lines, theoretical curves. Left: Standing volume of Norway spruce in natural boreal forest comprising initial seedling stands through to tree cohorts established and grown over time. Right: Volume increment of a Norway spruce stand on sites of medium fertility and moisture. Numbers refer to separate tree cohorts born, grown, and died over time from the newest one (number 1) to the oldest one (number 6)

tree populations/communities of varying ages (uneven-aged), which replace the tree populations/communities originally representing trees of the same age (even-aged) (e.g., Harmon 2002, 2009). The uneven-aged forest structure is possible for tree species tolerating a scarce supply of resources such as light as shade-tolerant Norway spruce. Scots pine and birch are shade-intolerant, with the natural forest structure of these species being even-aged rather than uneven-aged (Fig. 4.9; Box 4.2).

**Box 4.2 Carbon uptake in old-growth forest under natural conditions**

In old-growth forests or old natural forests, the net primary production is small relative to the carbon maintaining the ecosystem structure at different trophic levels. However, Luyssaert et al. (2008) found that "old growth forests with tree losses do not necessarily become carbon sources, as has been observed in even-aged plantations." Luyssaert et al. (2008) estimated that the net ecosystem production (*NEP*) in natural stands aged ≥200 years was 2.4 ± 0.8 Mg C ha$^{-1}$ yr.$^{-1}$. Similarly, Janisch and Harmon (2002), Law et al. (2001, 2003), Gough et al. (2008), and Law and Harmon (2011) also showed that old-growth forests intercept more carbon in photosynthesis than they emit in respiration. However, the carbon balance depends on site fertility, tree species, the maturity of dominant trees and the ingrowth of new trees, and growth of ground vegetation. Luyssaert et al. (2008) concluded that carbon dynamics in old-growth forests are driven by the age structure of the forest related to the mortality of mature trees and subsequent ingrowth of seedlings, as found by Sirén (1955) for northern boreal Norway spruce forests.

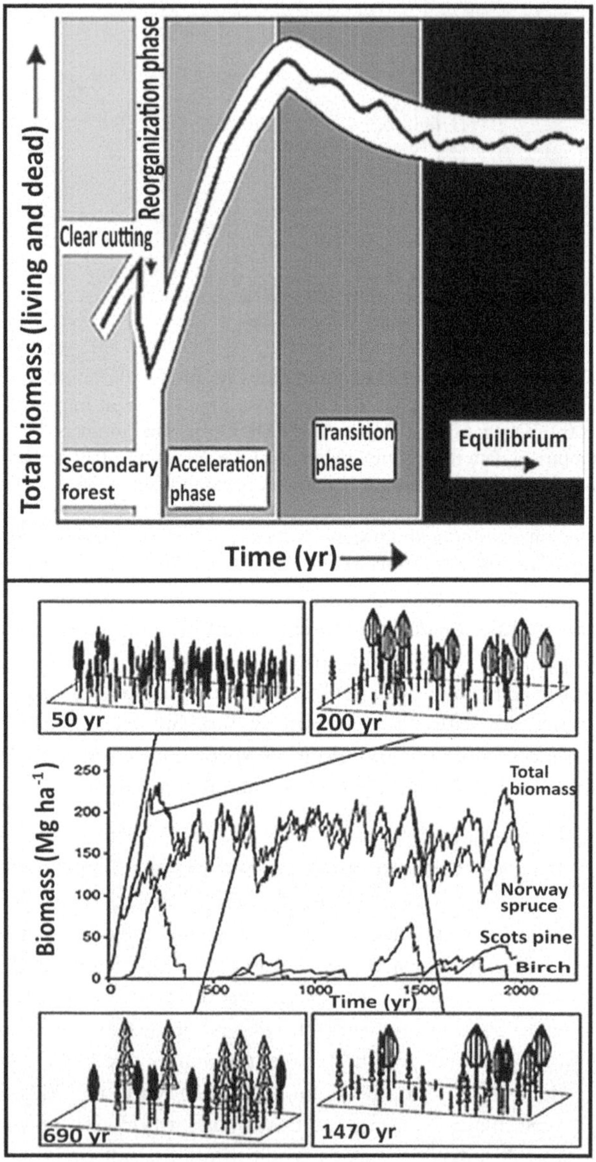


**Fig. 4.9** Mass and carbon dynamics in boreal and temperate forest chronosequences. Upper: Outline of how biomass (living, dead) is accumulated in a forest ecosystem over time in separate phases of the succession cycle after clear-cutting (Bormann and Likens 1979, p. 4). Permission of Cambridge University Press. Lower: Simulation of stem wood stocking with inserts of structural development of tree stand in a naturally growing and developing Norway spruce (*Picea abies*), Scots pine (*Pinus sylvestris*), and birch (*Betula* sp.) stand on moderately fertile site (*Myrtillus* type, MT) in the middle boreal forest zone (62° N). The long-term mean temperature in the area was 2.2 °C and precipitation 630 mm per year. The initial amount of litter and humus on the site was 67 Mg ha$^{-1}$. Simulations were based on the Sima model (Kellomäki 2005, p. 217). Compiled by the author of this book. Courtesy of the University of Eastern Finland

## 4.4 Carbon Dynamics at Patch and Chronosequence Scales: Ecophysiology

### *4.4.1 Carbon Dioxide, Energy, and Water Fluxes in Gaps, with Atmospheric Interactions*

Small- and large-scale disturbance modifies the carbon dynamics in forest landscapes from patches to ecosystems (sites, plant communities) at varying temporal scales. Depending on the developmental phase, a chronosequence may lose or absorb carbon. Carbon loss is most evident just after disturbance, which creates canopy openings (gaps), thus curbing the carbon uptake in trees. At the same time, the increasing decay of soil carbon (accumulated prior to the disturbance, i.e., legacy carbon), tends to increase carbon loss, thus reducing the contribution of legacy carbon to the total carbon stock in the ecosystem. On the other hand, canopy openings enhance the growth of ground vegetation (e.g., herbs, grasses, dwarf shrubs, mosses, lichens) on fertile sites, with effects on the biophysical and biochemical interactions between the forest and atmosphere.

According to Mamkin et al. (2016), clear-cutting immediately turned a Norway spruce-dominated boreal site into a $CO_2$ source. However, the $CO_2$ emissions decreased following the growth of ground vegetation (Fig. 4.10). In the main growing season, the daily *NEE* varied from a small source to a small sink depending on solar radiation, temperature, and soil moisture, with a zero balance between $CO_2$ uptake and emission. Over the period from April through to August, the total $CO_2$ emission from the clear-cut area (155 g C $m^{-2}$) indicated that the gap area was a $CO_2$ source before reforestation. The daily fluxes of sensible (*H*) and latent (*LE*) heat were also controlled by net radiation and the amount of ground vegetation in the gap area, where soil moisture was adequate through the whole growing season (Mamkin et al. 2016).

### *4.4.2 Carbon Storage and Fluxes in Secondary Succession of Coniferous Chronosequences*

In the seedling-, pole- and middle-aged phases of trees, carbon uptake in trees and ground vegetation in a chronosequence exceeds the carbon emissions from soil, whereas under the full maturity of trees, carbon uptake and losses tend to balance out. This phase-by-phase cycle of carbon loss and uptake also holds true for managed forest landscapes. In this context, Goulden et al. (2011) carried out year-round eddy covariance measurements in a chronosequence of boreal forest stands. The focus in measurements was to understand how ecosystem production and carbon stocks change during recovery from stand-replacing crown fires. The tree stands were 1, 6, 15, 23, 40, 74, and 154 years of age, and were in North-American boreal forests at 55–56° N and 98–99° W. The tree stands differed in the time since

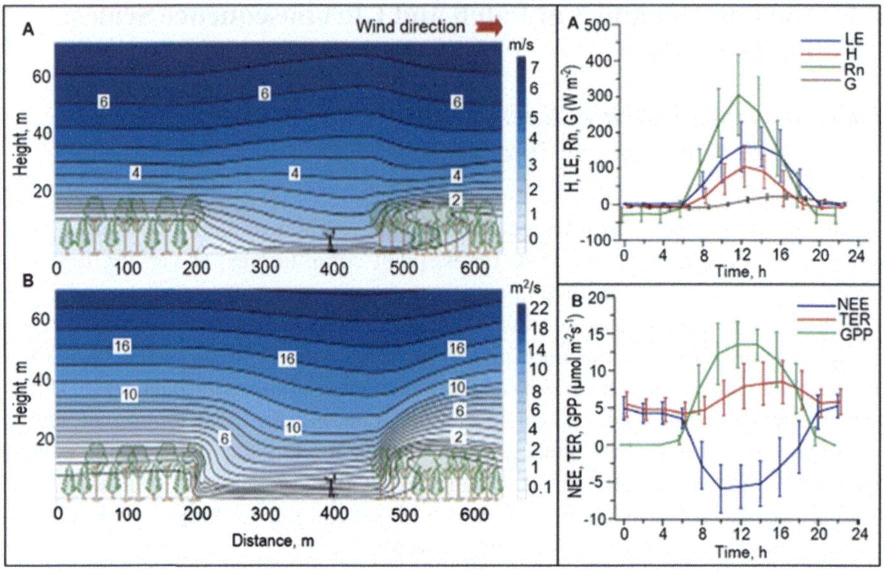


**Fig. 4.10** Carbon uptake and losses in a canopy opening induced by clear-cutting following timber harvest in a mixed spruce/birch/aspen forest in the southern boreal zone of European Russia (56° N, 33° E) (Mamkin et al. 2016). Open Access. Courtesy of IOP Publishing Ltd. Left/Upper: Modeled pattern of horizontal wind velocity. Left/Lower: Turbulent exchange along a south–north transect through the clear-cut area. Right/Upper: Diurnal variability of latent heat (*LE*), sensible heat (*H*), soil heat flux (*G*), net radiation ($R_n$). Right/Lower: Diurnal variability of net ecosystem exchange (*NEE*), total ecosystem respiration (*TER*), and gross primary production (*GPP*) from June through August

the last burn. The stands comprised coniferous trees (e.g., *Picea mariana*, *Pinus banksiana*) in middle, old, and mature phases. In the seedling and pole phase, the tree stands were coniferous, but mixed with deciduous trees (e.g., *Populus tremuloides*). Ground vegetation was representative of typical fertile upland sites and was sometimes mixed with mire vegetation. In Fig. 4.11, the column on the left, the net $CO_2$ exchange based covariance measurements over a chronosequence expanding from a one-year-old seedling stand to a 154-year-old tree stand at full maturity. The column on the right integrates the gross primary production (GPP), total ecosystem respiration ($R_{total}$), and net ecosystem production ($NEP_{EC}$) over the period from October 1 in 200 to September 30 in 2004.

In Fig. 4.11, the columns both on the left and right, $FCO_2$ presents (in the atmospheric sign convention) a positive flux indicating the net transfer of carbon from the ecosystem to the atmosphere. Similarly, GPP, $R_{total}$, and $NEP_{EC}$ present (in the ecological sign convention), a positive flux indicating the net transfer of carbon from the atmosphere to the ecosystem. According to Goulden et al. (2011):

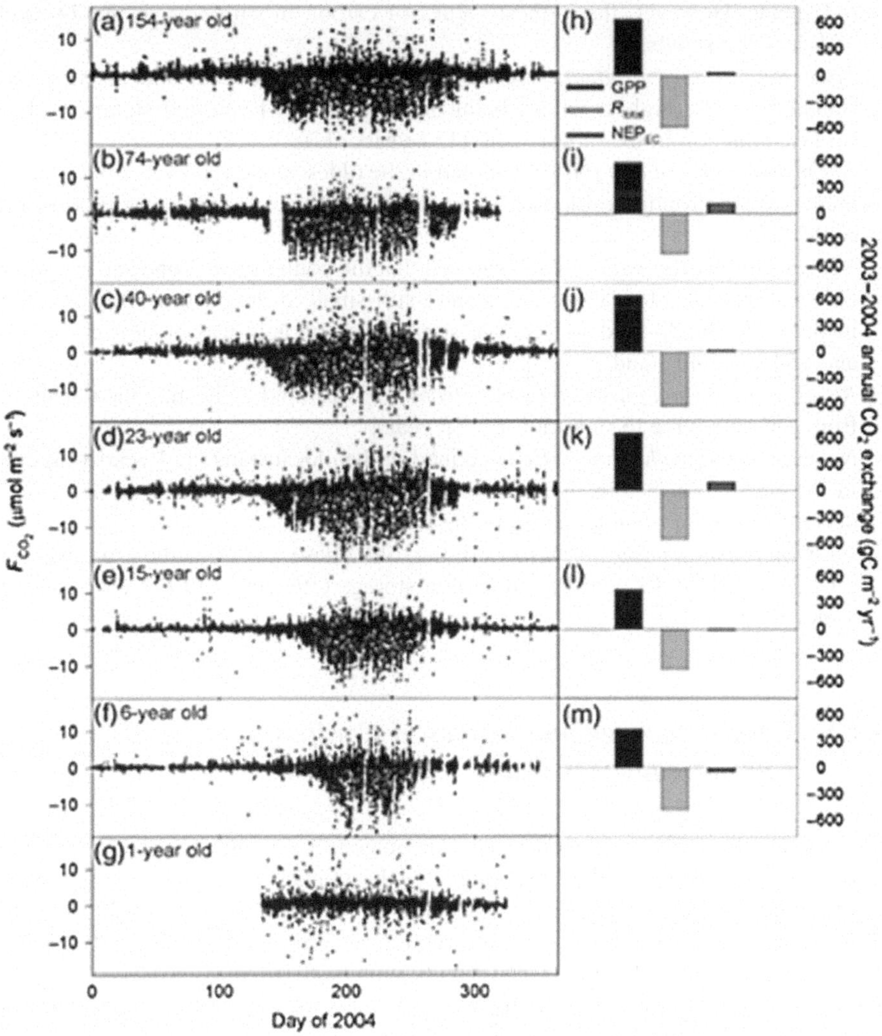


**Fig. 4.11** Measurements and analysis of carbon dynamics in a boreal forest ecosystem over a chronosequence from the seedling phase to full maturity of the tree stand upwards from below (Goulden et al. 2011). Open Access. Courtesy of Wiley. Column/Left/ (**a-g**): Net $CO_2$ exchange ($FCO_2$; 30 min covariances, mmol $m^{-2}$ $s^{-1}$) during 2004 at seven sites that differ in age since last burn. Column/Right/ (**h–m**): Gross primary production (GPP, g C $m^{-2}$ yr.$^{-1}$), ecosystem respiration ($R_{total}$, g C $m^{-2}$ yr.$^{-1}$), and net ecosystem production ($NEP_{EC}$, g C $m^{-2}$ yr.$^{-1}$) integrated from October 1, 2003, to September 30, 2004, at the six sites with year-round observations

- *Carbon stocking of living trees* was low in the seedling phase (i.e., in the one- and six-year-old stands), with stocking increasing logistically to the largest stocking in the 74- and 154-year-old stands.
- *Carbon stocks in the forest floor* ($C_{forest\ floor}$) and coarse woody debris ($C_{CWD}$) were high in the one-year-old stand but decreased in the six- through to 40-year-

old stands. However, the highest values of carbon in soil were in the 74- and 154-year-old stands.

- *Total net primary production* (TNPP) was low in the one- and six-year-old stands, with the highest values being in the 23- to 74-year-old stands, with a further reduction in the oldest stand (154 years old).
- *Net primary production* (NPP) declined in the oldest stand (154 years old) due to increased autotrophic respiration more than decreased gross primary production (GPP).
- *Net ecosystem production* (*NEP*) showed that the young stands (one and six years old) lost carbon. The carbon gains were still small in the 15-year-old stand but increased in the 23- and 74-year-old stands. Carbon gain was modest in the 40- and 154-year-old stand.
- *Recovery of carbon fluxes and stocks from fire* was rapid, indicating the transition from carbon source to sink took 11–12 years.
- *Net ecosystem production* (*NEP*) declined in the oldest stand (154 years old) due to increased losses from mortality of previously living trees and the increased decomposition of organic matter in the forest floor.
- *With increasing maturation* toward the old phase, *NPP*, carbon production efficiency (NPP/GPP), NEP, and carbon storage efficiency (*NEP/TNPP*) tended to decrease.

### 4.4.3 Carbon Storage and Fluxes in Coniferous Chronosequences Under Management

To analyze the carbon dynamics in a managed Scots pine (*Pinus sylvestris*) chronosequence, Kolari et al. (2004) estimated the annual *NEE* for a chronosequence representing four Scots pine stands in southern Finland during 2000–2002 using eddy covariance measurements. Concurrently, net ecosystem productivity (*NEP*) was estimated using growth measurements and modeled mass losses of woody debris. The stands were 4, 12, 40, and 75 years of age. The youngest stand represented a clear-cut area with small seedlings.

In the analysis, the four-year-old clear-cut area was a source of carbon throughout the year, combining a low gross primary productivity (*GPP*) with the total ecosystem respiration (*TER*) representing that in the forested stands. Annual *NEE* in the clear-cut area was 386 g C $m^{-2}$ $yr.^{-1}$ (Kolari et al. 2004) (Fig. 4.12). The tree growth was small and net ecosystem production (*NEP*) was −262 g C $m^{-2}$ $yr.^{-1}$, while the annual gross primary production (*GPP*) was 928–1072 C $m^{-2}$ $yr.^{-1}$. However, the total ecosystem respiration (*TER*) was variable at the stand scale, i.e., *TER* was 905 C $m^{-2}$ $yr.^{-1}$ for the 12-year-old stand and 616 C $m^{-2}$ $yr.^{-1}$ for the 75-year-old stand. The *NEE* in the 40-year-old stand varied between −179 and − 192 g C $m^{-2}$ $yr.^{-1}$, while *NEP* was between 214 and 242 g C $m^{-2}$ $yr.^{-1}$. The annual *NEE* of the 75-year-old stand was 323 g C $m^{-2}$ $yr.^{-1}$, and *NEP* was 252 g C $m^{-}$ $yr.^{-1}$. This indicated that there was no reduction in carbon sink strength with stand age (Kolari et al. 2007, 2009).

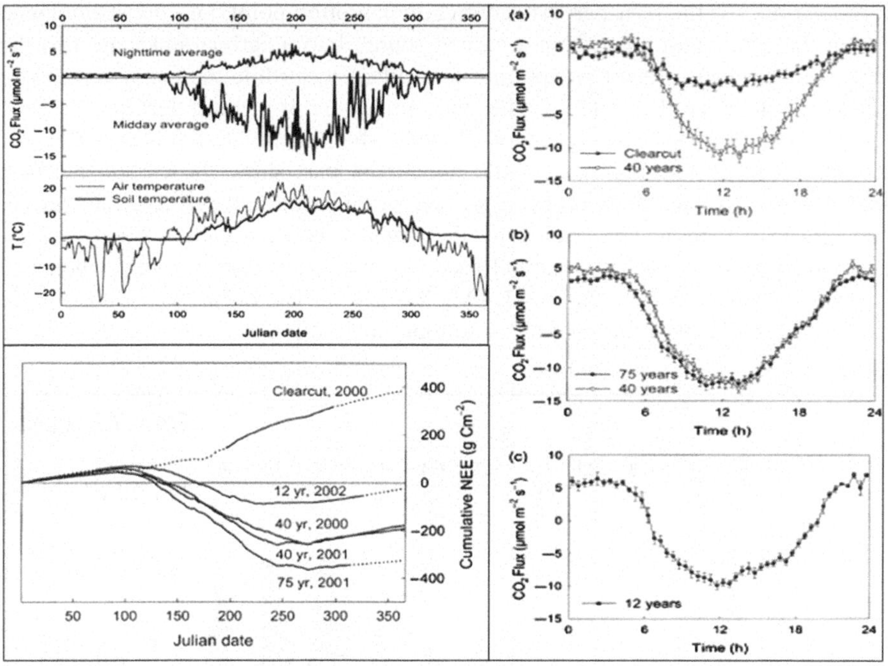


**Fig. 4.12** Carbon storage and fluxes in a coniferous chronosequence of stands (*Pinus sylvestris*) of varying age under management (Kolari et al. 2004). Permission of Wiley. Left/Upper and Middle: Annual pattern of $CO_2$ fluxes measured by eddy covariance, and soil temperature at 5-cm depth and daily mean air temperature at the 40-year-old site in year 2001. The flux graphs represent the average midday (10:00–14:00 hours) and midnight (22:00–02:00 hours) fluxes. Left/Below: Cumulative net ecosystem exchange *(NEE)* and the year of the measurements for Scots pine (*Pinus sylvestris*) stands of different ages. Solid lines indicate measured values and dotted lines the modeled values. Positive values indicate the loss of carbon and negative values the uptake of carbon. Right: Average diurnal courses of $CO_2$ fluxes for Scots pines stands of varying age in July–August: (**a**) year 2000, (**b**) year 2001, and (**c**) year 2002, as a function of tree stand age. The clear-cut site represents a four-year-old seedling stand. The error bars denote the standard error of the mean

## 4.5 Dynamics of Carbon Across Boreal Scots Pine Chronosequence: Biomass

### *4.5.1 Biomass, Growth, Carbon Fluxes, and Carbon Balance in Trees*

Uri et al. (2022) investigated through a chronosequence of Scots pine (*Pinus sylvestris*) the carbon dynamics in the hemiboreal conditions in the southern part of Estonia (58° N). Over the post-harvest recovery of carbon stocks and fluxes, study extended from a clear-cut area to a 110-year-old mature stand. The focus was on (i) recovery of carbon accumulation in the clear-cut area after harvest and regeneration; (ii) the post-harvest carbon compensation point, i.e., the stand age at which the ecosystem turns from a carbon source into a carbon sink; and (iii) the carbon payback period for recapturing the carbon lost while the ecosystem was a carbon source after clear-cutting. The study sites were of medium fertility.

Figure 4.13 shows the dynamics of the carbon stocks and fluxes across the chronosequence. The numbers in the uppermost part show the carbon balance, with the negative numbers referring to carbon emission and positive numbers to the uptake of carbon, thus indicating the negative/positive carbon budget of the ecosystem. In the clear-cut area and seedling phase, the carbon budget indicated carbon emission at a rate of up to $-1.9$ Mg ha$^{-1}$ yr.$^{-1}$. Thereafter, the carbon uptake increased rapidly from the pole phase to more mature phases, culminating at up to

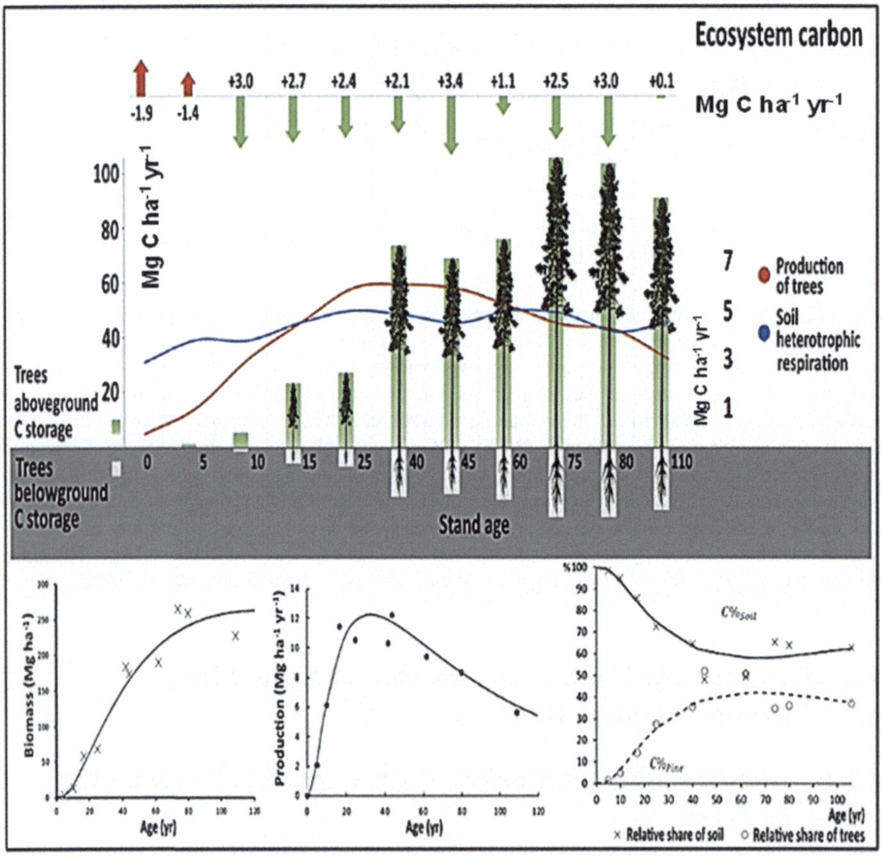


Fig. 4.13 Carbon storage, fluxes, and balance in a Scots pine (*Pinus sylvestris*) chronosequence (Uri et al. 2022). Permission of Elsevier, slightly modified by the author of this book. Upper: Dynamics of the carbon storage and fluxes in a chronosequence as a function of stand age. Lower/Left: Dynamics of biomass through chronosequence. Lowe/Middle: Dynamics of growth through chronosequence. Lower/Right: Dynamics of the relative share of ecosystem carbon in soil and trees through chonosequence

+3.4 Mg $ha^{-1}$ $yr.^{-1}$ in the phase of middle age at the pole phase. Toward the end of the chronosequence, the carbon balance tended to decline slightly, as indicated by the trend lines for the biomass production of trees and soil heterotrophic respiration in the figure. This decline was due to the further maturation of trees with reducing growth and increase of soil heterotrophic respiration.

Figure 4.13 further shows the dynamics of stem wood growth (*NPP*) and tree biomass over a chronosequence (Uri et al. 2022). *NPP* increased rapidly in young trees from the initial phase to 30–40 years of age, with an increase of up to 12 Mg carbon $ha^{-1}$ $yr.^{-1}$. Thereafter, *NPP* reduced gradually down to 5–6 Mg $ha^{-1}$ $yr.^{-1}$ toward the full maturity of trees, as previously found by Gao et al. (2018) for mature trees in old boreal forests. Along with the growth, the biomass of trees increased most rapidly in young and middle-aged trees and levelled off when the age of trees exceeded 75–80 years. Concurrently, the annual growth of shoots and needles ranged from 1.7 to 3.7 Mg $ha^{-1}$ $yr.^{-1}$ from trees in the seedling phase to trees in the pole phase. Thereafter, growth of needles and shoots was stable, balancing out with litterfall production of 1.9–4.2 Mg $ha^{-1}$ $yr.^{-1}$. The largest values represented the trees in the pole phase and the smallest trees in the seedling phase. The litterfall was strongly correlated with the age of trees and stabilized thereafter. Over the whole chronosequence, the mean litterfall was $1.99 \pm 0.14$ Mg $ha^{-1}$ $yr.^{-1}$.

The biomass accumulation in growing trees stabilized at up to 250 Mg biomass $ha^{-1}$ at full maturity (Fig. 4.13). The share of stem biomass (i.e., the above-ground biomass of the trees) increased up to 35–40% of the total carbon stock in trees and soil with increasing stand age. The share of the total carbon stock in soil was 55–60% when the tree stand was at full maturity. Uri et al. (2022) further found that the dry mass of the stump plus course roots with diameter > 10 cm ranged from 11.4 to 82.5 kg per tree over the chronosequence. The mass of stumps and coarse roots represented $41.0 \pm 1.5\%$ and $25.5 \pm 2.2\%$ of the total mass of the root systems. Consequently, the mean ratio between belowground/aboveground mass was $0.24 \pm 0.01$, with slight variability for trees of varied sizes. The understory vegetation further contributed 8–46% of total ecosystem *NPP*, with no clear age-related trend (Fig. 4.13).

Uri et al. (2022) further found that there was no clear correlation between stand age and total soil carbon stock to a depth of 50 cm, including the carbon in the soil organic layer (Fig. 4.13). Over the chronosequence, the soil carbon stock varied from 85 to 260 Mg C $ha^{-1}$, with the smallest value in the pole phase and the largest in the mature phase. In this context, the mean carbon stock in soil organic matter or the forest floor was 20% greater than the total soil carbon stock, ranging from 10 to 40%. Uri et al. (2022) found that the soil carbon stock had the major share of the total stock of ecosystem carbon in the young stands. Over time, the share of accumulated carbon increased, with the consequence that the carbon stocks in trees relative to soil carbon stock gradually balanced out in middle-aged trees (50–60 years of age). The largest and the smallest total carbon stocks over the chronosequence were 400 and 130 Mg C $ha^{-1}$, occurred in the pre-mature-phase stand and the pole-phase stand, respectively. According to Uri et al. (2022), there was a strong correlation between stand age and total carbon stock, which was related to the cumulative biomass of the trees.

Cumulative soil heterotrophic respiration ($R_h$) showed stable values along the chronosequence, varying between 3.8 and 5.4 Mg C ha$^{-1}$ yr.$^{-1}$. The $R_h$ flux of 2.9 Mg C ha$^{-1}$ yr.$^{-1}$ at the clear–cut site did not exceed the corresponding values in the stands. The net ecosystem production ($NEP = NPP - R_h$) along the chronosequence reflected the annual biomass production of the trees, peaking at the middle-aged phase and decreasing in the older stands. Thus, *NPP* was the main driver directing the dynamics of *NEP*. There was no significant correlation between $R_h$ and the dynamics of aboveground litter or fine root production. This can partly explain why no relationship was found between annual $R_h$ and stand age. The total ecosystem stocks followed the same trend as cumulative tree biomass. The post-harvest carbon compensation point was reached at the stand age of seven years, and carbon payback occurred at a stand age of 11–12 years. Thus, the carbon accumulating in the ecosystem was 2.5 Mg C ha$^{-1}$ yr.$^{-1}$, excluding the clear-cut areas and seedling stands. In the old stand, the carbon budget was almost balanced, with a modest annual accumulation of 0.12 Mg C ha$^{-1}$ yr.$^{-1}$.

### 4.5.2 Biomass, Growth, Carbon Fluxes, and Carbon in Ground Vegetation

According to Uri et al. (2022), the ground vegetation was mainly dominated by dwarf shrubs (e.g., *Vaccinium myrtillus*, *Vaccinium vitis-idaea*, *Calluna vulgaris*) and mosses (e.g., *Pleurozium schreberi, Hylocomium splendens*), while the role of herbs and grasses was negligible. The mean biomass was 1.5 ± 0.41 Mg ha$^{-1}$ for dwarf shrubs, and 2.4 ± 0.35 Mg ha$^{-1}$ for mosses over the whole chronosequence. There was no correlation between biomass in the ground vegetation and the age of trees (Table 4.1).

Over the chronosequence, the growth of fine roots in Scots pine increased from 0.47 Mg ha$^{-1}$ yr.$^{-1}$ in the seedling phase to 3.93 Mg ha$^{-1}$ yr.$^{-1}$ in the pole phase, with a slight decline along with further maturation of trees due to their declining growth. Concurrently, the growth of coarse roots of dwarf shrubs in the clear-cut area and in the seedling phase was 0.27 and 2.78 Mg ha$^{-1}$ yr.$^{-1}$, respectively, declining to 0.02 Mg ha$^{-1}$ yr.$^{-1}$ in the pole phase, and then increasing toward the end of the chronosequence up to 1.73 Mg ha$^{-1}$ yr.$^{-1}$. The growth of fine roots in dwarf shrubs ranged from 0.46 to1.08 Mg ha$^{-1}$ yr.$^{-1}$, with no clear trend from the clear-cut area to the fully-mature tree stand (Table 4.1).

**Table 4.1** Aboveground biomass and production of understory vegetation in a Scots pine (*Pinus sylvestris*) chronosequence as a function of the developmental phase (or development class, DC) (Uri et al. 2022)

| Aboveground biomass and production in the understory vegetation | | | | | | | | |
|---|---|---|---|---|---|---|---|---|
| | | Biomass (Mg ha$^{-1}$) | | | | Production (Mg ha$^{-1}$ yr.$^{-1}$) | | |
| Development class (DC) | Stand | Herbs | Dwarf shrubs | Mosses | Mean for DC | Dwarf shrubs | Mosses | Mean for DC |
| Clear-cut area | P0 | 0.18 | 0.53 | 3.55 | 4.26 | 0.20 | 1.21 | 1.59 |
| | P5 | 0.21 | 4.44 | 2.19 | | 2.09 | 0.74 | |
| Seedling phase | P10 | 0.34 | 3.36 | 3.71 | 5.59 | 1.29 | 1.26 | 2.35 |
| | P15 | 0.36 | 0.56 | 1.59 | | 0.21 | 0.54 | |
| Pole phase | P25 | 0.17 | 0.04 | 0.69 | 0.90 | 0.02 | 0.23 | 0.42 |
| | P40 | 0.03 | 1.09 | 1.01 | | 0.42 | 0.34 | |
| Middle-aged phase | P45 | 0.09 | 0.45 | 1.91 | 2.29 | 0.17 | 0.62 | 0.86 |
| | P60 | 0.02 | 0.64 | 3.95 | | 0.25 | 1.34 | |
| Pre-mature phase | P75 | 0.02 | 1.43 | 1.42 | 3.74 | 0.55 | 0.48 | 1.33 |
| | P80 | 0.03 | 1.94 | 3.56 | | 0.75 | 1.21 | 1.81 |
| Mature phase | P110 | 0.01 | 1.76 | 2.77 | 5.04 | 0.67 | 0.94 | |

### 4.5.3 *Biomass and Carbon in Roots of Trees and Ground Vegetation*

The biomass of fine roots ranged for Scots pine from 1.92 to 3.85 Mg ha$^{-1}$. Over the whole chronosequence, the mean fine root biomass for Scots pine was 2.90 ± 0.18 Mg ha$^{-1}$, and for the belowground parts of dwarf shrubs 1.32 ± 0.18 Mg ha$^{-1}$. There was a clear correlation between the mean fine root biomass per stand and stand age. In mature tree stands, the fine root biomass of dwarf shrubs was clearly larger than in younger tree stands. Over the whole chronosequence, the biomass of the fine roots of herbs was absent. On the other hand, the biomass of the coarse roots of dwarf shrubs was up to 6 Mg ha$^{-1}$ in the seedling phase, while decreasing clearly under trees in the pole- and early middle-aged phases. Thereafter, the mass of dwarf shrubs increased up to 4.5 Mg ha$^{-1}$ toward the end of the chronosequence. In general, the coarse root mass exceeded the above-soil mass, with a mean ratio of 1.64 ± 0.19 between the biomass in coarse roots and above-soil shoots based on single plants excavated from the soil (Table 4.2).

### 4.5.4 *Carbon Budget, Net Primary Production, and Net Ecosystem Production*

Uri et al. (2022) summarized their findings in Table 4.3. It shows the carbon budget for the chronosequence of Scots pine as a function of the developmental phase from

**Table 4.2** Belowground biomass and production of understory vegetation and Scots pine roots for the development classes (DC) through a choronosequence (Uri et al. 2022)

| Belowground biomass and production in the understory vegetation | | | | | | | | | | |
|---|---|---|---|---|---|---|---|---|---|---|
| | | Biomass (Mg ha$^{-1}$) | | | | | Production (Mg ha$^{-1}$ yr.$^{-1}$) | | | |
| | | | Dwarf shrubs | | | | | Dwarf shrubs | | |
| Development class (DC) | Stand | Pine fine roots | Coarse roots | Fine roots | Herbs, coarse and fine roots | Total for DC | Pine fine roots | Coarse roots | Fine roots | Total for DC |
| Clear-cut area | P0 | 0 | 0.70 | 0.92 | 0.21 | 1.83 | 0 | 0.27 | 0.46 | 0.94 |
| | P5 | 1.92 | 5.91 | 1.28 | 0.15 | | 0.47 | 2.78 | 0.64 | |
| Seedling phase | P10 | 3.26 | 4.70 | 1.31 | 0.75 | 8.33 | 2.41 | 1.80 | 0.66 | 4.52 |
| | P15 | 2.41 | 0.78 | 1.39 | 1.14 | | 1.78 | 0.30 | 0.70 | |
| Pole phase | P25 | 4.68 | 0.06 | 0.18 | 0.70 | 4.10 | 3.93 | 0.02 | 0.09 | 3.15 |
| | P40 | 3.85 | 1.70 | 1.90 | 0 | | 2.85 | 0.65 | 0.97 | |
| Middle-aged phase | P45 | 3.00 | 0.18 | 0.55 | 0.14 | 5.79 | 2.52 | 0.11 | 0.28 | 3.73 |
| | P60 | 2.35 | 0.99 | 1.31 | 0 | | 1.74 | 0.38 | 0.66 | |
| Pre-mature phase | P75 | 2.95 | 2.95 | 1.69 | 0 | 6.12 | 2.18 | 1.13 | 0.85 | 3.47 |
| | P80 | 3.38 | 4.02 | 1.86 | 0.27 | | 2.50 | 1.54 | 0.93 | |
| Mature phase | P110 | 2.73 | 4.51 | 2.16 | 0.06 | 9.50 | 2.02 | 1.73 | 1.08 | 5.07 |

**Table 4.3** Carbon budgets for a Scots pine chronosequence, with net primary production (*NPP*) and net ecosystem production (*NEP*) from a clear-cut site (P0) to full maturity (P110) (Uri et al. 2022)

| | | Clear-cut | Seedling phase | | | Pole phase | Middle-aged phase | | Pre-mature phase | | Mature phase | |
|---|---|---|---|---|---|---|---|---|---|---|---|---|
| Fluxes ($Mg\ ha^{-1}\ yr^{-1}$) | | P0 | P5 | P10 | P15 | P25 | P40 | P45 | P60 | P75 | P80 | P110 |
| Plant above ground | | | | | | | | | | | | |
| 1 | Tree biomass growth | 0.0 | 0.89 | 2.63 | 4.69 | 4.41 | 3.94 | 4.96 | 3.94 | 4.39 | 3.47 | 2.04 |
| 2 | Litter total | 0.0 | n.e. | 0.96 | 1.15 | 1.87 | 2.08 | 1.52 | 1.59 | 2.12 | 1.46 | 1.65 |
| 3 | Herbs and dwarf shrubs | 0.17 | 1.06 | 0.74 | 0.26 | 0.08 | 0.21 | 0.12 | 0.12 | 0.26 | 0.36 | 0.32 |
| 4 | Moss growth | 0.52 | 0.32 | 0.54 | 0.23 | 0.11 | 0.15 | 0.31 | 0.57 | 0.21 | 0.52 | 0.40 |
| Total aboveground carbon | | 0.69 | 2.27 | 3.91 | 5.13 | 4.60 | 4.30 | 5.39 | 4.63 | 4.86 | 4.35 | 2.76 |
| Plant below ground | | | | | | | | | | | | |
| 5 | Tree coarse root growth | 0.0 | 0.10 | 0.32 | 0.75 | 0.78 | 0.66 | 0.88 | 0.60 | 0.62 | 0.51 | 0.26 |
| 6 | Tree fine root growth | 0.0 | 0.24 | 1.21 | 0.89 | 1.97 | 1.43 | 1.26 | 0.87 | 1.09 | 1.25 | 1.01 |
| 7 | Dwarf shrub fine root growth | 0.09 | 0.13 | 0.12 | 0.13 | 0.00 | 0.18 | 0.05 | 0.12 | 0.16 | 0.17 | 0.20 |
| 8 | Dwarf shrub coarse root growth | 0.14 | n.e. | 0.90 | 0.15 | 0.01 | 0.33 | 0.13 | 0.19 | 0.57 | 0.77 | 0.86 |
| 9 | Herb roots and rhizomes | 0.10 | 0.08 | 0.35 | 0.52 | 0.27 | 0.05 | 0.06 | 0.05 | 0.06 | 0.12 | 0.03 |
| Total belowground carbon | | 0.33 | 0.55 | 2.90 | 2.44 | 3.03 | 2.65 | 2.38 | 1.83 | 2.50 | 2.82 | 2.36 |
| Output | | | | | | | | | | | | |
| 10 | Heterotrophic respiration | 2.90 | 4.21 | 3.78 | 4.93 | 5.27 | 4.81 | 4.37 | 5.39 | 4.87 | 4.13 | 5.00 |
| 11 | TOC leaching | n.e. | 0.02 | n.e. | n.e. | <0.01 | n.e. | 0.02 | n.e. | n.e. | n.e. | n.e. |
| 12 | Carbon input to soil = (2 + 3 + 4 + 6 + 7 + 9) | 0.88 | n.e. | 3.92 | 3.18 | 4.30 | 4.10 | 3.32 | 3.32 | 3.90 | 3.88 | 3.61 |
| 13 | Soil carbon exchange = (12–13) | −2.02 | n.e. | 0.14 | −1.75 | −0.97 | −0.71 | −1.07 | −2.07 | −0.97 | −0.25 | −1.39 |
| Productivity | | | | | | | | | | | | |
| 14 | NPP = (1 + 2 + 3 + 4 + 5 + 6 + 7 + 8 + 9) | 1.02 | 2.82 | 6.81 | 7.62 | 7.63 | 6.95 | 7.77 | 6.46 | 7.36 | 7.17 | 5.12 |
| 15 | NEP = (14–10) | −1.88 | −1.38 | 3.03 | 2.69 | 2.36 | 2.14 | 3.38 | 1.07 | 2.49 | 3.04 | 0.12 |

TOC leaching refers to total organic carbon lost in water flow from the ecosystem, and n.e. signifies no estimate available. Permission of Elsevier the clear-cut area to the stand with fully-mature trees. In the clear-cut area and in the stand with young seedlings, *NEP* was −1.9 and −1.4 Mg C $ha^{-1}$ $yr.^{-1}$, these sites thus being carbon sources. Other sites in the chronosequence were carbon sinks, with *NEP* ranging from +0.12 to +3.38 Mg C $ha^{-1}$ $yr.^{-1}$. The sink strength tended to be larger in the maturing seedling and pole phases than in more mature stands. In the most mature stand, the carbon uptake was small compared to that in young stands in the chronosequence.

## 4.6 Carbon Dynamics and Stocking of Trees with Species Richness in Ground Vegetation

### *4.6.1 Dry-Matter Production and Carbon Sequestration in Ground Vegetation*

In boreal forest ecosystems, ground vegetation constitutes up to 30% of the total biomass production depending on the site fertility in southern and middle Finland (Mälkönen 1974). This holds especially for poor sandy sites (*Calluna* site type, CT), where the annual growth of dwarf shrubs (*Calluna* and *Vaccinium* species), and mosses and lichens (*Pleurozium, Dicranum* and *Cladonia* spp.) is 200–800 kg $ha^{-1}$ $yr.^{-1}$ (e.g., Tamm 1953; Mälkönen 1974). On fertile sites, the share of the ground cover in the total dry-matter production in ecosystem is substantially smaller, mainly representing the growth of herbs and grasses (e.g., Sirén 1955; Persson 1975).

According to Uri et al. (2022), the growth and stocking of ground vegetation are linked to the dynamics (growth and stocking) of trees above ground vegetation. The dynamics of tree layer further effects on the richness of grown vegetation, e.g., how biomass of ground vegetation is distributed among different ground cover species. This holds both under natural (not managed and harvested) and commercial (managed and harvested) forests in different phase of chronosequences over the life span of trees from seedlings to full maturity. Kellomäki and Väisänen (1991) used this pattern in long-term simulations in studying (i) how the dry-mass growth/stocking in ground vegetation performs under no management vs. even-aged management and (ii) how biomass stocking of trees effects on the species richness related to the distribution of biomass among different species through a forest chronosequence under no management vs. even-aged management.

### *4.6.2 Performance of Tree Layer Above Soil with Impacts on Ground Vegetation*

The simulations were done for moderately fertile upland site (*Myrtillus* site type, MT) on moraine soil in middle boreal zone (62° N). In the site, the mean annual temperature sum was 1300 d.d., and the mean annual precipitation 500 mm (Kellomäki and Väisänen 1991). The natural deposition of nitrogen was 5 kg $ha^{-1}$ $yr.^{-1}$ throughout the simulations over 500 years. The properties of forest floor represented the situation after a hundred-year succession prior to the clear-cutting for initiating the simulations. The results of computations represented the mean values given by ten repeated simulations, with the values of the parameters for the death procedure being slightly varied randomly from run to run.

In the simulations, the initial tree stand was of natural origin representing a tree-species mixture regenerated naturally by Scots pine (*Pinus sylvestris*), Norway spruce (*Picea abies*) and birch (*Betula pendula* and *B. pubescens*) from nearby stands. The variability stem biomass of different species was related to their growth rate, and life span. The stem wood variability was further modified by annual regeneration success and mortality over chronosequence. Shade-tolerant Norway spruce became dominant over the whole simulation period, with the mass of stem wood fluctuating from 60 to 200 Mg $ha^{-1}$ depending on the regeneration, growth, and mortality affecting in different phases of the chronosequence (Fig. 4.14). However, the shade-intolerant Scots pine existed over hundred years through regeneration below gaps of dominating Norway spruce canopy. However, the contribution of Scots pine to the total ecosystem dry-mass production was small compared to that of Norway spruce. Birches were out-competed during the first 150 years, since the mean light conditions below the canopy (mainly $I/Io < 0.2$, ratio between light below tree canopy/light above tree canopy) was too small to allow its regeneration and growth under dominating Norway spruce canopy (Fig. 4.14). In the ecosystem scale, the accumulation and loss of carbon fluctuated following the balance between the gross primary production and ecosystem respiration. The long-term balance varied around one (i.e., $GPP/(R_A + R_H) \rightarrow 1$) from negative (carbon uptake or sink) to positive (carbon loss or source) in stable environmental conditions.

### 4.6.3 *Performance of Ground Vegetation Below Tree Layer Under No Management*

Figure 4.14 further details of the response of ground vegetation to changes in the varying dynamics and biomass stocking of trees under even-aged management. Under successional tree stand, the presence of different ground cover species in biomass is assumed to be dependent on the limits set by the environmental conditions affected by the properties and stocking of trees through the chronosequence of trees. After the clear-cut, the biomass of ground cover species was small, as under the full-stocked tree stand before the cut. Thereafter, biomass increased rapidly and culminated in a few years after the clear-cut. This was linked with the regeneration and growth of trees reducing the available light and nitrogen for the growth ground cover species.

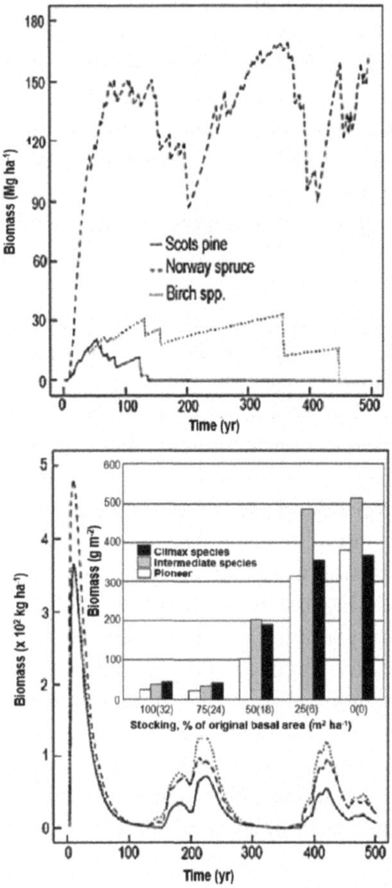


| Plant group, plant species | Mass, g 100 $m^{-2}$ | |
|---|---|---|
| | Before cut | 5 yr. after cut |
| **Pioneer species, total** | **52** | **825** |
| • *Festuca ovina* | | 412 |
| • *Carex globularis* | | 206 |
| • *Stellaria graminea* | | 52 |
| • *Carex digitata* | | 28 |
| • *Calamagrostis* | | 2 |
| • *Hypericum maculatum* | 26 | 13 |
| • *Cetraria islandica* | 13 | |
| • *Cladonia fimbriata* | 7 | |
| • *Rumex actosella* | 3 | |
| • *Epilobium angustifolium* | 2 | 6 |
| • *Avenella flexuosa* | 1 | 103 |
| • *Cladonia arbuscula* | | 3 |
| **Intermediate species, total** | **72** | **537** |
| • *Hieracium murorum* | 37 | 1 |
| • *Peltigera apthosa* | 18 | |
| • *Peltigera malacea* | | 135 |
| • *Maianthemum bifolium* | 9 | |
| • *Othilia secunda* | 5 | 269 |
| • *Melampyrum pratense* | 2 | 8 |
| • *Equisetum sylvaticum* | | 17 |
| • *Drypteris dilata* | 1 | 67 |
| • *Polypodium vulgare* | | 34 |
| • *Hepatica nobilis* | | 4 |
| • *Viola riviniana* | | 2 |
| **Climax species, total** | **65** | **104** |
| • *Pleroziun schreberi* | 33 | 42 |
| • *Ptilium crista-castrensis* | 16 | 3 |
| • *Dicranum elatum* | 8 | |
| • *Hylocomium splendens* | 4 | 21 |
| • *Pyrola chlorantha* | 2 | 5 |
| • *Diopteris carthusiana* | 1 | 13 |
| • *Lycopodium clavatum* | 1 | 10 |
| • *Lycopodium annotinum* | | 7 |
| • *Oxalis acetosella* | | 1 |
| • *Moneses uniflora* | | 2 |
| **Grand total** | **189** | **1,4666** |

**Fig. 4.14** Simulations of successional dynamics of tree species composition, changes in ground species composition under dominating trees, and an overview of the development of species richness after clear-cutting over time (Kellomäki and Väisänen 1991). Courtesy of Elsevier. Left/Upper: Simulations of biomass dynamics of Scots pine, Norway spruce, and birch under no management. Left/Below/excluding the Insert: Simulations of biomass dynamics of ground vegetation representing pioneer species (thin dotted line), intermediate species (thick dotted line) and climax species (solid black line). Left/Below/including the Insert: Simulations of biomass ground vegetation representing pioneer species (white column), intermediate species (grey column) and climax species (black column) as a function of the percentage of basal area of remaining trees in harvest. The percentage of remaining trees is 100% excluding any cutting (with the basal area 32 $m^2$ $ha^{-1}$). Concurrently, the percentage of remaining trees is 0% representing clear-cutting (with the basal area 0 $m^2$ $ha^{-1}$ after cutting). The columns within this range represent the effects of thinning intensity of remaining stocking from 75, 50, and 25% of trees on the biomass of ground cover compared to that before thinning. Right: Occurrence of different plant species in ground cover and their combined biomass divided in the pioneer, intermediate, and climax species in the biomass before cutting and five years after cutting

In early succession, *pioneer species* are dominant under substantial supply of light and nitrogen, but in late succession they are partly replaced by *climax species* capable of surviving under limited supply of light and nitrogen. The occurrence of ground cover species between the dominance of pioneer and climax species, there are many ground cover species through the chronosequence with no clear dominance. By the definition, these *intermediate species* survived in any phase of the chronosequence. In successional processes through a chronosequence, the occurrence and abundance of pioneer, intermediate, and climax species are modified by the availability of light and nitrogen in the forest floor forming a replacement (or reoccupation) process changing the species composition between species in the species classes representing pioneer, intermediate, and climax species. This process changes allocation of area for different species, thus modifying the species richness driven by the dynamics of tree biomass and gaps in canopy above ground vegetation as formulated below (Kauppi et al. 1978) (Box 4.3).

**Box 4.3 Growth and development of ground vegetation under growth and development of tree stand**

In analyzing the growth of development of ground cover plant communities, Kauppi et al. (1978) divided the stand area ($X_{1n}$) into not occupied by species in the year $n$. The area may be occupied by pioneer ($X_{2n}$), intermediate ($X_{3n}$) and climax ($X_{4n}$) species referred to the occurrence of the species in different phases of the chronosequence. The reoccupation ratio ($a_{ij}$) is the ratio between the area occupied by the species class $i$ from the species class $j$ and the area of species class $j$. Consequently, the area reoccupied by species $i$ from species $j$ is $a_{ij} \times X_{jn-1}$. This implies that the climax species increase their area at the expense the other species:

$$X_{4n} = a_{41} \times X_{1n-1} + a_{42} \times X_{2n-1} + a_{43-1} \times X_{3n-1} + X_{4n-1} \tag{4.1}$$

The area of intermediate species can be increased at the expense of non-occupied areas and pioneer species while losing area to climax species:

$$X_{3n} = a_{31} \times X_{1n-1} + a_{32} \times X_{2n-1} + X_{3n-1} - a_{43} \times X_{3n-1} \tag{4.2}$$

The pioneer species can increase their coverage at the expense of non-occupied areas and lose it to intermediate and climax species:

$$X_{2n} = a_{21} \times X_{1n-1} + X_{2n-1} - a_{32} \times X_{2n-1} - a_{42} \times X_{2n-1} \tag{4.3}$$

The non-occupied area only decreases:

(continued)

**Box 4.3** (continued)

$$X_{1n} = X_{1n-1} - a_{21} \times X_{1n-1} - a_{31} \times X_{1n-1} - a_{41} \times X_{1n-1} \quad (4.4)$$

The equations are recursive, i.e. the situation for one year is calculated with the help of the situation in the previous year. For their use in the matrix calculations, see Kauppi et al. (1978).

In initialization, the biomass of each species group and the proportion of individual species in each group were random one derived from the empirical material for middle boreal forests in Finland (61–62° N) based on the third National Forest Inventory of Finnish forests (1951–1953) (Kauppi et al. 1978). The sample plots formed a successional series over one hundred years from clear-cutting areas to trees stands of full maturity. The structure of ground vegetation was descried by using the coverage of each species related to total area of the observation plot. Vascular plants (field layer), and bryophytes (bottom layer) were treated separately in the inventory as done in the analysis.

The plant species in the ground cover were divided into pioneer, intermediate, and climax species as described above. Thus, the initial composition of ground vegetation was reduced to three species groups facilitating several tens of species being treated in the computations. The species groups were used as basic computational units. The growth of species in each group was related to the available light, soil water and nitrogen with the distinction between vascular and non-vascular plants as controlled by the gap dynamics of tree canopy. Light and nitrogen had clear effects on growth, while the role of water was negligible, thus excluded from the further model-based analyses. The growth of ground cover species in each group was computed separately in the same manner as in the case of trees using the parameter values specific for each species group affecting potential growth (Kellomäki and Väisänen 1991):

$$G_0(t) = e(bdgro \times M(t) \quad (4.5)$$

$$\frac{dG}{t} = G_0(t) \times G_l(t) \times G_N(t) \quad (4.6)$$

where $G_o$ (g m$^{-2}$ yr.$^{-1}$) is the potential growth; $M$ (g m$^{-2}$) is the current biomass of a particular species group; $G_l$ and $G_N$ are the growth multipliers dependent on the availability of light and nitrogen; bdgro is a parameter; and $t$ time in years.

The current biomass of a particular species group in any year is computed as a balance of the growth and death rates integrated over the preceding years. The death rate of each species group was assumed to be 5% of the standing crop (Mälkönen 1974), which was converted to litter and, subsequently, to humus. Litter and humus were decomposed in the same way as litter and humus originating from trees. The biomass of each species group was converted into individual species belonging to this group. *First*, any species present in the initial set of species was present in any phase of the succession. *Second*, the biomass of the species group was allocated to individual species of this group through a random procedure, which was repeated every year. Consequently, the biomass growth of a particular species within the species group varied from year to year.

The biomass of ground vegetation below the fully closed canopy was only a few grams per square meter, represented by dwarf shrubs and mosses (e.g., *Vaccinium myrtillus, Vaccinium vitis-idaea, Hylocomium splendens, Pleurozium schreberi, Dicranum sp.*) (Fig. 4.14). The mass of herbs and grasses was also small (e.g., *Avenella flexuosa, Maianthemum bifolium*). Following the clear-cutting or thinning (removal of basal area with consequent stem wood mass), the mass of these pioneer species increased in few years up to several hundred grams per square meter. This also held for climax species, but their share of the mass in ground cover decreased compared to that before cutting with the reduction of canopy cover. The increase of the number and mass of herbs and grasses was related to increases in available light and nitrogen. However, the dominance of herbs and grasses decreased, and the dominance of dwarf shrubs and mosses increased, if the canopy closed again related to the ongoing growth and development of trees.

### *4.6.4 Performance of Ground Vegetation Below Tree Layer Under Management*

The sensitivity of the ground cover to the dynamics of the tree layer was further studied with the help of simulations where the gap size varied in otherwise stable conditions, i.e., in the stand at the age of 80 years. The density of the tree layer in the model stand was thinned by 0, 25, 50, 75, and 100% from the prevailing one (the insert in Fig. 4.14). Thereafter the stand was allowed to grow and develop 20 years up to the age of 100 years. Biomass accumulation responded slightly to thinning at a rate lower than 50% of removal stem wood stocking. At a higher thinning rate (> 50%), the response of all the groups of species was vigorous, but the response of intermediate and climax species levelled off at a lower thinning rate than did that of pioneer species. This was expected based on the response pattern of different species in relation to the availability of light and nitrogen (Kellomäki and Väisänen 1991).

## 4.7 Richness of Heterotrophic Species Related to Presence of Dead Wood

Richness of plants (tree species, ground cover species) indicates the variability of niches for heterotrophic species through forest chronosequences. Especially, the presence of dead wood increases the abundance and richness of heterotrophic organisms in forest soil. In this context, Sandström et al. (2019) investigated how the presence of dead wood affects the abundance and richness of species in different taxon groups in selected temperate and boreal protected forests. Based on a meta-analysis, the study included saproxylic insects, ground insects, wood-inhabiting fungi, lichens, reptiles, and cavity-nesting birds. The meta-analysis (Box 4.4) included 37 peer-reviewed studies representing the stand-scale biodiversity of the study sites to identify how changes in dead wood (dead wood manipulation) may affect stand-scale biodiversity. The dead wood manipulation included (i) creation of dead wood from living trees at the site; (ii) addition of dead wood originating outside the sites; and (iii) treatment of dead wood in prescribed burning at the sites.

**Box 4.4 Overview of meta-analyses**

In the meta-analysis, Sandström et al. (2019) combined the results of multiple scientific studies that address the same question. The analysis is based on statistics, aiming to derive a pooled estimate closest to the unknown common truth based on how this error is perceived. The response ratio (*RR*) indicates the magnitude of the effect of selected factors (e.g., thinning or fertilization) on selected processes and components (e.g., carbon uptake/emission or carbon stocks above/below the soil surface) (e.g., Chen et al. 2020):

$$RR = ln\frac{X_t}{X_c} = lnX_t - lnX_c \qquad (4.7)$$

where $X_t$ and $X_c$ are the mean values for the impacts on the selected factor and on the control group against which the impacts are compared. Meta-analyses provide an estimate of the unknown effect size, and, further, the capacity to contrast results from different studies and identify: (i) patterns among different studies; (ii) sources of disagreement among those results; (iii) or other interesting relationships that may become known with multiple studies.

Sandström et al. (2019) found that the abundance and richness of saproxylic insects and fungi increased along with the increasing amount of dead wood (Fig. 4.15), while changes in dead wood had no effect on ground insects and cavity-nesting birds. However, the findings showed that an increasing amount of dead wood tended to increase the abundance and richness of species dependent on dead wood. Burning of dead wood, as with other dead wood manipulation, increased the abundance of saproxylic insects. However, burning tended to reduce the amount of dead wood such as snags, with subsequent reduction of biodiversity, for example, of cavity-nesting birds. Sandström et al. (2019) concluded that manipulating dead wood is an effective conservation management approach for supporting biodiversity in protected areas.

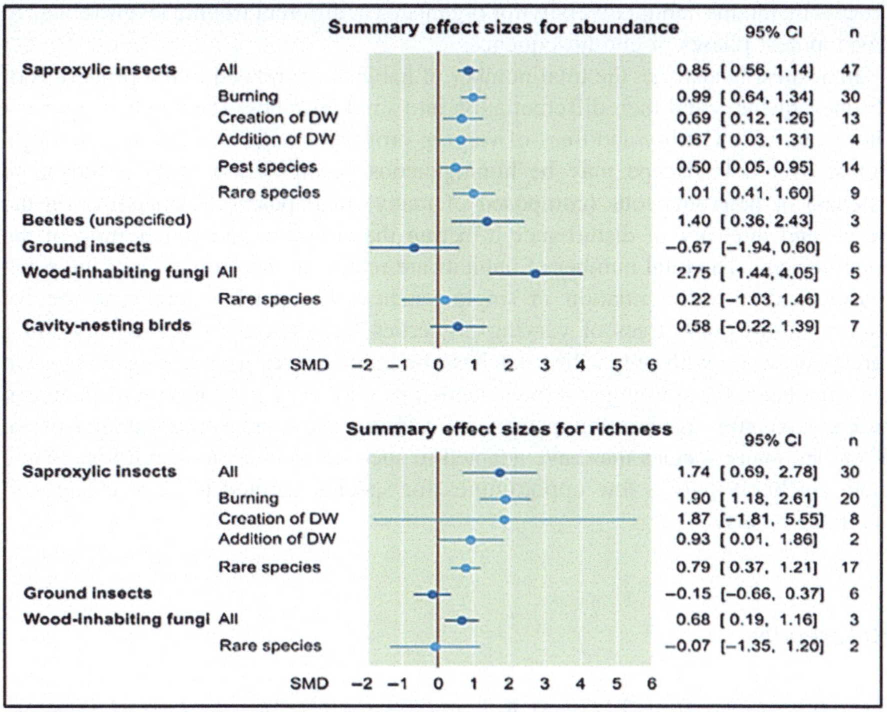


**Fig. 4.15** Response of the abundance and richness of different taxon groups to dead wood manipulation (Sandström et al. 2019). Open Access. Permission of Wiley. Species abundance is the number of individuals per species, while relative abundance refers to the evenness of distribution of individuals among species in a community. Species richness is the number of varied species present in a site, ecological community, landscape, or region. SMD = standardized mean difference in relation to sites with no manipulation of dead wood; 95% CL = mean and 95% confidence interval of the summary effect; n = sample size, referring to the number of comparisons

## 4.8 Concluding Remarks

The carbon dynamics of forest ecosystems are closely linked to the structure and mass of trees and ground vegetation. Ecosystem dynamics control several feedbacks affecting the carbon sink/source dynamics, and further nutrient, and hydrological cycles. Dynamics of a forest ecosystem is biologically related to the amount of biomass (living, dead) of trees and ground vegetation occupying the site. The mass in the ecosystem buffers the impacts of energy (radiation, gravity, heat, wind force, water flow) affecting the energetic, hydrological, and biogeochemical factors controlling the functioning of the ecosystem. Habitat and species diversity reflect ecosystem energetics, where the gross primary production and respiration losses (autotrophic and heterotrophic respiration) converge with each other toward the old-growth phase (climax phase) through a forest chronosequence. This long-term process maintains habitat diversity for organisms at different trophic levels related to development phases of chronosequences.

In a forest landscape, the total number of habitats are related to the properties of chronosequences and their differentiation into single patches. The patches represent sites of varying fertility and trees of varying properties (e.g., species, age, density). Forest over a landscape may be homogeneous (composed of only a few large patches) or heterogeneous (composed of many small patches) depending on the extent and intensity of disturbance affecting the structure and functioning of the patch mosaic. The total number of habitats are related to the properties of the patch mosaic and its differentiation in single patches. The patches represent sites of varying fertility and trees of varying properties (e.g., species, age, density). In a forest landscape with only a few patches, the habitat diversity changes slowly. On the other hand, the splitting of a forest landscape into many indicates rapid change in habitat diversity representing patches of varying ages and species composition. There are many species that have adapted to such heterogeneous conditions, while there are likely only a few opportunities for species adapted to slow changes in habitat diversity.

## References

Amiro BD, Barr AG, Black TA, Brown R, Brown MN, Chen J, Clark KL, Davis KJ, Desai AR, Dore S, Engel V, Fuentes JD, Goldstein AH, Goulden ML, Kolb TE et al (2010) Ecosystem carbon dioxide fluxes after disturbance in forests of North America. J Geophys Res/Biogeosci 115:1–28

Bateman IJ, Mace GM, Fezzi C, Atkinson G, Turner K (2011) Economic analysis for ecosystem service assessments. Environ Resource Econ 48:177–217

Bhatti JS, van Kooten GC, Apps MJ, Laird LD, Campbell ID, Campbell C, Turetsky MR, Yu Z, Banfield E (2003) Carbon balance and climate change in boreal forests. In: Burton PJ, Messier C, Smith DW, Adamowicz WL (eds) Towards sustainable management of the boreal forest. NRC Research Press, Ottawa, pp 799–855

Bolton NW, D'Amato AW (2011) Regeneration responses to gap size and coarse wood debris within natural disturbance-based silvicultural systems in north-eastern Minnesota, USA. For Ecol Manag 262:2015–1222

Bormann FH, Likens GE (1979) Pattern and process in a forested ecosystem. Springer, New York

Busing RT (1994) Canopy cover of tree regeneration in old-growth cove forests of the Appalachian Mountains. Vegetatio 115:19–27

Chen LF, Xiang YZ, He ZB, Lin PF, Zhu Z (2020) A meta-analysis of the impacts of forest logging on soil $CO_2$ efflux. Sci Cold Arid Regions 12(3):165–179

Framstadt E, de Wit H, Mäkipää R, Larjovaara M, Vesterdal L, Karltun E (2013) Biodiversity, carbon storage and dynamics of northern forests. TemaNord 2013(507):1–130

Gao B, Taylor AR, Searle EB, Kumar P, Ma Z, Hume AM, Chen HYH (2018) Carbon storage declines in old boreal forests irrespective of succession pathway. Ecosystems 21:1168–1182

Gough CM, Vogel CS, Schmid HP, Curtis CP (2008) Controls on annual forest carbon storage: lessons from the past and predictions for the future. Bioscience 58(7):609–622

Goulden ML, McMillan AMS, Winston GC, Rocha AV, Manies KL, Harden JW, Bond-Lamberty BP (2011) Patterns of NPP, GPP, respiration, and NEP during boreal forest succession. Glob Chang Biol 17:855–871

Harmon ME (2002) Carbon sequestration in forests: addressing the scale question. J For 99(4): 24–29

Harmon ME (2009) Woody detritus mass and its contribution to carbon dynamics of old-growth forest: the temporal context. In: Wirth C, Gleixner G, Heimann M (eds) Old-growth forests: function, fate, and value, vol 207. Ecological Studies, pp 159–190

Janisch JE, Harmon ME (2002) Successional changes in live and dead wood carbon stores: implications for net ecosystem productivity. Tree Physiol 22:77–89

Jentsch A, Beierkuhnlein C, White PS (2002) Scale, the dynamic stability of forest ecosystems, and the persistence of biodiversity. Silva Fennica 36(1):393–400

Kauppi P, Hari P, Kellomäki S (1978) A discrete time model for succession of ground cover communities after clear cutting. Oikos 30:100–105

Kazimirov NM (1971) Spruce forests of Karelia. Nauka, Leningrad

Kellomäki S (2005) Metsäekologia. Joensuun Yliopistopaino, Joensuu

Kellomäki S, Väisänen H (1991) Application of a gap model for the simulation of forest ground vegetation in boreal conditions. For Ecol Manag 42:35–47

Kolari P, Pumpanen J, Rannik Ü, Ilvesniemi H, Hari P, Berninger F (2004) Carbon balance of different aged Scots pine forests in Southern Finland. Glob Chang Biol 10:1106–1119

Kolari P, Lappalainen HK, Hänninen H, Hari P (2007) Relationship between temperature and the seasonal course of photosynthesis in Scots pine at northern timberline and southern boreal zone. Tellus 59B:542–552

Kolari P, Kulmala L, Pumpanen J, Launiainen S, Ilvesniemi H, Hari P, Nikinmaa E (2009) $CO_2$ exchange and component $CO_2$ fluxes of a boreal Scots pine forest. Boreal Environ Res 14:761–783

Law BE, Harmon ME (2011) Forest sector carbon management, measurement and verification, and discussion of policy related to climate change. Carbon Manage 2(1):73–84

Law BE, Thornton PE, Irvine J, Anthoni PM, Svan TS (2001) Carbon storage and fluxes in ponderosa pine forests at different developmental stages. Glob Chang Biol 7:755–777

Law BE, Sun OJ, Campbell J, van Tuyl S, Thornton PE (2003) Changes in carbon storage and fluxes in a chronosequence of ponderosa pine. Glob Chang Biol 9:510–524

Liu S, Bond-Lamberty B, Hicke JA, Vargas R, Zhao S, Chen J, Edburg SL, Hu Y, Liu J, McGuire AD, Xiao J, Keane R, Yuan W, Tang J, Luo Y, Potter C, Oeding J (2011) Simulating the impacts of disturbances on forest carbon cycling in North America: processes, data, models, and challenges. J Geophys Res 116:G00K08

Luyssaert S, Schulze E-D, Börner A, Knohl A, Hessenmöller D, Law DBE et al (2008) Old-growth forests as global carbon sinks. Nature 455(11):213–215

Mälkönen E (1974) Annual primary production and nutrient cycled in some Scots pine stands. Communicationes Instituti Forestalis Fenniae 84(5):1–87

Mamkin V, Kurbatova J, Avilov V, Mukhartova Y, Krupenko A, Ivanov D, Levashova N, Olchev A (2016) Changes in net ecosystem exchange of $CO_2$, latent and sensible heat fluxes in a recently clearcut spruce forest in western Russia: results from an experimental and modeling analysis. Environ Res Lett 11(12):125012

McKinley DC, Ryan MG, Birdsey RA, Giardina CP, Harmon ME, Heath LS, Houghton RA, Jackson RB, Morrison JF, Murray BC (2011) A synthesis of current knowledge on forests and carbon storage in the United States. Ecol Appl 21(6):1902–1924

Muscolo A, Bagnato S, Sidari M, Mercurio R (2014) A review of the roles of forest canopy gaps. J For Res 25(4):725–735

Oliver CD, Larson BC (1990) Forest stand dynamics. McGraw Hill, New York

Persson H (1975) Dry matter production of dwarf shrubs, mosses, and lichens in some Scots pine stands at Ivantjärnheden, Central Sweden. Swed. Conifer. For. Proj. Tech. Report, 2/1975

Pickett STA, Rogers KH (1997) Patch dynamics: the transformation of landscape structure and function. In: Bissonette JA (ed) Wildlife and landscape ecology: effects of pattern and scale. Springer, New York, pp 101–127

Pickett STA, Collins SL, Armesto JJ (1987) Models, mechanisms, and pathways of succession. Bot Rev 53:335–371

Sandström J, Bernes C, Junninen K, Lõhmus A, Macdonald E, Müller J, Jonsson BG (2019) Impacts of dead wood manipulation on the biodiversity of temperate and boreal forests: a systematic review. J Appl Ecol 56:1770–1781

Shorohova E, Kuuluvainen T, Kangur A, Jõgiste K (2009) Natural stand structures, disturbance regimes and successional dynamics in the Eurasian boreal forests: a review with special reference to Russian studies. Ann For Sci 66:1–20

Sing L, Ray D, Watts K (2015) Ecosystem services and forest management. Research Note FCRN020. Forestry Commission, Bristol

Sirén G (1955) The development of spruce forest on raw humus sites in Northern Finland and its ecology. Acta Forestalia Fennica 62(4):1–408

Tamm CO (1953) Growth, yield, and nutrition in carpets of forest mosses (*Hylocomium splendens*). Medd Skogsfordskningsinst 43(1):1–140

Uri V, Kukumägi M, Aosaar J, Varik M, Becker H, Aun K, Lõhmus K, Soorsaar K, Astover A, Uri M, Buht M, Sepaste A, Padari A (2022) The dynamics of the carbon storage and fluxes in Scots pine (*Pinus sylvestris*) chronosequence. Sci Total Environ 817:152973

White PS, Pickett STA (1985) Natural disturbance and patch dynamics: an introduction. In: Pickett STA, White PS (eds) The ecology of natural disturbances and patch dynamics. Academic Press, San Diego, pp 3–13

White PS, Harrod J, Walker JL, Jentsch A (2000) Disturbance, scale, and boundary in wilderness management. USDA Forest Service Proceedings RMRS-P-15-VOL2:27–42

# Chapter 5
# Carbon Dynamics with Links to Growth and Mortality of Trees and Ground Vegetation

## Contents



 
S. Kellomäki, *Forest Management for Timber Production and Climate Change Mitigation*, Managing Forest Ecosystems 44,
https://doi.org/10.1007/978-3-031-71575-4_5

**Abstract** The functioning of boreal forest ecosystems is based on the ecosystem structure and functioning, which involves carbon cycling through its uptake in photosynthesis to retention in biomass and soil organic matter. Atmospheric carbon is bound in the ecosystem through the growth of biomass, which synthesizes new tissue in trees and ground vegetation. Soil carbon input includes litter, exudates from roots and mycorrhizal fungal mycelia, and bodies of the soil biota. In decay, litter is converted into partly decayed organic matter and further humus. The difference between gross primary production (*GPP*) and autotrophic respiration ($R_A$) yields the net primary production (*NPP*), the rate at which energy and carbon are accumulated and maintained in the form of mass in trees and other green plants. *NPP* declines in aging trees, i.e., *GPP* declines more rapidly than the increase in $R_A$. Finally, carbon departs the ecosystem in carbon dioxide emitted into the atmosphere and in dissolved in surface and ground water flows.

## 5.1 Carbon in Boreal Forest Ecosystems Under Management

### *5.1.1 Carbon Fluxes and Stocks*

The functioning of boreal forest ecosystems is based on the cycle of carbon, which involves in the uptake, retention in biomass and soil organic matter, and eventual emission of carbon to the atmosphere or dissolution in surface and ground water flows (Fig. 5.1). Emissions can be both autotrophic and heterotrophic, the latter representing the decay of dead soil organic matter. The carbon cycle is further driven by the hydrological and nutrient cycles, in that the availability of soil water and nutrients control carbon uptake and the use of assimilated carbon for regeneration and growth of trees and ground vegetation. Carbon allocation to the different organs of trees and ground vegetation further affects the amount and properties of soil organic matter (SOM) and carbon stocks on the soil and within the soil profile. The functioning and structure of forest ecosystems produce different ecosystem services, including regulating (e.g., mitigation of climatic warming), provisioning (e.g., timber and biomass), and cultural (e.g., amenity values) services. The harvesting of biomass and timber further cycle forest carbon through the manufacturing industry and wood-based products, which are eventually abandoned and burned/decayed, emitting carbon to the atmosphere.

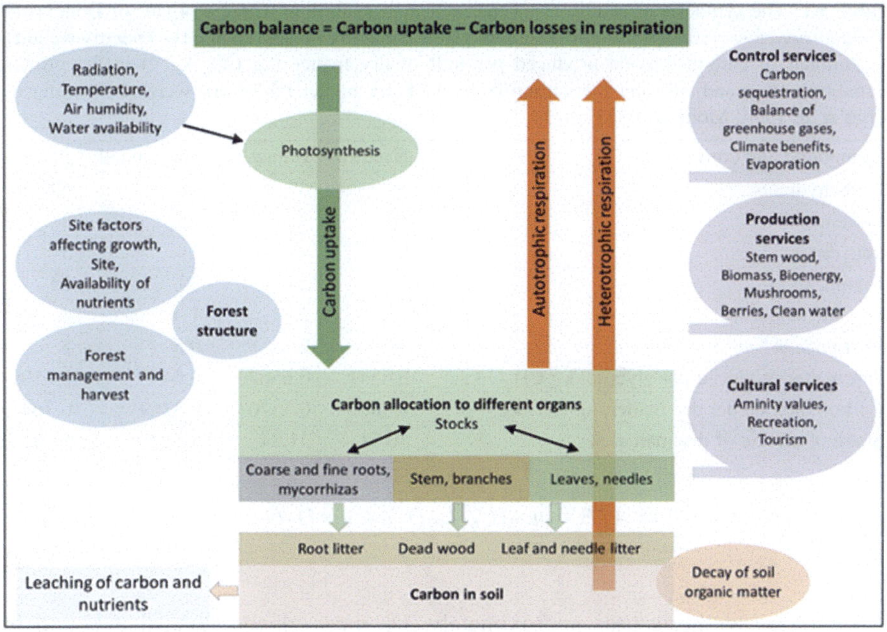


**Fig. 5.1** Overview of the dynamics of carbon fluxes and stocks in boreal forest ecosystems, which cycle carbon through ecosystems and to the atmosphere, affecting the provisioning of ecosystem services (Launiainen et al. 2015). Courtesy of Natural Resources Institute Finland

### *5.1.2 Carbon Bound in Trees and GroundVegetation*

Atmospheric carbon is bound in the ecosystem in the biomass growth, which synthesizes new tissue in trees and ground vegetation. Biomass growth is a function of the photosynthates and of maintenance and growth respiration: $\Delta W = P - (R_m + R_g)$, where $\Delta W$ is the biomass growth, $P$ is the gross amount of carbon available for maintenance and growth in the whole tree or in each organ (allocated to organ). $R_{\mathrm{m}}$ is the maintenance respiration and $R_{\mathrm{g}}$ the growth respiration, driving the maintenance of existing living tissues and the growth of new tissues, respectively. Biomass growth is manifested in growth across the dimensions of trees such as height and diameter, the former representing the primary and the latter, the secondary growth. It is also manifested in the mass of tree organs (foliage, branches, stem, coarse roots, fine roots).

Table 5.1 contains an overview of the chemical composition of matter in Douglas fir (*Pseudotsuga menziesii*), with breakdown of the proportion of carbohydrates and other chemicals used in the growth of its organs. In growth, glucose, and other initial substances (e.g., other carbohydrates, fats, etc.) are converted into new molecules such as cellulose, lignin, proteins, and fats. The amount of energy and respiration is related to the chemical composition of the plant material.

**Table 5.1** The synthesis of cellulose: (i) chemical composition of selected organs of Douglas fir (*Pseudotsuga menziesii*); (ii) the conversion of glycose into dry matter (kg of $CH_2O$ per dry weight), the amount of carbon dioxide produced per unit of dry matter (kg $CO_2$ $kg^{-1}$) in the growth respiration ($R_g$); and (iii) and the carbon content of dry matter (% of dry weight) (Penning de Vries et al. 1974; Mohren 1987)

| Compound and conversion | Needles | Branches | Stem wood | Roots |
|---|---|---|---|---|
| Carbohydrates, % | 60 | 65 | 66 | 62 |
| Proteins, % | 10 | 3 | 0.5 | 3 |
| Fats, % | 8 | 4 | 2 | 5 |
| Lignin, % | 15 | 24 | 30 | 24 |
| Organic acids, % | 4 | 2 | 1 | 4 |
| Nutrients, % | 3 | 2 | 0.5 | 2 |
| Dry matter produced per glycose, kg $CH_2O$ $kg^{-1}$ | 0.6541 | 0.6562 | 0.6443 | 0.6509 |
| Carbon dioxide per dry matter, kg $CO_2$ $kg^{-1}$ | 0.3866 | 0.3526 | 0.3469 | 0.3644 |
| Share of carbon of dry matter, % | 50.61 | 51.34 | 52.62 | 51.51 |

$$n \times (C_6H_{12}O_6) + 2n \times ATP \rightarrow n \times (C_6H_{10}O_5) + n \times H_2O + 2n \times ADP + 2n \times P. \tag{5.1}$$

Cellulose synthesis uses approximately 7% more glucose than that bound in cellulose itself. In the synthesis of proteins, fats, and lignin, 66%, 20%, and 26% more glucose, respectively, is used for ATP than is bound in these compounds themselves. In this example, the carbohydrate content in different tree organs varies in the range of 60–66% of organ dry matter, the highest values representing sapwood and the lowest, needles. In needles, the protein content (10%) is three times higher than in other organs (<3%). In needles, the fat (8%) and nutrient contents are also higher than in other organs, while the lignin content is lower. The amount of glucose used to produce dry matter is 0.65 kg per kg of dry matter regardless of organ. Similarly, the growth respiration is similar in different organs (0.35–0.39 kg $CO_2$ per kg of dry matter) for different organs. The carbon content of organs ranges from 51 to 53% of dry matter, the lowest value being for needles and the highest for sapwood.

### 5.1.3 Carbon Distributed in Boreal Forest Ecosystems

The whole mass of plants (trees or other plants) is the total mass of the living organs. The whole mass is related to the allocation and turnover of mass in different organs. In trees, for example, the diameter growth of the stem downwards from the stem apex is linearly related to the concurrent accumulation of foliage downwards. The total foliage mass ($W$, kg) is a function of the stem mass: $W = A \times (dbh^2 \times H)^b$, where $dbh$ is the breast-height diameter, $H$ is the tree height, and $A$ and $b$ are parameters.

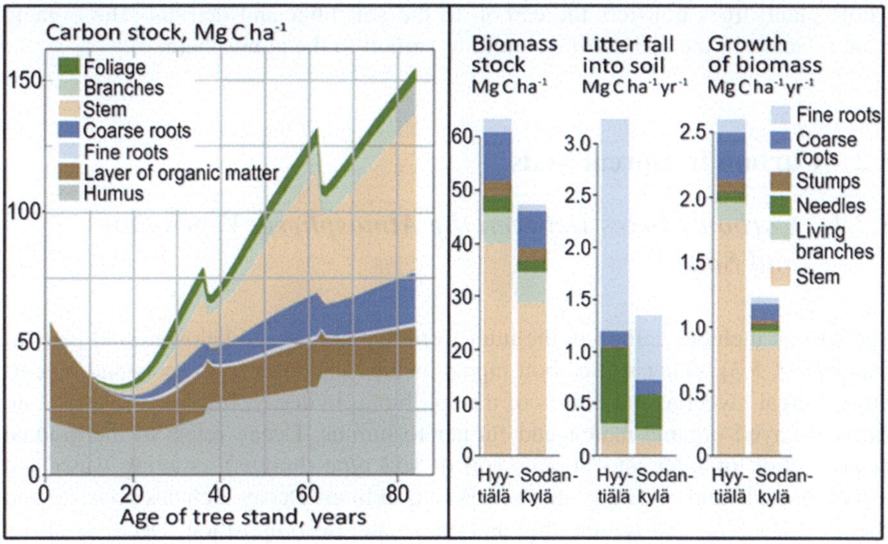


**Fig. 5.2** Distribution of mass and growth in forest ecosystems and between tree organs in southern and northern boreal forests (Launiainen et al. 2015). Courtesy of Natural Resources Institute Finland. Left: Distribution of carbon (Mg ha$^{-1}$) in different organs of Scots pine over 90 years in southern boreal forests (Hyytiälä, 61° N) under even-aged management. Right: Net biomass growth, biomass, and litter production in uneven-aged Scots pine forests in southern (Hyytiälä, 61° N) and northern (Sodankylä, 67° N) boreal forests

Similar allometric relations can be compiled for other organs (e.g., Mälkönen 1974, 1977).

Figure 5.2 shows how the mass of different organs in stands of 45-year-old Scots pine individuals is distributed in the southern (62° N) and northern (67° N) boreal zones. In the south, the whole mass of carbon is 65 Mg C ha$^{-1}$, while in the north, it is 47 Mg C ha$^{-1}$ (about 70% of that in the south). This difference implies differences in growing conditions and management history affecting the mean annual growth rate, which is about half of that in the north as in the south (about 2.6 Mg ha$^{-1}$ yr.$^{-1}$). The distribution of mass between organs is similar at both sites. However, mass distribution becomes increasingly dominated by stem mass as trees mature. In mature 90-year-old trees, the stem wood mass is 60–70 Mg C ha$^{-1}$, while the combined mass of foliage, branches, and coarse and fine roots are 80–90 Mg C ha$^{-1}$ (Fig. 5.2). The share of these organs also increases over time, but their increase of carbon stock is small in relation to that in stem wood (e.g., Launiainen et al. 2015).

Figure 5.2 also shows that soil carbon (litter and humus) decreases from 50 down to 20 Mg C ha$^{-1}$ 15–20 years after the clear-cutting of mature trees in the previous production cycle. Thereafter, the carbon in the litter layer and humus increases with the growth and development of trees planted in the clear-cut area. Harvest residues and fine and coarse roots remaining from removed trees further increase the litter and humus in soil. In this process, the mortality and subsequent litterfall of living plant organs (from trees, vascular, and nonvascular plants in ground vegetation) and whole plants/trees transfers the carbon to the soil litter and detritus. The organic matter decays in the soil detritus, emitting carbon to the atmosphere.

## 5.2 Carbon in Boreal Soils

### *5.2.1 Carbon Fluxes Between the Atmosphere, Vegetation, and Soil*

Soil carbon includes carbon in the litter from the above- and belowground parts of plants (Fig. 5.3). Other soil carbon inputs include exudates from roots and mycorrhizal fungal mycelia, and bodies of the soil biota. In decay, litter is converted into partly decayed organic matter and further to humus. Decay refers to the gradual disintegration of litter into organic soil or soil type (humus), emitting water and carbon dioxide and releasing nutrients bound in litter. Decay includes physical and chemical leaching and weathering, and enzymatic decomposition. Decay is also a biological process in which heterotrophic organisms obtain energy from dead plant material. Heterotrophic organisms include vertebrates, invertebrates, bacteria, and fungi, representing different taxonomic groups and trophic levels. Soil fauna and microorganisms make combined nutrients available and make soil organic matter more susceptible to leaching and erosion (Mayer et al. 2020).

During decay, the mass of litter and humus declines and the structure of compounds becomes simpler (Fig. 5.3). Eventually, some humus turns into humus colloids mixed with mineral particles in the soil (Mason 1976; Persson et al. 1980). Over time, the organic matter on mineral surfaces reduces the decay of carbon particles. This stabilizes the dynamics of soil carbon by slowing decay, thus protecting carbon stocks in the soil profile. In mineral soil, carbon particles occur in micropores at a depth where the organic matter is less accessible to microbial decay. Microbes convert most of the carbon in plant biomass into carbon dioxide, and the minor share into stable soil carbon. Thus, soil microbes play the double role of rapidly decaying carbon in litter and retaining carbon originating from the carbon uptake in plants (Bradford et al. 2016).

### *5.2.2 Carbon in the Soil Profile*

In northern Europe, the majority of boreal upland forest soils are covered by an organic layer (organic horizon or O-horizon). The organic layer includes litter and partly decayed litter from trees and ground vegetation. Further down, humus contains the decayed organic material below the litter. The origin of the litter is still identifiable in the upper parts of the litter, but the identifiability declines as decay proceeds to form peat-like matter below the identifiable litter (Fig. 5.4). The decay of

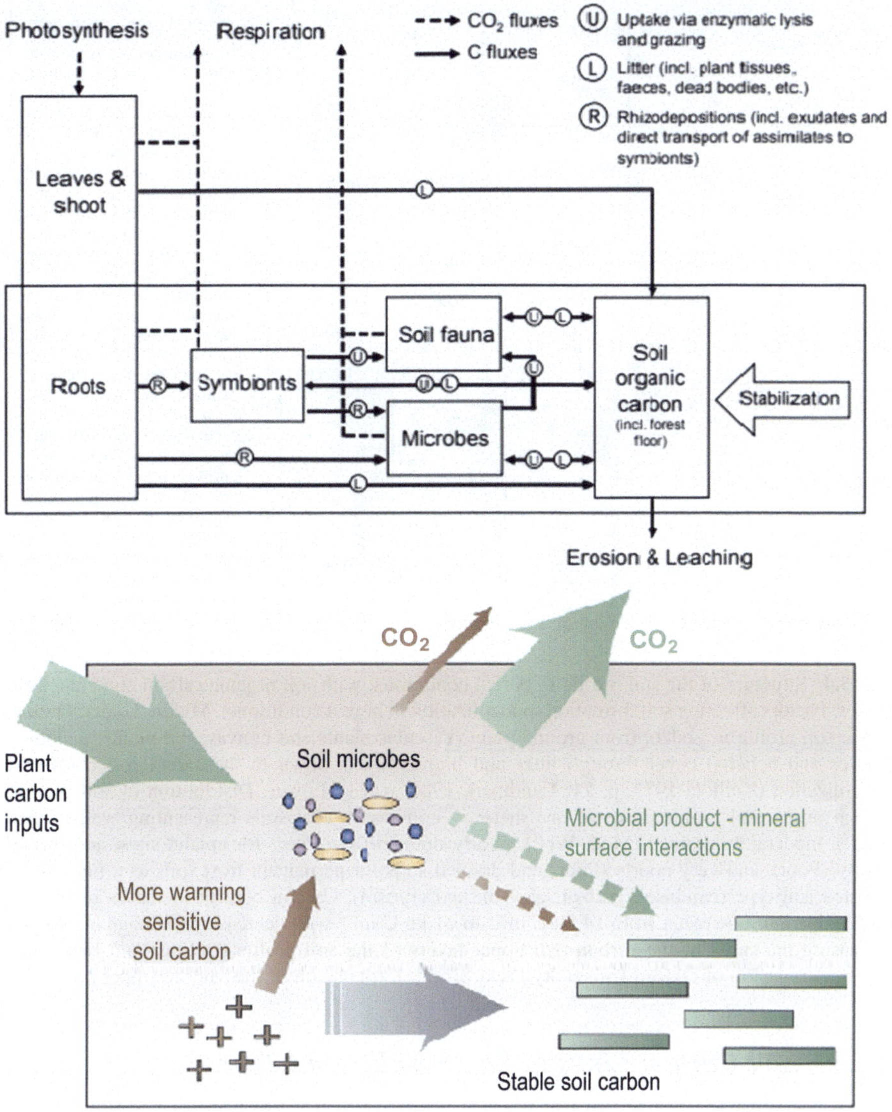


**Fig. 5.3** Carbon fluxes between the atmosphere, vegetation, and soil. Upper: Carbon fluxes between the atmosphere, vegetation, and soil (Mayer et al. 2020). Permission of Elsevier. Lower: Dual role of soil microbes in the decay and stabilization of soil carbon (Bradford et al. 2016). Permission of Springer Nature

humus is related to the activity of fungi, bacteria, and invertebrates, which use litter and humus to obtain energy. Litter, humus, humus mixed with upper mineral soil, and humus particles leached lower into the soil profile from soil organic matter (SOM). In peatlands, mineral soil is covered by peat, which is the partly decayed debris of ground vegetation (e.g., *Sphagnum* sp., *Carex* sp.) and litter from trees growing in waterlogged sites.

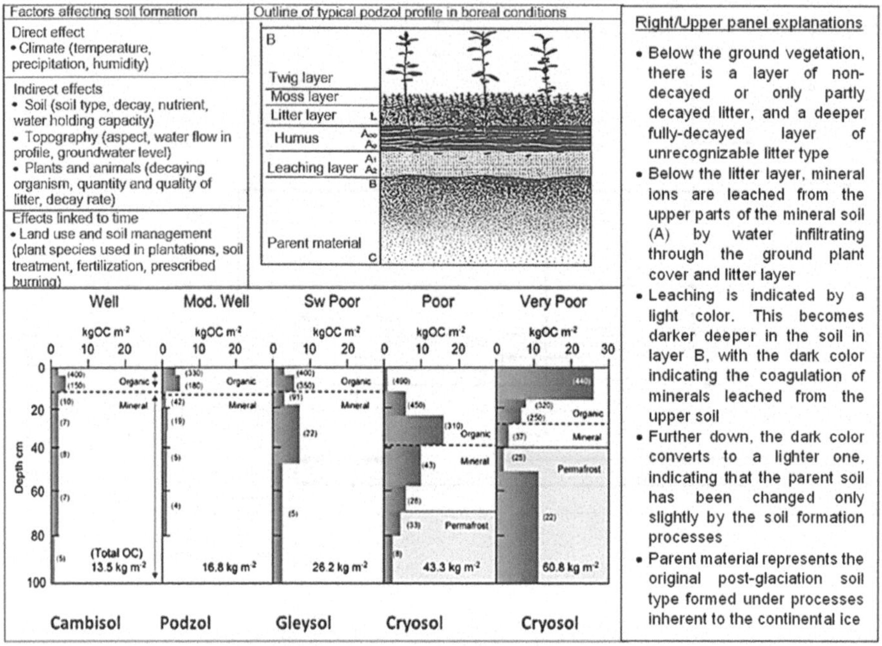


**Fig. 5.4** Structure of the soil profile in boreal conditions, with soil organic carbon contents. Left/ Upper: Factors affecting soil formation (podzolization in boreal conditions). Middle/Upper: Outline of the soil profile in podzol from ground plant (vascular plants and nonvascular plants, including mosses and lichens) cover through litter and humus layers down to the parent soil below the podzolization (Kalliola 1973, p. 71; Lundmark 1986, p. 73). Lower: Distribution of soil organic carbon in the soil profile down to one meter in common boreal soils representing well-drained (Well), moderately drained (Mod. Well), poorly drained (Sw. Poor) for upland sites, and further poorly (Poor), and very poorly (Poor) and drained soils for permanent frost soils as a function of selected soil type (cambisol, podzol, gleysol, and cryosol). Carbon content increases in the soil profile down to one meter from 14 kg C m$^{-2}$ to 61 kg C m$^{-2}$ with decreasing drainage of soil and increasing thickness of the carbon-rich upper layers of the soil profile (DeLuca and Boisvenue 2012). Permission of Oxford University Press. Right: Explanation of the Middle/Upper panel

## 5.2.3 *Soil Carbon Links to Site Fertility and Type*

Carbon uptake and storage in trees, ground vegetation, and soil are connected to site fertility. Site fertility refers to the ability of forest soils to sustain plant growth with specific rate in forest ecosystems, especially tree growth, which is the main focus of forest management. Site types can be used to identify the fertility of forest sites, as in Finland, and its links with carbon at the site. The concept of forest types refers to the presence of varied species (trees, other woody plants, grasses, herbs, mosses, lichens) that indicate the degree of site fertility, or site type. In the plant community, different species compete among one another, with the consequence that the community structure becomes regular over time (*E*) with regard to site fertility. Such

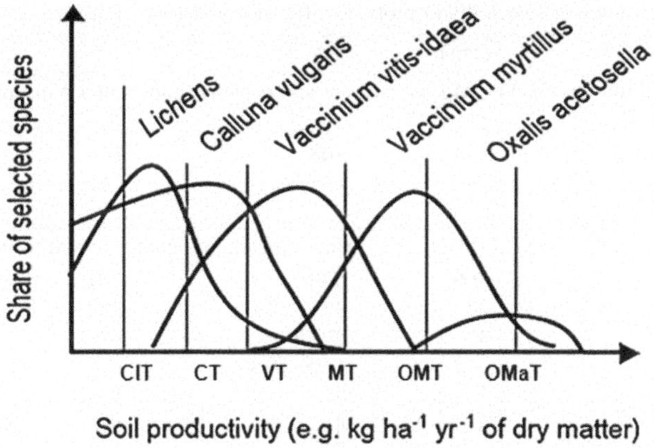


**Fig. 5.5** Schematic depiction of how some species in boreal ground vegetation are distributed in relation to the potential productivity of upland forest sites, defined by site types modified from Kauppi et al. (1976, p. 26). Compiled by the author of this book. Legend for site types: ClT = *Cladonia* type (extremely poor), CT = *Calluna* type (very poor), VT = *Vaccinium* type (moderately poor), MT = *Myrtillus* type (moderately fertile), OMT = *Oxalis–Myrtillus* type (fertile), OMaT = *Oxalis–Maianthemum* type (very fertile)

plant communities deviate from each other depending on site fertility ($P$), thus indicating the variability in fertility among sites (Cajander 1909):

$$P = f(E) \rightarrow P = f(\text{climatic factors, edaphic factors}) \rightarrow P = f(\text{plant community}) \tag{5.2}$$

Sites with similar ground vegetation (i.e., the same species and species composition) are of the same site type. This is the clearest for mature tree stands with a closed canopy, where the ground cover community is in equilibrium with the prevailing growing conditions. However, any site with ground cover deviating temporarily from that in the mature stand belongs to the same site type. The ground vegetation (e.g., *Myrtillus* site type, MT) may be grass- and herb-dominated in a young tree stand, but the ground vegetation shifts to a dominance of dwarf shrubs and mosses when the tree layer matures and the canopy closes. Thus, the long-term development of ground cover converges with the typical ground cover of the *Myrtillus* site type. As a young tree stand develops into a mature stand, the dominance of different species changes but the same plant species persist over time (Fig. 5.5).

Site types represent a continuous variability of fertility, which is divided into fertility classes, where the internal variability is smaller than that between fertility classes. The variability of fertility is related to the variability in climatic and edaphic conditions, which create a gradient in the productivity between site types, as shown in Table 5.2 for southern- and middle-boreal forests. Productivity declines from the

**Table 5.2** Mean values of selected soil properties for different forest types

| Property | OMT | MT | VT | CT |
|---|---|---|---|---|
| Mass of litter and humus (SOM), Mg ha$^{-1}$ (layer and organic matter mixed in mineral soil) | | | | |
| Humus mass<br>Humus in mineral soil (0–30 cm)<br>Total SOM | 23<br>124<br>147 | 29<br>108<br>137 | 30<br>98<br>128 | 27<br>84<br>111 |
| Nutrients in litter and humus (SOM), kg ha$^{-1}$ (layer of organic matter mixed in mineral soil) | | | | |
| N | 500 | 530 | 420 | 330 |
| P | 34 | 37 | 32 | 24 |
| K | 33 | 37 | 30 | 27 |
| Ca | 175 | 160 | 130 | 80 |
| Mg | 34 | 26 | 20 | 16 |
| Humus pH | 5.2 | 4.8 | 4.6 | 4.2 |
| Nutrients in mineral soil below humus, kg ha$^{-1}$ (excluding humus mixed in mineral soil) | | | | |
| N | 2800 | 2000 | 1500 | 1200 |
| P | 6 | 7 | 9 | 9 |
| K | 56 | 41 | 35 | 33 |
| Ca | 340 | 162 | 120 | 56 |
| Mg | 58 | 27 | 18 | 12 |
| Field capacity, mm | 125 | 100 | 75 | 15 |
| Wilting point, mm | 75 | 45 | 5 | 5 |
| Typical soil type | Fine moraine | Medium fine moraine | Coarse moraine | Coarse sand |

fertile (OMT, *Oxalis–Myrtillus* site type and more fertile sites) through to the moderately fertile (MT, *Myrtillus* site type) and moderately poor (VT, *Vaccinium* site type) to poor (CT, *Calluna* site type and less fertile) sites, including the majority of upland forested land. Fertile site types represent sites with fine-grained soils containing a mixture of varying sizes of soil particles (e.g., clay moraine), while poor sites are common on coarse-grained soils (e.g., sandy soils). Increasing site fertility indicates an increasing amount of soil organic matter (SOM), with an increasing amount of nitrogen and calcium, and a reduction in soil acidity. Increasing site fertility is further linked to the availability of water with increasing nutrient availability at a given site (e.g., Urvas and Erviö 1974; Westman 1990; Pardossi et al. 2009; Launiainen et al. 2022).

The nutrient content in litter decreases in SOM in the early phase of succession, but the amount of litter and humus with nutrient storage increases over the long term (Table 5.2). The accumulation of nitrogen is particularly large. For example, in mature Scots pine stands the amount of nitrogen bound in litter and humus may exceed 1000 kg ha$^{-1}$ in moderately fertile sites (*Myrtillus* site type, MT) in the southern boreal zone. In such conditions, clear-cutting substantially increases the mineralization of nitrogen: five years after the clear-cutting cut, the nitrogen bound in SOM is likely 20% lower than that before the cut. For other nutrients, the effects of clear-cutting are smaller than that for nitrogen. In litter and humus, the amount of phosphorus may be 70–100 kg ha$^{-1}$, for potassium 100–200 kg ha$^{-1}$, and for calcium 100–250 kg ha$^{-1}$.

## 5.3 Carbon in Organic Matter and Ecosystem Energetics

### *5.3.1 Carbon Dynamics in Forest Ecosystems*

Vegetation (trees, shrubs, and ground vegetation species) forms a production system that intercepts energy into the ecosystem for use at different trophic levels of food webs. Production requires nutrients (nitrogen, mineral nutrients) originating from (i) the exchange between different parts of plants and trees (internal nutrient cycle); (ii) the exchange between trees and plants and their surrounding environment (within-community nutrient cycle); and (iii) the exchange between surrounding communities (between-community nutrient cycle). These cycles combined form the nutrient cycle in an ecosystem, which is further linked to the nutrient cycles in surrounding ecosystems. Energy and nutrients move from the production system to the detritus system via litter, which is decayed along with humus and releases nutrients for reuse (decomposition system) (Fig. 5.6).

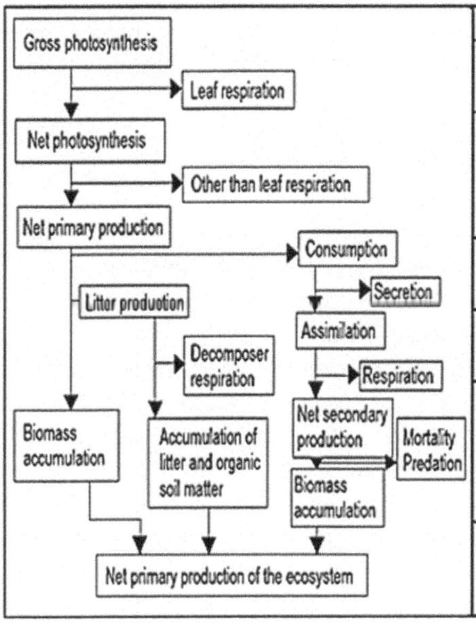


| Concepts | Definition |
|---|---|
| Gross primary production *GPP* [g m$^{-2}$ yr$^{-1}$] | Gross carbon fixation in forest ecosystems |
| Autotrophic respiration *RA* [g m$^{-2}$ yr$^{-1}$]<br>• Maintenance $R_m$<br>• Growth $R_g$ | Loss rate of carbon through autotrophic respiration rate $RA = R_m + R_g$ |
| Net primary production *NPP* [g m$^{-2}$ yr$^{-1}$] | Gross production minus autotrophic respiration $NPP = GPP - RA$ |
| Heterotrophic respiration *RH* [g m$^{-2}$ yr$^{-1}$] | Loss rate of carbon through decay and respiration of secondary producers |
| Net ecosystem production *NEP* or net ecosystem exchange *NEE* [g m$^{-2}$ yr$^{-1}$] | Gross primary production minus total respiration or net primary production minus heterotrophic respiration, $NEE = GPP - RA - RH$ or $NEE = NPP - RH$ |
| Net biome production *NBP* [g m$^{-2}$ yr$^{-1}$] | Net biome production $NBP = NEE$ plus carbon lost in disturbances, e.g., harvest, fires |

**Fig. 5.6** Carbon flow and stocks, and definitions of the main concepts used for energy flow and carbon dynamics in forest ecosystems. Compiled by the author of this book. Left: Schematic depiction of the structure and functioning of carbon flows through a forest ecosystem based on Aber and Melillo (1991, p. 35). Right: Concepts with definitions widely used to describe the carbon dynamics in forest ecosystems

The mass of organic matter and its dynamics are linked with the ecosystem energetics, which involve the use of energy bound in gross primary production (*GPP*) to renew and maintain organisms at different trophic levels of food webs. The difference between *GPP* and autotrophic respiration ($R_A$) yields the net primary production (*NPP*), the rate at which energy is accumulated and maintained in the form of mass in trees and other plants (Fig. 5.6). NPP declines in aging trees primarily driven by slowing *GPP*. This decreases more rapidly with increasing age than $R_A$ does, but the ratio of *NPP/GPP* remains approximately constant within a biome (Tang et al. 2014). Mass is further used in secondary production, which involves higher trophic levels in t*he ecosystem energetics, as indicated by the heterotrophic respiration* ($R_H$)*. The difference between GPP* and the sum of $R_A$ and $R_H$ indicates the overall net production in the ecosystem (*NEP*), representing the mass accumulated in the ecosystem ($NEP = GPP - R_A - R_H = NPP - R_H$). Furthermore, net biome production (*NBP*) is used to link the uptake and loss of carbon and nitrogen with carbon losses through natural (e.g., fire) and anthropogenic disturbance (management and harvest).

## 5.3.2 Carbon Uptake and Stocks in Trees

At the stand/community scale, the mass of trees ($M(t)$) or plants at a given moment depends on the past growth and development ($M(t\text{-}1)$), and the current regeneration ($S(t)$), growth ($G(t)$), and mortality ($D(t)$):

$$M(t) = M(t-1) + G(t) + S(t) - D(t). \tag{5.3}$$

Furthermore,
$M(t) = n(t-1) \times \overline{m}(t) + n(t-1) \times \overline{g}(t) + h(t) \times \overline{s}(t) - k(t) \times (\overline{m}(t) + \overline{g}(t))$, where $n$ is the density of trees, $\overline{m}$ is the mean mass of trees, $\overline{g}$ is the mean growth of trees, $h$ is the density of seedlings born in regeneration, $\overline{s}$ is the mean mass of seedlings, and $k$ is the mortality (density of dying trees).

In crop physiology, regeneration and growth represent the NPP when energy is used for growth in the mass of different organs and seed crops. Subtracting $R_A$ from GPP gives the gross biomass growth ($NPP = GPP - R_A$). The mass of litter or death of whole trees and other plants subtracted from the gross growth implies the net biomass growth accumulating over time, yielding the stocking (Fig. 5.7). The accumulation of mass in trees is dependent on the growth rate, which is linked to site fertility. The growth rate or mass accumulation rate follows a culminating function related to tree species, stand age, and site fertility. Regardless of tree species, the slow growth of stem wood in early succession is followed by rapid growth until the growth slows down and gradually declines in late succession. In this respect, there are pronounced differences in growth between sites of different fertility. In fertile sites (*Oxalis–Myrtillus* site type, OMT), the growth rate of Scots pine is nearly three times larger than that in poor sites (*Cladonia* site type, ClT). On

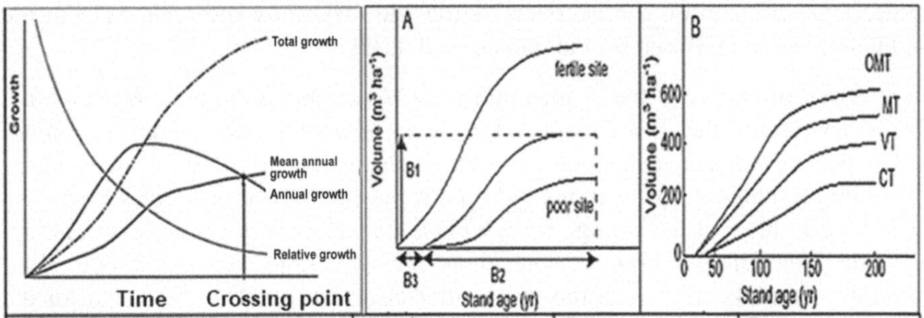


| Site type | $B_1$, m³ ha⁻¹ | $B_2$, yr | $B_3$, yr |
|---|---|---|---|
| *Oxalis-Myrtillus* (OMT) | 600 | 10 | 135 |
| *Myrtillus* (MT) | 520 | 15 | 160 |
| *Vaccinium* (VT) | 400 | 25 | 175 |
| *Calluna* (CT) | 300 | 40 | 185 |
| *Cladonia* (CIT) | < 200 | >60 | >190 |

**Fig. 5.7** Schematic depiction of growth and yield in boreal forests as a function of time/age and site fertility, based on growth and yield studies. Compiled by the author of this book. Left/Upper: Schematic depiction of stem wood growth over time, featuring widely used concepts in growth and yield studies, i.e., total growth, annual growth, mean annual growth, and relative growth (percentage of annual growth in relation to stem wood stock) over time. Middle/Upper: Overview for calculating the growth of stem wood as a function of time in sites of varying fertility using the equation: $M(t) = B_1 \times \left(1.0 - e^{B_3 \times (t - B_2)}\right)$. The parameter $B_1$ ($m^3\ ha^{-1}$) is the maximum value of stem wood, $B_2$ is the year to achieve rapid growth since establishment, and $B_3$ is the time to obtain the maximum value of stem wood. Right/Upper: Growth of Scots pine as a function of time and site fertility in the southern boreal zone, when using the values of parameters estimated based on the growth and yield tables (Koivisto 1962). Lower: Parameter values per site type used in calculations

the other hand, the growth rate of Scots pine and birch in early succession is faster than that of Norway spruce. The rapid growth of Norway spruce continues for longer than in Scots pine and birch (Koivisto 1962).

### 5.3.3 *Carbon Uptake and Stocks in Ground Vegetation*

Ground vegetation is broadly divided into field and bottom layers. The field layer includes vascular plants (herbs, grasses, and dwarf shrubs) and the bottom layer, non-vascular plants (mosses and lichens). After clear-cutting, the growth of ground vegetation tends to accelerate, especially on fertile sites occupied with herbs and grasses. In such conditions, the enhancing growth of ground vegetation retains nutrients in increasing biomass before the reforestation, thus reducing the wash of nutrients released in the enhancing decay of soil organic matter. In this respect, Palviainen et al. (2005) investigated the aboveground growth and mass, nutrient contents (N, P, K, and Ca), and species composition of ground vegetation one year before and seven years after clear-cutting of a mixed forest dominated by Norway spruce (*Picea abies*) on a moderately fertile site (*Myrtillus* site type, MT) in the middle-boreal zone. Based on Palviainen et al. (2005):

- In terms of biomass, the feather mosses (*Pleurozium schreberi*, *Hylocomium splendens*) and the dwarf shrubs (*Vaccinium myrtillus*, *Vaccinium vitis-idaea*) lost their dominance in the ground vegetation in the mature forest following clear-cutting. Excluding *Hylocomium splendens*, the occurrence of species recovered 3–5 years after clear-cutting. *Hylocomium splendens* was sensitive to cutting, while *Pleurozium schreberi*, *Vaccinium myrtillus*, and *Vaccinium vitis-idaea* returned rapidly to their initial levels after clear-cutting. This held also for the nutrient contents in ground vegetation.
- The biomass of hair-grass *(Deschampsia flexuosa)* increased immediately after clear-cutting, and the biomass of *fireweed (Epilobium angustifolium)* increased 3–5 years after cutting. These species retained nutrients efficiently because of their rapid biomass growth and high nutrient content in biomass. The total biomass and nutrient contents in such ground vegetation exceeded those of the pre-cutting levels. However, the share of biomass and nutrient contents in mosses decreased after cutting.

For stand-replacing disturbance events such as clear-cutting under even-aged management, the growth or the mass accumulation of vascular plants is partly inversely related to the increasing stocking of trees, from clear-cutting to the regeneration and full maturation of trees (e.g., Muukkonen and Mäkipää 2006). The mass of herbs and grasses reduces when the tree canopy closes, allowing dwarf shrubs and mosses to increase in mass relative to herbs and grasses. Under fully mature trees with a partially closed canopy, the mass of herbs, grasses, and dwarf shrubs tends to increase slightly in fertile sites. This likely holds also for mosses. Based on Palviainen and Finér (2007):

- The mass of dwarf shrubs increases up to 100–300 kg ha$^{-1}$ in 20 years after clear-cutting (e.g., with maximum biomass of up to 500–1000 kg ha$^{-1}$, mainly of bilberry (*Vaccinium myrtillus*) and cowberry (*Vaccinium vitis-idaea*).
- The mass of herbs and grasses increases 3–5 years after clear-cutting, with a maximum mass of 2000–3000 kg ha$^{-1}$ depending on the species composition.
- The maximum biomass of mosses (e.g., combined communities of *Pleurozium schreberi*, *Hylocomium splendens*, *Dicranum* species) may be up to 400–1000 kg ha$^{-1}$.

The amount and distribution of biomass among different species varies in relation to the properties of maturing trees (Table 5.3). Under a partially closed canopy, the biomass of ground vegetation is large, including herbs and grasses, when the spacing of trees is wide on fertile sites. The opposite scenario occurs in stands with narrow spacing, where the total biomass of ground vegetation is small and mainly represented by dwarf shrubs and mosses. The tree species composition further affects the properties of the ground vegetation. In fertile sites, the dominance of deciduous trees increases the growth of herbs and grasses due to moderate shading and beneficial temperature and moisture conditions. On the other hand, the

**Table 5.3** Biomass of ground vegetation in southern boreal forests by site fertility as a function of site type and the age of tree stand

| Site type and species group | Parameters of the model shown in the footnotes for species groups per for different site | | | Biomass, kg ha$^{-1}$ per species groups as a function of age of Norway spruce stand on *Myrtillus* site type over the period of 120 years since clear cut and planting | | | |
|---|---|---|---|---|---|---|---|
| | $B_1$ (kg ha$^{-1}$) | $B_2$ (yr) | $B_3$ (yr) | Stand age (yr) | Dwarf shrubs | Herbs, grasses | Total |
| | | | | | Biomass, kg ha$^{-1}$ | | |
| *Oxalis-Myrtillus* site type (OMT) | | | | | | | |
| • Dwarf shrubs | 600 | 2 | 10 | 0 | 380 | 50 | 440 |
| • Grasses, herbs | 3000 | 1 | 3 | 5 | 1520 | 2840 | 4360 |
| *Myrtillus site* type (MT) | | | | 10 | 1590 | 2350 | 3940 |
| • Dwarf shrubs | 1000 | 2 | 12 | 15 | 990 | 1040 | 2030 |
| • Grasses, herbs | 2000 | 3 | 4 | 20 | 730 | 220 | 950 |
| *Vaccinium* site type (VT) | | | | 30 | 650 | 30 | 680 |
| • Dwarf shrubs | 1500 | 3 | 15 | 40 | 640 | 20 | 660 |
| • Grasses, herbs | 500 | 3 | 5 | 50 | 640 | 20 | 660 |
| *Calluna* site type (CT) | | | | 60 | 630 | 20 | 650 |
| • Dwarf shrubs | 1000 | 3 | 17 | 70 | 490 | 20 | 510 |
| • Grasses, herbs | 100 | 4 | 6 | 80 | 350 | 20 | 370 |
| *Cladonia* site type (ClT) | | | | 90 | 350 | 30 | 380 |
| • Dwarf shrubs | 400 | 5 | 20 | 100 | 370 | 50 | 420 |
| • Grasses, herbs | < 20 | 5 | 20 | 120 | 380 | 50 | 440 |

Left: Mass was calculated with the equation $M(t) = B_1 \times \left(1.0 - e^{B_3 \times (t - B_2)}\right.$, where $M(t)$ is the mass of dwarf shrubs, herbs, and grasses as a function of site fertility and the age ($t$) of the tree stand (e.g., Mälkönen 1974; Napola 1982)

Middle: Parameter $B_1$ (kg $ha_{-1}$) indicates the maximum value of mass in early succession after clear-cutting and before slight shading from tree seedlings, $B_2$ the years to achieve rapid growth after clear-cutting, and $B_3$ the time to achieve the maximum value of biomass

Right: Mass and structure of ground cover in a Norway spruce (*Picea abies*) stand in a site of medium fertility (*Myrtillus* site type, MT) in the southern boreal forests as a function of the age of tree stands in simulations using the Forcyte model (Kimmins and Scoullar 1984). The initial stand density was 4000 seedlings ha$^{-1}$

dominance of conifers with pronounced shading tends to increase the growth of dwarf shrubs and reduces that of herbs and grasses for canopy shading with shading-limited soil moisture and temperature. In poor sites, the influence of trees on ground vegetation is small, and the dominance of dwarf shrubs, mosses, and lichens remains even in clear-cut areas, whereas the role of herbs and grasses is marginal.

## 5.4 Litterfall Related to Tree Growth

### *5.4.1 Stem Wood Growth and Litterfall in Trees and Stands*

Regarding trees, death may be divided into litterfall and the death or mortality of the whole tree. Litterfall includes the annual death of a part of the mass of different tree organs (foliage, branches, bark, coarse roots, and fine roots) in living trees, while mortality refers to the death of whole trees, including all the organs. Litter may further be divided into non-woody (e.g., needles/leaves, fine roots) and woody litter (e.g., shoots/branches, coarse roots).

In boreal coniferous trees, non-woody litter mainly comprises dying needles and fine roots, which represent the turnover (annual share of foliage to die) of the mass of these organs. This also holds for deciduous trees, whose foliage (leaves) totally dies and renews following annual or seasonal cycles of climate and weather. In both coniferous and deciduous trees, a portion of the fine roots survives to the next growing season, representing the species-specific turnover and the dependence on site conditions. Consequently, the foliage/fine root mass generated in a given year decreases depending on the maturity of trees and, in conifers, on the survival rate related to the age of needles. In both coniferous and deciduous trees, the mass generated yearly, and the mass distributed to different age classes, are allometrically related to the size (maturity) of the tree.

In general, there are close links between the biomass growth and mortality of different organs and of the whole tree. Both in Scots pine and Norway spruce, the annual death of needles is directly related to the growth of stem wood in such a way that the death of branches is initiated with a several-year delay compared to that in needles and fine roots. In this regard, Fig. 5.8 summarizes how the fall of non-wood

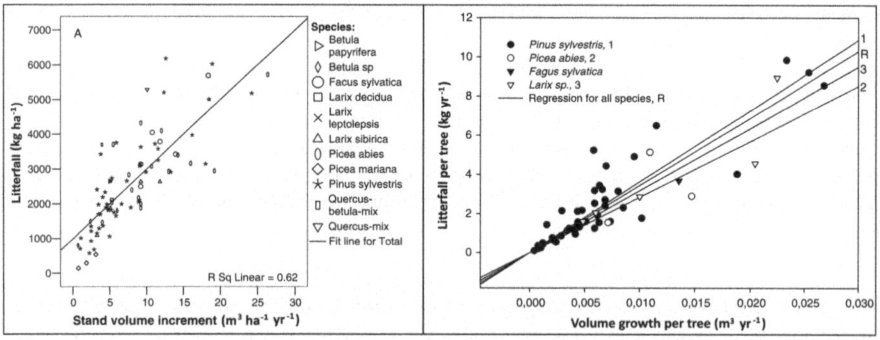


**Fig. 5.8** Needle/leaf litterfall related to stem wood growth for selected tree species common in boreal and temperate zones based on different studies (Matala et al. 2008). Permission of Taylor & Francis. Left: Needle/leaf litterfall as a function of stem wood growth of selected tree species over all the species included in the study. Right: Needle/leaf litterfall as a function of stem wood growth of selected trees. The symbol (R) refers to the regression for all the tree species selected to the study shown on the left, and the symbol (1) refers to Scots pine; (2) to Norway spruce; and (3) to larch species

coniferous/deciduous litter (foliage, shoots) is linked to the stem wood growth of single trees and stand-based measurements for several boreal and temperate tree species. In this case, the litter also includes other dead material from living trees. These experimental studies show a linear dependence of litter on the stem wood growth, thus demonstrating the allometric relationship between the growth of stem wood and foliage. Relationship between stem wood growth ($\Delta V(t)$, $m^3$ $yr.^{-1}$) and litterfall ($LF(t)$, kg $yr.^{-1}$) may be presented for a single tree as follows (Matala et al. 2008):

$$LF(t) = \beta \times \Delta V(t) \tag{5.4}$$

The combined value of the parameter $\beta$ was 343.2 for all species, 362.0 for Scots pine, 283.9 for Norway spruce (*Picea abies*), and 316.6 for larches (*Larix* sp.) (Matala et al. 2008). This implies that Scots pine produces 22% more litter per stem wood growth than Norway spruce, thus indicating tree-species differences affecting SOM and carbon in soil detritus. This is also evident in the left panel in Fig. 5.8, which shows the litterfall being 2000–3000 kg $ha^{-1}$ over a range of 5–10 $m^3$ $ha^{-1}$ $yr.^{-1}$ of stem wood growth. On the other hand, the litterfall is inversely related to the northward increase of latitude; that is, in southern-boreal Finland, the litterfall is 2000–3000 kg $ha^{-1}$, and in northern-boreal Finland, up to 1000 kg $ha^{-1}$.

### 5.4.2 *Litter Type and Share of Mortality Related to Organ Mass*

Woody litter is litter from the stem, including bark, branches/shoot axis, and coarse roots, while non-woody litter represents litter from foliage, regenerative organs (i.e., flowers, cones, seeds etc.), and fine roots. The amount of litter from different organs is related to the mass of organs and the turnover of mass representing the percentage or fraction of litter per year in relation to the mass of the living organs. The turnover of wood litter is very low (<1%), while the turnover of non-woody litter in conifers is high, with up to 30% for needles and up to 85% for fine roots. In deciduous trees, the turnover of foliage is up to 80% in the phase when leaves turn yellow (Table 5.4).

### 5.4.3 *Turnover of Biomass and Litterfall in Trees*

The allometric distribution of growth between different organs establishes the quantitative relations between the mass of the stem, branches, foliage, coarse roots, and fine roots (e.g., Mälkönen 1974). This implies a regular allocation of biomass relative to the share of biomass present in the different organs at a given

**Table 5.4** Litter types and turnover for selecwted boreal tree species, compiled by Lehtonen et al. (2016) based on different studies

| Tree species | Biomass component | Mass loss per (year, %) | Reference |
|---|---|---|---|
| Scots pine | Leaves | 27 (18–34) | Ťupek et al. (2015) |
| Norway spruce | Leaves | 13 (9–13) | Ťupek et al. (2015) |
| Broad-leaved | Leaves | 79 | Ťupek et al. (2015) |
| Scots pine | Branches | 2 | Lehtonen et al. (2004) |
| Norway spruce | Branches | 1.25 | Muukkonen and Lehtonen (2004) |
| Broad-leaved | Branches | 1.35 | Lehtonen et al. (2004) |
| Scots pine | Bark | 0.3 | Mälkönen (1977) |
| Norway spruce | Bark | – | Mälkönen (1977) |
| Broad-leaved | Bark | 0.01 | Mälkönen (1977) |
| Scots pine | Coarse roots | 2 | Lehtonen et al. (2004) |
| Norway spruce | Coarse roots | 1.25 | Muukkonen and Lehtonen (2004) |
| Broad-leaved | Coarse roots | 1.35 | Muukkonen and Lehtonen (2004) |
| Scots pine (1) | Fine roots | 85 | Kleja et al. (2008) |
| Norway spruce (1) | Fine roots | 85 | Kleja et al. (2008) |
| Broad-leaved | Fine roots | 85 | Kleja et al. (2008) |
| Scots pine (2) | Fine roots | 50 | Leppälammi-Kujansuu et al. (2014a, b) |
| Norway spruce (2) | Fine roots | 50 | Leppälammi-Kujansuu et al. (2014a, b) |
| Broad-leaved (2) | Fine roots | 50 | Leppälammi-Kujansuu et al. (2014a, b) |

Legend: In the column on the Left: (1) refers to the north, and (2) to the south

moment. Thus, the mass of different organs represents the difference between growth and mortality, with the annual turnover of mass in different organs referring to the fraction of mass lost per year in any tree organ. This holds also for other vascular plants, and for mosses and lichens. According to Muukkonen and Lehtonen (2004) and Muukkonen (2005), the turnover of biomass such as that of needles ($r_j$, fraction yr.$^{-1}$) for Scots pine and Norway spruce can be calculated as:

$$r_j = \frac{\sum_{i=0}^{n-1} (b_i) - b_{i+1}) \times w_i \times d_i}{\sum_{i=0}^{n-1} (b_i \times w_i)}. \tag{5.5}$$

where $n$ refers to the number of age classes of needles, $b$ is the percentage survival of mass cohorts ($i$) of needles, $w$ is the weight factors indicating the weight of needles over time, and $d$ is the weight loss of needles when dying. Over time, the litterfall ($l(t)_i$) from trees and ground vegetation can be calculated by multiplying the total biomass of specific organs by the turnover: $l(t)_i = r_i \times w_i$ (Muukkonen 2005).

### 5.4.4 Litter from Woody Organs of Trees: Stem and Branches

In trees, woody litter includes dead shoots, branches, the bark of stem, and coarse roots. In Norway spruce, the turnover of total branch biomass is dependent on stand density, tree size, and tree stocking (Fig. 5.9). This holds also for Scots pine and birches, with the branch turnover being 0.020 $yr.^{-1}$ for Scots pine and 0.0135 $yr.^{-1}$ for birches. Lehtonen et al. (2004) further showed that the share of branch litter from the branch mass in small Scots pines (diameter < 5 cm) was 4–6% and declined down to 1–2% in mature trees (diameter > 20 cm) along with the higher crown base. The pattern was similar in Norway spruce, but the turnover rate (0.0125 $yr.^{-1}$) was about half of that of Scots pine, and the death of branches was less sensitive to the size and stocking of trees, especially for widely spaced trees. Mälkönen (1977), and later Lehtonen et al. (2016), estimated that the turnover rate of bark is likely 0.003 $yr.^{-1}$ in Scots pine and 0.0001 $yr.^{-1}$ in deciduous trees. In coarse roots, the turnover rate is also small: 0.02 $yr.^{-1}$ for Scots pine, 0.0125 $yr.^{-1}$ for Norway spruce, and 0.0135 $yr.^{-1}$ for deciduous trees.

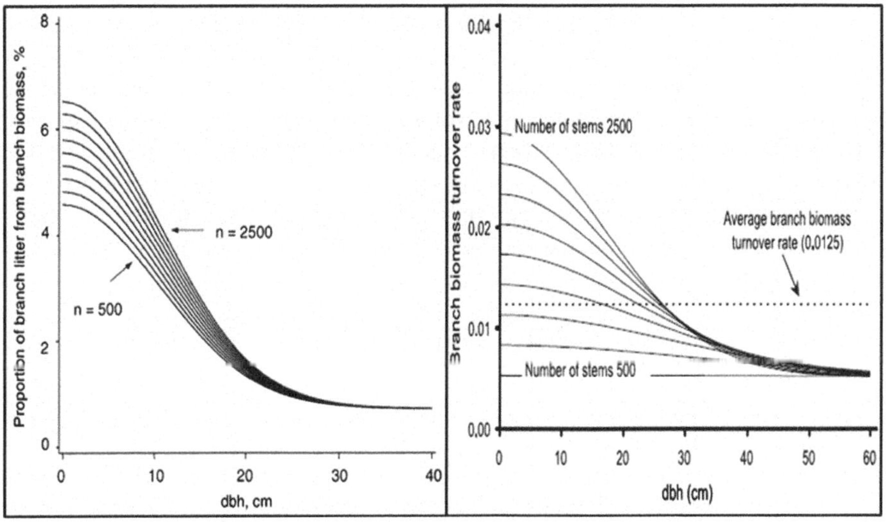


**Fig. 5.9** Dynamics of branches in crown of boreal Scots pine (*Pinus sylvestris*) and Norway spruce (*Picea abies*) related to the stem diameter of trees and stand density indicated by the number of trees per hectare. Left: Proportion of branch litter from branch biomass in Scots pine (Lehtonen et al. 2004). Permission of Elsevier. Right: Turnover rate of branch biomass in Norway spruce (Muukkonen and Lehtonen 2004). Open Access. Permission of NRC Canada

### 5.4.5 Litter from Non-woody Organs of Trees: Foliage

Litter from foliage and fine roots comprises the majority of carbon inputs to the soil, with a smaller contribution from the carbon in woody litter and regenerative organs. According to Muukkonen and Lehtonen (2004), the mean annual turnover rate for needles of Norway spruce is 0.10 $yr.^{-1}$. In any needle cohort, half of the needles are shed within 5.5 years, and all the needles within 12 years. Muukkonen and Lehtonen (2004) further found that the survival rate of Scots pine needles from their generation to the age of four years was up to 80–85% but declined linearly to 10–15% as a function of needle age from four to six years (Table 5.4). The mean maximum survival of needles in boreal Scots pine is 4–6 years, with an annual turnover of 0.25–0.50 $yr.^{-1}$ in southern and 0.10–0.25 $yr.^{-1}$ in northern boreal forests (e.g., Pensa and Sellin 2002; Muukkonen and Lehtonen 2004; Muukkonen 2005; Vanninen and Mäkelä 2005; Lehtonen et al. 2008, 2016; Ťupek et al. 2015). In terms of temperature, the life span of needles of Scots pine in the northern boreal forests is five to six years under annual temperatures of −2 to −4 °C, while it is three to four years under annual temperatures of +2 to +6 °C in the southern boreal forests (e.g., Reich et al. 1996, 2014). The maximum survival of needles of boreal Norway spruce is 8–12 years, with a turnover of 0.10–0.15 $yr.^{-1}$ and is longer on poor and northern sites than on southern boreal and fertile sites (Lehtonen et al. 2016) (Table 5.4).

There is biogeographic variation in needle turnover in boreal and temperate conifers, as evident in Scots pine and Norway spruce (Fig. 5.10). In both species, the pattern is similar, but the life span of needles is longer in Norway spruce than in Scots pine. Regardless of species, the life span of needles is longer at high latitudes and in sites with low mean annual temperatures than at low latitudes and under moderate mean annual temperatures (Reich et al. 2014). In Scots pine, the death of needles over four years from birth to death is:

$$d(t) = (1 - p(1)) \times mn(0) + (1 - p(2)) \times mn(1) + \ldots + (1 - p(5)) \times mn(4)) \tag{5.6}$$

where $d(t)$ is the mass of needle litter in the year $t$, and $mn$ (0) . . .$mn$ (4) is the mass of needles up to four years of age with a probability $p$ (1) . . .$p$ (5) of surviving to the next year. This pattern holds for Norway spruce across the boreal zone, but the longevity of needles in Norway spruce is double that in Scots pine both in the south and north.

In Scots pine, the annual total needle litter is in the range of 500–4500 kg $ha^{-1}$ $yr.^{-1}$ (e.g., Jalkanen 1998; Niinemets and Lukjanova 2003) across the boreal zone depending on the growth, density, and maturity of tree stands. In Norway spruce, the needle litterfall varies from 500 to 4500 kg $ha^{-1}$ $yr.^{-1}$, covering more than 50% of the aboveground litter in the sites dominated by this tree species (Muukkonen and Lehtonen 2004). In both species, the needle litter yield is larger in the southern and fertile sites than in the northern and poor sites (e.g., Ukonmaanaho 2013). In Norway

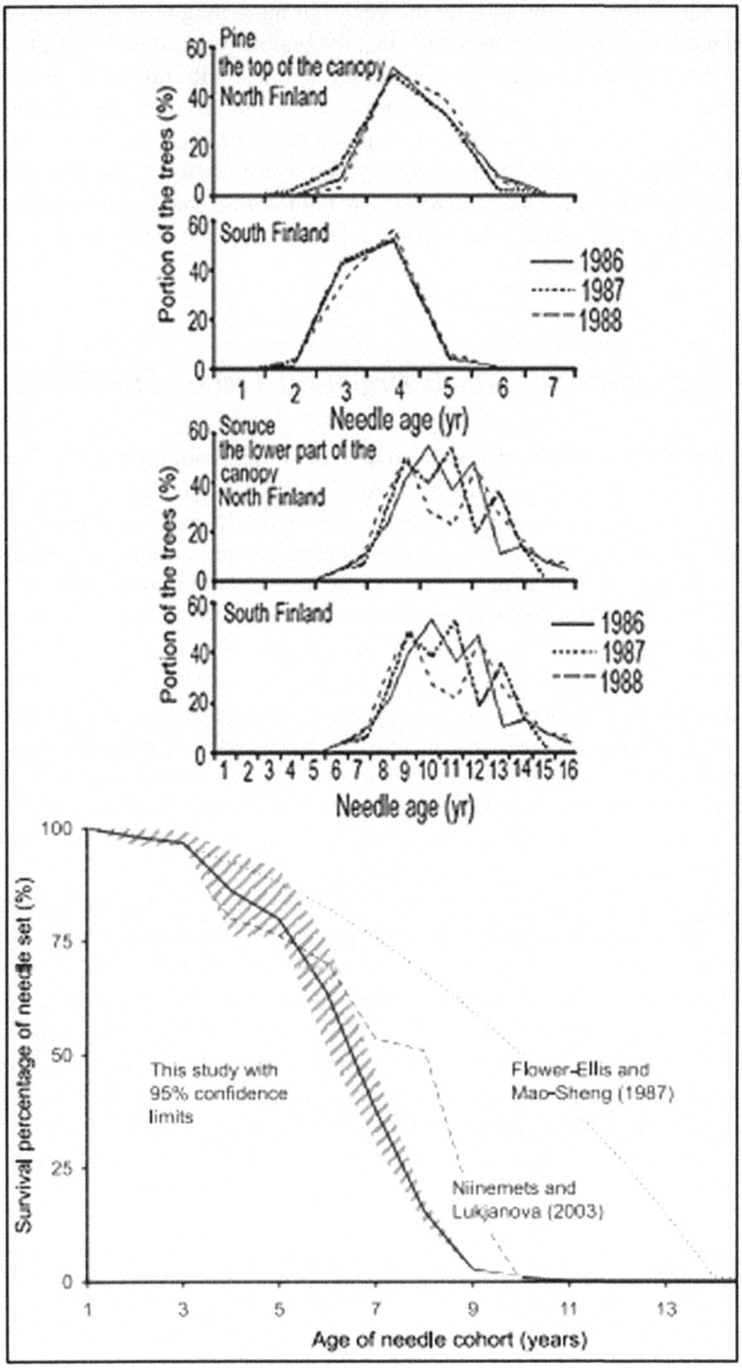


**Fig. 5.10** Life span of needles of Norway spruce and Scots pine in southern and northern boreal conditions. Upper: Number of needle cohorts in Scots pine and Norway spruce in southern and northern boreal conditions (Jukola-Sulonen et al. 1990. Courtesy of Springer Nature. The needle

spruce, needle litter falls throughout the year, but more than 60% of annual litterfall occurs in the growing season. In Scots pine, the majority of needle litterfall occurs in autumn after growth cessation. In birch, the average litterfall is in the range of 1500–3500 kg ha$^{-1}$ yr.$^{-1}$, the lower value referring to the northern and the higher to southern sites (e.g., Ukonmaanaho et al. 2008; Ukonmaanaho 2013). In mature birch stands, the annual litterfall is slightly larger in Pubescent (*Betula pubescens*) than in Pendula birch (*Betula pendula*). In both cases, the foliage litter is about 70–80% of the total annual aboveground litterfall by the time the foliage turns yellow (Niemistö et al. 2004).

### 5.4.6 Litter from Non-woody Organs of Trees—Fine Roots

In the dynamics of boreal forest ecosystems, fine roots (with diameter $< 2$ mm) are important contributors of carbon in the soil profile, in which they mix with soil particles down to the parent material. Fine roots regulate the biogeochemical cycles of ecosystems, with impacts on carbon storage in ecosystems, which mitigates climatic warming. In this respect, Yuan and Chen (2010) summarized the findings of a combined dataset of 218 studies on boreal forests. The mean fine root biomass was 5.28 Mg ha$^{-1}$ (Fig. 5.11), with fine root production of 2.82 Mg ha$^{-1}$ yr. $^{-1}$, representing 32% of the annual NPP of the boreal forest. The mean turnover of fine roots was 1.07 yr.$^{-1}$. The production, biomass, and turnover rate of fine roots in boreal forests increased along with increasing mean annual temperature and precipitation, while the biomass of fine roots in the boreal forest decreased along with the availability of nitrogen and phosphorus in the soil. Throughout the maturation of tree stands, fine root biomass increased until the stand age of 100 years. Thereafter, the fine root biomass leveled off or even reduced.

The amount and mortality of fine roots in boreal conifers may be related to the amount and mortality of needles. Vanninen and Mäkelä (1999) divided the fine roots into small (diameter $\leq 2$ mm) and large (diameter 2–5 mm) roots. They found that in a chronosequence of boreal Scots pine, the mass of total fine roots increased from 1500 kg ha$^{-1}$ in a young stand to 3000 kg ha$^{-1}$ in a mature stand. They further found that the ratio between the total mass of fine roots and needle mass varied from 0.6 to1.0 on a poor site (*Calluna* type, CT) and from 0.5 to 0.6 on a moderately fertile site (*Myrtillus* type, MT). Furthermore, the fine root mass was linearly related to the basal area of trees. This held also for Finér et al. (2007), who found the fine root

**Fig. 5.10** (continued) distribution can be modelled with Eq. 5.6, e.g., for boreal Scots pine (Oker-Blom et al. 1988). Lower: Percentage survival of needles of Norway spruce with the 95% confidence limits for the age of the needle cohort (Muukkonen and Lehtonen 2004). Permission of NRC Canada. The findings are compared with those of Flower-Ellis and Mao-Sheng (1987) and Niinemets and Lukjanova (2003)

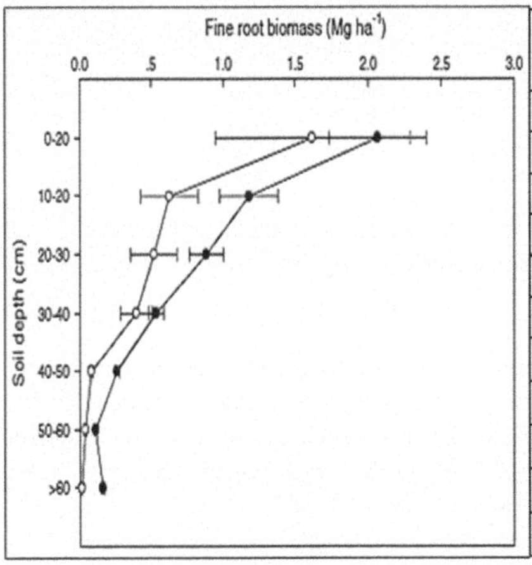


| Life trait | Biomass, Mg ha$^{-1}$ | Growth rate Mg ha$^{-1}$ yr$^{-1}$ | Turnover, yr$^{-1}$ |
|---|---|---|---|
| Broad-leaved | 2.12±0.10 | 2.01±0.17 | 1.15±0.23 |
| Needle-leaved | 2.92±0.36 | 2.31±0.12 | 0.76±0.06 |
| Early-successional | 2.91±0.14 | 2.62±0.23 | 0.86±0.15 |
| Late-successional | 2.81±0.12 | 1.94±0.18 | 0.80±0.06 |
| | | | |
| Genus | | | |
| • *Abies* | 2.83±0.50 | 4.39±0.62 | 0.61±0.12 |
| • *Betula* | 4.15±0.59 | 1.85±0.33 | 1.22±0.56 |
| • *Picea* | 2.78±0.13 | 1.80±0.18 | 0.84±0.07 |
| • *Pinus* | 2.52±0.13 | 2.80±0.27 | 0.61±0.17 |
| • *Populus* | 4.80±0.55 | 1.96±0.47 | 1.28±0.26 |

**Fig. 5.11** Distribution of fine root biomass through the soil profile and the mass, growth, and turnover of fine roots for selected tree genera common in boreal forests (Yuan and Chen 2010). Permission of Taylor & Francis. Left: Vertical distribution of fine roots of early-successional (closed circle) and late-successional (open circle) trees in boreal forests. Right: Biomass, growth rate, and turnover of fine roots (diameters <2 mm) in boreal forests

(<2 mm) mass increasing from 2000 kg ha$^{-1}$ in juvenile Scots pine stands to 8000 kg ha$^{-1}$ in mature ones.

Lehtonen et al. (2016) found that the turnover of fine roots is 0.5 yr.$^{-1}$ for Scots pine, Norway spruce, and deciduous trees in the northern boreal zone, and 0.85 yr.$^{-1}$ in the southern boreal zone. The latter value is close to that found by Brunner et al. (2013) for European forests representing both native and introduced tree species: 0.88 yr.$^{-1}$. On the other hand, Persson (1980) reported that the fine root turnover in a 20-year-old Scots pine stand is substantially faster (2.0 yr.$^{-1}$) than in a 120-year-old stand (0.7 yr.$^{-1}$). This declining trend for Scots pine in fine root turnover was also found by Helmisaari et al. (2002) for Nordic boreal tree species and Yuan and Chen (2010) for several Canadian boreal tree species.

In boreal forests, the majority of carbon input into the soil originates from the renewal of fine roots and aboveground litterfall, as shown by Leppälammi-Kujansuu et al. (2014a, b). They investigated the belowground carbon input from the fine root litter of trees and understory vegetation in relation to the aboveground litterfall. The study concerned two Norway spruce (*Picea abies* L.) stands located in northern (66° N) and southern (61° N) Finland, with comparison to published data of the species along the south-north climate gradient in Sweden (Fig. 5.12). The fine root biomass per basal area of trees was twice as high in the north as in the south, i.e., 170 kg m$^{-2}$ in the north to a soil depth of 30 cm and 90 kg m$^{-2}$ in the south to a soil depth of 20 cm. On the other hand, the fine roots of trees in the north persisted significantly longer (97 ± 2 weeks) than of those in the south (89 ± 2 weeks) or of understory plants at both sites. The annual production of tree foliage litter was higher in the southern stand, but the total amount of litter (including trees and understory, above-

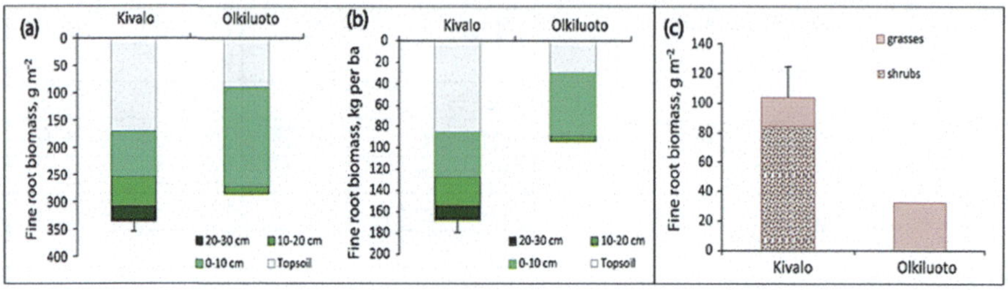


**Fig. 5.12** Mass of fine roots of Norway spruce and understory plants in the northern (Kivalo, 66° N) and southern (Olkiluoto, 60° N) boreal conditions (Leppälammi-Kujansuu et al. 2014b). Courtesy of Elsevier. Left (**a**): Mass of fine roots (g $m^{-2}$) in the soil profile depending on the depth in soil. Middle (**b**): Mass of fine roots (kg $m^{-2}$) of Norway spruce per basal area (ba) of stand in the soil profile. Right (**c**): Mass of fine roots of understory plants (g $m^{-2}$)

and belowground) was similar at both sites, as was the ratio between the above- and belowground litter production. The role of understory vegetation was greater in the northern site, where it was responsible for 33% and 23% of above- and belowground litter production, respectively, compared to 15% and 11% in the south. Both above- and belowground understory carbon input is substantial and should be considered in ecosystem carbon cycle models.

### 5.4.7 *Biomass Allocation and Whole-Tree Mass Turnover*

Based on several sources, Muukkonen (2006) estimated the biomass turnover of different organs of Scots pine over the boreal zone (60–70° N) in Finland (Fig. 5.13). The turnover values were the same in southern and northern Finland except for needle mass, which was two times larger in the south than in the north. In coniferous trees, needle mortality is gradual, following the cohort-based aging and death of needles and eventually turning yellow and losing weight, as in deciduous foliage. The turnover of needles is faster in the south than in the north; for example, 0.10 $yr.^{-1}$ in the south and 0.05 $yr.^{-1}$ in north for Scots pine, and 0.22 $yr.^{-1}$ in the south and 0.10 $yr.^{-1}$ in the north for Norway spruce. In deciduous trees, the annual litterfall was pronounced, with the foliage turnover of 0.78 $yr.^{-1}$ representing the litterfall when foliage turns yellow in late summer and autumn (Muukkonen 2006) (Fig. 5.13).

### 5.4.8 *Litter from Trees Through Different Successional Phases*

In boreal conditions, litterfall takes place throughout the year, but its temporal variability is large as it depends on the litter type, tree species, and season. In

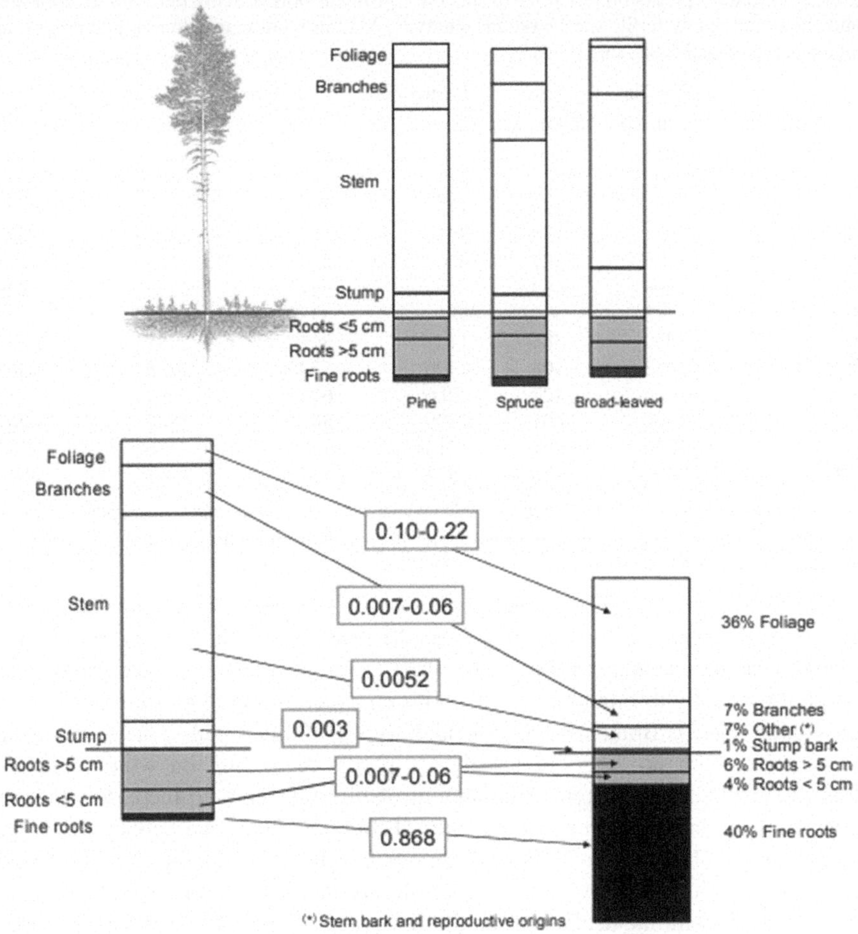


| Organ | Norway spruce | Scots pine | Deciduous trees |
|---|---|---|---|
| | Turn over $yr^{-1}$ | | |
| Foliage | 0.10 (south) 0.05 (north) | 0.22 (south) 0.10 (north) | 0.78 (south and north) |
| Branches/coarse roots | 0.0125 | 0.007-0.06 | 0.0135 |
| Stump bark | 0.0125 | 0.0030 | 0.0001 |
| Reproductive organs/stem bark | 0.0027 | 0.0052 | 0.0029 |
| Fine roots | 0.811 | 0.868 | 0.800 |

**Fig. 5.13** Biomass and turnover of different organs in Norway spruce, Scots pine, and deciduous trees over boreal forests in Finland (60–70° N) (Muukkonen 2006). Courtesy of Finnish Society of Forest Science. Upper: Mean relative biomass distribution in Scots pine, Norway spruce, and deciduous trees. Middle: Mean relative biomass and turnover (arrows with labels, fraction per year) and percentage litter of the total litterfall (rightmost) in the main organs of Scots pine. Below: Summary of mean turnover per organ for Norway spruce, Scots pine, and deciduous trees

**Table 5.5** Annual production of litter in naturally growing boreal Scots pine (*Pinus sylvestris*) stands in a moderately fertile site (*Myrtillus* site type, MT) in relation to different tree organs and stand age, based on Mälkönen (1974)

| | Needles | Bark | Branches | Fine roots | Heartwood | Total |
|---|---|---|---|---|---|---|
| Stand age (yr) | Litter production, kg ha$^{-1}$ yr.$^{-1}$ | | | | | |
| 5 | 101 | <1 | <1 | 4 | <1 | 105 |
| 10 | 200 | <1 | 66 | 11 | <1 | 278 |
| 15 | 428 | 18 | 122 | 39 | 88 | 697 |
| 20 | 765 | 138 | 321 | 122 | 638 | 1986 |
| 30 | 1636 | 337 | 630 | 323 | 1765 | 4693 |
| 40 | 1809 | 483 | 880 | 541 | 3225 | 6940 |
| 50 | 1713 | 397 | 981 | 538 | 3025 | 6656 |
| 60 | 1812 | 388 | 1124 | 565 | 2955 | 6846 |
| 70 | 1883 | 408 | 1206 | 572 | 2735 | 6807 |
| 80 | 1965 | 427 | 1286 | 611 | 2794 | 7085 |
| 90 | 1460 | 401 | 1299 | 543 | 2502 | 6207 |
| 100 | 1016 | 397 | 1313 | 480 | 2336 | 5543 |
| 120 | 900 | 420 | 1317 | 367 | 1807 | 4814 |

The Forcyte model (Kimmins and Scoullar 1984) was used in the simulations over a period from an open area with plantings to a stand of fully mature trees grown under no management. The initial stand density was 4000 seedlings ha$^{-1}$

Scots pine, for example, the majority of needle litter falls in the late part of the main growing season and autumn, but some other types of litter (e.g., cones, branches, bark) fall gradually throughout the year with no exact timing (e.g., Flower-Ellis and Mao-Sheng 1987). Branch litter and other types of litter with a large mass fall occasionally, with no seasonal variability, partly depending on wind force and snow load. Regardless of type, the amount of litterfall tends to increase when the mass of organs increases in maturing trees (Table 5.5). In boreal Scots pine, on the other hand, the share of needle litter decreases in relation to the litter mass of cones and branches. Concurrently, the share of slowly decaying litter (branches, cones, bark) increases with the decline of rapidly decaying needle litter. This likely applies for the share of bark litter, but its amount and dynamics are poorly understood in relation to the total litter. In boreal Scots pine, dead bark is retained for several decades regardless of site fertility. The amount of bark litter may exceed 500 kg ha$^{-1}$ yr.$^{-1}$ in moderately fertile sites in southern boreal conditions (Mälkönen 1974).

The death of branches is related to the growth and development of the crown and the canopy of the tree stand. The death of branches is accelerated when the canopy closes (i.e., the amount of foliage >95% of full canopy closure), and branches in the lower crown fall deeper into the shading canopy.

Dead branches retained in trees for years or even decades until their strength declines, and they fall due to wind force and/or snow load forming branch litter. In southern boreal conditions, dead branches in young Scots pine occur from the fifth to the sixth whorl from the stem apex downwards, and living branches occur only seldom from the 12th whorl downwards depending on stand density and spatial distribution of trees. The retention time is longer in poor sites than in fertile sites, and longer in narrowly than in widely spaced stands (Heikinheimo1953). In southern boreal conditions, the amount of branch litter exceeds 1000 kg ha$^{-1}$ yr.$^{-1}$ in moderately fertile sites (*Myrtillus* site type, MT).

The mortality of fine roots is more rapid than that of other organs. For example, in boreal Scots pine, the turnover of fine roots by 140–150% per year represents the continuous mass growth of current-year roots and roots from previous years (Persson et al. 1980; Makkonen and Helmisaari 1998). However, the growth and mortality of fine roots is still poorly understood, and they are widely related to the growth and mortality of foliage. This is because it is assumed that fine roots are like foliage in terms of their role in resource uptake for whole-tree metabolism, growth, and development (e.g., Kimmins and Scoullar 1984). In southern boreal conditions, the amount of fine root litter in Scots pine, for example, seems to be less than 1000 kg ha$^{-1}$ yr.$^{-1}$ (Makkonen and Helmisaari 1998), which is markedly less than that of the needle litter. It is evident that a large number of fine roots survive for a period longer than one year or less.

The death of stem wood and coarse (woody) roots is related to the formation of heartwood when sapwood ages. The maximum age of sapwood varies in the range of 10–20 years and is shorter in fertile sites and longer in poor sites. Dying sapwood loses its capacity to transfer water, and the structure and color of sapwood cells change. Furthermore, the nitrogen content in heartwood (0.10–0.15%) is smaller than in sapwood (0.2–0.3%) (Kramer and Kozlowski 1979, p. 306). Heartwood in the stem and coarse roots remains a part of trees and provides mechanical support against wind force and snow loading. Nutrients bound in heartwood in the stem and coarse roots return to the nutrient cycle when the whole tree falls to the soil and begins decaying. In southern boreal conditions, the annual production of heartwood in Scots pine in moderately fertile sites (*Myrtillus* site type, MT) is 2000 kg ha$^{-1}$ yr.$^{-1}$.

## 5.5 Litter from Ground Vegetation

### *5.5.1 Turnover of Mass*

As the litter from trees drives the within-stand nutrient cycle, so does the litter produced by the ground vegetation. In vascular plants, the turnover concerns above- and belowground biomass dependent on species with producing litter. In non-vascular plants, the litter production concerns only aboveground biomass. In vascular and non-vascular plants, the biomass turnover with litter production is widely based on the share of the biomass dying annually. In this context, the litter production in ground vegetation is expressed in the percentage of dying stocking of different plants in field/bottom layers or in different classes of vascular/non-vascular plants. Until now, the litter production concerns few common plants, and the litter production is based on estimates related to the annual life span of few species.

Regarding vascular plants, the share of annually dying shoots and roots of dwarf shrubs is 25%, while the dying share of shoots in herbs and grasses is 25–100%. The mortality of roots in vascular plants is likely around 25% of the mass, depending on species. This implies that in dwarf shrubs, the turnover of shoot mass is 0.25 $yr.^{-1}$ and in root mass 0.33 $yr.^{-1}$. In herbs and grasses, the turnover of aboveground biomass is 0.25–1.0 $yr.^{-1}$ and in the belowground 0.25 $yr.^{-1}$ (e.g., Mälkönen 1974, 1977). The aboveground part of herbs dies at the end of the growing season, while one third of roots survives over winter to the next growing season. In grasses, a part of aboveground and belowground biomass survives through winter like in hair-grass *(Deschampsia flexuosa)*. In non-vascular plants (lichens and mosses), the share of annually dying biomass is 30% (e.g., Mälkönen 1974, 1977; Palviainen and Finér 2007). In this context, Muukkonen (2006) estimated that the turnover for mosses is 0.33 $yr.^{-1}$ and for lichens 0.1 $yr.^{-1}$.

The impacts of ground vegetation on soil carbon are dependent on the production and properties of litter. In dwarf shrubs like bilberry (*Vaccinium myrtillus*) the litter production is important. The nitrogen content of bilberry litter is high, which makes bilberry litter decay rapidly in contrast that to cowberry (*Vaccinium vitis-idaea*) and black crowberry (*Empetrum nigrum*) litter. In litter of black crowberry, the content of polyphenolic compounds is high, which reduce the decay of litter. Moss litter (e.g., *Pleurozium schreberi, Hylocomium splendens*) decays slowly and enhances the accumulation of carbon in soil surface. On the other hand, moss litter maintains ground cover moisture, thus increasing the decay of soil detritus below litter layer (Wardle et al. 2003; Nilsson and Wardle 2005).

### 5.5.2 Litter Over Time

Over a boreal chronosequence, litter production in ground vegetation depends on the site fertility, tree species composition and successional phase. Table 5.6 demonstrates how the litter production in ground vegetation is dependent on the site fertility through a Scots pine chronosequence on moderately fertile site (*Myrtillus* site type, MT). The litter originating from ground vegetation increases during the early phase of succession and maximizes when the tree cover closes in the seedling/pole phase before commercial thinning. In this phase, herbs and grasses are partly replaced by dwarf shrubs and mosses, and the total amount of litter from ground vegetation decreases under the tree cover. The litter from ground vegetation remains smaller than that in gaps (or clear-cut areas), while the accumulation of soil organic matter increases. This is due to the slow decay of litter from dwarf shrubs and mosses compared to that from herbs and grasses. In southern boreal conditions, for example, the litter production in mesic sites is 100–2000 kg $ha^{-1}$ $yr.^{-1}$, the larger values being obtained a few years after clear-cutting and in young stands in fertile sites, while the values are smaller in mature stands. In poor and dry sites, the litter production per year may vary from a few tens of kilograms up to 1500 kg $ha^{-1}$ $yr.^{-1}$.

**Table 5.6** Estimates of the amount of litter production in ground vegetation in southern boreal conditions as a function of site type and age of natural Scots pine stands based on a Forcyte model (Kimmins and Scoullar 1984)

| | CT | VT | MT | OMT |
|---|---|---|---|---|
| Stand age (yr) | Litter production in ground vegetation, kg $ha^{-1}$ $yr.^{-1}$ | | | |
| 5 | 971 | 1192 | 1481 | 1688 |
| 10 | 1243 | 1566 | 1845 | 1839 |
| 15 | 1370 | 1524 | 874 | 738 |
| 20 | 1185 | 892 | 306 | 214 |
| 30 | 308 | 237 | 164 | 126 |
| 40 | 187 | 63 | 162 | 125 |
| 50 | 135 | 37 | 162 | 124 |
| 60 | 100 | 22 | 159 | 124 |
| 70 | 71 | 23 | 126 | 124 |
| 80 | 54 | 20 | 91 | 124 |
| 90 | 43 | 28 | 87 | 124 |
| 100 | 37 | 97 | 87 | 124 |
| 120 | 45 | 192 | 87 | 124 |

Simulations covered a period from the regeneration area with small seedlings to the full maturation of trees grown and developed under no management. In the simulations, the initial stand density was 4000 seedlings $ha^{-1}$

### 5.5.3 *Spatial Distribution of Ground Cover Litter Across the Boreal Zone*

Under even-aged management, harvest and management operations drive the succession of forest ecosystems, with changing uptake and stocks of carbon in ground vegetation from one developing phase to the next one over multiple production cycles. To date, the biomass of ground vegetation and related litter has been excluded from ordinary forest inventories. Therefore, there is no clear understanding of how ground vegetation affects carbon uptake and stocks in managed boreal forests. This is especially problematic when assessing how forestry and forest management affect the role of forest ecosystems in mitigating climatic warming in boreal conditions.

For boreal forests, the spatial estimates of the role of ground vegetation in the carbon cycle in unmanaged and managed forests are based on model simulations. Such models are parametrized based on field measurements as that used by Lehtonen et al. (2016). The simulations show that the amount of litter from trees and from ground vegetation is inversely related across the boreal zone as found by Lehtonen et al. (2016). They indicate that the total litterfall combined from both sources is four times as large in southern boreal sites than in northern boreal sites across the south/north temperature gradient. The model-based estimate further shows the mean amount of carbon in litter from trees is approximately 2000 kg C $ha^{-1}$ in the south and 900 kg C $ha^{-1}$ in the north (Fig. 5.14). Concurrently, the carbon in litter from ground vegetation is approximately 470 kg C $ha^{-1}$ in the south and 860 kg C $ha^{-1}$ in

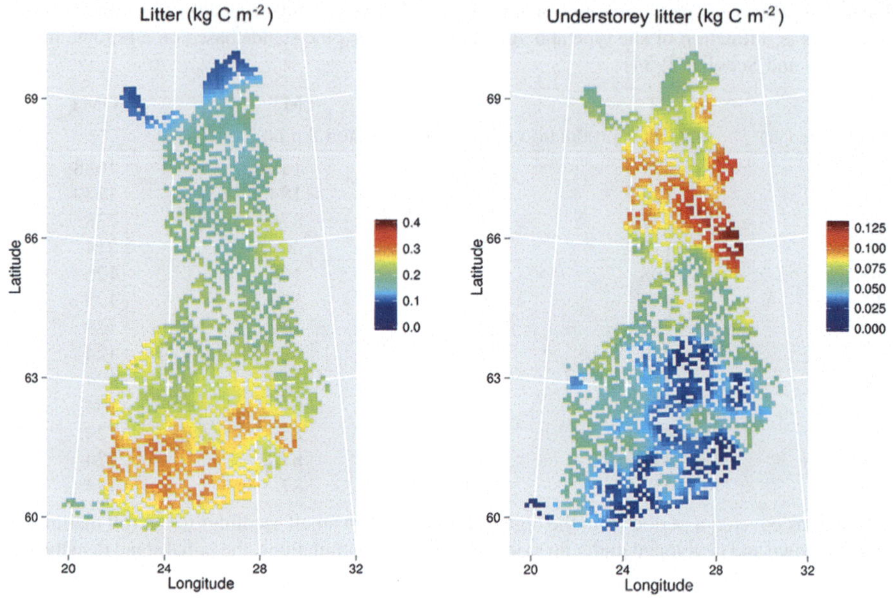


**Fig. 5.14** Model-based estimate of spatial distribution of litter across the boreal zone in Finland (Lehtonen et al. 2016). Open Access. Courtesy of European Geosciences Union. Left: Total carbon input in litter combining litter from trees and ground vegetation. Right: Carbon input in litter from ground vegetation

the north. This implies that the mean carbon amount in the litter from trees and ground vegetation is equal. Lehtonen et al. (2016) further emphasize that the litter from ground vegetation compensates the litter from trees in the ecosystem carbon cycle under wide stand spacing and nutrient-limited conditions. This pattern reflects the difference in the primary growth of trees between southern and northern boreal sites. In general, the distribution of total litter and that from ground vegetation follows the mean distribution of temperature and precipitation across boreal forests over Finland.

The yield and properties of litter are site-specific and depend on the site fertility, properties of dominating tree species, structure of the tree stand (spacing, stocking, maturity), and plant species in the ground vegetation. These factors affect the spatial distribution of litter and consequent distribution of soil organic matter (Fig. 5.14). According to Lehtonen et al. (2016) and Ukonmaanaho et al. (2008), the share of needle litter from the total tree litter (i.e., litter from above-ground from different organs ((e.g., branches, stem etc.) including also needles) is larger in the southern boreal zone than in the northern boreal zone. In the south, the litter layer also contains more twigs and cones than in the north, while the share of needles from the total amount of tree litter is larger in the north than in the south (Ukonmaanaho et al. 2008; Hilli 2013).

Tree species dominance further effect on the amount and properties of litter produced by ground vegetation as found by Hilli et al. (2010) and Hilli (2013). Concurrently, the litter production in ground vegetation was larger in Scots pine- than in Norway spruce-dominated sites across the boreal zone, especially in the north. Stendahl et al. (2010) also showed that ground vegetation produces more litter under Scots pine than under Norway spruce in the north related to the differences in ground vegetation. Concurrently, the moss litter under Scots pine canopies decreases opposite to the increasing share of dwarf shrubs, with the increase of slowly decaying dwarf shrub litter in the total litterfall. This likely increases the carbon accumulation on soil, especially in mature Scots pine stands, with moderately wide spacing (Stendahl et al. 2010).

## 5.6 Mortality of Trees Over Time

### *5.6.1 Life Span of Selected Boreal and Temperate Tree Species*

Tree mortality refers to the death of the whole tree. Trees may die in any phase of their life span due to a shortage of resources and other reasons unrelated to their increasing mass and maturation. A shortage of resources in relation to demand may induce subsequent causes of mortality. For example, declining physiological activity (primary cause) may trigger insect and fungal attacks (secondary cause), increasing the mortality risk in any phase of growth and development. This is especially the case in mature trees, where the increasing demand of photosynthates for maintenance makes trees susceptible to insect and fungal pests. Fire, strong winds, and heavy snow load further damage trees, and increase the risk of insect and fungal attacks even on undamaged trees. In the long term, the damage induced by insect and fungal attacks seems to be the main reason for trees to die over the successional cycle from seedlings to old-growth trees, with links to self-thinning (Fig. 5.15).

Tree mortality may also be related to the growth and mass of trees at a given moment. This indicates that a large share of trees in the original tree cohort has died by that moment. This implies an inverse relationship between the mass or mean mass of trees and population/stand density, as formulated by Reineke (1933). This rule was further developed by Yoda et al. (1963), whose self-thinning (natural thinning) model and its modifications are also widely used in modelling the dynamics of populations/stands of trees (Fig. 5.15):

$$Y = C \times \rho^{k} \rightarrow \overline{Y} = C \times \rho^{k-1}. \qquad (5.7)$$

where $Y$ is the total stocking of stem wood in trees, $\overline{Y}$ is the mean volume/mass of single trees, $C$, and $k$ (dimensionless) are parameters, and $\rho$ is the density of the population/stand (trees per unit area). The self-thinning model is applicable if the

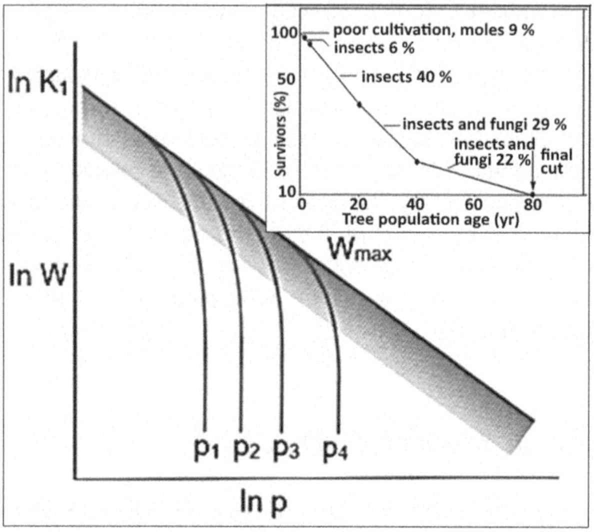


**Fig. 5.15** Schematic depiction of the concurrent development of the mean mass ($W$) in trees as a function of varying initial spacing ($p_i$) modified based on Landsberg (1986, pp. 39–42), where $W_{max}$ indicates the maximum mean mass of trees in a population attained by self-thinning. The curves $p_i$ show that the mass in a population increases until the within-population competition (shaded area) tends to reduce growth and initiate mortality. This trigger self-thinning when $W$ converges with $W_{max}$ ($p_i$), with a consequent decline of population density. Insert: Example of biotic causes of tree death; survival of a cohort of silver pine (*Pinus monticola*) over a longer period (Silvertown 1982, pp. 67–71). Permission of Springer Nature

following assumptions hold: (i) any trees in an even-aged population/stand have nearly the same geometric structure regardless of developmental phase and site fertility; (ii) self-thinning in the population/stand occurs only if the canopy cover exceeds the full occupation (100%) of the site; and (iii) the full canopy cover remains after the self-thinning events.

The self-thinning model between the population is geometric rather than causal. Even in old-growth trees, meristems are capable of renewal, as exemplified by trees of several thousands of years old. The mechanism of the mortality of whole trees is poorly understood, but the death of trees depends on their age and the environmental conditions (e.g., reduced water availability under long drought period) acting through the growth. In the case of age-dependent mortality, a tree may die at any age. This implies that a given share of trees in the original cohort never achieves the maximum age of the species (AGEMAX, yr), with the probability ($p(n)$) of dying in the year $n$ (Botkin et al. 1972):

$$p(n) = 1 - (1 - \varepsilon)^n \tag{5.8}$$

where $\varepsilon$ is the yearly rate of mortality. When assuming that 2% of trees in a cohort achieve the maximum age, then $(1-\varepsilon)^{\mathrm{AGEMAX}} = 2\times10^{-2}$ gives the approximate value $\varepsilon = 4.0/AGEMAX$. The probability of tree death may further be related to age-independent causes like declining growth. Botkin et al. (1972) assumed that diameter growth <0.01 cm yr.$^{-1}$ indicates a reduced capacity of a tree to grow and compete in the population/stand. This scenario allows only 1% probability of survival in the consecutive 10 years. A tree with such low growth in the two previous years is expected to die with a probability of 0.368 in any year. Trees may also die for episodic reasons. For example, high winds, heavy snow load, wildfires, and intensive insect and fungal attacks increase the age-dependent and age-independent risks of dying (Fig. 5.15).

### *5.6.2 Mortality of Trees in Relation to Total Tree Growth*

The approach of Botkin et al. (1972) was used in a gap-type model Sima (Kellomäki et al. 2008), which Mazziotta et al. (2014) used to investigate the dependence of the mortality of selected tree species depending on site fertility under no management. Over a 120-year period, 20–40% of the tree biomass may die, depending on tree species, site type, and the density of the initial stand (Table 5.7), with higher mortality in dense stands than sparse stands. The results agreed with the findings in growth and yield studies (Koivisto 1962) and in simulations based on the growth and yield model Motti (Hynynen et al. 2002). In the latter case, the mortality partly utilized the inverse relationship between stand density and the volume of trees (e.g., Reineke 1933; Yoda et al. 1963).

The length of the life span of trees is species-specific; nevertheless, it is difficult to determine. The death of trees belonging to a given cohort is gradual. This implies that the length of the life span refers to the point from the generation of the cohort to the point when a given percentage of the original number of trees still survives. If one accepts 1% survival, the maximum life span for Scots pine is 400–450 years, for Norway spruce, 300–350 years, and for birch, 100–150 years (e.g., Shorohova et al. 2011) in southern and mid boreal forests. Nevertheless, the life span of individual trees may be longer, as shown in Table 5.8: the oldest Scots pine may live from 400 to 800 years and the oldest Norway spruce, from 300 to 400 years. In general, the life span of deciduous trees is shorter: from 100 to 150 years for Pendula birch, from 100 to 120 years for Pubescent birch, 100 years for aspen, and 50 years for grey alder. The age of individual trees varies, but the life span proceeds more slowly in poor than in fertile sites, indicating that the lifespan of different species tends to be longer in poor than in fertile sites, and longer in the northern than in southern boreal zone.

**Table 5.7** Dead wood share (%) of total production calculated using the results of the gap-type Sima model and the growth and yield model Motti (Mazziotta et al. 2014) over 120 years, with the comparison to the values representing the finding for growth and yield studies concerning Scots pine, Norway spruce and birch (Koivisto 1962)

| | | Share of dead wood, southern boreal | | | Share of dead wood, northern boreal | | |
|---|---|---|---|---|---|---|---|
| Species and site type | Initial stand density, trees $ha^{-1}$ | Sima model, % | Motti model, % | Growth and yield tables, % | Sima model, % | Motti model, % | Growth and yield table, % |
| Norway spruce MT | 1200 | 21.9 | 5.5 | | 45.4 | 5.5 | – |
| | 2400 | 28.1 | 21.6 | 22 | 48.3 | 12.8 | |
| | 4800 | 33.6 | 34 | | 54.3 | 23.3 | |
| OMT | 1200 | 23.9 | 10.5 | | 50.4 | 4.6 | – |
| | 2400 | 30.5 | 24.9 | 16.6 | 55.9 | 13.5 | |
| | 4800 | 36.2 | 34 | | 58.5 | 26.4 | |
| Scots pine VT | 1200 | 18.9 | 15.6 | | 24.9 | 13.7 | |
| | 2400 | 26.8 | 26.8 | 31 | 31.3 | 23.8 | 35.7 |
| | 4800 | 35.8 | 36 | | 41 | 32.1 | |
| MT | 1200 | 28.7 | 24.7 | | 41.3 | 15 | |
| | 2400 | 37.8 | 36.5 | 29.2 | 45.7 | 31.8 | 41.2 |
| | 4800 | 40.3 | 44.8 | | 49.6 | 40.1 | |
| OMT | 1200 | 27.3 | 6.9 | | – | – | – |
| | 2400 | 34.2 | 33.1 | 27.6 | | | |
| | 4800 | 40.6 | 39.7 | | | | |
| Birch MT | 1200 | 39.6 | 33.7 | | 44.4 | 7.7 | – |
| | 2400 | 38.5 | 43.2 | 32.5 | 41.8 | 17.4 | |
| | 4800 | 38.6 | 47.5 | | 46.7 | 26.5 | |
| OMT | 1200 | 44.8 | 39.6 | | 55 | 9.8 | – |
| | 2400 | 43 | 47.9 | 35.5 | 49.5 | 23.2 | |
| | 4800 | 44 | 52.2 | | 51.7 | 32 | |

The simulations are for the southern (N 62°) and northern boreal conditions (N 66°) as a function of site type and initial density of seedling stands

### *5.6.3 Longevity of Mass in Standing and Fallen Dead Trees*

Tree death is related the gradual deterioration of physiological and ecological functions over time (Fig. 5.16). According to Leikola (1969), the death of Scots pine unexposed to any external forces occurs in the northern boreal forest at the age 420–450 years. Thereafter, it takes 35–40 years until the stem is debarked, but dead trees remain standing for 200–500 years if not exposed to turning forces like wind or snow load. Standing dead trees provide unique niches for many vertebrates and invertebrates that require specific sites and substrates to successfully complete their life cycles (Niemelä et al. 2002). Tree death initiates decay and changes in the properties of niches and substrates until the dead tree falls. Decay of tree biomass continues in trees lying on the ground until the dead biomass decomposes in humic

**Table 5.8** Estimated mean maximum life span of selected European and North American coniferous and deciduous trees (Sarvas 1964)

| Tree species | Life span, yr |
|---|---|
| European coniferous trees | |
| • European fir (*Abies alba*). | 300–400 |
| • Norway spruce (*Picea abies*). | 300–400 |
| • Scots pine (*Pinus sylvestris*). | 400–600 |
| North-western American coniferous trees | |
| • Sitka spruce (*Picea sitchensis*). | 800–900 |
| • Douglas fir (*Pseudotsuga menziesii*). | 1300–1500 |
| • Giant sequoia (*Sequoia gigantea*). | 3000–4000 |
| European deciduous trees | |
| • Birch (*Betula* sp.) | 100–150 |
| • Elm (*Ulmus* sp.) | 500–600 |
| • Linden (*Tilia* sp.) | 800–1000 |

substances, emitting carbon dioxide and releasing nutrients for uptake in the nutrient cycle. Dead trees lying on the ground provide niches for many specialized fungi mostly due to their slow rate of decay and the gradual changes in substrate for the survival of fungi and invertebrates specialized for different decay phases of dead wood (Niemelä et al. 2002).

Decay affects the physical and chemical properties of wood until the wood loses its original structure. The decay of stem wood results in a decline in wood density ($Y$, mg $cm^{-3}$). This is shown in Fig. 5.16 for Douglas fir, where wood density is declining as a function of the residence time of a stem on the forest floor. Density followed the equation: $Y = B \times e^{k \times t}$, where $B$ (350 mg $cm^{-3)}$ and $k$ (0.0063 $yr.^{-1}$) are parameters and $t$ are time (Means et al. 1985). In the decaying stems, the wood lost about 70% of its original density within 300 years. The decay rate was related to the chemical properties of wood: lignin decayed substantially more slowly ($k = 0.0039$ $yr.^{-1}$) than the acid detergent-soluble fraction ($k = 0.0094$ $yr.^{-1}$) and cellulose ($k = 0.0109$ $yr.^{-1}$).

### 5.6.4 *Decay of Standing and Lying Dead Trees*

Standing dead trees and dead trees lying on the ground are important for the diversity of forest ecosystems as they provide varying habitats for many vertebrates and especially for invertebrates (Fig. 5.17). Maser et al. (1979) used a five-grade classification to describe the decay process, with changes in the properties of dead trees or snags standing or lying on the ground. In classes 1–2, the bark is intact, while some part of the bark remains on the stem in class 3, and none remains in classes 4–5. Branches <3 cm are still attached to the stem in decay class 1, but not in classes 2–5. At the same time, the stem wood turns to being soft and powdery, and its color changes from its original color to light yellow or grey. Typically, a newly fallen stem remains above the ground due to the support branches, stones, etc. Throughout the decay process, the whole stem comes to lie on the ground.

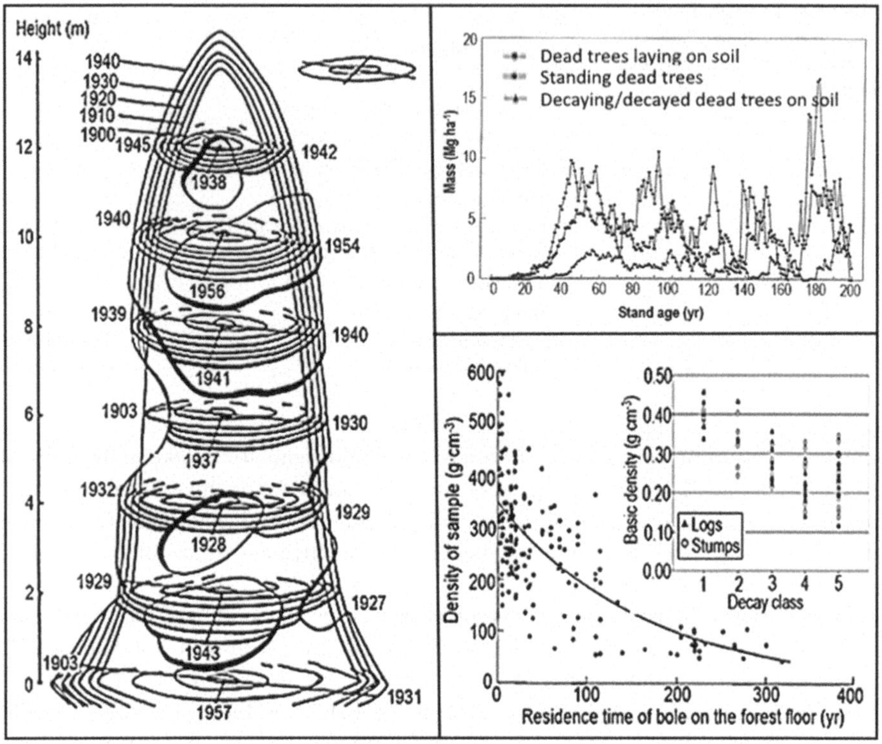


**Fig. 5.16** Growth and longevity in old-growth boreal Scots pine and of mass in standing and fallen dead trees exposed to decomposition. Left: An example demonstrating how the diameter growth distributes in different parts of the stem as northern-boreal Scots pine gradually matures and dies (Leikola 1969). Courtesy of Finnish Society of Forest Science. Right/Upper: Output of simulations on the amount of standing and fallen dead trees in boreal Scots stands after 30 years, and the amount of decayed stem wood still recognizable on the ground after 200 years. The simulations were performed with an earlier version of the gap-type Sima model for the southern boreal zone (61° N) on a moderately fertile site (*Myrtillus* site type, MT). The simulations were done under the current climate, assuming natural growth and development, and excluding any management. The initial seedling stand was 4500 seedlings per hectare (Kellomäki 1994). Compiled by the author of this book. Right/Lower: Density of Douglas fir (*Pseudotsuga menziesii*) stem wood as a function of the residence time (years) of the bole on the forest floor (Means et al. 1985). Courtesy of Canadian Science Publishing

**Box 5.1 Decomposition and its effects on wood properties of dead trees over time**

Trees newly fallen on the forest floor may be living or dead (Fig. 5.17). The fall of trees is due to abiotic disturbance (e.g., high wind, snow load), while snags are usually trees that are deteriorated and collapsed. Maser et al. (1979)

(continued)

**Box 5.1** (continued)
classified standing trees and snags into log classes representing different log decay phases. According to Ausmus (1977), the stages of wood decay can be characterized as the following stages of decay from the log of a living tree to a fully decayed log:

- Standing dead trees: trees or snags that are erect or leaning but have no photosynthetic capacity.
- Sound felled trees: logs recently windthrown or struck by snow load or lighting.
- Diseased and felled: timber only recently felled, exhibiting obvious pathological conditions but with invertebrate communities absent.
- Colonization and conditioning: wood moist but with no channelizing invertebrate present; diplopods and other surface-inhabiting invertebrates, blue-stain fungi, and mycelia of imperfect fungi present; little discoloration.
- Channelization: presence of and colonization by channelizing invertebrates.
- Succession and secondary colonization: colonization by channelizing invertebrates and large numbers of invertebrates without channelizing mouthparts, soil-dwelling microarthropods, predators, earthworms; large amount of faecal material and soil; high diversity of the microbial community within channels.
- Maceration: extensive fungal rhizomorphs, root invasion, soil and faecal material clogging, chambers, secondary colonizing invertebrates, obvious fungal dominance.
- Incorporation: wood no longer recognizable by species, 20–30% soil mixed into wood, mass of mycelia, fibrous and woody root systems colonizing throughout.

In different phases, slow-decaying wood debris has an important role in the mineral cycle, nutrient immobilization, nitrogen cycle, spread of fire, and habitats for ground-dwelling fauna.

Figure 5.17 also demonstrates simulated probabilities for the length of time that Scots pine, Norway spruce, and Pendula birch remain standing after death. According to Mäkinen et al. (2006), the probabilities are high in the first few years after death, especially for birch. However, the probabilities for birch decline more rapidly than in Scots pine and Norway spruce. Birches in the snag phase disappeared within 40 years after dying, while in Scots pine and Norway spruce the probabilities of trees remaining in the snag phase of 40 years after dying were 0.05–0.10. The snag phases for Norway spruce and Scots pine extended from 60 to 70 year. The remaining fraction of stem wood mass followed the same pattern as that for the probability for snags of Norway spruce and Scots pine. In birch, the fraction of

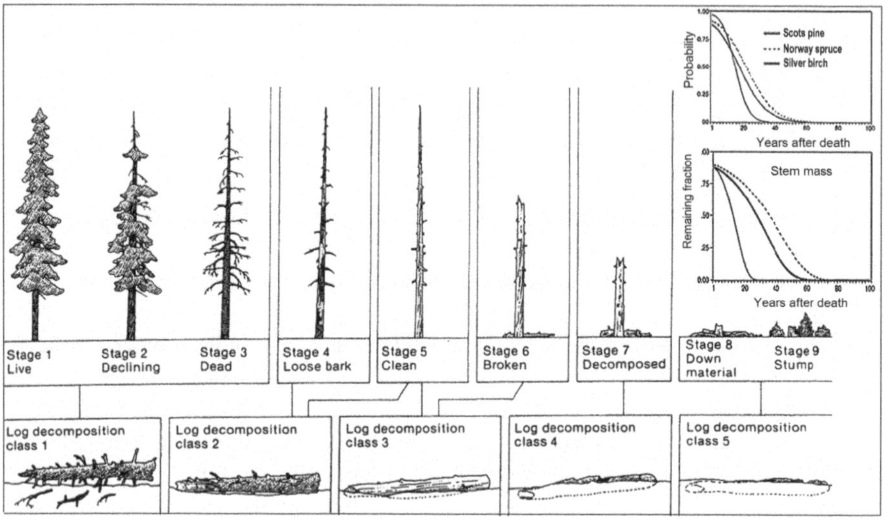


**Fig. 5.17** Decay of standing and lying dead trees, with further details in Box 5.1 for decomposition phases and their effects on wood properties of dead trees over time. Left: Schematic depiction of the life cycle of a tree through its declining growth to death and fall, and further through different phases of decay to its full decomposition into humic substances and release of nutrients and carbon dioxide (Maser et al. 1979). Courtesy of USDA Forest Service. Right/Upper: Predicted probabilities for a Scots pine (*Pinus sylvestris*), Norway spruce (*Picea abies*), and Pendula birch (*Betula pendula*) being a snag (standing dead tree) in the years after death (Mäkinen et al. 2006) Permission of Wiley. Right/Middle: Predicted remaining stem biomass in long-term thinning experiments in southern and mid-boreal forests in Finland (Mäkinen et al. 2006). Permission of Wiley

remaining stem wood declined rapidly, converging toward zero when 25–30 years had elapsed after death (Mäkinen et al. 2006).

# 5.7 Decay and Accumulation of Organic Matter in Soil

## *5.7.1 Mechanisms of Litter Decay*

During decomposition, the mass of litter and humus declines through heterotrophic respiration, and the structure of compounds becomes simpler. Eventually, some humus turns into humus colloids, which increases the capacity of soil to bind water and nutrients usable for plants (Mason 1976; Persson et al. 1980). Heterotrophic respiration represents the soil detritus cycle, where litter, harvest residues, and humus are decayed further humus particles. Decay refers to the gradual disintegration of litter into organic soil, or soil type (humus), emitting water and carbon dioxide and releasing nutrients bound in litter. Decomposition may be divided into three separate processes, namely leaching, weathering, and decay (Fig. 5.18):

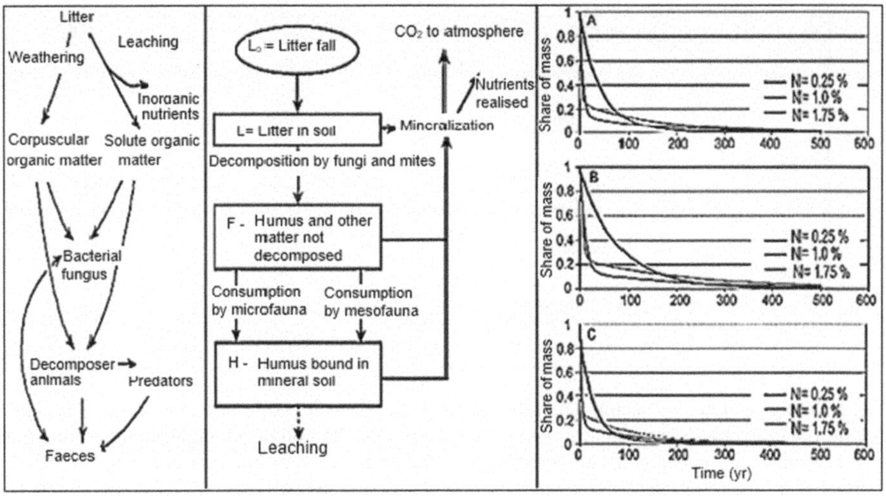


**Fig. 5.18** Outlines of processes driving the decomposition of soil organic matter (SOM). Left: Schematic depiction of the different phases of litter and humus decomposition based on Mason (1976) and Persson et al. (1980). Middle: Schematic depiction of the decomposition of SOM from the litter to the humus phase, based on Chertov et al. (2001). Courtesy of Elsevier. Right: Decay of needle litter in a southern boreal Scots pine stand as a function of the nitrogen content of litter under A: current temperature; B: a temperature reduction of 4 °C < present, and C: a temperature increases of 4 °C > present for boreal conditions (62° N). Simulated results were generated using a process-based model (FinnFor) (Kellomäki and Väisänen 1997)

- Leaching and weathering are physical and chemical processes in which several organic and inorganic substances are released. Decay may further be divided into enzymatic and physical decomposition.
- Decay is also a biological process in which heterotrophic organisms obtain energy from dead plant material. Heterotrophic organisms include vertebrates, invertebrates, bacteria, and fungi. Biological decay involves fungal and bacterial activity, but several invertebrates also graze on soil organic matter (SOM).
- These organisms represent different taxonomic groups and trophic positions, divided into saprotrophic, biotrophic, and necrotrophic organisms. Saprophytes mainly include fungi and bacteria. Saprophytes also include vertebrates and invertebrates that consume litter. Under boreal conditions, the decay rate is mainly controlled by temperature and, to a lesser extent, by moisture. High nitrogen content in litter indicates a high decay rate.

According to Chertov et al. (2001), litter falling in soil ($L$) forms the layer where organic matter decays (Fig. 5.19). Decay is related to the activity of fungi and mites converting litter to humus and other undecayed matter ($F$). Litter is consumed by micro- and macrofauna that produce humus ($H$), which is partly bound in the upper mineral soil. Finally, humus in the mineral soil partly leaches into the soil profile. In different phases, decay releases different levels of nutrients and carbon dioxide depending on the temperature, moisture, and nitrogen content in organic matter. A

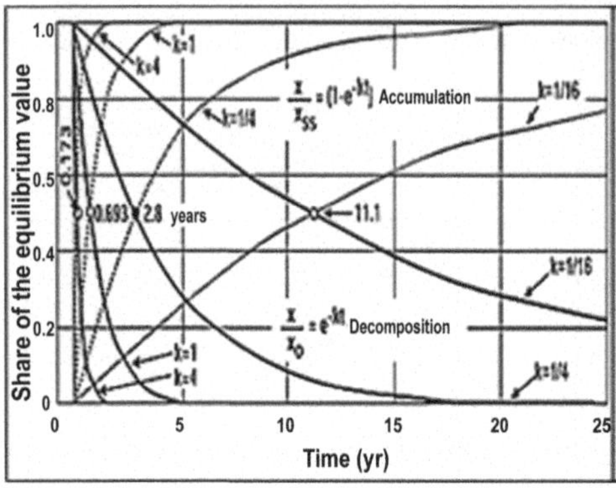


| Parameter k = X/Xss | 1/k = Xss/X | Time for 50% mass loss, 0.693/k | Time for 95% mass loss, 2/k |
|---|---|---|---|
| 4 | 0.25 | 0.173 | 0.75 |
| 2 | 0.5 | 0.346 | 1.50 |
| 1 | 1.0 | 0.693 | 3.0 |
| 0.693 | 1.442 | 1.000 | 4.33 |
| 0.5 | 2 | 1.386 | 6 |
| 0.25 | 4 | 2.772 | 12 |
| 0.125 | 8 | 5.544 | 24 |
| 0.0625 | 16 | 11.09 | 48 |
| 0.0312 | 32 | 22.21 | 96 |
| 0.0156 | 64 | 44.42 | 192 |
| 0.01 | 100 | 69.31 | 300 |
| 0.003 | 333 | 232.3 | 1000 |

**Fig. 5.19** Decay rate and accumulation of carbon in soil (Olson 1963). Permission of Wiley. Left: Fraction of litter remaining from the original weight and the fraction of soil organic matter from that at the steady state under constant litterfall as a function of decay rate (*k*) and time (*t*). Right: Examples of time required for 50% and 95% mass loss of decaying matter at a given decay rate (*k*) assuming a steady state of litterfall

temperature increase tends to increase the decay rate in an exponential manner in the temperature range of 0–30 °C, above which a higher temperature tends to decrease the decay rate. High nitrogen content in litter implies a high decay rate, especially in the early phase of litter decay. According to Berg et al. (2001), the rate of soil organic matter (SOM) build-up is related to the amount of foliar litterfall and the nitrogen concentration in foliar litter. The nitrogen-rich litter of Norway spruce is associated with a significantly higher accumulation of humus compared to that for Scots pine.

The decay rate of any litter type is further related to the lignin (*L*) and nitrogen (*N*) contents. The decay rate is low when the lignin content is high, or when the nitrogen content is low. As decay progresses, the chemical composition of the litter changes. The content of soluble and fast-decaying compounds reduces rapidly, while the content of lignin and other slowly decaying compounds increases. Organic matter accumulates in soil particularly fast if the lignin content of the original litter is high (Mikola 1960; Melillo et al. 1982; Aber et al. 1990). The full decay of nutrient-rich litter such as that of deciduous foliage takes several decades, while the full decay of woody litter requires up to one hundred years. In boreal/temperate conditions, the average time for the full decay per selected litter type in years (i): 150 for stem wood, 100 for bark, 50 for branches, 20–50 for leaves/needles, and 50 for roots; (ii) in dwarf shrubs: 20 for leaves 20 and 50 for stems, shoots, and roots; and (iii) in herbs and grasses: 25 for shoots and 40 for roots (Kimmins and Scoullar 1984).

In general, soil organic matter (litter and humus, SOM) is linked to the balance between the carbon input from litterfall and the carbon output through decay. Let *X*(*t*) refer to the amount of SOM per unit of area, and Δ*X* to the change (increase/decrease) in SOM (Olson 1963):

$$\frac{\Delta X}{\Delta t} = \text{increase of matter} - \text{decrease of matter.} \tag{5.9}$$

$L(t)$ refers to the amount of litter falling on soil (amount of annual litter). The change in SOM may be written in the form:

$$\frac{dX}{X} = L(t) - k \times X(t). \tag{5.10}$$

During decay, the matter declines at the rate $k \times X(t)$, which is dependent on its amount in soil ($X(t)$) and the decay rate ($k$). When $X(t)$ represents the mass of a single litter cohort, then:

$$\frac{dX}{X} = -kdt. \tag{5.11}$$

The mass of a given litter cohort ($X$ (0)) reduces at a constant rate, which gives the mass fraction $X(t)$ in relation to $X$ (0) at a given moment. According to Olson (1963), the remaining mass fraction of the litter cohort and the mass fraction of accumulated litter compared to the amount of carbon at the steady state is:

$$\text{Fraction of litter still remaining from original weight} = \frac{X}{X_o} = e^{-k \times t} \tag{5.12}$$

$$\text{Fraction of SOM of that at steady state} = \frac{X}{X_{ss}} = 1 - e^{-k \times t} \tag{5.13}$$

where is $X_o$ is the mass of litter at the fall ($t = 0$), $X$ is the remaining mass in the year $t$, and $X_{ss}$ is the mass of accumulated litter at the steady state (Fig. 5.19).

The decay and accumulation trajectories under a given decay rate are inversely related. The time for the decay of matter to half of the original value is obtained from: $0.5 = e^{-k \times t}$, $t = -\ln(0.5)/k = 0.693/k = 0.693/0.0625 = 11.1$ years. For example, at the decay rate $k = 0.25$ yr.$^{-1}$, the mass is reduced by 50% in 2.8 years, whereas at the rate $k = 0.0625$ yr.$^{-1}$, it takes 11.1 years for the same mass loss. The period $2/k$ is required for a 95% reduction of the original mass.

Depending on tree species, the decay rate ($k$) for coniferous foliage is 0.20–0.40 yr.$^{-1}$ and for deciduous foliage, 0.30–0.60 yr.$^{-1}$ (e.g., Valachovic et al. 2004). In the former case, 95% of the mass would be lost in 10–15 years and in the latter case, 5–10 years, depending on tree species.

- The decay rate for Scots pine twigs and branches may vary in the range of 0.128–0.188 yr.$^{-1}$ (95% mass loss in 15–25 years) and 0.066–0.127 yr.$^{-1}$ (95% mass loss in 25–45 years): the lower limit of the $k$ value represents the northern boreal (66° N), and the upper limit, the southern boreal (61° N) and hemiboreal (59° N) conditions (Vávrová et al. 2009).
- In coniferous coarse woody debris (e.g., Scots pine, Norway spruce), the decay rate varies between 0.010 and 0.035 yr.$^{-1}$ (e.g., Laiho and Prescott 2004;

Vávrová et al. 2009). A period of 80–300 years are required to achieve 95% mass loss.

- Krankina and Harmon (1995) estimated that the decay rate for Norway spruce in north-western Russia (59° N) is 0.0033–0.034 yr.$^{-1}$: 80–90 years are required to achieve 95% mass loss, which is close to that suggested by Mäkinen et al. (2006).
- Russell et al. (2014) estimated that for the temperate forest in eastern North America, the residence time of dead wood is 57–120 years for conifers and 40–70 years for hardwoods. This implies a decay rate of 0.024–0.040 yr.$^{-1}$ for conifers, and 0.043–0.064 yr.$^{-1}$ for hardwoods.

### *5.7.2 Carbon Sequestration Rate in the Boreal Zone by Tree Species*

Carbon sequestration refers to the process of carbon uptake from the atmosphere and storage in carbon pools or stocks in the ecosystem. In boreal forests, carbon sequestration varies depending on the site properties, the structure and functioning of the ecosystem, and the management and harvesting of forests aiming at timber and biomass production. Concurrently, the variability of carbon stocks in trees is especially pronounced, while the carbon stock in soil is stable even under intensive soil management.

Berg et al. (2009) conducted direct measurements to determine the rate of carbon sequestration in the humus layer across the boreal zone through Sweden. The study covered the podzol soils based on a dataset of 41 years of humus inventories, including the mean humus layer thickness in 82,513 plots. The data represented three forest types: (i) all combinations of tree species; (ii) forests dominated (>70%) by Norway spruce (*Picea abies*); and (iii) forests dominated (>70%) by Scots pine (*Pinus sylvestris*). For each forest type, the intersections in 25 × 25 km grids were used in kriging interpolation to relate changes in the humus layer thickness to land area. At each intersection, mean humus thickness for each year was calculated and regressed against time to obtain the rate of change. The rate of change, humus bulk density, and humus carbon concentration enabled the calculation of carbon sequestration rates. Berg et al. (2009) found that the mean carbon sequestration rate was 251 kg C ha$^{-1}$ yr.$^{-1}$. The sequestration rate was positively correlated with the temperature sum but was also affected by forest management. In the Scots pine-dominated forest, the mean sequestration rate was 283 kg C ha$^{-1}$ yr.$^{-1}$, and in the Norway spruce-dominated forests, 239 kg C ha$^{-1}$ yr.$^{-1}$. Under similar site conditions, the carbon sequestration rate was larger for Scots pine than for Norway spruce, with the difference being 71 kg C ha$^{-1}$ yr.$^{-1}$, highlighting the importance of tree species in storing carbon in boreal upland sites.

### 5.7.3 Carbon Accumulated in Boreal Forests by Site Fertility

Regardless of site type, carbon density increases with increasing productivity from northern to southern Finland, along with increasing growth and litterfall (Liski and Westman 1997a, b). On upland soils, the carbon density is 2.2 kg C $m^{-2}$ in the organic soil layer at the most fertile sites (*Oxalis–Myrtillus* site type, OMT) and sites with higher fertility. At the moderately fertile and moderately poor sites (*Myrtillus* and *Vaccinium* site types, MT, and VT), the carbon density is 1.9 kg C $m^{-2}$, and at the poor sites (*Calluna* site type, CT, and even poorer sites), 1.2 kg C $m^{-2}$. These values are of the same magnitude as that found by Lukina et al. (2020) for boreal upland sites in north-western Russia. In Sweden, the carbon density in the organic soil layer varies from 2 to 10 kg C $m^{-2}$ from upland mineral soils from northern (65° N) to southern (60° N) Sweden, with a clear influence of thermal and hydrological conditions on carbon accumulation (Olsson et al. 2009).

Based on Liski and Westman (1997a, b), the organic layer (litter and humus) in the Finnish sites contained 28% of the total carbon in soil, while the corresponding values of carbon in the mineral soil down to one meter was 68%, and below that 4%. The total amount of carbon in soil was more than double that in trees. The increasing temperature increased the carbon density 0.266 kg C $m^{-2}$ per increase of 100 d.d. in the temperature sum. Along with the site fertility, the temperature variability explained 36% and 70% of the variability of carbon in the organic and mineral layers, respectively. Based on Finnish National Forest Inventory data, Liski and Westman (1997b) estimated that the total carbon storage in the upland sites throughout Finland falls in the range of 1109–1315 Tg.

Soil carbon stock comprises nearly half of the carbon in boreal upland sites, but the carbon accumulation is dependent on site-specific biomass production, the properties of tree species and populations, and also on climatic conditions, as shown in Fig. 5.20. Based on the thickness of the soil organic matter layer, the mean current amount of soil organic matter (SOM) in the mineral soil of southern boreal forests may exceed 60–90 Mg $ha^{-1}$ on fertile sites (*Myrtillus*-type sites and more fertile sites), while on poor sites (*Vaccinium*-type sites and less fertile sites), the mean amount of soil organic matter may be less than half of this. In northern boreal forests, the amount of SOM is 80–90% of that in the southern boreal forests, regardless of site type (Kellomäki et al. 2005, 2008). The northward decline in the amount of organic matter is linked to the concurrent decline of forest growth in this direction. However, the spatial variability of soil organic matter is large, reflecting the variability in site fertility, thermal conditions, tree species, management, harvest intensity and frequency and disturbance history.

Climatic warming increases tree growth, suggesting the higher accumulation of soil organic matter (SOM) despite its increasing decay under such conditions (Fig. 5.20). Based on model simulations, Kellomäki et al. (2005) showed that climatic warming may increase SOM in southern and mid-boreal forests by 5% even in the short term (the period 2001–2020). The same trend occurs later in the north, where the amount of SOM increases from 2001 to 2050 by as much as 15% in

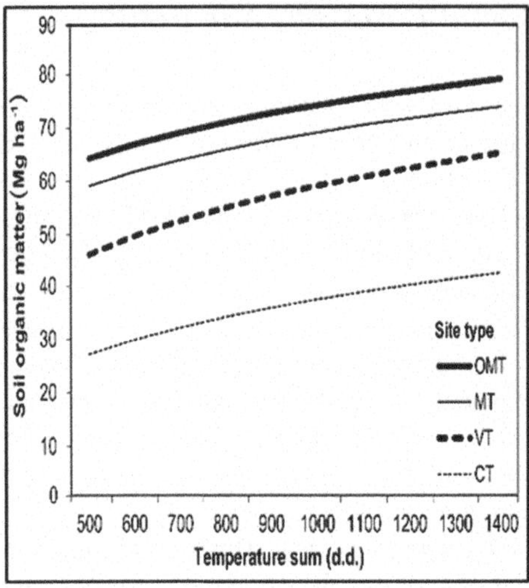


| Region and period | Organic matter, Mg ha$^{-1}$ (% of that under current climate) | Available nitrogen, kg ha$^{-1}$ (% of that under current climate) |
|---|---|---|
| Southern Finland | | |
| • Current | 80.3 | 43.2 |
| • 1990–2020 | 80.3 (0) | 44.5 (+3) |
| • 2021-2050 | 104.0 (+29) | 55.4 (+28) |
| • 2071–2099 | 133.5 (+66) | 63.7 (+47) |
| Northern Finland | | |
| • Current | 63.2 | 26.2 |
| • 1990–2020 | 63.1 (-0.3) | 27.1 (+4) |
| • 2021-2050 | 74.2 (+17) | 43.4 (+65) |
| • 2071- 2099 | 88.6 (+40) | 46.7 (+78) |
| Total | | |
| • Current | 73.4 | 36.3 |
| • 1990–2020 | 73.3 (<0) | 37.4 (+3) |
| • 2021-2050 | 91.9 (+25) | 50.4 (+39) |
| • 2071- 2099 | 115 (+57) | 56.8 (+56) |

**Fig. 5.20** Carbon accumulated in boreal upland in mineral soils forests in relation to site fertility (Kellomäki et al. 2005). Compiled by the author of this book. Left: Mean amount of soil organic matter (litter, humus) in forest sites on mineral soils as a function of temperature sum and site fertility (site type), based on Finnish National Forest Inventory data for the thickness of the soil organic layer. The temperature sum represents from the northern (e.g., 600 d.d.) to southern (e.g., 1400 d.d.) latitudes across the boreal zone in Finland. Right: Mean amount of soil organic matter (litter, humus) and available nitrogen in soil currently and in the period 1990–2099 under climate change between southern and northern Finland. The percentage change in relation to the current values is shown in the parentheses. Southern Finland refers to the forests below 63° N and northern Finland to the forests above 63° N

some locations. During 2071–2100, the amount of SOM is further increased in the north, occasionally even more than 30%. In the south, the amount of SOM also increases locally but even reduces by 20–30%. By the end of this century, climatic warming may increase the mean amount of SOM by 40% in the north and 60% in the south (Fig. 5.20). This difference between the southern and northern sites follows the difference in growth and the consequent litterfall, which strongly controls the accumulation of organic matter. In the north, low temperature increases further the accumulation of SOM due to low decay.

In boreal forests, the nitrogen cycle is temperature-dependent on the decay of SOM, whereas low water supply only occasionally limits the decay of SOM. A gradually warming climate is likely to increase the decay of litter and humus, making the nitrogen cycle faster and increasing the amount of nitrogen available for enhancing growth (Fig. 5.20). Based on Kellomäki et al. (2005), the increase is likely several percentage points even in the period 2001–2020, further 30–60% in 2021–2050, and 50–80% in 2070–2100. The increase of available nitrogen is pronounced in the north, where the percentage changes exceed those in the south. However, the amount of available nitrogen is still greater in the south than in the north under warming climate by 2070-2100.

### 5.7.4 Dynamics of Accumulation Soil Carbon Related to Site Fertility

The amount of litter and humus in soil varies in relation to the site fertility and structure of tree stands (species, density, growth, maturity, and developmental phase) (Table 5.9). In mesic sites, for example, the amount of litter and humus declines substantially after clear-cutting due to low litterfall and increased decay rate when herbs and grasses dominate the ground vegetation. This developmental phase is followed by a phase in which trees and shrubs invade the site, increasing the litterfall. At the same time, tree canopies close, and the decay rate of litter declines when the mean soil temperature remains lower than that in the clear-cut area and in young seedling stands. The amount of litter and humus in the soil is largest in mature stands. In the southern boreal zone, for example, the amount of litter and humus may be larger than 100 Mg ha$^{-1}$ in mature Scots pine stands in moderately fertile sites (*Myrtillus* site type, MT), depending on the properties of the tree stands.

The nutrient content in litter decreases in soil organic matter (SOM) in the early phase of succession, but the amount of litter and humus with nutrient storage increases in the long term (Table 5.9). The accumulation of nitrogen is particularly large in mature Scots pine stands. Nitrogen is mainly bound in litter and humus exceeding 1000 kg ha$^{-1}$ in moderately fertile sites (*Myrtillus* site type, MT) in the southern boreal zone. In such conditions, clear-cutting substantially increases the mineralization of nitrogen. In five years after the cut, the nitrogen bound in SOM is likely 20% lower than that before the clear-cutting. For other nutrients, the effects of cutting are smaller than that on nitrogen. In litter and humus, the amount of phosphorus may be 70–100 kg ha$^{-1}$, potassium, 100–200 kg ha$^{-1}$, and calcium, 100–250 kg ha$^{-1}$.

**Table 5.9** Estimated amount of litter and humus in Scots pine (*Pinus sylvestris*) stands in the southern boreal zone as a function of tree stand age and site fertility (site type)

| | CT | VT | MT | OMT |
|---|---|---|---|---|
| Stand age, yr | Litter and humus, Mg ha$^{-1}$ | | | |
| 5 | 89 | 96 | 127 | 126 |
| 10 | 77 | 78 | 97 | 96 |
| 15 | 69 | 66 | 81 | 84 |
| 20 | 63 | 59 | 74 | 81 |
| 30 | 55 | 58 | 75 | 84 |
| 40 | 56 | 62 | 101 | 111 |
| 50 | 63 | 77 | 109 | 118 |
| 60 | 68 | 86 | 119 | 125 |
| 70 | 71 | 93 | 123 | 130 |
| 80 | 76 | 99 | 128 | 136 |
| 90 | 83 | 102 | 128 | 141 |
| 100 | 92 | 104 | 126 | 132 |
| 120 | 101 | 92 | 109 | 120 |

Estimates are based on the Forcyte model (Kimmins and Scoullar 1984). Simulations were performed over a period from the open area to the full maturation of trees grown under no management. The initial stand density was 4000 seedlings ha$^{-1}$

## 5.8 Dead Wood in Boreal Forests

### *5.8.1 Biodiversity vs. Dead Wood*

In general, the relative supply of biodiversity (primary and secondary diversity) is larger in naturally developing forests than in managed forests. In the latter case, dying and dead trees are regularly harvested to prevent economic losses and reduce insect attacks on the remaining trees (e.g., Essen et al. 1992, 1997). In close-to-nature management (i.e., uneven-aged management), this tends to increase the structural complexity of the forest, but reduce the continuity of dead wood, because large-sized trees are harvested in selective cuttings thus reducing the formation of large-sized dead wood.

The loss of secondary biodiversity further increases under the increasing management intensity of both even-aged and uneven-aged forests. Under even-aged forest, the primary diversity tends to increase with the increasing share of early successional phases in forest landscapes due to the variability of stand developmental phases from clear-cut areas to mature stands (Sing et al. 2018). This trend is not under uneven-aged managed forests, where the stand-based variability in different developmental phases over the forest landscape is missing. Under even-aged management, decaying wood may increase when single trees or tree groups and large-sized retention trees are left in clear-cut areas to fall in the wind and left to decay on soil (Fig. 5.21). Under even-aged management, secondary biodiversity may also be enhanced if clear-cut areas with fallen trees are burnt in soil management, in contrast to uneven-aged management (e.g., Granath et al. 2018).

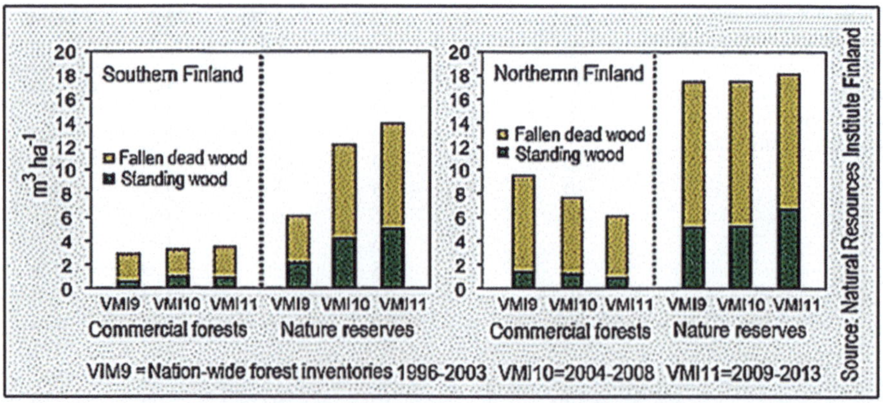


**Fig. 5.21** Standing and fallen dead wood in southern and northern Finnish commercial forests and nature reserves, compiled from several nation-wide forest inventories since 1996 https://www.luonnontila.fi/fi/elinymparistot/metsat/me6-lahopuun-maara). Courtesy of Natural Resources Institute Finland

### *5.8.2 Dead Wood Across Boreal Forests*

In northern Finland, the amount of dead wood is generally larger than in southern Finland, while in the north, the amount has declined due to an increase in cuttings (Fig. 5.21). Currently, dead wood amounts to 3–4 $m^3$ $ha^{-1}$ in the southern and 6 $m^3$ $ha^{-1}$ in the northern boreal forests (e.g., Ihalainen and Mäkelä 2009). In the natural forests, the values are clearly larger: 14 $m^3$ $ha^{-1}$ in the south and 18 $m^3$ $ha^{-1}$ in the north. According to the Finnish National Forest Inventory in 2009–2011 (VMI11), the increase of standing dead trees was especially large. Since the previous inventories, the deadwood in nature reserves had doubled in southern Finland, while in the commercial forests, the increase was 25%. However, the amount of dead wood in the natural forests varies depending on the fertility and geographical location of the site. The largest dead wood resources are in the south, where the amount of dead wood is up to 50–120 $m^3$ $ha^{-1}$ in old natural Norway spruce stands. In such cases, the amount of dead wood might have exceeded several hundred cubic meters per hectare immediately after wind or fire catastrophes.

### *5.8.3 Forest Structure and Dead Wood Effect on Boreal Forest Diversity*

Siitonen and Martikainen (1994) provide an example of how the mosaic structure of forests affects the number of beetles and their diversity in southern boreal conditions in an area either side of the border between Finland and Russia (Fig. 5.22). On the

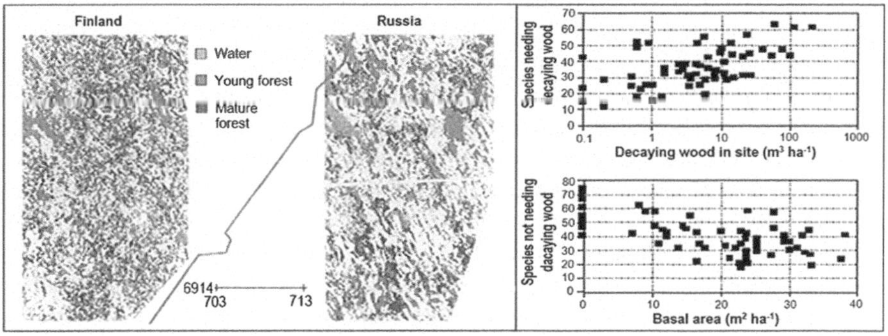


**Fig. 5.22** Effect of forest structure and occurrence of decaying wood on successes of invertebrate species in boreal forests (Siitonen and Martikainen 1994). Courtesy of Natural Resources Institute Finland. Left: Landsat image taken on May 27, 1992, on a segment of forest either side of the border between Finland and Russia in the southern/mid-boreal zone, showing how the varying management and harvest activities affect forest structure at the landscape scale. Water is indicated by pale grey, young forests by middle grey, and mature forest by dark grey. Right/Above: Relationship between the amount of dead wood and the number of beetle species obligately dependent on dead wood. Right/Below: Relationship between the stocking of living trees and the number of beetle species not obligately dependent on dead wood

**Table 5.10** Number of beetles and beetle species found in an investigation conducted either side of the border between Finland and Russia in the Karelian region

| Species group | Number of beetle species on the Finnish side of the border | Number of beetle species on the Russian side of the border | Total number of beetle species on both sides of the border |
|---|---|---|---|
| Number of beetles found | | | |
| • Decaying wood primarily needed | 3645 | 4964 | 8609 |
| • Decaying wood secondarily needed | 1893 | 2202 | 4095 |
| • Others | 8657 | 7663 | 16,320 |
| • Total | 14,195 | 14,829 | 29,024 |
| Number of beetle species | | | |
| • Decaying wood primarily needed | 179 | 217 | 253 |
| • Decaying wood secondarily needed | 55 | 58 | 64 |
| • Others | 245 | 232 | 307 |
| • Total | 479 | 507 | 623 |

Beetles were further classified into those primarily requiring decayed wood and those secondarily requiring decayed wood (Siitonen and Martikainen 1994)

Finnish side, the forest is split into small patches indicative of long-term management, intensive timber harvest, and the dominance of small forest holdings owned by private forest owners. On the Russian side, the forest is also split into patches, but the mean patch size is substantially larger than on the Finnish side. On the Russian side, for example, the rotation length is long, and wildfires have affected the forest more than on the Finnish side. This further implies that the amount of dead wood on the Russian side is substantially larger than on the Finnish side, where commercial thinning has been used regularly, in contrast to the Russian side. Such differences in management and timber harvest are clearly reflected in the beetle fauna, as shown in Table 5.10.

Siitonen and Martikainen (1994) also found that the number of beetle species correlated with the amount of dead wood (Fig. 5.22). The number of species primarily needing dead wood increased more rapidly than linearly along with the amount of dead wood. On the other hand, the number of species not primarily needing dead wood decreased along with the increasing stocking of trees. In mature stands with full tree cover, the number of species was one quarter of that in clear-cut areas. It was evident that the differences in the amount of beetle species between Finland and Russia reflected the differences in the patch mosaic of forests, tree species composition, and amount of dead wood in forests. There were more species preferring dead wood on the Russian side (213 species) than on the Finnish side (179 species). In total, 17 of the 20 most common species preferring dead wood were found on the Russian side, most of which preferred decaying birch wood. The majority of the species were present on both sides of the border, but the rarest species were mainly collected on the Russian side (Table 5.10).

## 5.9 Concluding Remarks

The functioning of boreal forest ecosystems is based on the cycle of carbon. This involves in the uptake, retention in biomass and soil organic matter in forest ecosystem in an interaction with atmosphere. Forest ecosystem has a double role for being a carbon sink in carbon uptake in assimilative and growth processes from atmosphere, and a carbon source for emitting carbon into the atmosphere in decaying processes. Growth and litter fall are accumulating carbon in trees and ground vegetation and further in organic matter (SOM) on soil surface and through upper soil profile. Carbon allocation to the different organs of trees and ground vegetation affects the growth and further litterfall, which transfers carbon from living trees and ground vegetation for accumulating on site surface and further deeper through soil profile below litter and organic matter on soil surface. Litterfall links primary production with the decay of litter and detritus, thus cycling carbon from the atmosphere through the ecosystem with links to sink/source process in carbon cycle. This affects the amount and properties of soil organic matter, with impacts on mitigating climatic warming.

Carbon dynamics in forest ecosystem is further affect by the mortality of trees. Trees may die in any phase of their life span due to resource shortages and other reasons unrelated to their increasing mass and maturation. A shortage of resources in relation to demand may induce subsequent causes of tree death: declining physiological activity (primary cause) can trigger insect and fungal attacks (secondary cause), increasing the mortality risk in any phase of growth and development. This is especially the case in mature trees, where the increasing demand of photosynthates for maintenance makes trees susceptible to insect and fungal pests. Fire, strong wind, and heavy snow load further damage trees, and increase the risk of insect and fungal attacks even on undamaged trees. In the long term, the damage induced by insect and fungal attacks seems to be the main cause of tree death over the successional cycle from seedlings to old-growth trees, with links to self-thinning. Standing dead trees and dead trees lying on the ground are important for the diversity of forest ecosystems, providing varying habitat for many vertebrates and invertebrates.

## References

Aber JD, Melillo JM (1991) Terrestrial ecosystems. Saunders College Publishing, Philadelphia

Aber JD, Melillo JM, McClaugherty CA (1990) Predicting long-term patterns of mass loss, nitrogen dynamics, and soil organic matter formation from initial fine litter chemistry in temperate forest ecosystems. Can J Bot 68:2201–2208

Ausmus BS (1977) Regulation of wood decomposition rates by arthropod and annelid populations. In: Lohm U, Persson T (eds) Soil organisms as components of ecosystems. Ecological Bulletin, vol 25. Swedish Natural Science Research Council, Uppsala, pp 180–192

Berg B, McClaugherty C, De Santo AV, Johnson D (2001) Humus build-up in boreal forests: effects of litter fall and its N concentration. Can J For Res 31:988–998

Berg B, Johansson M-B, Nilsson Å, Grundersen P, Norell L (2009) Sequestration of carbon in the humus layer of Swedish forests – direct measurements. Can J For Res 39:962–975

Botkin DB, Janak JF, Wallis JR (1972) Some ecological consequences of a computer model of forest growth. J Ecol 60:849–872

Bradford MA, Wieder WR, Bonan GB, Fierer N, Raymond PA (2016) Managing uncertainty in soil carbon feedbacks to climate change. Nat Clim Chang 6:751–758

Brunner I, Bakker MR, Björk GB, Hirano Y, Lukac M, Aranda X, Børja I, Eldhuset TD, Helmisaari HS, Jourdan C, Konôpka B, López BC, Pérez CM, Persson H, Ostonen I (2013) Fine-root turnover rates of European forests revisited: an analysis of data from sequential coring and ingrowth cores. Plant Soil 362:357–372

Cajander AK (1909) Über Waldtypen. Fennia 28(2):1–175

Chertov OG, Komarov AS, Nadporozhskaya M, Bykhovets SS, Zudin SL (2001) ROMUL – a model of forest soil organic matter dynamics as a substantial tool for forest ecosystem modeling. Ecol Model 138:289–308

DeLuca TH, Boisvenue C (2012) Boreal forest soil carbon: distribution, function, and modelling. Forestry 85(2):162–184

Essen P, Ehnsröm A, Eriksson L, Sjöberg K (1992) Boreal forest – the focal habitats of Fennoscandia. In: Hansson L (ed) Ecological principles of nature conservation: applications in temperate and boreal environments. Elsevier Applied Science, London, pp 252–325

Essen PA, Ehnström B, Ericson L, Sjöberg K (1997) Boreal forests. Ecol Bullet 46:16–47

Finér L, Helmisaari HS, Lõhmus K, Majdi H, Brunner I, Børja I, Eldhuset T, Godbold D, Grebenc T, Konôpka B, Kraigher H, Möttönen MR, Ohashi M, Oleksyn J, Ostonen I, Uri V, Vanguelova E (2007) Variation in fine root biomass of three European tree species: beech (*Fagus sylvatica* L.), Norway spruce (*Picea abies* L. Karst.), and Scots pine (*Pinus sylvestris* L.). Plant Biosyst 141:394–405

Flower-Ellis JGK, Mao-Sheng Y (1987) Vertical and cohort life-tables for needles of Norway spruce from northern Sweden. Sveriges Lantbruksuniversitet. Institut Skogsekologie, Skoglig Marklära Rapport 57:1–24

Granath G, Kouki J, Johanson S, Heikkala O, Rodrígues A, Strengbom J (2018) Trade-offs in production and biodiversity under prescribed burning and retention regimes in boreal forests. J Appl Ecol 55:1658–1667

Helmisaari HS, Makkonen K, Kellomäki S, Valtonen E, Mälkönen E (2002) Below- and above-ground biomass, production and nitrogen use of Scots pine stands in eastern Finland. For Ecol Manag 165:317–326

Hilli S (2013) Significance of litter production of forest stands and ground vegetation in the formation of organic matter and storage of carbon in boreal coniferous forests. In: Merilä P, Jortikka S (eds) Forest condition monitoring in Finland – national report. The Finnish Forest Research Institute. htt://www.metla.fi/metinfo/forest conditions/projects/

Hilli S, Stark S, Derome J (2010) Litter decomposition rates in relation to litter stocks in boreal coniferous forests along climatic and soil fertility gradients. Appl Soil Ecol 46:200–208

Hynynen J, Ojansuu R, Hökkä H, Salminen H, Siipilehto J, Haapala P (2002) Models for predicting stand development in MELA system. Finn For Res Inst Res Paper 835:1–116

Ihalainen A, Mäkelä H (2009) Kuolleen puuston määrä Etelä- ja Pohjois-Suomessa 2004–2007. Metsätieteellinen aikakauskirja I/2009:35–56

Jalkanen R (1998) Fluctuation in the number of needle sets and needle shed in *Pinus sylvestris*. Scand J For Res 13:284–291

Jukola-Sulonen EL, Mikkola K, Salemaa M (1990) The occurrence of abiotic damage and its relation to defoliation (needle loss) of conifers in Finland (1985–1988). In: Kauppi P, Anttila P, Kenttämies K (eds) Acidification in Finland. Springer-Verlag, Berlin, pp 523–560

Kalliola R (1973) Suomen kasvimaantiede. Werner Söderström Oy, Porvoo

Kauppi P, Kellomäki S, Saastamoinen O (1976) Metsäaapinen. Luonto-Liitto ry and Suomen Luonnonsuojeluliitto ry. Luonnonsuojelujulkaisuja Sarja A 3:1–160

Kellomäki S (1994) Kestävän kehityksen edellytykset Suomessa: metsätalous. Imatran Voiman juhlavuoden julkaisu. Kustannusosakeyhtiö Tammi, Helsinki, pp 417–482

Kellomäki S, Väisänen H (1997) Modelling the dynamics of the boreal forest ecosystems for climate change studies in the boreal conditions. Ecol Model 97(1, 2):121–140

Kellomäki S, Strandman H, Nuutinen T, Peltola H, Korhonen KT, Väisänen H (2005) Adaptation of forest ecosystems, forests, and forestry to climate change. Finn Environ Inst, FinAdapt Working Paper 4:1–50

Kellomäki S, Peltola H, Nuutinen T, Korhonen KT, Strandman H (2008) Sensitivity of managed boreal forests in Finland to climate change, with implications for adaptive management. Philos Trans R Soc B 363:2341–2351

Kimmins JP, Scoullar KA (1984) Forcyte-10: a user's manual. Environment Canada, Canadian Forestry Service, Ottawa

Kleja DB, Svensson M, Majdi H, Jansson P, Langvall O, Bergkvist B, Johansson M, Weslien P, Truusb L, Lindroth A (2008) Pools and fluxes of carbon in three Norway spruce ecosystems along a climatic gradient in Sweden. Biogeochemistry 89:7–25. https://doi.org/10.1007/s10533

Koivisto P (1962) Kasvu- ja tuotostaulukoita. Growth and yield tables. Communicationes Instituti Forestalis Fenniae 65(8):1–49

Kramer PJ, Kozlowski TT (1979) Physiology of woody plants. Academic Press, New York

Krankina ON, Harmon ME (1995) Dynamics of the dead wood carbon pool in Northwestern Russian boreal forests. Water Air Soil Pollut 82:227–238

Laiho R, Prescott CE (2004) Decay and nutrient dynamics of coarse woody debris in northern coniferous forests: synthesis. Can J For Res 34:763–777

Landsberg JJ (1986) Physiological ecology of forest production. Academic Press, London

Launiainen S, Lehtonen A, Laurén A (2015) Boreaalisten metsien hiilivirrat ja varastot. In: Salo K (ed) Metsä. Monikäyttö ja ekosysteemipalvelut. Luonnonvarakeskus (Luke), Helsinki, p 328

Launiainen S, Kieloaho A-J, Lindroos A-J, Salmivaara A, Ilvesniemi H, Heiskanen J (2022) Water retention characteristics of mineral forest soils in Finland: impacts for modeling soil moisture. Forests 13(1797):1–20

Lehtonen A, Sievänen R, Mäkelä A, Mäkipää R, Korhonen KT, Hokkanen T (2004) Potential litterfall of Scots pine branches in southern Finland. Ecol Model 180:305–315

Lehtonen A, Lindholm M, Hokkanen T, Salminen H, Jalkanen R (2008) Testing dependence between growth and needle litterfall in Scots pine – a case study in northern Finland. Tree Physiol 28:1741–1749

Lehtonen A, Linkosalo T, Peltoniemi M, Sievänen R, Mäkipää R, Tamminen P, Salemaa M, Nieminen T, Ťupek B, Heikkinen J, Komarov A (2016) Forest soil carbon stock estimates in a nationwide inventory: evaluating performance of the ROMULv and Yasso07 models. Geosci Model Dev 9:4169–4183

Leikola M (1969) Havaintoja männyn paksuuskasvun loppumisesta ja puiden keloutumisesta Inarin Lapissa. Summary: on the termination of diameter growth of Scots pine in old age in northernmost Finnish Lapland. Silva Fennica 3(1):50–61

Leppälammi-Kujansuu J, Salemaa M, Kleja DB, Linder S, Helmisaari H-S (2014a) Fine root turnover and litter production of Norway spruce in a long-term temperature and nutrient manipulation experiment. Plant Soil 374:73–88

Leppälammi-Kujansuu J, Aro L, Salemaa M, Hansson K, Kleja DB (2014b) Fine root longevity and carbon input into soil from below- and aboveground litter in climatically contrasting forests. For Ecol Manag 326:79–90

Liski J, Westman CJ (1997a) Carbon storage in forest soil of Finland: 1. Effect of thermoclimate. Biogeochemistry 36:239–260

Liski J, Westman CJ (1997b) Carbon storage in forest soil of Finland: 2. Size and regional patterns. Biogeochemistry 36:261–274

Lukina N, Kuznetsova A, Tikhonova E, Smirnov V, Danilova M, Gornov A, Bakhmet O, Kryshen A, Tebenkova D, Shashkov M, Kryazeva S (2020) Linking forest vegetation and soil carbon stock in Northwestern Russia. Forests 11:979

Lundmark JE (1986) Skogsmarkens ekologi. Ståndortsanpasst skogsbruk. Del 1 – Grunder. Fälths Tryckeri, Värnmo

Mäkinen H, Hynynen J, Siitonen J, Sievänen R (2006) Predicting the decomposition of Scots pine, Norway spruce, and birch stems in Finland. Ecol Appl 16:1865–1879

Makkonen K, Helmisaari HS (1998) Seasonal and yearly variations of fine-root biomass and necromass in a Scots pine (*Pinus sylvestris*) stand. For Ecol Manag 102:283–290

Mälkönen E (1974) Annual primary production and nutrient cycle in some Scots pine stands. Seloste: vuotuinen primäärituotos ja ravinteiden kiertokulku männikössä. Communicationes Instituti Forestalis Fenniae 84(5):1–87

Mälkönen E (1977) Annual primary production and nutrient cycle in a birch stand. Seloste: Vuotuinen primäärituotos ja ravinteiden kiertokulku eräässä koivikossa. Communicationes Instituti Forestalis Fenniae 91(5):1–28

Maser C, Anderson RG, Cromack Jr K, Williams JT, Martin RE (1979) Dead and down wood material. In: Thomas JW (ed) Wildlife habitats in managed forests: the Blue Mountains of Oregon and Washington. USDA Forest Service, Agriculture Handbook No. 553, Portland, pp 78–95

Mason CF (1976) Decomposition. Edwald Arnold Press, Southampton

Matala J, Kellomäki S, Nuutinen T (2008) Litterfall in relation to volume growth of trees: analysis based on literature. Scand J For Res 23:194–202

Mayer M, Prescott CE, Abaker WEA, Augusto L, Cécillon L, Ferreira GWD, James J, Jandl R, Katzensteiner K, Laclau JP, Laganière J, Nouvellon Y, Paré D, John A, Stanturf JA, Vanguelova EI, Vesterdal L (2020) Influence of forest management activities on soil organic carbon stocks: a knowledge synthesis. For Ecol Manag 466:118127

Mazziotta A, Mönkkönen M, Strandman H, Routa J, Tikkanen OP, Kellomäki S (2014) Modeling the effects of climate change and management on the dead wood dynamics in boreal forest plantations. Eur J For Res 133:405–421

Means JE, Cromack K Jr, MacMillan PC (1985) Comparison of decomposition models using wood density of Douglas fir logs. Can J For Res 1:1092–1098

Melillo JM, Aber JD, Muratore JF (1982) Nitrogen and lignin control of hardwood litter composition dynamics. Ecology 63(3):621–626

Mikola P (1960) Comparative experiment on decomposition rates of forest litter in southern and northern Finland. Oikos 11:161–166

Mohren GMJ (1987) Simulation of forest growth, applied to Douglas fir stands in the Netherlands. PhD thesis, University of Wageningen

Muukkonen P (2005) Needle biomass turnover rates of Scots pine (*Pinus sylvestris* L.) derived from the litter-shed dynamics. Trees 19:273–279

Muukkonen P (2006) Forest inventory-based large-scale forest biomass and carbon budget assessment: new enhanced methods and use of remote sensing for verification. Dissertationes Forestales 30:1–49, plus seven separate papers. Finnish Society of Forest Science, Helsinki

Muukkonen P, Lehtonen A (2004) Needle and branch biomass turnover rates of Norway spruce (*Picea abies*). Can J For Res 34:2517–2527

Muukkonen P, Mäkipää R (2006) Empirical biomass models of understorey vegetation in boreal forests according to stand and site attributes. Boreal Environ Res 11:355–369

Napola ML (1982) Puuston kehityksen vaikutus kuusikon pintakasvillisuuteen. Metsänhoitotieteen laudaturtyö. MSc thesis. Helsingin Yliopiston metsänhoitotieteen laitos

Niemelä T, Wallenius T, Kotiranta H (2002) The kelo tree, a vanishing substrate of specified wood-inhabiting fungi. Polish Bot J 47(2):91–101

Niemistö P, Hokkanen T, Varama M (2004) Karikemäärän muutokset 1982–2001 ja puiden kunto lumi- ja hallamittariesiintymisen vaivaamissa koivikoissa Noormarkussa. Metsätieteen aikakauskirja I(2004):21–41

Niinemets Ü, Lukjanova A (2003) Needle longevity, shoot growth and branching frequency in relation to site fertility and within-canopy light conditions in *Pinus sylvestris*. Ann For Sci 60: 195–208

Nilsson MC, Wardle DA (2005) Understory vegetation as a forest ecosystem driver: evidence from the northern Swedish boreal forest. Front Ecol Environ 3:421–428

Oker-Blom P, Kellomäki S, Valtonen E, Väisänen H (1988) Structural development of Scots pine stands with varying initial density: a simulation model. Scand J For Res 3:114–129

Olson JS (1963) Energy storage and the balance of producers and decomposers in ecological systems. Ecology 44(2):322–331

Olsson MT, Erlandersson M, Lundin L, Nilsson T, Nilsson Å, Stendahl J (2009) Organic carbon stocks in Swedish podzol soils in relation to soil hydrology and other site characteristics. Silva Fennica 43(2):209–222

Palviainen M, Finér L (2007) Ravinteiden pidättyminen kasvillisuuteen päätehakkuun ja maanmuokkauksen jälkeen. Metsätieteen aikakauskirja 3:287–290

Palviainen M, Finér L, Mannerkoski H, Piirainen S, Starr M (2005) Responses of ground vegetation species to clear-cutting in a boreal forest: aboveground biomass and nutrient contents during the first 7 years. Ecol Res 20(6):652–660

Pardossi A, Incrocci L, Incrocci G, Malorgio F, Battista P, Bacci L, Rapi B, Marzialetti P, Hemming J, Balendonck J (2009) Root zone sensors for irrigation management in intensive agriculture. Sensors:9809–2835

Penning de Vries FWT, Brunstin A, Laar HH (1974) Products, requirements, and efficiency of biosynthesis; a quantitative approach. J Theor Biol 45:339–377

Pensa M, Sellin A (2002) Needle longevity of Scots pine in relation to folia nitrogen content, specific leaf area, and shoot growth in different forest types. Can J For Res 32:1225–1231

Persson T, Bååth E, Clarholm M, Lundqvist H, Söderström BE, Sohlenius B (1980) Trophic structure, biomass dynamics and carbon metabolism of soil organisms in Scots pine forest. Ecol Bull 32:419–462

Reich BR, Oleksyn J, Modrzynsky J, Tjoelker MG (1996) Evidence that longer needle retention of spruce and pine populations at high elevations and high latitudes is largely a phenotypic response. Tree Physiol 16:643–647

Reich PB, Rich RL, Lu X, Wang Y-P, Oleksyn J (2014) Biogeographic variation in evergreen conifer needle longevity and impacts on boreal carbon cycle projections. PNAS 111:10703 13708

Reineke H (1933) Perfecting a stand-density index for even-aged forests. J Agric Res 36:627–638

Russell MB, Woodall CW, Fraver S, D'Amato AW, Domke GM, Skog EK (2014) Residence times and decay rates of downed woody debris biomass/carbon in eastern US forests. Ecosystems 17: 765–777

Sarvas R (1964) Havupuut. Werner Söderström Oy, Porvoo

Shorohova E, Kneeshaw D, Kuuluvainen T, Gaunthier S (2011) Variability and dynamics of old-growth forests in circumboreal zone: implications for conservation, restoration, and management. Silva Fennica 45:785–806

Siitonen J, Martikainen P (1994) Occurrence of rare and threatened insects living on decaying *Populus tremula*: a comparison between Finnish and Russian Karelia. Scand J For Res 9(2): 185–191

Silvertown JA (1982) Introduction to plant population ecology. Longman, London, and New York

Sing L, Metzger MJ, Paterson JS, Ray D (2018) A review of the effects of forest management intensity on ecosystem services for northern European temperate forests with a focus on the UK. Forestry 91:151–164

Stendahl J, Johansson MB, Eriksson E, Nilsson Å, Langvall O (2010) Soil organic carbon in Swedish spruce and pine forests – differences in stock levels and regional patterns. Silva Fennica 44:5–21

Tang J, Luyssaert S, Richardson AD, Kutsch W, Janssens IA (2014) Steeper decline in forest photosynthesis than respiration explains age-driven decrease in forest growth. PNAS 111(24): 8856–8860

Ťupek B, Mäkipää R, Heikkinen J, Peltoniemi M, Ukonmaanaho L, Hokkanen T, Nöjd P, Nevalainen S, Lindgren M, Lehtonen A (2015) Foliar turnover rates in Finland – comparing estimates from needle-cohort and litterfall-biomass methods. Boreal Environ Res 20:283–304

Ukonmaanaho L (2013) Litterfall production and quality at Level II sites. In: Merilä P, Jortikka S (eds) Forest condition monitoring in Finland: national report. The Finnish Forest Research Institute

Ukonmaanaho L, Merilä P, Nöjd P, Nieminen TM (2008) Litterfall production and nutrient return to the forest floor in Scots pine and Norway spruce stands in Finland. Boreal Environ Res 13 (suppl. B):67–91

Urvas L, Erviö R (1974) Metsätyypin määräytyminen maalajin ja maaperän kemiallisten ominaisuuksien perusteella. Summary: Influence of the soil type and the chemical properties of soil on the determining of the forest type. J Sci Agric Soc Finland 46(3):307–319

Valachovic YS, Caldwell BA, Cromack K Jr, Griffiths RP (2004) Leaf litter chemistry controls on decomposition of Pacific Northwest trees and shrubs. Can J For Res 34:2131–2147

Vanninen P, Mäkelä A (1999) Fine root biomass of Scots pine stands differing in age and soil fertility in southern Finland. Tree Physiol 19:823–830

Vanninen P, Mäkelä A (2005) Carbon budget for Scots pine trees: effects of size, competition, and site fertility on growth allocation and production. Tree Physiol 25:17–30

Vávrová P, Penttilä T, Laiho R (2009) Decomposition of Scots pine fine woody debris in boreal conditions: implementations for estimating carbon pools and fluxes. For Ecol Manag 257:401–412

Wardle DA, Nilsson MC, Zackrisson O, Gallet C (2003) Determinants of litter mixing effects in a Swedish boreal forest. Soil Biol Biochem 35:827–835

Westman KJ (1990) Metsämaan fysikaaliset ja fysikaalis-kemialliset ominaisuudet CT-OMaT kasvupaikkasarjassa. Summary: Soil physical and physico-chemical properties of Finnish upland sites. Silva Fennica 24(1):141–158

Yoda K, Kira K, Ogawa H, Hozumi K (1963) Intraspecific competition among higher plants: XI Self-thinning in overcrowded pure stands under cultivated and natural conditions. J Biol Osaka City Univ D14:107–129

Yuan ZY, Chen YH (2010) Fine root biomass, production, turnover rates, and nutrient contents in boreal forest ecosystems in relation to species, climate, fertility, and stand age: literature review and meta-analyses. Crit Rev Plant Sci 29(4):204–221

# Part III
# Management of Boreal Forests for Different Ecosystem Services

# Chapter 6
# Management for Timber Production and Carbon Sequestration

## Contents




**Abstract** Forest management can be divided into two basic strategies: uneven-aged and even-aged management. Uneven-aged management combines cutting removal through selective cutting from above with natural regeneration and ingrowth of seedlings. Uneven-aged management excludes the specific rotation and stand-based forest structure of landscapes. This is contrast to even-aged management based on sites with tree stands, using thinning from below or above. In even-aged management, reforestation is mainly based on clear-cutting and seeding/planting or through natural regeneration using shelterwood or seed tree cuttings. Both even- and uneven-aged management seem to generate similar mean growth, carbon stocks, and





S. Kellomäki, *Forest Management for Timber Production and Climate Change Mitigation*, Managing Forest Ecosystems 44,
https://doi.org/10.1007/978-3-031-71575-4_6

timber yield under the currently recommended management patterns. Using even-aged management, carbon sequestration is sensitive to the thinning intensity in removal of trees and extension of rotation time. A moderate decrease of thinning intensity and extension of rotation length increase the capacity and duration of carbon retention in forest ecosystems. Concurrently, the changes in soil carbon stocks in response to thinning and extension of rotation are delayed and less sensitive than the changes of carbon stocks in trees. Varying combinations of different management operations affect given timber yield and climate benefits due to the complex interaction between management and the retention of carbon in forest ecosystems.

## 6.1 Boreal Forests and Greenhouse Gas Emission Reduction

### *6.1.1 Mitigating Climate Change: Global Scale*

Sustainable forestry involves ecological dimensions as well as economic and social acceptability. Ecological dimensions mainly involve the links of carbon dynamics to the production of biomass and conservation of biodiversity. In this context, sustainability requirement deals with ecological processes such as (i) the fixation of carbon and energy in the ecosystem (energy); (ii) the conversion of carbon and energy to chemical energy and organic matter; and (iii) the distribution of chemical energy and organic matter between different species (diversity). Forestry is sustainable if the energy fixation and the production and distribution of organic matter between different species follow the natural dynamics of forest succession under management and harvest cycles selected for producing different ecosystem services.

In the global scale, the role of forests and forestry is important in mitigating climate change. Appropriate management and harvest can maintain global forest area at sufficiently elevated levels for sequestering and storing atmospheric carbon, thus slowing climatic warming. According to the rules of the Kyoto Protocol (Box 6.1), and those of the United Nations Framework Convention on Climate Change (UNFCCC), forests provide a sink for carbon dioxide ($CO_2$), and their maintenance thus meets the national commitments to reduce greenhouse gas (GHG) emissions. The Kyoto Protocol is an international agreement linked to the UNFCCC that **commits** the Parties by setting internationally binding targets for reducing atmospheric greenhouse gases. The Kyoto Protocol implements the objective of the UNFCCC to reduce the onset of global warming based on the principle of common but differentiated responsibilities. In this respect, it acknowledges that individual countries have different capabilities of combating climate change depending on their economic development. The Protocol contains country-specific obligations to reduce current GHG emissions, because many countries are more historically responsible than others for the current levels of GHG gases in the atmosphere.

The Kyoto Protocol operationalizes the UNFCCC by committing industrialized countries and countries with economies in transition to limit and reduce GHG emissions in accordance with agreed individual targets. The Convention itself only asks those countries to adopt policies and measures on GHG emissions and to report periodically. To this end, proper land use and management can be used to increase the carbon sequestration and terrestrial carbon stocks in forest ecosystems (e.g., Jandl et al. 2007). The Kyoto Protocol came into force in 2005. Thereafter, its content has been specified in a series of Conferences of the Parties in UNFCCC (COP) following the increasing knowledge on the likely changes in climate.

**Box 6.1 The Kyoto Protocol**

The Kyoto Protocol establishes legally binding targets and timetables for reducing greenhouse gas (GHG) emissions in developed countries. The GHG reduction commitments for the first commitment period (2008–2012) for the key boreal signatories of the protocol were (expressed as a percentage of 1990 emissions levels): Canada 94%, Norway 101%, Sweden 92%, Finland 92%, Russian Federation 100%.

The Protocol recognizes that land-use change, and forestry affect the net GHG balance (sources and sinks). Articles 3.3 and 3.4 provide forest-related mechanisms for compliance with the above targets based on forest management practices since 1990.

- Article 3.3 requires an accounting of carbon stock changes resulting from land-use change (Afforestation, Reforestation, and Deforestation (ARD)) since 1990.
- Article 3.4 allows countries to include, if they choose to do so, carbon gains/losses.
- Each Party must stipulate by 2006 whether they will include forest management practices in their reporting.
- If included, sinks due to forest management can be used to offset any net ARD source up to a cap of 9 Tg C year$^{-1}$. Any remaining sink is then credited up to the country-specific cap (Canada's cap is 12 Tg C years$^{-1}$).

Definitions

- Afforestation: The direct human-induced conversion of land that has not been forested for a period of at least 50 years to forested land through planting, seeding, and/or human-induced promotion of natural seed sources.
- Reforestation: The direct human-induced conversion of non-forested land to forested land through planting, seeding, and/or human-induced promotion of natural seed sources, on land that was forested but has been converted to non-forest land. For the first commitment period, reforestation

(continued)

**Box 6.1** (continued)

activities will be limited to reforestation occurring on those lands that did not contain forest on 31 December 1989.

- Forest management: A system of practices for stewardship and use of forest land aimed at fulfilling relevant ecological (including biological diversity), economic, and social functions of forests in a sustainable manner.

Additional information on the Kyoto Protocol, including recent developments and negotiations, can be found on the United Nations Framework Conversion on Climate Change website, http://unfccc.int/

### *6.1.2 Management of Carbon in the Successional Dynamics of Forest Ecosystems*

The forest ecosystem includes biotic (living organisms, including trees and other organisms) and abiotic (site, including climate and soil) components, which form the ecosystem structure (Fig. 6.1). The biotic and abiotic components interact through

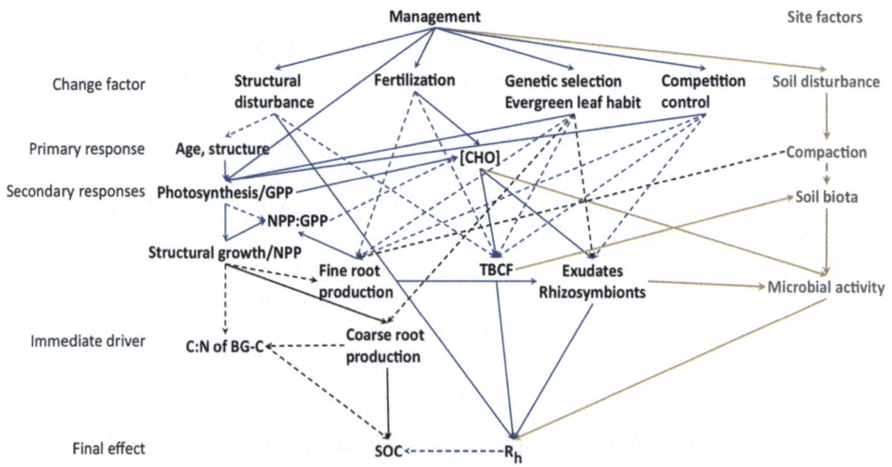


**Fig. 6.1** Schematic depiction of how selected management measures affect the interaction between genotype and environment, and how these factors affect the production of ecosystem goods and services alone or in combination (Noormets et al. 2015). Permission of Elsevier. The arrows and abbreviations outline the effects of management on forest productivity and carbon sequestration as follows: (i) solid arrows indicate a positive, and dashed arrows a negative effect; (ii) brown arrows mark processes affecting the recalcitrance of soil carbon; (iii) orange arrows mark processes operating through soil disturbance; (iv) *GPP*: gross primary productivity; (v) *NPP*: net primary productivity; (vi) *[CHO]*: carbohydrate concentration; (vii) *DTBCF*: total belowground carbon flux; (viii) *C/N*: ratio of carbon to nitrogen; (ix) *BG-C*: belowground carbon; (x) *SOC*: soil organic carbon; and (xi) $R_h$: heterotrophic respiration

biological, geochemical, and physical processes (ecosystem functioning) that take place between the components. This leads to a flow of energy and carbon that is dependent on trophic structure, biotic diversity, and material cycles. Concerning environmental, economic, and social benefits, ecosystem services are linked to ecosystem structure and functioning such as carbon sequestration, with concurrent mitigation of climatic warming.

Following the forest management and land-use principles defined in the Kyoto Protocol (Box 6.1), management includes a set of operations (measures) that may be used, alone or in combination, to reroute succession dynamics to produce different goods and services based on natural ecosystem dynamics. Succession is directed mainly by modifying the structure and subsequent functioning of populations and communities, representing both primary (forming organic matter from inorganic matter) and secondary (micro- and macro-organisms consuming primary production) production (Fig. 6.1). Management operations may be combined in different ways depending on the properties of forest domains, and on the edaphic and climatic conditions. This emphasizes the role of accepted strategies in management of forest ecosystems rather than the role of single management operations.

Under management, boreal forests mainly comprise tree species with long life spans. In forestry, the length of rotation, for example, is typically in the range of 60–150 years for Scots pine (*Pinus sylvestris*) and Norway spruce (*Picea abies*) in northern Europe depending on the prevailing climatic and edaphic conditions. This makes these forests important for long-term carbon storage. Carbon is taken in uptake through primary production, including gross primary production and net primary production of the above- and belowground biomass, while carbon loss occurs through autotrophic and heterotrophic respiration. Both uptake and loss of carbon contribute to the carbon balance and concurrent carbon stocks in trees and soil over forest chronosequences from the seedling phase to mature trees. In this context, the mitigation of climatic warming is related to the cycling of carbon between the ecosystem and atmosphere (Fig. 6.1). In management, these eco-physiological processes are, for example, affected by the reforestation combined with genotype selection/tree species choice, competition control such as the density control of tree stands in different phase of chronosequence, as emphasized by Noormets et al. (2015). Management also affects the properties of soil, with impacts on soil biota and microbiologic activity, and subsequent effects on soil organic matter and the decay of soil detritus through soil preparation and fertilization.

### 6.1.3 Carbon in Managed Forests, with Links Between and Beyond Ecosystems

In forestry, the net carbon exchange in timber production includes (i) the carbon taken up and bound in biomass; (ii) carbon emissions from decaying litter and humus; and (iii) carbon emissions from management and harvest operations.

Forestry has further links outside the forest ecosystem through the carbon bound in timber and biomass used in the manufacturing and energy industries. The recycling of carbon to the atmosphere also occurs in the carbon released from the combustion of biomass and degradation of wood-based products no more in use. Consequently, the net carbon exchange ($C_{net}$) can be calculated as (Kilpeläinen et al. 2011):

$$C_{net} = C_{seq} + C_{man} + C_{decomp} + C_{harv} \tag{6.1}$$

where $C_{seq}$ is the sequestered carbon (carbon bound in the ecosystem), $C_{man}$ is the carbon emitted in management operations, $C_{decomp}$ is the carbon emitted in decomposition of litter humus and decomposition and burnt wood-based material, and $C_{harv}$ is the carbon emitted in harvest operations. The links between different management operations and further carbon sink/source dynamics are demonstrated in Fig. 6.2 based on Kilpeläinen et al. (2011).

Kilpeläinen et al. (2011) applies the Eq. 6.1 to estimate the net $CO_2$ exchange of forest production for boreal Norway spruce (*Picea abies*) in southern (62° N) and northern (66° N) boreal forests. The sites were of medium fertility (*Myrtillus* site type, MT). In the initialization of simulations, the amount of soil organic matter (SOM with litter and humus) was 67 Mg $ha^{-1}$ for the south and 65 Mg $ha^{-1}$ for the north, representing SOM prior to clear-cutting (old carbon, or legacy carbon). The calculations were performed under the current (CU) and changing climate (CC). The latter one based on the A2 climate change scenario (IPCC 2007), which involves the rise of atmospheric $CO_2$ concentration up to 840 ppm by 2100. Under this scenario, the mean January temperature was likely to rise by 7.6 °C, and mean July temperature by 3.4 °C, over the whole of Finland. Precipitation increased by 30% in the winter and 10% in summer by 2100.

Kilpeläinen et al. (2011) applied in their study the management recommendations for privately owned forests. In both southern and northern sites, the initial stand density was 2500 seedlings per hectare, with an initial diameter of 2 cm (Fig. 6.2). In the energy wood thinning (EWT, pre-commercial thinning), the stand density was reduced to 1000–1200 trees per hectare. Thereafter, thinning rules based on the development of basal area and dominant height of trees were used in the specification of thinning interventions throughout the rotation. The final-cut (FF) was performed at the end of 80-year rotation. The simulations were performed separately for traditional timber production (TP) and for the integrated production of timber and bioenergy (BP). In the latter case, the energy biomass was harvested in (i) the energy wood harvest (EWT; (ii) whole-tree harvest in precommercial thinning); (iii) and final-cut (FF, timber, and logging residues). In thinning between the EWT and FF, no energy biomass was harvested.

Table 6.1 shows the carbon dynamics in the simulations when applying the Eq. 6.1 to calculate net carbon exchange. It includes both ecological processes in forests and anthropogenic processes such as management, and manufacturing of materials and their use outside the forest ecosystem. The results are shown for traditional timber production (pulpwood and saw logs) with integrated timber and

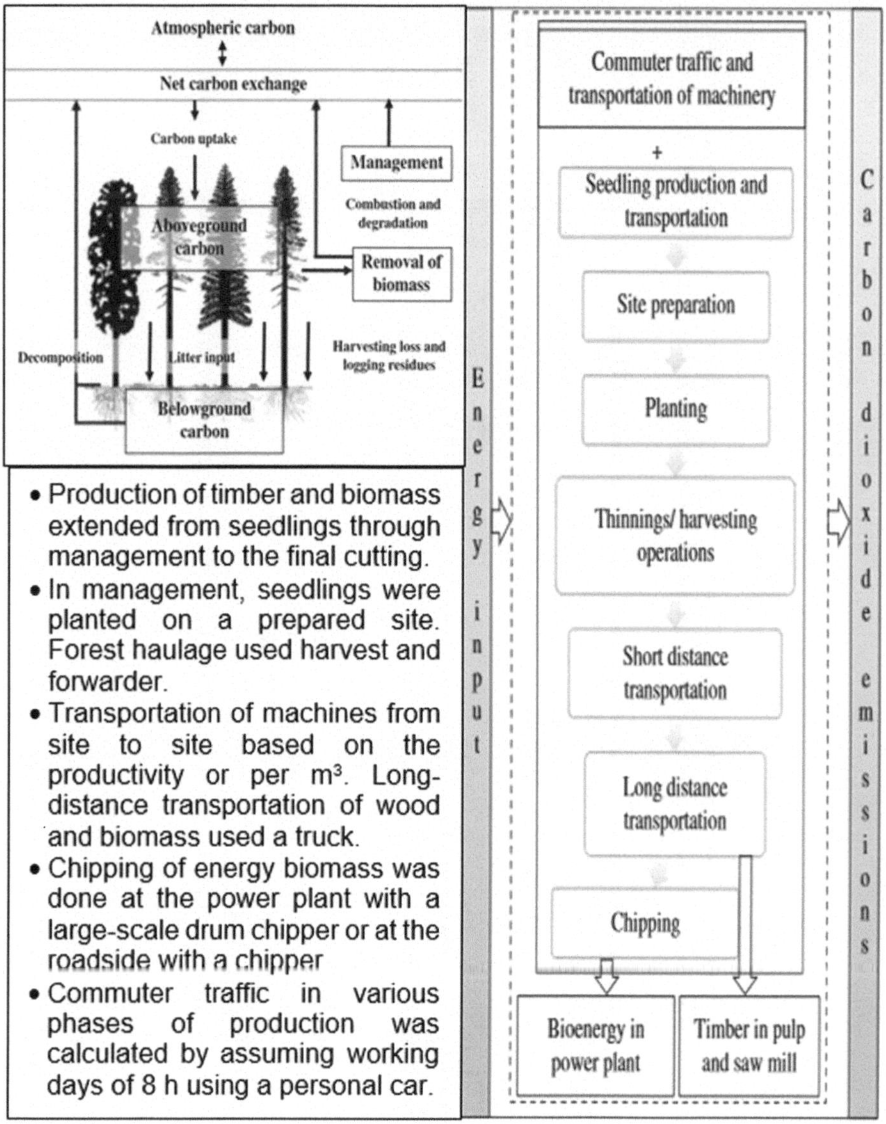


**Fig. 6.2** Carbon in managed forests, with links between ecosystem structure and functioning and further to the use of timber and energy biomass in manufacturing and energy industries (Kilpeläinen et al. 2011). Permission of Wiley. Left:/Upper: Carbon exchange of forest production. Carbon uptake = photosynthesis – plant respiration. Left/Lower: Summary of main points explaining the management and production of timber and biomass. Right: Forest production system boundaries used in the calculations

**Table 6.1** Mean values (g $CO_2$ $m^{-2}$ $yr.^{-1}$) of carbon flows calculated using the Eq. 6.1 to define the main carbon flows in the carbon dynamics in an 80-year rotation of Norway spruce (*Picea abies*) in southern and northern boreal forests under different management regimes and current and warming climate (Kilpeläinen et al. 2011)

| Management regimes and climate with explanations below | Mean values for carbon flows over an 80-year rotation for separate flows, g $CO_2$ $m^{-2}$ $yr.^{-1}$ | | | | | | |
|---|---|---|---|---|---|---|---|
| | Carbon sequestration | Carbon decomposition | Carbon in pulp wood | Carbon in saw logs | Carbon in energy biomass | Carbon emitted in management and harvest | Net carbon balance |
| Southern Finland | | | | | | | |
| • TP, CU | 1138 | 728 | 70 | 15 | 0 | 5 | −319 |
| • BT, CU | 1112 | 725 | 26 | 22 | 222 | 7 | −110 |
| • TP, CC | 1040 | 791 | 73 | 15 | 0 | 4 | −157 |
| • BT, CC | 1053 | 751 | 29 | 8 | 211 | 6 | −49 |
| Northern Finland | | | | | | | |
| • TP, CU | 974 | 617 | 71 | 13 | 0 | 4 | −214 |
| • BT, CU | 941 | 616 | 40 | 0 | 228 | 6 | −51 |
| • TP, CC | 1198 | 753 | 81 | 22 | 0 | 5 | −337 |
| • BT, CC | 1134 | 721 | 50 | 18 | 217 | 7 | −120 |

Legend for the column on the left: *TP* traditional timber production regime, *BT* integrated timber and biomass production regime, *CU* current climate, *CC* changing climate bioenergy production (logging residues, stumps, and roots) for Norway spruce. The magnitude of management-related emissions on net carbon exchange was smaller compared with the total ecosystem fluxes, with emission being the largest from decomposition. In addition, the effects of management options and climate were higher on the decomposition of new humus (de novo carbon accumulated since clear-cutting) compared with old humus (legacy carbon). The results also showed that the increased biomass growth due to climate warming could not compensate for carbon loss related to decomposition and biomass combustion in the south. Consequently, the carbon emissions due to the use of biomass for energy were 172 and 188 kg $CO_2$ $MWh^{-1}$ under the current and warming climate, respectively. In the north, the carbon emissions from biomass combustion were 199 and 157 kg $CO_2$ $MWh^{-1}$, respectively.

## 6.2 Management Related to Interactions Between Ecosystems Processes

### *6.2.1 Primary and Secondary Succession*

Long-term dynamics of forest vegetation is a successive one called the succession, which refers to the temporal changes in growth and development of vegetation (trees, shrubs, ground vegetation) in selected spatial scale as in trees stands forming a chronosequences. Succession is divided into primary and secondary ones. In both cases, the carbon sequestration and stocking are determined by the prevailing site conditions, the properties of the vegetation (trees, shrubs, ground vegetation), and the decay of organic matter in soil detritus.

*Primary succession* refers to the colonization of vegetation in sites previously unoccupied by plant cover, such as after expansion of a shoreline following land uplift (Fig. 6.3). This also holds for the reforestation of arable land that has lost its forest-based carbon (or legacy carbon) when converted from original forest into arable land. The reforestation initiates the primary succession and the accumulation of de novo carbon in site and vegetation. Under repeated disturbances or management cycles, the primary succession is replaced by secondary succession, with the declining soil legacy carbon being gradually replaced by the increasing accumulation of de novo carbon.

*Secondary succession* refers to the dynamics of carbon sequestration and stocking in sites previously occupied by vegetation as in forest ecosystems (Fig. 6.3). Secondary succession is driven by natural disturbances (e.g., wildfire, high wind etc.) and management and harvest measures in producing timber and biomass. In secondary succession, legacy carbon represents the carbon that remains on a site following a stand-replacing disturbance event, such as clear-cutting area, with links inherent to carbon sequestration and stocking in the previous production cycle. This is opposite to the primary succussion, whose processes have no links

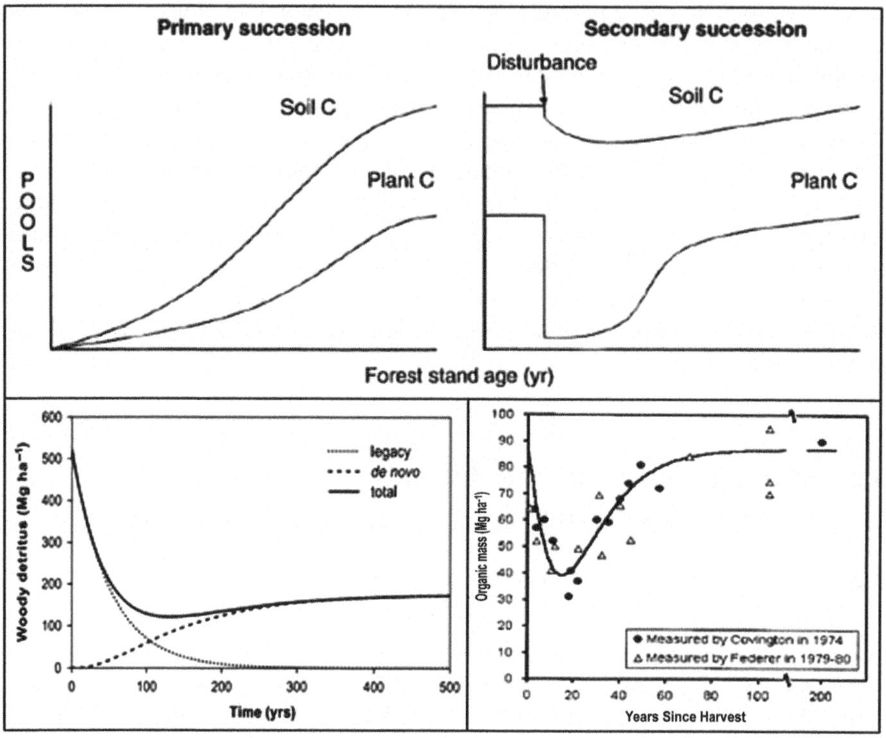


**Fig. 6.3** Schematic depiction of carbon dynamics in primary and secondary succession, the latter induced by disturbance with impacts on legacy and de novo carbon in the forest ecosystem. Left/Upper: Overview of carbon accumulation trends in soil and vegetation in primary succession (no management). Courtesy of Springer Nature. Right/Upper: Secondary succession following disturbance such as clear-cutting, with the impacts on carbon in soil and vegetation. Courtesy of Springer Nature. Left/Lower: Overview of modelled total carbon divided into legacy and de novo carbon in woody detritus mass through secondary succession following disturbance and afforestation under no management (Harmon 2001). Permission of Oxford University Press. Right/Lower: Ecosystem carbon combining legacy and de novo organic mass in trees and soil as a function of time from harvesting based on field measurements and model calculations for secondary succession under management, as shown by Yanai et al. (2003) adapted from Covington (1983). Courtesy of Springer Nature

to previously carbon sequestration and stocks. In secondary succession, the previous interactions affect the recovery of vegetation, as due to vegetation and tree seedlings existing in areas prior to cutting. Consequently, legacy carbon decreases in relation to increasing de novo carbon through a new production cycle toward the full maturity of trees. Secondary succession is driven by ecological processes (regeneration, growth, and mortality) and natural disturbances and management/harvest measures in producing timber and biomass, and further sequestrating and stocking carbon in forest ecosystems.

### 6.2.2 *Succession of Forest Ecosystems, with Management Implications*

In forestry perspective, the long-term dynamics of forest ecosystem refers to the secondary succession, as addressed in this book. In this context, management and harvest control the production of timber and biomass, with sequestrating and stocking of carbon in different successional phases through stand-based chronosequences or forest landscapes. Management and harvest are sustainable if the energy fixation and the production and distribution of organic matter between different species follow the natural dynamics of forest succession. Concurrently, sustainability deals with ecological processes such as addressed above: (i) the fixation of carbon and energy in the ecosystem (energy); (ii) the conversion of carbon and energy to chemical energy and organic matter (structure); and (iii) the distribution of chemical energy and organic matter between different species (diversity). Succession of forest ecosystem can be divided into autogenic (or endogenous) and allogenic (or exogenous) ones depending on how the disturbances or management/harvest affect ecological processes, which drive the growth and development of trees and tree stands through management and harvest cycles in forestry.

*Autogenic succession* refers to mortality of single trees or groups of trees in minor disturbances (e.g., wind load, snow load, insect, and fungi attacks) in different phases of life cycle of trees (Fig. 6.4) Death of single trees and the consequent gaps in the canopy release resources and enhance regeneration and growth of remaining trees. Autogenic succession creates a forest landscape, with no clear grouping of trees in separate stands. Such a landscape is homogenous opposite to the landscspe created by *allogenic succession*. In this case, the dynamics of forest ecosystems is related to by major disturbances (e.g., large-scale wildfire, high wind) with large-scale damages, which modify the ecosystem structure over stands/communities (Fig. 6.4). In major disturbances, trees die at the same time over a large area. This release of space for trees to regenerate and grow as a population/stand, with individuals of the same age. Major disturbances create a forest landscape with a heterogenous mosaic of tree populations/communities (stands) of varying character. This forest structure leads to even-aged management through the chorosequences from the reforestation phase to mature phase, which is followed the regenerative cutting for the next rotation.

In the scale of landscape, trees in gaps and trees in patch-based stands shift from one developmental phase to another. During the shift, trees in separate gaps/patches represent structural differentiation related to the life cycle of single trees in gaps or tree stands in patches of several trees occupying the same site. Wind or fire, for example, turn the development of single gaps or patches or several patches back to some earlier successional phase. The structure and functioning of the forest over the whole region vary through time following the successional dynamics of each gap and patch from the reorganization after disturbance to the stability. The mass (trees and soil, living and dead) over a region oscillates following the succession of gaps and patches in landscape mosaic (shifting-mosaic in steady state). In this context,

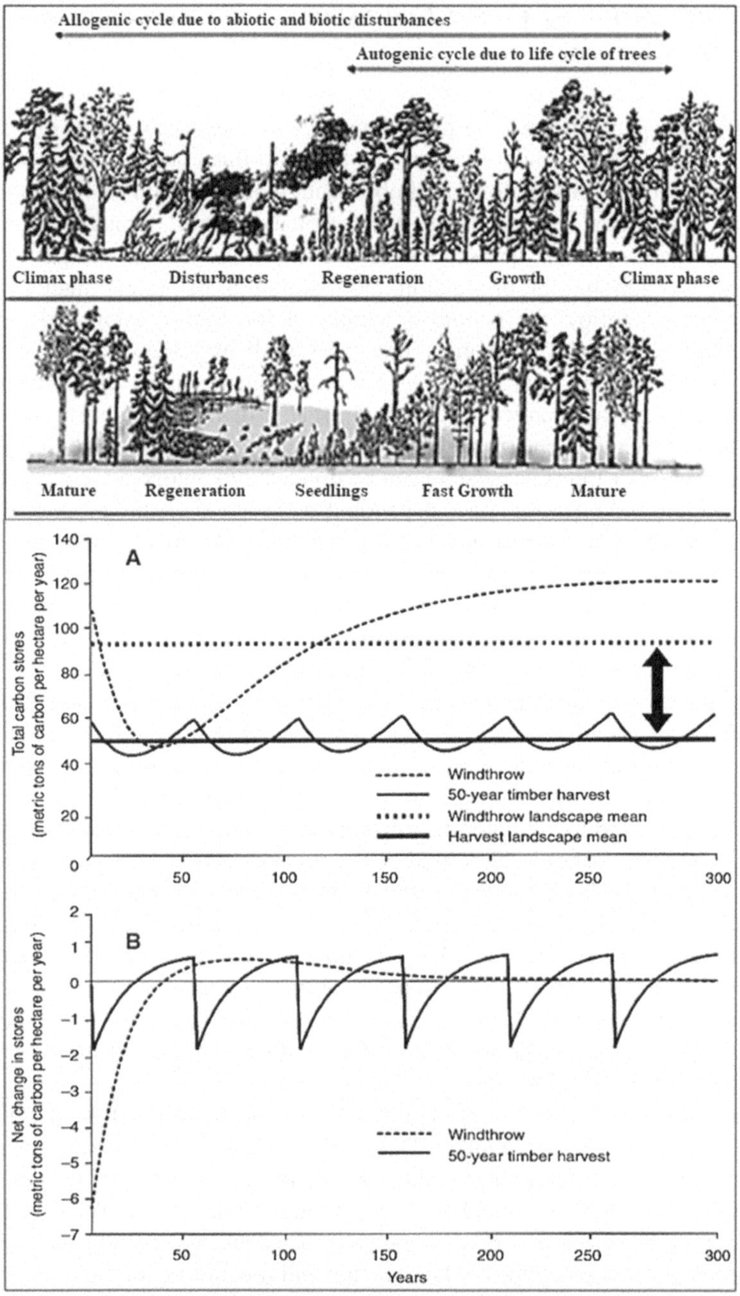


**Fig. 6.4** Forest dynamics under natural succession, and effects of management on carbon stocks in the forest landscape related to wind-induced disturbance and regular management under selected rotation in management. Upper: Schematic depiction natural forest development under allogenic and autogenic disturbance. Compiled by the author of this book. Upper/Middle: Even-aged management in different phases of the production cycle and the timing of selected management measures in controlling the rotation cycle, assuming stand-by-stand management. Complied the

successional dynamics emphasizes the spatial patterning at a variety of scales, including single trees, populations/stands of trees, communities of trees, and forest landscapes (e.g., Bormann and Likens 1979, pp. 164-176).

### 6.2.3 Changes of Carbon Stocks in Relation to Disturbance in the Landscape

Figure 6.4 further outlines how disturbance such as the impacts of windthrow and selected management on the total carbon stock and the net change of carbon stock in the forest landscape. In this theoretical analysis, Harmon (2001) combined the temporal changes in mass and net carbon sequestration rates to complete windthrow with a return period of 300 years, and timber harvest with a mean rotation of 50 years. In this scenario, the timber harvest covered 65% of the living biomass in boles, converting the timber to products with a mean lifespan of 50 years. The horizontal lines indicate the mean landscape-scale stock of carbon for these two different disturbance regimes (windthrow, management). The mean net change in carbon stock for both disturbance regimes is zero over the 300-year period, i.e., both systems are in balance at the landscape scale in terms of carbon sequestration. In this regard, the black arrow indicates the potential net change in carbon stocks caused by moving from one disturbance regime (natural/management) to another; that is, from a single windthrow to regular harvesting over time. Under regular harvests, the total carbon stock over the landscape is less sensitive than that after a single wind-induced disturbance. In the long term, disturbance scenarios produce a similar carbon balance over the forest landscape; that is, higher over time after a single wind throw but lower for regular harvesting over several rotation periods.

### 6.2.4 Management Links to Ecosystems, Impacts on Structure and Functioning

Under even-aged management, the carbon balance and accumulation of carbon stocks can be increased through proper management, reducing the time needed for the rapid establishment of the tree stand for carbon uptake in ecosystem processes and structures (Fig. 6.5). In reforestation, this is achieved by measures such as site preparation, and planting. The duration of net carbon loss is further reduced if more

**Fig. 6.4** (continued) author of by this book. Middle: Total carbon stocks in the ecosystem and net change in carbon stocks affected by windthrow, a 50-year cycle of timber harvest, and the mean over the windthrow-affected forest landscape (Harmon 2001). Permission of Oxford University Press. Lower: Net exchange of carbon storage affected by a windthrow with recovery and a 50-year cycle of timber harvest (Harmon 2001). Permission of Oxford University Press

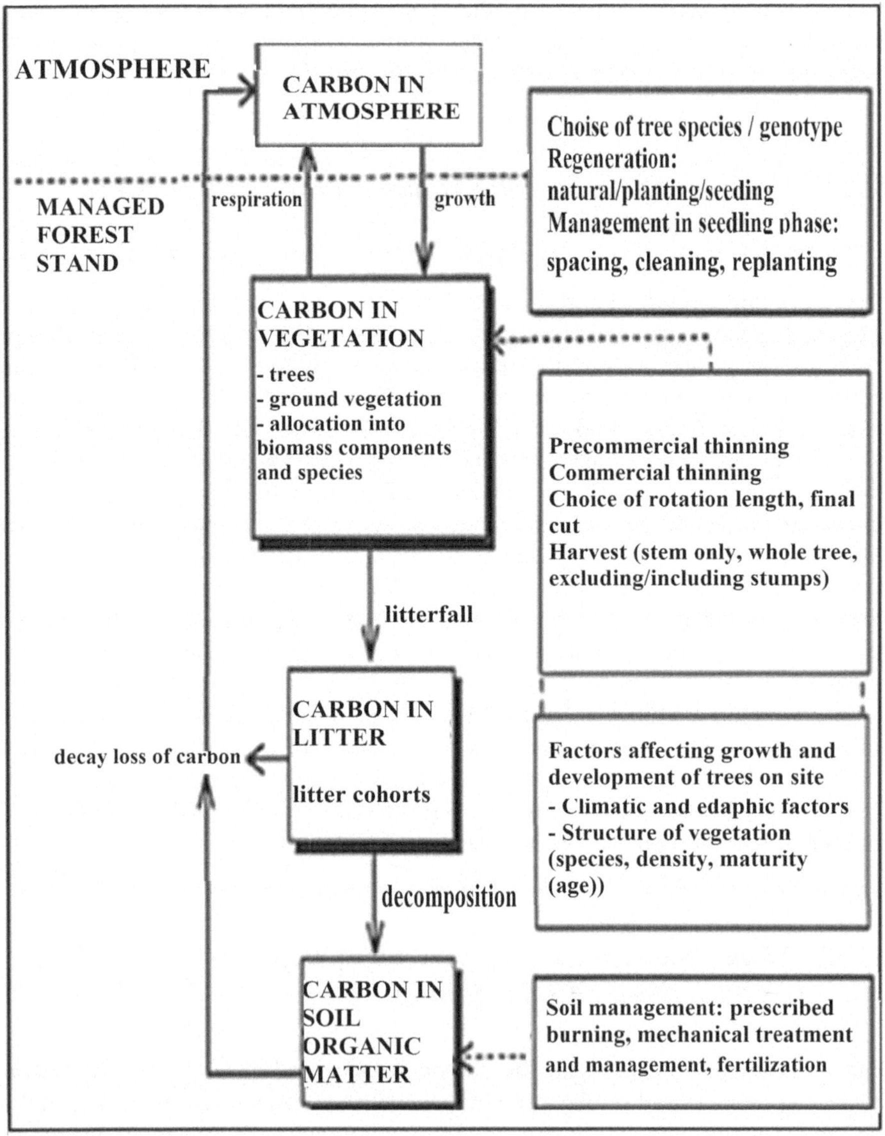


**Fig. 6.5** Major management measures used in even-aged management, with links to the structure and functioning of boreal forest ecosystems (Karjalainen 1996)

productive tree species or fast-growing species, or their most productive genotype seeds/seedlings (provenances), are used in reforestation. This will yield highest carbon accumulation rates in the early years after clear-cutting. Carbon losses can also be reduced if excess soil preparation and damage to residual trees are avoided in even-aged management. Carbon losses increase more under whole-tree harvest, while stem-only harvest leaves the carbon-containing logging residues in the site, thus increasing the carbon stock in soil detritus.

Management may combine one or several management measures. They are directed toward soil and trees, representing allogenic and autogenic measures, respectively, in relation to the selected management strategy (Karjalainen 1996) (Fig. 6.5). More specifically, allogenic measures involve the soil and ground vegetation, while autogenic measures involve trees in single-species populations or in communities of several species. Allogenic measures are mainly driven by external forces, with long-term effects on site fertility depending on the physical and chemical properties of soil, and also on the dynamics of carbon stocks in the decay of soil organic matter. Autogenic measures include harvest and regeneration (clear-cut with planting or shelterwood/seed tree cutting with natural regeneration), as well as the management of trees in seedling and thinning phases through the early production cycle. Furthermore, autogenic measures have key effects on the carbon uptake and stocks in vegetation (trees, ground vegetation) and removal of carbon outside the forest ecosystem in harvest.

## 6.3 Management Strategies and Operations Affecting Forest Ecosystem Dynamics

In boreal forestry, management can be divided into two mainstream strategies: even-aged and uneven-aged management, comprising different management operations. The distinction between even-aged (clear-cutting and seeding/planting, retaining seed trees, and using shelterwood/seed tree methods of natural regeneration) and uneven-aged management (group or single-tree selection, coppicing) is based on the degree of canopy influence in different developmental phases as listed in Table 6.2. In boreal conditions, the even-aged management is based on the sexual regeneration of main coniferous and deciduous tree species, while vegetative regeneration of commercial tree species occur for few deciduous species like European aspen (*Populus tremula*) with effective regeneration through root buds. The regenerative differences effect on the management practices from reforestation through chronosequence to final cutting for utilizing these differences in management phases of production cycle.

In *even-aged management*, the production cycle may include (Fig. 6.5; Table 6.2) (i) the soil management prior to regeneration (natural regeneration, planting/seeding); (ii) precommercial management with management of seedling stands (replanting, spacing, cleaning); (iii) pre-commercial and commercial thinning; (iv) fertilization; (v) pruning; and (vi) terminal cutting for regeneration. The frequency and intensity of each operation are based on the environmental conditions (e.g., site fertility) and stand structure (tree species composition, maturity, stocking), rotation length, and harvest intensity (stem-wood-only harvest, whole-tree harvest). The production cycle (or rotation) integrates several operations in a sequence, which

**Table 6.2** Outline of selected management systems applicable in boreal and temperate forests, modified slightly from Coates and Burton (1997)

| Management system | Overall description | Microclimate in area as affected by present/ nearby forest canopy | Canopy opening, ha |
|---|---|---|---|
| Even-aged management, with sexual reproduction | | | |
| Clear-cutting, with seeding or planting | All trees removed from a partial or full patch, or several patches | Open conditions, with limited canopy effect of surrounding trees | > 0.3–0.5 |
| Seed-tree, natural regeneration | Small number of seed-producing trees (15–50 parent trees $ha^{-1}$) on cleared area for a short time | Open conditions, with limited canopy effect of surrounding trees | > 0.5–1.0 |
| Shelterwood, natural regeneration | Successive regenerative cuts, with a partial cover of parent trees through the regenerative phase | Protective cover during regenerative phase | |
| • Uniform | Even distribution of canopy openings, young trees are even-aged | Protective canopy declining from initiation to full removal of parent trees | < 0.01–0.1 |
| • Group | Canopy openings are scattered, young trees are even-aged | Gaps close to full cover in 20–40 years | < 0.01–0.1 initially, larger over time |
| • Irregular | Canopy openings are irregularly and gradually distributed, young trees uneven-aged | Gaps close in 20–40 years, but few trees may be retained | < 0.01–0.1 initially, larger over time |
| • Strip | Canopy openings represent well-defined strips, regeneration forms even-aged young tree stands | Some canopy shading during regeneration, later removed | < 2 times height of nearby trees |
| Uneven-aged management, with sexual reproduction | | | |
| Selection | Canopy cover retained over the area | Canopy influence dominated over the area | |
| • Single-tree, natural regeneration | Trees periodically removed individually across the whole area | Continuous forest cover | < 0.01 |
| • Group, natural regeneration | Trees periodically removed in small groups | Continuous forest cover, with openings | 0.01–0.1 |
| Even-aged/uneven-aged management, with vegetative reproduction | | | |
| Coppicing, regeneration through stump and root buds | Trees generate from root- and stump-buds immediately or gradually after cutting | Open conditions exceed canopy effects | |

follows the growth and development of tree populations or communities from the seedling phase to full maturity.

In even-aged management, soil management is tailored to the given tree-species for enhancing the success of both natural and artificial (seeding/planting) reforestation. In seeding/planting, the appropriate selection of genotype (tree species or provenance of tree species) for a given site is of priority importance in reforestation and pre-commercial management. The precommercial management may include the tending of seedling populations and communities (pre-commercial thinning and spacing, cleaning, removal of competing ground vegetation, replanting) and thinning (commercial) repeated in maturing stands. The production of timber and biomass may be increased by reducing the rotation length under intensive management and by increasing the harvest intensity, such as through whole-tree harvesting (the harvest of stem wood, branches, foliage, and even stumps and coarse roots) instead of stem-wood-only harvesting. The rotation length and harvest intensity have a strong influence on the sustainability and the probability of realizing non-timber objectives combined with timber-oriented management.

*Uneven-aged management* is the most applicable for tree species of climax type, such as Norway spruce, which is tolerant to shading (Fig. 6.5; Table 6.2). In uneven-aged management, the regeneration phase and growth phase are combined, with a focus on individual trees of different age occupying the site. The length of the production cycle is related to the length of the interval between cuttings. These interventions involve selective cutting of individual trees (single-tree selection) and/or of small groups of trees (group selection) removed in patches. In selective cutting, larger (dominant, codominant) trees are harvested but partly left, as in the case of smaller trees (from seedling to selection). No terminal cut is performed because the selective cutting and regeneration are combined. In uneven-aged management, the natural regeneration and ingrowth of seedlings occur in canopy gaps between the trees remaining after the previous selection. Ingrowth refers to a process in which seedlings/small trees grow from a given canopy stratum into the next stratum, and so forth (Lundqvist 2017). In uneven-aged management, the age and size structure distribution of trees is retained, with large numbers of suppressed seedlings/trees dominated by a few large (dominating, codominant) trees, with a gradually decreasing number of trees over the diameter distribution (e.g., Saksa and Valkonen 2011; Eerikäinen et al. 2014; Ekholm et al. 2023).

## 6.4 Management for Timber and Carbon: Model-Based Experiment

### *6.4.1 Overview of Model Simulations*

In northern European boreal forestry, even-aged management and harvesting is widely used. To date, experimental studies on the success of uneven-aged

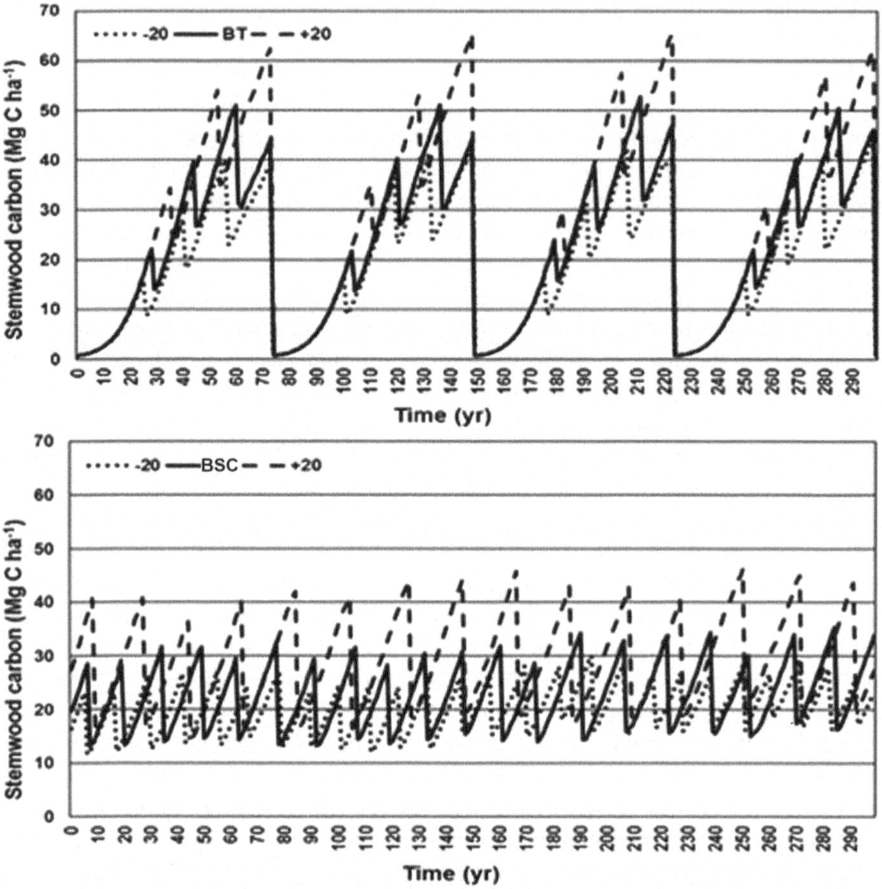


**Fig. 6.6** Carbon in stem wood under even-aged and uneven-aged management as a function of harvest/selection intensity (Kellomäki et al. 2019). Permission of Oxford University Press. Numbers in the x-axis refers to the years in the simulation period from the years 225 to 525 used in the data analysis. Upper: Mean stem wood carbon under even-aged management, when using the baseline thinning intensity (BT) and reduced (BT+20) and increased (BT-20) thinning intensity over 300 years. Below: Mean stem wood carbon under uneven-aged management over 300 years, when using the baseline selection intensity (BSC) and reduced (BSC+20) and increased (BSC-20) selection intensity

management are still limited compared to model-based findings, in contrast to even-aged management. Despite being in an emergent state, uneven-aged management is gaining in popularity for environmental reasons. The majority of the practice, economics, and technology in implementing uneven-aged management is still mostly based on model simulations.

Kellomäki et al. (2019), for example, carried out simulation experiments to analyse the potential of uneven-aged management in producing timber and sequestrating and stocking carbon compared to even-aged management (Fig. 6.6). In these simulations, a gap-type ecosystem model was used for a boreal Norway spruce forest on a medium-fertile upland site (*Myrtillus* site type, MT) in the mid-boreal zone (62° N). Under even-aged management, planting of 1800 seedlings $ha^{-1}$ with butt diameters of 2.5 cm (unimproved seedlings, no breeding gain assumed) was used to initialize the simulations at the clear-cut area. In the baseline thinning (BT) (Äijälä et al. 2014, p. 147), thinning from below was done when the dominant height was 12–22 m and the basal area 24–28 $m^2$ $ha^{-1}$. After thinning, the remaining basal area was 15–20 $m^2$ $ha^{-1}$, depending on the dominant height. In two optional thinning modes were further used, with the basal area being ±20 per cent higher/lower (BT + 20 and BT − 20) than in the BT over a rotation. Clear-cutting was done at the end of the 75-year rotation. The simulations were extended over 525 years, using seven rotations (7 × 75 years). The three first 75-year rotations were excluded from the analysis in order to stabilize the effects of the initial stand conditions on the model output. Four last rotations of 75 years, between years 225 and 525 (i.e., 300 years), were included in the data analysis.

Under uneven-aged management, the simulations were initiated in the same way as for even-aged management (Fig. 6.6). Consequently, thinning from below was done up to the year 75 from the initiation, but the clear-cutting was not done, allowing natural seeding to occur. Thereafter, single-tree selective cutting was done, one after another, following the given interval, thus gradually switching the even-aged management to an uneven-aged system. Selective cuttings were done whenever the basal area of a stand was ≥20 $m^2$ $ha^{-1}$, by reducing the basal area to 11 $m^2$ $ha^{-1}$. In addition to the baseline selective cutting (BSC) Äijälä et al. 2014, p. 120), two optional selective cutting modes were used, representing ±20 per cent higher/lower basal area thresholds (BSC + 20 and BSC − 20) than those used in the BSC over a production cycle. The simulations were extended over 525 years. The simulation output between years 225 and 525 (i.e., 300 years) was included in the comparative analysis in order to stabilize the effects of the initial stand conditions on the model output.

### 6.4.2 Distribution of Stem Wood Stock in Relation to Management

Under even-aged management, thinning occurred 20–30 years after the initiation of each cycle (Fig. 6.6): earlier for increasing (BT-20) and later for decreasing (BT+20) thinning intensity (Kellomäki et al. 2019). Under uneven-aged management, the recommended basal area limit in initializing selective cutting was 30 $m^2$ $ha^{-1}$ in the BSC. In this case, the basal area after cutting was 11 $m^2$ $ha^{-1}$. In selection with increased intensity (BSC-20), the upper basal area limit was 25 $m^2$ $ha^{-1}$ and the lower limit again 11 $m^2$ $ha^{-1}$. When reduced removal (BSC+20) was performed, the upper limit was 40 $m^2$ $ha^{-1}$ or even more, while the basal area after cutting was 20 $m^2$ $ha^{-1}$. In uneven-aged management, the BSC yielded a cutting interval of

15 years, while the 20% increase/decrease of section intensity gave a mean cutting interval of 10 and 20 years, respectively.

### 6.4.3 Distribution of Stem Diameter of Trees in Relation to Management

Under even-aged management, the mean diameter distribution of trees over 300 years (4 × 75-year rotations) spanned from seedlings just planted to trees of 35–40 cm diameter just before the terminal cut (Fig. 6.7). Under even-aged management, the density of large trees (e.g., diameter ≥ 25 cm) was approximately 500–600 trees per hectare over 300-year period (Kellomäki et al. 2019). Under uneven-aged management, the mean density of large trees (diameter ≥ 25 cm) was lower, at up to 180–200 trees per hectare, while the density of small trees was substantially higher than under even-aged management. Under uneven-aged management, the reduced selective intensity (BSC+20) tended to increase the density of small seedlings (diameter < 5 cm) more compared to that under increased selective intensity (BSC-20). Under uneven-aged management (BSC), the seed crop had a clear effect on the number of seedlings germinating annually and also on the mean

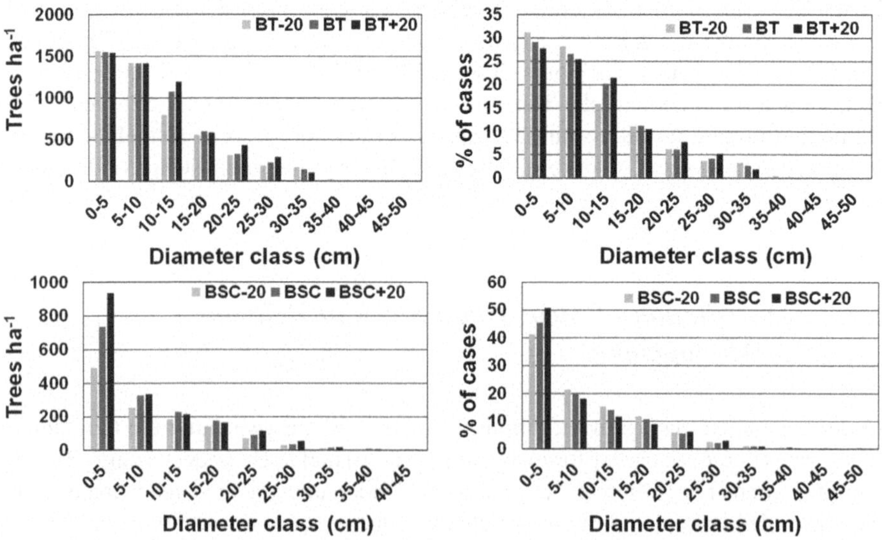


**Fig. 6.7** Absolute and relative density distribution of trees in relation to diameter using the mean values over 300 years (Kellomäki et al. 2019). Permission of Oxford University Press. Upper/Left/Right: Even-aged management using baseline thinning (BT) intensity and reduced (BT+20) and increased (BT-20) thinning intensity over a rotation of 4 × 75 years. Lower/Left/Right: Uneven-aged management for a 15-year selection interval over the period of 1 × 300 years, with the baseline selection (BSC) intensity and reduced (BSC+20) and increased (BSC-20) selection intensity

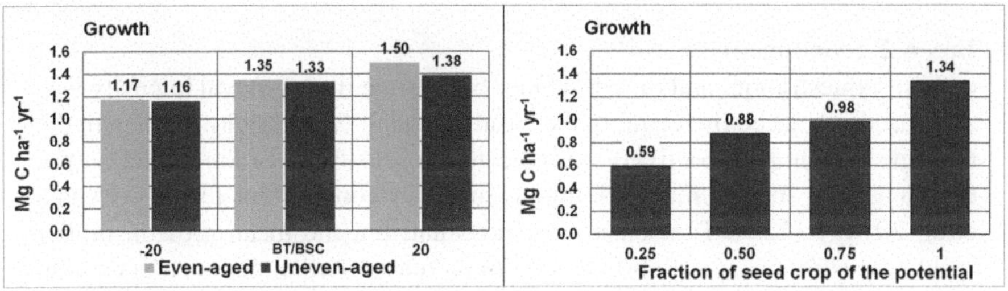


**Fig. 6.8** Growth of stem wood in relation to management regime, and the effect of seed crop on growth under uneven-aged management (Kellomäki et al. 2019). Permission of Oxford University Press. Left: Mean annual growth of stem wood carbon (including natural mortality) under even-aged and uneven-aged management as a function of baseline thinning/selection (BT, BSC) with reduced (BT+20 and BSC+20) and increased (BT-20 and BSC-20) removal intensity. Right: Mean annual growth of stem wood carbon (including natural mortality) under uneven-aged management with baseline selection (BSC) as a function of the fraction of full potential of the seed crop

diameter distribution of trees over 300 years. For example, the reduction of yearly seed crop to 25–75% of the potential reduced the number of small seedlings (diameter < 5 cm) to 20–50% of that under no reduction in seed crop.

### 6.4.4 Growth of Stem Wood in Relation to Management

Regardless of management regime, the mean annual growth of stem wood carbon was nearly the same when the baseline removal (BT and BSC) was performed (Fig. 6.8; Box 6.2). When the target basal area was reduced (BT-20 and BSC-20), the mean annual growth declined about 15% in relation to the baseline removal under both management regimes (Kellomäki et al. 2019). In response to an increased target basal area (BT+20 and BSC+20), the mean annual growth increased about 10% under even-aged and about 5% under uneven-aged management compared to the baseline. The growth remained about 9% lower for uneven-aged than even-aged management. Under uneven-aged management, the annual mean seed crop had a clear effect on the growth of stem wood carbon. For example, the reduction of yearly seed crop to 25–75% of the potential reduced the mean annual growth to 44–73% of that when seed crop was not reduced.

**Box 6.2 Sensitivity of timber production and carbon sequestration to management strategy**

Under even-aged and uneven-aged management, the recommended thinning intensities (Äijälä et al. 2014, p. 120) were used as the reference management (baseline thinning) to indicate the impacts of management regimes on growth,

(continued)

**Box 6.2** (continued)
carbon sequestration, and timber. Under both strategies, removal intensity was decreased/increased by retaining the basal area at 20% higher/lower than in the baseline thinning, respectively, to indicate the sensitivity of the model output to changes in management commonly used in commercial forestry (Äijälä et al. 2014, p. 120). The sensitivity was small but the mean growth, timber yield, and carbon sequestration tended to increase more under even-aged than under uneven-aged management when removal intensity was reduced. However, management strategy had little to no effect on the mean stem wood growth (1.17–1.50 Mg C $ha^{-1}$ $yr.^{-1}$). This is equivalent to 4.5–6.0 $m^3$ $ha^{-1}$ $yr.^{-1}$, around the same magnitude as that found by Laiho et al. (2011) for uneven-aged Norway spruce in the southern boreal zone. Similarly, Lundmark et al. (2016) found even-aged and uneven-aged management to have only minor differences in climatic benefits when the same growth, harvest, and use of timber for wood-based products were used in the simulations.

### *6.4.5 Carbon Uptake and Stocks*

Carbon sequestration refers to the carbon uptake of whole trees in the net primary growth (Fig. 6.9). Litterfall transfers carbon into the soil organic matter representing previous years' litter plus humus, which are decayed, emitting carbon to the atmosphere (Kellomäki et al. 2019). In the simulation of this cycle, the mean transfer of carbon in each phase tended to be slightly larger under even-aged than under uneven-aged management. Similarly, the carbon stock in trees, soil, and the whole ecosystem tended to be slightly larger under even-aged than uneven-aged management, excluding the carbon stock in trees when maintaining either higher or lower stocking in management (3% higher).

### *6.4.6 Carbon Exchange and Retention*

The mean annual net ecosystem exchange (*NEE*) indicates the balance between the carbon net uptake bound in the growth of whole trees (sequestration) and the carbon emission in the decay of litter and humus in soil. In the simulation, *NEE* tended to be larger under uneven-aged than under even-aged management, implying that carbon uptake was relatively higher than carbon emission in the former than in the latter management regime (Fig. 6.10). Under uneven-aged management, the value of *NEE* was 4–6% higher under BSC and the increased selection intensity (BSC-20) than under even-aged management and 2% lower under reduced selection intensity (BSC +20) than under even-aged management (Kellomäki et al. 2019).

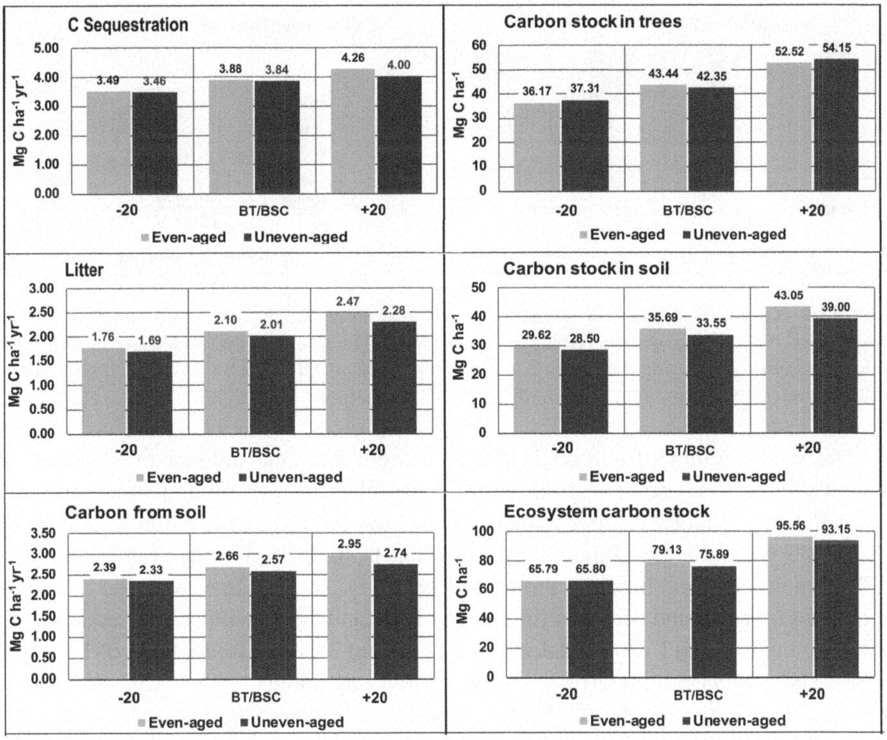


**Fig. 6.9** Mean carbon fluxes and stocks in forest ecosystem under even-aged (BT, BT+20, BT−20, with rotation lengths of 75 years) and uneven-aged (BSC, BSC+20, BSC−20) management with full seed crop potential (seed crop fraction = 1), under baseline management and with higher and lower basal area thresholds for management interventions. Permission of Oxford University Press. Left/Upper: Mean annual carbon uptake in trees. Left/Middle: Mean annual carbon in litterfall originating from different tree organs. Left/Below: Mean annual carbon emission from soil (litter and humus). Right/Upper: Mean carbon stock in trees (including stems, branches, foliage, and roots). Right/Middle: Mean carbon stock in soil (litter and humus). Right/Lower: Mean ecosystem carbon in trees and soil

In both cases, the retention of lower stocking under even-aged and uneven-aged management reduced the values of *NEE*, in contrast to the retention of higher stocking (Fig. 6.10). However, the carbon in trees and soil (the whole ecosystem) was retained over 28–36 years depending on management regime. The mean carbon retention time was a few percentage points (1–5%) longer under even-aged than uneven-aged management regardless of stocking. However, lower stocking or more intensive thinning tended to reduce the carbon retention, while higher stocking increased it, as was also the case in the net ecosystem exchange (*NEE*). Under uneven-aged management, a smaller seed crops substantially reduced *NEE* in relation to that under even-aged management. At the same, time, the carbon retention time increased substantially (5–16%) regardless of the size of the seed crop (Kellomäki et al. 2019).

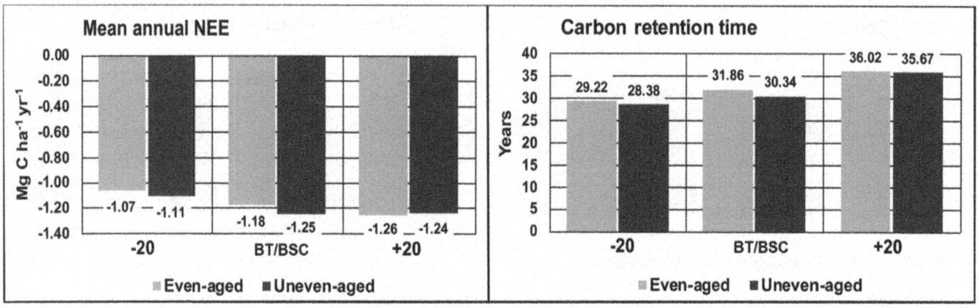


**Fig. 6.10** Carbon exchange and retention related to thinning under even-aged and to selection under uneven-aged management (Kellomäki et al. 2019). Thinning was performed from below, and selection from above, using the baseline management intensity (BT and BSC) and ±20% thinning/selection intensity. Permission of Oxford University Press

The total amount of carbon in the ecosystem was also related to the rotation length under even-aged management and to the selection interval under uneven-aged management. Under even-aged management, the increase of rotation length from 75 to 100 years increased the carbon retention from 33 to 37 years (an increase of 12%) when using the baseline thinning (Fig. 6.11). At the same time, the retention increased up to 36 and 42 years (increases of 9% and 14%) with a 20% reduction in thinning intensity (BT+20) and decreased to 30 and 33 years (reductions of 11% and 9%, respectively) with a 20% increase of thinning intensity (BT-20) (Kellomäki et al. 2019). Longer rotation time increased carbon retention, but this increase eventually declined; evidently, the carbon sequestration in maturing trees decreased in relation to carbon emission. The increase of rotation length from 75 to 100 years increased the carbon in the soil, trees, and ecosystem by 7%, 32%, and 21% under baseline thinning (BT) intensity, and 16%, 39%, and 29% under reduced thinning intensity (BT + 20), and further 6%, 37%, and 23% under increased thinning intensity (BT-20), respectively.

It is evident that stand-level $CO_2$ fluxes can be controlled with the intensity of thinning or selection, including the $CO_2$ soil efflux (e.g., Kumpu et al. 2018). Under uneven-aged management, the baseline selection intensity with a 15-year interval retained carbon in the ecosystem for 32 years (Fig. 6.12). At the same time, a 20% reduction in the remaining basal area (BSC-20) reduced the selection interval to 11 years, with carbon retained in the ecosystem for 28 years. Similarly, a 20% increase of remaining basal area (BSC+20) increased the selection interval to 23 years, with ecosystem carbon retention of 35 years. Further increases in the selection interval (or the remaining basal area) linearly increased carbon retention, but regeneration is likely to reduce at the same time, thus reducing the increase in carbon retention. The mean duration of carbon retention in the soil, trees, and ecosystem increased by 16%, 27%, and 22% under reduced selection intensity (BSC+20) and 15%, 12%, and 13% under increased selection intensity (BSC-20), respectively (Kellomäki et al. 2019).

Over different thinning/selection intensities, carbon was retained in the ecosystem for 3% longer under even-aged than uneven-aged management. The increase of

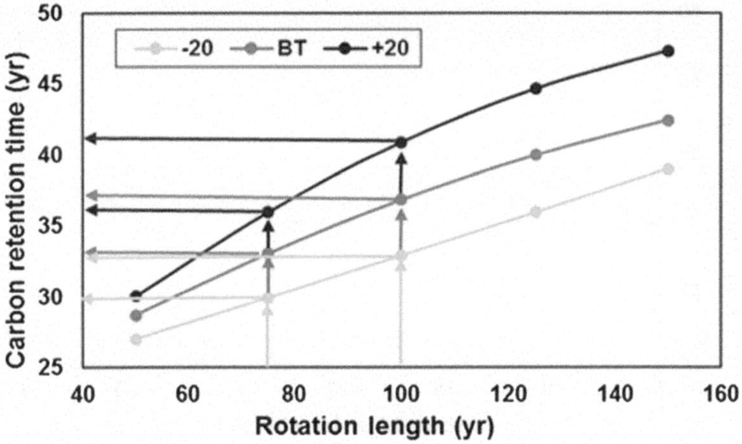


| Rotation and thinning intensity | Carbon retention time (yr) | Carbon retention, % change of BT |
|---|---|---|
| Rotation, 75 yr | | |
| • BT | 33 | - |
| • BT + 20% | 36 | +9 |
| • BT - 20% | 30 | -11 |
| Rotation, 100 yr | | |
| • BT | 37 | +12 |
| • BT + 20% | 42 | +14 |
| • BT - 20% | 33 | -9 |

**Fig. 6.11** Carbon retention time in a forest ecosystem under even-aged management (Kellomäki et al. 2019). Permission of Oxford University Press. Upper: Retention time as a function of rotation length and thinning intensity under Baseline thinning (BT) and increasing/decreasing thinning intensity (±20% change compared to that under BT). Lower: Mean carbon retention as a function of rotation length and thinning intensity under Baseline thinning (BT) and increasing/decreasing thinning intensity (±20% change compared to that under BT)

rotation length to 100 years yielded 16% longer carbon retention than under the recommended uneven-aged management operations as summarized over the thinning/selection intensities. Furthermore, the carbon retention under a 100-year rotation of even-aged management was still 9% larger that under a 23-year selection interval with reduced selection intensity under uneven-aged management. This implied 6–16%, 37–39%, and 23–29% increases of carbon retention time in the soil, trees, and ecosystem, respectively, depending on the removal intensity.

### *6.4.7 Stem Wood Stocking, Harvest, and Timber*

The results of simulations of stem wood stocking, harvest quantity, and the distribution of harvest into saw logs and pulpwood are illustrated in Fig. 6.13. Regardless

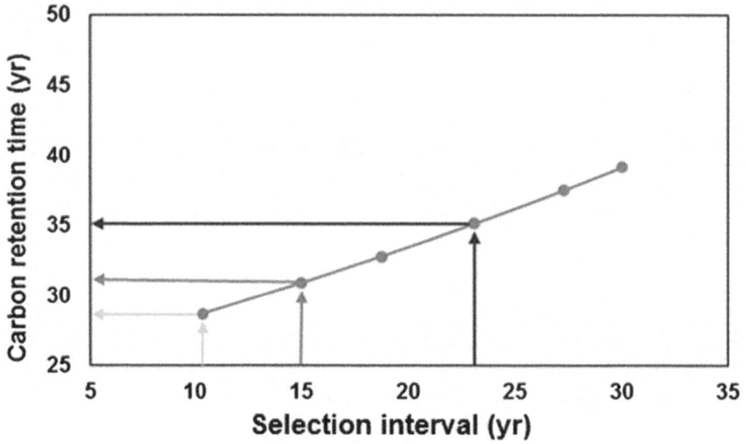


| Selection intensity | Selection interval (yr) | Carbon retention (yr) | Carbon retention, % change of BSC |
|---|---|---|---|
| • BSC | 15 | 32 | - |
| • BSC + 20% | 23 | 35 | +9 |
| • BSC - 20% | 11 | 28 | -9 |

**Fig. 6.12** Carbon retention time in a forest ecosystem under uneven-aged management. (Kellomäki et al. 2019). Permission of Oxford University Press. Upper: Retention time as a function of selection interval. Lower: Mean carbon retention as a function of selection interval and under Baseline selection (BSC) and increasing/decreasing selection intensity (±20% change compared to that under BSC)

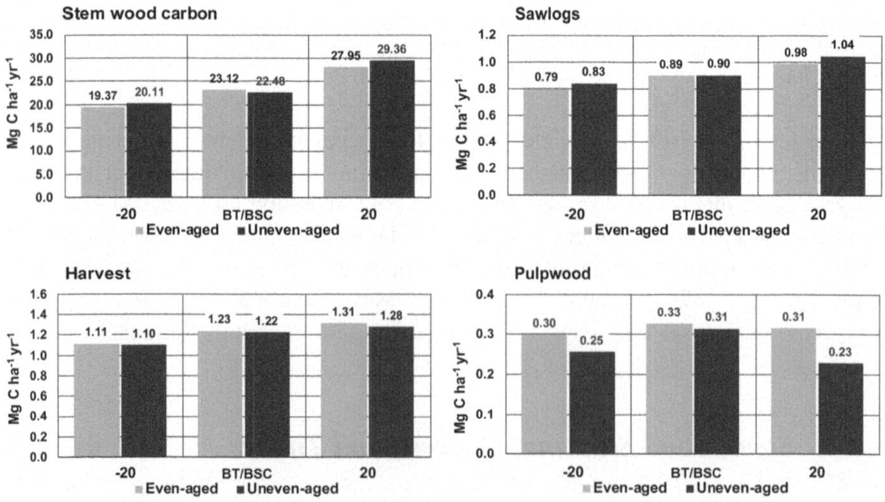


**Fig. 6.13** Carbon in stem wood stocking, harvested wood, and its timber assortments (data from Kellomäki et al. 2019). Permission of Oxford University Press

of management regime, reduced thinning/selection intensity (BT+20/BSC+20), increased the stem wood carbon stock up to 15%, while intensive thinning/selection (BT-20/BSC-20) reduced the carbon stock up to 13% compared to baseline thinning/selection. The amount of carbon in stem wood stock was slightly (up to 3%) larger under even-aged than uneven-aged management. Under uneven-aged management, both the reduction and increase of selection intensity slightly increased (up to 5%) stem wood carbon compared to even-aged management with thinning.

For the mean annual harvest, the differences between management regimes and thinning/selection intensities were small (Fig. 6.13). This held especially for the yield of saw logs, while the mean pulp wood yield tended to be higher (up to 30%) under even-aged management. This was especially the case under reduced removal intensity. Under uneven-aged management, the size of the seed crop had nearly no effect on stem wood carbon, whereas a reduction of seed crops substantially reduced the harvest. For example, the harvest decreased to 46% when the mean annual seed crop was 25% of the potential annual seed crop. The reduction of harvest further implied decreasing yields of saw logs and pulpwood. This reduction was more sensitive for pulpwood than for saw logs, but even in the latter case, the impact of reducing the seed crop was clear (Kellomäki et al. 2019).

Carbon in timber yields was further converted into wood in cubic meters and divided into saw logs and pulpwood to calculate the net present value (NPV) for timber production (Table 6.3). The differences between the management regimes and thinning/selection intensities were small. However, the amount of saw logs was slightly larger and of pulpwood slightly smaller under uneven-aged management. Consequently, the NPV under even-aged management was smaller than that of uneven-aged management when the production cycle under even-aged management was initiated by planting on bare land. In this case, NPV was 1640 € $ha^{-1}$ under baseline thinning under even-aged management, versus 7392 € $ha^{-1}$ under uneven-aged management (Kellomäki et al. 2019).

**Table 6.3** Total and annual timber yield over 300 years in cubic meters for saw logs and pulpwood under both even-aged and uneven-aged management using baseline thinning (BT) and selection (BSC) intensity, respectively, as well as reduced (BT+20 and BSC+20) and increased (BT-20 and BSC-20) thinning/selection intensity

| | Even-aged | | | Uneven-aged | | |
|---|---|---|---|---|---|---|
| Management | Saw logs | Pulpwood | Total | Saw logs | Pulpwood | Total |
| Total over 300 years, $m^3$ $ha^{-1}$ | | | | | | |
| BT-20 | 1190 | 450 | 1640 | 1252 | 382 | 1634 |
| BT | 1338 | 489 | 1827 | 1345 | 469 | 1815 |
| BT+20 | 1475 | 467 | 1942 | 1562 | 341 | 1902 |
| Mean, $m^3$ $ha^{-1}$ $yr.^{-1}$ | | | | | | |
| B-20 | 3.97 | 1.50 | 5.47 | 4.17 | 1.27 | 5.45 |
| BT | 4.46 | 1.63 | 6.09 | 4.48 | 1.57 | 6.05 |
| BT+20 | 4.92 | 1.56 | 6.47 | 5.21 | 1.14 | 6.34 |

Under even-aged management, the total timber yield is for a 4 × 75-year period and under uneven-aged management, one 300-year period (Kellomäki et al. 2019)

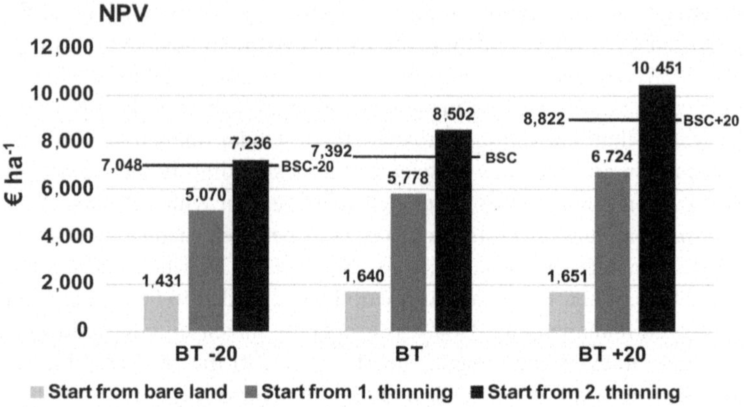


**Fig. 6.14** Net present value (NPV, 3% interest rate) for total timber yield under even-aged and uneven-aged management (Kellomäki et al. 2019). Permission of Oxford University Press. Under even-aged management, baseline thinning (BT) and decreased (BT+20) and increased (BT-20) thinning intensities were used over the 75-year rotation. For initialization of the calculations for even-aged management, three optional initial cases were used: initialization of a seedling stand on bare land (scenario 1), initialization of a young stand in the first thinning phase (scenario 2), or initialization of a mature stand in the thinning phase (scenario 3). Under uneven-aged management (horizontal lines), baseline selection (BSC) and decreased (BSC+20) and increased (BSC-20) basal area thresholds for selection were used, with full seed crop potential (seed crop fraction: 1.0). Under both management systems, the simulation period was 300 years. Under even-aged management, NPV calculations were initiated: (i) for a clear-cut area on which seedlings were planted in the first year (scenario planting on base soil); (ii) at the first thinning phase (scenario 1); and (iii) at the second thinning phase (scenario 3)

When the NPV calculation was initiated under even-aged management in previously established stands in the first thinning phase, the difference between even-aged and uneven-aged management decreased substantially (Table 6.3; Fig. 6.14). Under BT, NVP was 5778 € $ha^{-1}$, which was 28% lower than that for uneven-aged management (7392 € $ha^{-1}$). When initializing the thinning in the mature thinning stand, NPV for even-aged management tended to be higher than that for uneven-aged management. In this case, NPV for BT under even-aged management was 8502 € $ha^{-1}$, which was 15% higher than that for uneven-aged management. Under BT + 20, NPV for even-aged management was 10,451 € $ha^{-1}$, which was 18% higher than that for uneven-aged management (8822 € $ha^{-1}$). Under even-aged management with using BT-20 thinning option, the NPV was quite the same as that under uneven-aged management (BSC) with 7236 vs. 7048 € $ha^{-1}$. In general, the lower basal area reduced NPV (13–14%), while increasing basal area increased NPV up to 16%, under even-aged management. The pattern was different under uneven-aged management, with a NPV reduction of 5% with the reduced basal area and an increase of 19% with the increased basal area (Kellomäki et al. 2019).

## 6.5 6.5 Optimized Management for Production of Multiple Services

### 6.5.1 *Combining Ecosystem Services with Synergistic and Trade-Off Relations*

There may be trade-offs between different ecosystem services when combining the production of different services in the same forest landscape. Trade-offs refer to tactical or strategic choices made based on a full understanding the advantages and disadvantages of forest management and harvest decisions for the forest in each landscape. However, a particular management decision may increase other services synergistically. For example, synergistic relations exist between the growth of trees, which controls erosion and sustains soil fertility, thus driving further tree growth. The interaction may also be an inverse one, like that between the biomass stock of trees and of berry crops in the ground vegetation, which tends to reduce with increasing shading below the tree canopy.

Trade-offs between ecosystem services may take place at different spatial and temporal scales or in the short or long term over a given forest landscape. In this context, Triviño et al. (2015, 2017) examined the trade-offs between provisioning (revenues from timber selling) and regulating (carbon storage and sequestration) ecosystem services under several management options in a large mid-boreal forest landscape (62° N) in Central Finland. These studies aimed to highlight management regimes that simultaneously improve revenues from timber selling and carbon sequestration in ecosystem carbon stocks, with impacts on biodiversity. Triviño et al. (2015, 2017) used multi-objective optimization to identify how to combine varying management regimes in an optimal way for the production of timber, carbon storage, and forest biodiversity (deadwood, habitat availability) in a set of management plans feasible for a forest landscape. The acceptable management plan is a combination of several management regimes applicable to stands comprising the forest landscape. If there is any conflict between the management objectives, their maximal values are not obtainable at the same time for all the objectives.

Triviño et al. (2015) compiled management plans for a forest area through management regimes determined that maximized harvest revenue (NPV) from timber production and the carbon services by solving the following bi-objective optimization problem:

$$\text{maximize } (f_1(x), f_2(x)) \qquad (6.2)$$
$$\text{subject to } x \in X$$

where $X$ is the set of optional management regimes, $f_1(x)$ is the net present value (NPV), and $f_2(x)$ is the objective function for the carbon services over a 50-year planning period for applying the management plan $x \in X$. In multi-objective optimization, both objective functions are maximized simultaneously to define the

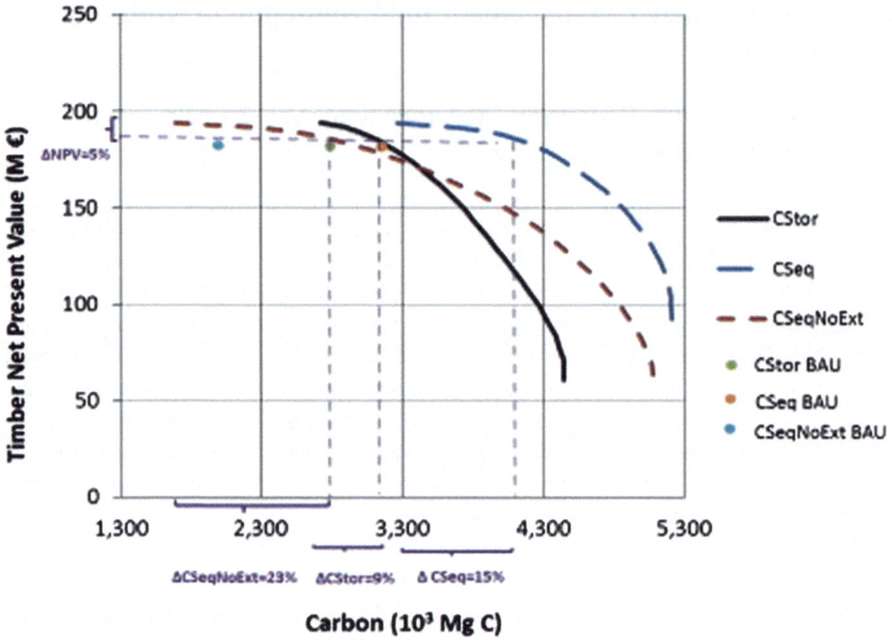


**Fig. 6.15** Curves representing Pareto-optimal plans describe the trade-offs between (i) carbon services, i.e., carbon storage and carbon sequestration; Mg of carbon, averaged across every 5 years); and (ii) harvest revenues (net present value, NPV) for the three different carbon services (Triviño et al. 2015). Permission of Elsevier. The reference line for a change of 5% in the NPV is included (dashed and light violet-colored line) and the changes for each carbon service, where the dashed line intercepts with each of them. The three points show the positions for the exclusive application of the current management regime (business as usual, BAU). Legend in Figure: $C_{Stor}$ = carbon storage; $C_{Seq}$ = carbon sequestration; $C_{StorNoExt}$ = carbon sequestrated, non-extracted; $C_{Stor}$ Bau = carbon storage, recommended management; $C_{Seq}$ Bau = recommended management; $C_{StorNoExt}$ Bau = carbon sequestrated but non-extracted, recommended management

management plans that are Pareto-optimal, including different trade-offs among the objectives.

Figure 6.15 shows the set of Pareto-optimal plans, trade-offs between the harvest revenues and the carbon services (carbon storage (Cstor)), carbon sequestration (Cseq), and carbon sequestered but non-extracted (CseqNotExt). The relationship was nonlinear when different management plans were connected. The slopes of the curves were flat for low values of carbon, implying the first increments of carbon storage or sequestration being inexpensive. According to Triviño et al. (2015), maximizing the total carbon sequestered, and carbon sequestered but not extracted yielded large NPV losses. The loss was even more expensive for maximizing the total amount of carbon storage, as the curve was steeper.

### 6.5.2 *Timber, Carbon Sequestration, and Biodiversity at the Forest Landscape Scale*

In Triviño et al. (2015, 2017), the stand-based management included seven options, ranging from the currently recommended management (business as usual) to the setting-aside of stands (Fig. 6.16). At the landscape-scale, multi-objective optimization was used to analyze the trade-offs between timber production and carbon sequestration and carbon stocks. In this respect, the analysis allowed for identification of (i) when current management regimes are inefficient for providing timber or carbon storage; and (ii) where carbon storage may increase with no or only negligible reductions in the production of timber, or vice versa. In this context, Triviño et al. (2015, 2017) addressed the following specific questions: (i) "What is the potential of the forest landscape and the optimal combinations of management regimes to simultaneously produce economic revenues and regulate climate? (ii) Is

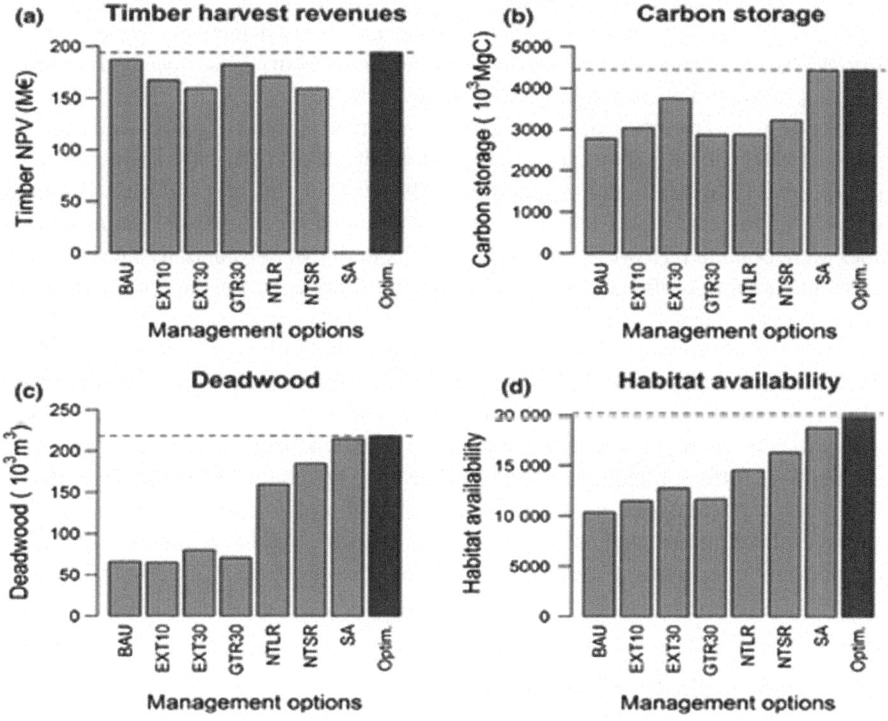


**Fig. 6.16** Timber, carbon stock, dead wood, and habitat availability over the study area for different management regimes, the intensity of which reduces from left to right (Triviño et al. 2017). Free Access. British Ecological Society. Legends for management options: Business as usual = BAU; Set aside = SA; Extended rotation by ten years = EXT10; Extended rotation by thirty years = EXT30; Green tree retention = GTR30; No thinning with long rotation = NTRL

there a difference between the two carbon measures regarding to their trade-offs with timber production?"

In optimizing, each pair of objectives (biobjective optimization) or all three objectives (triobjective optimization) were used. Biobjective optimization indicated the severity of trade-offs between pairs of objectives, and triobjective optimization how the pairwise trade-offs change while a third objective is also considered. The trade-offs between carbon storage and biodiversity changed when different levels of timber harvest were reached, ranging from maximal timber harvest through moderate losses to no pre-set constraints for timber. A single compromise (joint-production) solution was a management plan that provided a given amount of timber, with the smallest losses in carbon storage and biodiversity related to their maximum values.

Over the whole forest region, the mean maximum return (timber harvest in NPV) was 5800 € $ha^{-1}$, while the mean maximum carbon stock was 103 Mg C $ha^{-1}$. The mean maximum amount of dead wood was 5.1 $m^3$ $ha^{-1}$ and of habitat availability, 0.47 $ha^{-1}$. Timber harvest tended to decrease slightly when management intensity decreased. This contrasted with carbon stock, which was largest under no management or harvest. Similarly, the amount of dead wood and habitat availability increased substantially when the management and harvest intensity were reduced (Fig. 6.16). The trade-off between timber and biodiversity was clear and stronger than that between the carbon stock and biodiversity (Triviño et al. 2015, 2017).

According to Triviño et al. (2015, 2017), any given amount of timber had a strong trade-off effect with carbon stock and biodiversity (Fig. 6.16). When maximizing timber (99.9% of the maximal potential), 39–46% of the maximum dead wood, 61–64% of the maximum habitat availability, and 65% of the maximum carbon stock were obtained. However, small reductions in timber substantially increased the multifunctionality of the landscape. The impact of timber harvesting on biodiversity was especially strong: a reduction of 1–5% in the timber harvest increased the maximum dead wood by 47–90% and habitat availability by 65–88%. The effect on carbon stock was small, being 66–67% of that under maximum timber production. No management regime could maximize timber yield, carbon stock, or biodiversity in combination because of the conflicts between different services. The trade-offs between different management objectives may be reduced when management is diversified over the forest landscape. The elongation of rotation and the increasing setting aside of unmanaged patches likely improves the parallel production of timber and non-timber services in commercial forests (Triviño et al. 2015, 2017).

## 6.6 6.6 Concluding Remarks

Forest ecosystems include biotic (living organisms, including trees and other organisms) and abiotic (site, including climate and soil) components, which form the ecosystem structure. The biotic and abiotic components interact through biological, geochemical, and physical processes (ecosystem functioning) that occur between them. The organisms in a given area interact with the physical environment, leading to a flow of energy and carbon depending on trophic structure, biotic diversity, and material cycles. These components provide environmental, economic, and social benefits or ecosystem services linked to ecosystem structure and functioning. In this regard, management involves a set of operations that may be used, alone or in combination, to reroute the succession dynamics to produce in sustainable ways different goods and services. In forestry, the sustainability considers several production objectives to yield simultaneously selected services aimed at by management. There may be a strong synergetic effect between different services, such as between timber supply and carbon sequestration. However, the strong synergy between timber and carbon sequestration tends to imply a strong trade-off in relation to biological diversity for the managed boreal forests. Trade-offs imply the consideration for differentiation of ecosystem services at temporal and spatial scales for sustainable land use and management.

## References

Äijälä O, Koistinen A, Sved J, Vanhatalo K, Väisänen P (eds) (2014) Hyvän metsänhoidon suositukset – METSÄNHOITO. Tapio Oy, Helsinki

Bormann FH, Likens GE (1979) Pattern and process in a forested ecosystem. Springer, New York

Coates KD, Burton PJ (1997) A gap-based approach for development of silvicultural systems to address ecosystem management objectives. For Ecol Manag 99:337–354

Covington WW (1983) Changes in the forest floor organic matter and nutrient content following clear cutting in northern hardwoods. Ecol 62:41–8

Eerikäinen K, Valkonen S, Saksa T (2014) Ingrowth, survival, and height growth of small trees in uneven-aged *Picea abies* stands in southern Finland. Forest Ecosyst 1:5

Ekholm A, Lundqvist L, Axelsson EP, Egnell G, Hjältén J, Lundmark T, Sjögren J (2023) Long-term yield and biodiversity in stands managed with the selection system and the rotation forestry system: a quality review. For Ecol Manag 537:120920

Harmon ME (2001) Carbon sequestration in forests: addressing the scale question. J For 99(4): 24–29

IPCC (2007) Climate Change 2007: impacts, adaptation, and vulnerability. In: Working Group II contribution to the Fourth Assessment Report of the Intergovernmental Panel on Climate Change. Cambridge University Press, Cambridge

Jandl R, Vesterdal L, Olsson M, Bens M, Badeck F, Rock J (2007) Carbon sequestration and forest management. CABI Reviews 2007 2(17)

Karjalainen T (1996) Dynamics of carbon flow through forest ecosystem and the potential of carbon sequestration in forests and wood products in Finland. University of Joensuu, Faculty of Forestry Research Notes 40:1–30

Kellomäki S, Strandman H, Peltola H (2019) Effects of even-aged and uneven-aged management on carbon dynamics and timber yield in boreal Norway spruce stands: a forest ecosystem model approach. Forestry 92:635–647

Kilpeläinen A, Alam A, Strandman H, Kellomäki S (2011) Life cycle assessment tool for estimating net $CO_2$ exchange of forest production. GCB Bioenergy 3(6):461–471

Kumpu A, Mäkelä A, Pumpanen J, Saarinen J, Berninger F (2018) Soil $CO_2$ efflux in even-aged and uneven-aged Norway spruce stands in southern Finland. iForest Biogeosci Forestry 11(6): 705–712

Laiho O, Lähde E, Pukkala T (2011) Uneven- vs. even-aged management in Finnish boreal forests. Forestry 84(5):547–556
Lundmark T, Bergh J, Nordin A, Fahlvik N, Poudel BC (2016) Comparison of Carbon balances between continuous-cover and clear-cut forestry in Sweden. Ambio 45:203–213
Lundqvist L (2017) Tamm review: selection system reduces long-term volume growth in Fennoscandic uneven-aged Norway spruce forests. For Ecol Manag 391:362–375
Noormets A, Epron D, Domec JC, McNulty SG, Fox T, Sun T, King JS (2015) Effects of forest management on productivity and carbon sequestration: a review and hypothesis. For Ecol Manag 355:124–140
Saksa T, Valkonen S (2011) Dynamics of seedling establishment and survival in uneven-aged boreal forests. For Ecol Manag 261:1409–1414
Triviño M, Juutinen A, Mazziotta A, Miettinen K, Podkopaev D, Reunanen P, Mönkkönen M (2015) Managing a boreal forest landscape for providing timber, storing, and sequestering carbon. Ecosyst Serv 14:179–189
Triviño M, Pohjanmies T, Mazziotta A, Juutinen A, Podkopaev D, Le Tortorec E, Mönkkönen M (2017) Optimizing management to enhance multifunctionality in a boreal forest landscape. J Appl Ecol 54(1):61–70
Yanai RD, Currie WS, Goodale CL (2003) Soil carbon dynamics after forest harvest: an ecosystem paradigm reconsidered. Ecosyst 6:197–212. https://doi.org/10.1007/s10021-002-0206-5

# Chapter 7
# Soil Management Impacts on Soil Carbon

## Contents




**Abstract** Soil represents long-term interactions between the ground (mineral material and organic soil matter), climate, trees, ground vegetation, soil animals, and microbial communities. These interactions modify the structure and functioning of soil, including the physical and chemical properties related to its stocks of water, nutrients, and organic matter with carbon. In natural succession, wildfires and the uprooting of trees disturb the surface and even deeper layers of soil through the rooting zone. Disturbance of forest soils maintains their long-term fertility based on changes in primary and secondary production over space and time. The response of soil carbon is also affected by the different impacts of forest management on the structure and dynamics of carbon stocks in the ecosystem. Soil management can be divided into prescribed burning and mechanical management. Both of these





S. Kellomäki, *Forest Management for Timber Production and Climate Change Mitigation*, Managing Forest Ecosystems 44,
https://doi.org/10.1007/978-3-031-71575-4_7

processes affect the soil carbon bound in ground vegetation, soil litter, soil detritus, and mineral soil affecting the release of nutrients and carbon bound in soil organic matter. The consequent mixture of soil layers with patches of open mineral soil provides seedbeds for germination of seeds and rooting of initial seedlings improving the success of natural regeneration. Soil management also makes manual seeding and planting in reforestation easier, thus reducing the required workforce and enhancing regeneration success.

## 7.1 Soil Management Links to Succession of Forest Ecosystems

### *7.1.1 Ecosystem Dynamics and Soil Properties*

In forest ecosystems, soil includes mineral soil (parent material, basic soil components) in layers through the soil profile, and further layers of litter and humus layer on mineral soil. Concurrently, vegetation alters soil physical and chemical properties, with impacts on soil biology and chemistry and on the growth and success of plants. Soil represents long-term interactions between mineral particles, climate, trees, ground vegetation, soil animals, and microbial communities. These interactions modify the structure and functioning of soils, including the stocks of water, nutrients, and organic matter with carbon. They affect nutrient fluxes and availability for uptake in physiological processes driving autotrophic and heterotrophic production in the ecosystem. The properties of soil carbon and the soil environment control ecosystem processes such as nutrient uptake, decomposition of soil organic matter (SOM) with cycling of nutrients, and availability of water. Boreal forest soil includes large carbon pools of new (de novo) and old (legacy) carbon, with a significant role in the global carbon cycle (e.g., DeLuca and Boisevenue 2012).

In natural succession, vegetation alters the chemical and physical properties of soil, with impacts on the regeneration, growth, development, and mortality of trees. Wildfires and the uprooting of trees in high wind disturb the surface and even deeper layers of soil through the rooting zone. In uprooting, the root-soil plate is turned upright, thus disturbing the soil profile representing the main location of the roots in soil (e.g., Ulanova 2000). Over time, roots decay and soil and expected root-soil plates or soil mounds represent fully/partly open mineral soil patches below canopy gaps in places of uprooted trees. They provide favourable microsites in which seedlings can establish. Lyford and MacLean (1966) emphasize that uprooting with soil disturbance enhances the regeneration of both coniferous and deciduous trees in temperate and boreal conditions, as occurs in soil management used in forestry for enhancing regeneration success.

### 7.1.2 *Response of Soil Carbon to Harvesting and Management of Harvest Residues*

In forestry, soil management impacts on soil carbon stock through successive production cycles. The response of soil carbon varies depending on soil type and different harvesting strategies and management operations. In this regard, James, and Harrison (2016) conducted a meta-analysis on the effect of timber harvesting on soil carbon, accounting for the inherent variability in boreal and temperate forests. This meta-analysis used a database of 945 soil carbon responses to harvesting collected from 112 publications around the world, representing temperate forests.

Harvesting and management and soil management tended to reduce soil carbon stocks (James and Harrison (2016) (Fig. 7.1):

- Harvesting reduced soil carbon by 11.2% on average, with a 95% confidence interval of [14.1%, 8.5%].
- There was substantial variation between responses at different soil depths, with the largest losses found in the litter layer on the forest floor O horizon (30.2%).
- Much smaller losses (3.3%) occurred in soil carbon pools 0–15 cm in depth. In deep soil (60–100 cm), carbon loss of 17.7% was found after harvesting.
- After harvesting, soil carbon recovered over several decades depending on soil type, even up to 75 years.

Management strategies and operations further affect the structure and dynamics of forest ecosystems with impacts on soil carbon (James and Harrison (2016) (Fig. 7.1):

- There was only a small difference in the loss of soil carbon between harvesting through thinning and clear-cutting.
- Loss of soil carbon under clear-cutting was 9.3% lower than that under thinning.
- Harvest intensity affected soil carbon: under whole-tree harvest, soil carbon may reduce soil carbon by 13.3% compared to that under bole-based harvest.
- Intensive site preparation with raised planting/seeding beds tended not to reduce soil carbon, especially in the mineral soil. This held for natural reforestation using seed-tree methods.
- Removal of logging residues resulted in no significant additional losses of soil carbon.

Figure 7.1 further shows that prescribed burning after the harvest reduced soil carbon up to 15.2% more compared to soil not burnt for reforestation (James and Harrison 2016):

- Effect was severe in the organic surface layer (O horizon) that contained litter, with 40.9% additional soil carbon loss compared to that in the unburnt sites.
- Additional 8.3% loss occurred in the mineral soil below the organic surface layer in burnt sites.

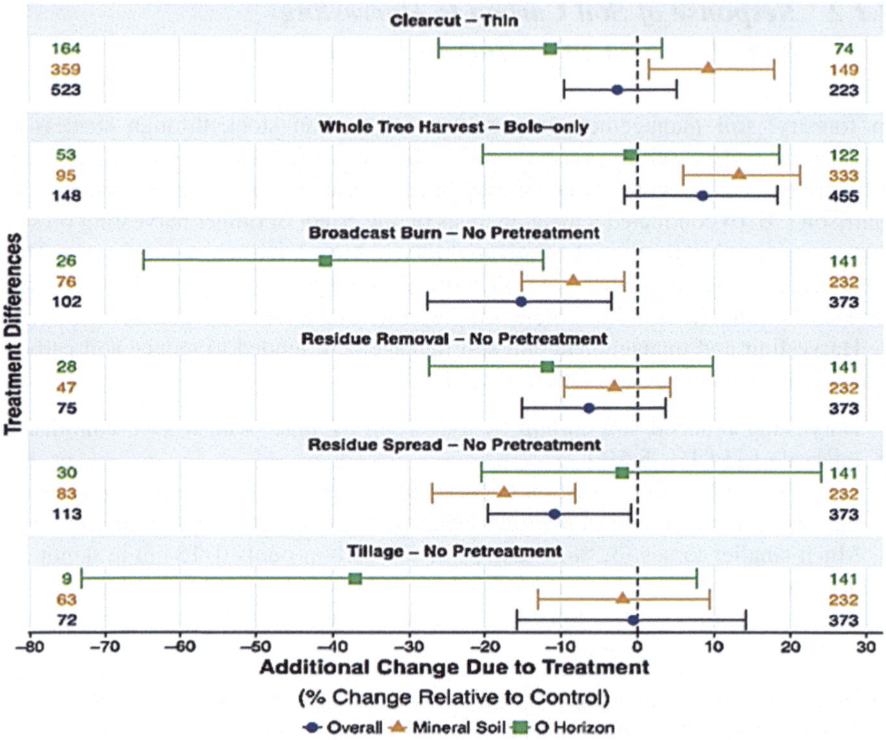


**Fig. 7.1** Differences in the response of soil carbon to harvesting and soil management between treatment strategies (James and Harrison 2016). Open Access. Courtesy of MDPI. Differences are calculated through subtraction: [more intensive treatment] minus [less intensive treatment], with positive differences representing reductions in carbon loss with increasing treatment intensity and vice versa. Mean effect differences with confidence intervals overlapping the dashed line (0%) show no significant difference in soil carbon responses between the harvesting, residual management, or pre-treatment strategies. Number of response ratios (*k*) for the intensive treatment in each comparison appears on the left and for the less intensive treatment on the right. The number of response ratios depends on the number of experimental conditions imposed on the number of samples taken by depth

- Spreading of logging residues such as branches across harvested sites resulted in an additional loss of soil carbon (10.9%), with the extra losses occurring mostly in the mineral soil (17.5%).

## 7.2 Soil Management Impacts on Soil Properties

In natural succession, disturbance of forest soils maintains their long-term fertility based on changes in primary and secondary production over space and time. This "natural" soil management can be replaced or enhanced with intentional soil

**Table 7.1** Overview of selected methods of managing soil for forest plantations

| Methods | Impact mechanisms | Impacts on soil | Impacts on vegetation | Preferable target site |
|---|---|---|---|---|
| Prescribed burning | High temperature | Release of nutrients, warmer soil, reduction of soil carbon | Aboveground vegetation is destroyed but the roots of vascular plants partly remain in the soil profile | *Vaccinium* site types (VT) and more fertile sites |
| Tilt ploughing | Mechanical opening of the soil surface and mixing of deeper soil layers | Release of nutrients, patchy warming of soil, clear reduction of soil carbon | Aboveground vegetation is destroyed but the roots of vascular plants remain in the soil profile between ploughing tracks | Water-logged sites on fine-grained moraine soils |
| Disk tilt ploughing | Mechanical opening of the soil surface and mixing of deeper soil layers | Release of nutrients, patchy warming of soil, reduction of soil carbon | Aboveground vegetation is destroyed but the roots of vascular plants partly remain in the soil profile | Sites on sand-rich moraine soils |
| Patching | Mechanical opening of the soil surface and mixing of deeper soil layers | Release of nutrients, patchy warming of soil, reduction of soil carbon in patches | Aboveground vegetation is destroyed but vegetation remains between patches | *Vaccinium* site type (VT) on sand-rich soils and less fertile sites |
| Mounding | Mechanical opening of the soil surface and mixing of deeper soil layers | Release of nutrients, patchy warming of soil, slight reduction of soil carbon likely | Aboveground vegetation is destroyed but vegetation remains between mounds | Water-logged sites on fine-grained moraine soils |

disturbance (Table 7.1). In intentional management, soil management can be divided into prescribed burning and mechanical management. These accelerate the regeneration and growth of trees based on several mechanisms, including (i) improving the soil temperature, moisture conditions, and nutrient supply; (ii) reducing the compet ing vegetation (ground vegetation, bushes, non-priority trees abandoned in the terminal cut); and (iii) reducing attacks by small mammals, insects (e.g., weevils), and fungi on seedlings. Both prescribed burning and mechanical soil management have similar effects. However, burning of logging residues, litter, and the soil surface reduces the potential for uncontrolled reproduction of insects and fungi that would damage seedlings. Both prescribed burning and mechanical soil management make manual and mechanical seeding and planting easier, thus reducing the required workforce, and enhancing regeneration success.

Mechanical site preparation encompasses a variety of soil cultivation practices with different intensities of disturbance of the soil surface and upper parts of the soil profile. These include scarification from disc-trenching (blade-scarification), pit-mounding, ploughing, and bedding (Table 7.1). Scarification involves removing the ground vegetation and the uppermost organic soil layer. Disc-trenching/harrowing creates continuous furrows in which the organic layer is removed and turned over to form ridges (Piirainen et al. 2015). Mounding creates deep furrows, either continuous (bedding, ploughing) or in the form of pits and elevated mounds with an inverted or mixed soil layer. Mechanical management operations minimizing soil disturbances tend to conserve and increase the accumulation of carbon in the soil profile (Powlson et al. 2014).

## 7.3 Prescribed Burning and Soil Carbon

### *7.3.1 Impacts of Wildfires on Boreal Forests*

Wildfires are common in boreal forests in the continental parts of the Northern Hemisphere. Wildfires refer to forest-disturbing fires with no management intention. In purely natural conditions, lightning is a cause of forest fires. However, most forest fires are triggered by the careless use of fire by people. Long- and short-term variability in precipitation, temperature, and air humidity affects the moisture of fuel such as litter, and the subsequent probability of forest fire occurrence. Climate and weather also affect the moisture of fuel, impacting on the subsequent fire risk and intensity of fire, spreading rate of fire, and the area being burnt. The spreading rate of fire is also affected by topography, wind velocity, and the availability of fuel (Amiro et al. 2001).

In Finland, the mean annual precipitation varies in a range of 300–700 mm in such a way that summertime precipitation is evenly distributed through the main growing season. In these conditions, the evaporation is low with a consequence that even low precipitation keeps litter and humus moist. The general fire risk is low under the current climate, but climatic warming with longer summers and a consequent increase in drought events is likely to increase fire risk in northern boreal conditions. For example, Zackrisson (1977) found that wildfires are typical in northern summers when the temperature in June and July is above the long-term mean. Furthermore, changes in the amount and seasonal distribution of precipitation may alter the risk of damaging wildfires. Early snow melt combined with only a minor increase of precipitation in the spring and early summer is likely to increase fire risk and damages even in humid boreal conditions.

Fire damage is any fire-induced damage to the ground (ground vegetation), soil organic matter (SOM), and trees, including the death of whole trees (Fig. 7.2), with improving natural reforestation in boreal forests (e.g., Vanha-Majamaa et al. 1996). However, intensive burning tends to slow the natural regeneration because of higher temperatures and enhanced water loss in surface soil following burning. In northern Europe, the natural mean length of a fire return period ranges from 30 to 300 years and is shorter in sites of poor fertility (dry) than in sites of high fertility (moist) (e.g., Shorohova et al. 2008). However, the length of fire return periods varies depending on the site conditions, ground vegetation, and tree species.

According to Oris et al. (2014), carbon fluxes from brunt site and vegetation are affected by (i) the carbon bound in photosynthesis in net primary production (*NPP*);

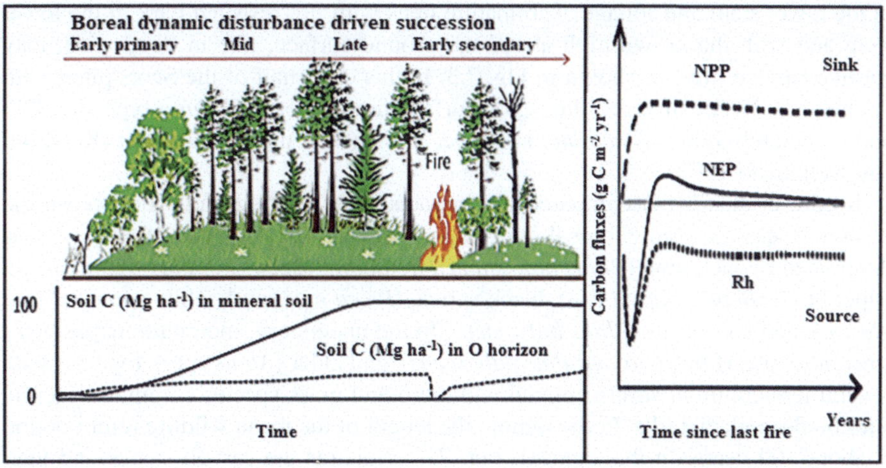


**Fig. 7.2** Schematic depicting how wildfires affect ecosystem carbon and its sink/source dynamics of carbon as a function of the time since the last fire. Left: Schematic depicting how the accumulation of carbon in soil and in litter and humus on soil during primary succession is disturbed by wildfire, with impacts on carbon in the soil and soil profile (DeLuca and Boisevenue 2012). Permission of Oxford University Press. Right: Conceptual net ecosystem production (*NEP*), net primary production (*NPP*), and heterotrophic respiration ($R_h$) after a boreal forest fire during secondary succession (Oris et al. 2014). Courtesy of NRC Research Press

and (ii) released through autotrophic respiration ($R_a$) and soil (heterotrophic respiration ($R_h$), with carbon loss from the ecosystem. These factors are the main drivers in the post-fire restoration of boreal forest carbon stocks. In fire, photosynthesis is terminated through the mortality of trees, or declines in damaged trees exposed to high temperature. This holds for autotrophic respiration ($R_a$) and for microbial soil respiration ($R_h$) under severe fire. Figure 7.2 demonstrates that boreal forest ecosystems remain carbon sources for several years as a result of increased soil respiration, decomposition of dead wood, and the absence of vegetative cover. Over time after fire, net primary productivity ($NPP$ = photosynthesis − $R_a$) increases in young forests (seedling and pole phases of trees, with increasing growth of ground vegetation) and culminates before declining toward the full maturity of trees. However, old-growth forests are not necessarily a carbon source, but they may have the potential to sequester atmospheric carbon (Luyssaert et al. 2008). After wildfires, net ecosystem productivity ($NEP = NPP - R_h$) tends to return to the pre-fire level after 10–40 years depending on forest type, ground vegetation, and burning intensity.

Wildfires are most detrimental when the fire spreads in tree crowns, damaging the foliage and thus killing whole trees (Figs. 7.1 and 7.2). Fire spreading on the ground surface is not as detrimental as it only damages vascular tissues in the lower stem. This holds if trees are sufficiently tall that their crown and foliage extend above the damaging flames. However, surface fires are detrimental for seedlings and young trees in the precommercial and pole phases, with serious damages to vascular tissue in the lower stem and foliage. Tall mature trees with fire-resistant bark in the lower stem and with the crown high above the ground surface, like in Scots pine, may survive surface fires, as shown in Fig. 7.3. In this case, half of the Scots pines with diameter $> 20$ cm survived through a surface fire on poor (*Calluna* type site, CT) and moderately poor (*Vaccinium* site type, VT) sites in the boreal forest (Kolström and Kellomäki 1993).

Seedre et al. (2014) compared the impacts of harvesting and wildfires on the carbon stocks in young boreal forests in Canada (49° N). The study site was dominated by jack pine (*Pinus banksiana*), trembling aspen (*Populus tremuloides*), paper birch (*Betula papyrifera*), black spruce (*Picea mariana*), white spruce (*Picea glauca*), and balsam fir (*Abies balsamea*). In the understory, mountain maple (*Acer spicatum*), dwarf raspberry (*Rubus pubescens*), and alder (*Alnus* spp.), together with several less-common shrubs, mixed with herb and grass species common in fertile sites on deep glacial tills. In the region, the length of the mean wildfire return period is about 100 years. In this context, Fig. 7.4 compares the carbon stocks and their distribution following disturbance of the forest ecosystem through harvesting via clear-cutting and through fire, over a 27-year-old chronosequence, with impacts on carbon dynamics, including carbon stocks in the living biomass, deadwood, forest floor, and mineral soil.

According to Seedre et al. (2014), the distribution of total ecosystem carbon varied between carbon stocks based on the disturbance induced by clear-cutting and fire. Carbon in living biomass was clearly higher after harvesting than after fire due to the living residual trees and regeneration of trees and shrubs following harvesting. Most living biomass was lost in fire; that is, postfire stands contained higher stocks of carbon in deadwood than in living biomass.

Carbon in snags and stumps peaked and was highest, immediately after fire. However, the carbon in snags and stumps reduced rapidly within eight years of fire that caused the downfall of dead trees, thus contributing carbon to the downed woody debris on soil. Carbon in the forest floor was clearly lower shortly after fire than after harvesting. However, the carbon stocks in the forest floor after fire and after harvesting converged eight years after disturbance.

For the total ecosystem carbon stock, there was no significant difference between harvesting and fire in early phases of stand development. The maximum ecosystem carbon stock occurred 27 years after fire or harvesting, with $185.1 \pm 18.2$ Mg ha$^{-1}$ for harvesting and $163.6 \pm 8.0$ Mg ha$^{-1}$ for fire. Seedre et al. (2014) concluded that the differences between individual carbon pools were clear, but the total carbon stock was similar, after wildfires and clear-cutting in boreal forests.

The findings of Kolström and Kellomäki (1993) indicated that the survival of trees exposed to wildfires is species-specific. In boreal forests, climatic warming is shifting the wildfire disturbance regime to one with more frequent fires that burn more deeply into organic soils, releasing sequestered carbon to the atmosphere. This effects a long-term change in the dominance of different tree species in the boreal forests, as shown by Mack et al. (2021), with impacts on carbon sequestration at the ecosystem scale. They found that severe burning of organic soils shifted tree dominance from slow-growing black spruce (*Picea mariana*) to the dominance of

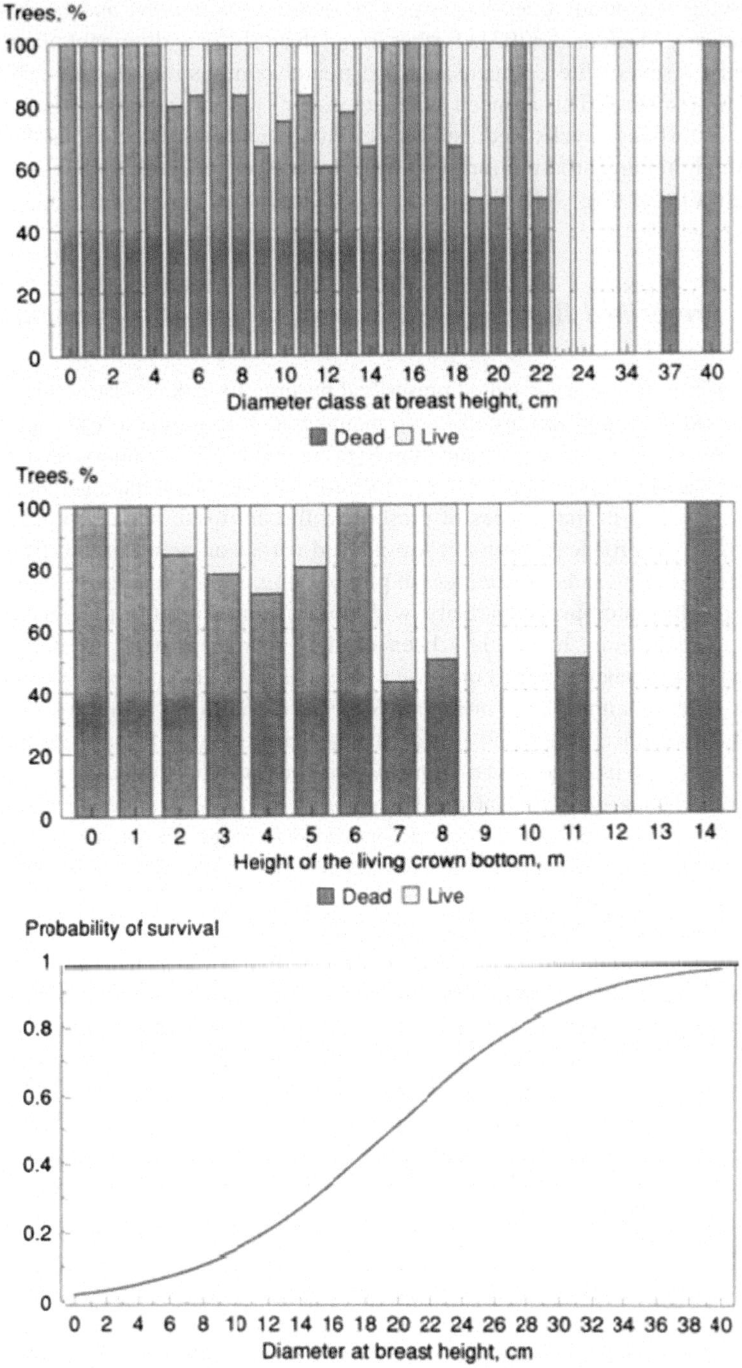


**Fig. 7.3** Mortality and survival of boreal Scots pines (62° N) exposed to wildfire as a function of tree stem diameter and crown-base height (Kolström and Kellomäki 1993). Courtesy of the Finnish Society of Forest Science. Upper: Share of dead and living Scots pines exposed to fire as a function

fast-growing deciduous trees like aspen (*Populus tremuloides*) and birch (*Betula neoalaskana*). Mack et al. (2021) further found that "shifts in dominant plant species catalyzed by severe fire compensated for greater combustion of soil carbon over decadal time scales." This induced a net increase in carbon storage by a factor of five over the disturbance cycle. Reduced fire activity due to future deciduous-dominated forests might retain carbon in forest landscapes, further reducing the fire risks under climatic warming (Fig. 7.5).

### *7.3.2 Prescribed Burning vs. Wildfires in Soil Management*

The use of fire in forestry is called prescribed burning to make a distinction between wildfires and the organized use of fire in management. In boreal forestry, prescribed burning is commonly used for managing logging residues and unharvested trees left in clear-cut areas. This practice is applicable only in even-aged management. In land management, two primary types of prescribed fire are used, namely broadcast and pile burning. In broadcast burning, fire is used across an area ranging from a few hectares up to thousands of hectares. In pile burning, separate piles of fuels such as logging residues are burnt separately, with fire not spreading from pile to another.

Prescribed burning has widely been used in management to (i) improve soil conditions for regeneration; (ii) increase nutrient availability and reduce competition between seedlings and plants in ground vegetation; and (iii) improve soil hygiene conditions to reduce insect and fungal attacks on seedlings. Prescribed burning is further used in the management of conservation areas to improve habitats of rare and threatened species with special dead and burnt wood requirements. Hence, prescribed burning is used under desired conditions to meet specific objectives, such as to restore fire regimes in adapted ecosystems or to limit the amount of dry brush in an area prone to wildfires.

In the context of forest soil, prescribed burning affects the biological, chemical, and physical attributes of soils. The impact of fire is dependent on the (i) fire intensity; (ii) duration and recurrence of fire; (iii) fuel load; and (iv) soil properties (Fig. 7.6) (Agbeshie et al. 2022). Low-intensity fires change the chemical properties of the soil surface through ash, including by increasing the availability of nutrients and reducing acidity (increasing pH). On the other hand, high-intensity fires may totally burn the organic matter on the soil surface (litter, upper humus layers), with severely harmful impacts: (i) evaporate nutrients; (ii) destroy the aggregate stability of soil; (iii) increase soil bulk density, and the hydrophobicity of soil particles; and

**Fig. 7.3** (continued) of tree stem diameter. Middle: Share of dead and live Scots pine exposed to fire as a function of the height of the living crown. Lower: The probability of Scots pine survival is a function of tree stem diameter ($x$): prob.($x$) = $e^z/(1 + e^z)$, where $z = \beta_0 + \beta_1 x X_1 + \beta_2 x X_2 + \ldots + \beta_n x X_n$, $_0 + \beta_1 x X_1 + \beta_2 x X_2 + \ldots + \beta_n x X_n$, and $\beta_o \ldots \beta_n$ are the coefficients for the factors $X_1 \ldots X_n$ affecting survival

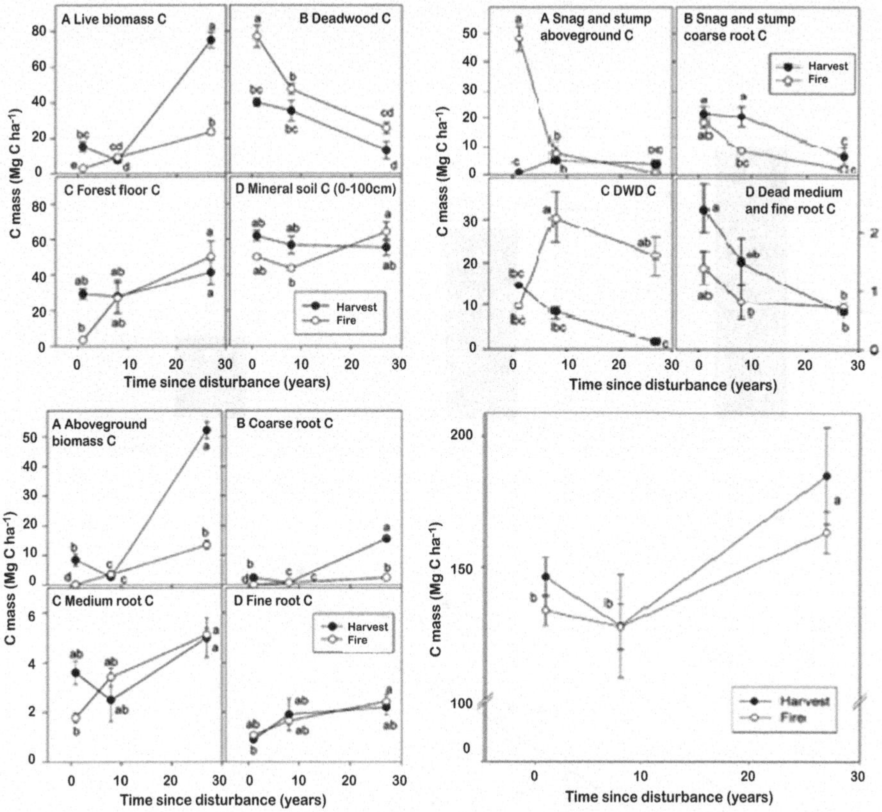


**Fig. 7.4** Recovery of ecosystem carbon in boreal forests from seedling phase to pole phase (Seedre et al. 2014). Permission of Springer Nature. Left/Upper: Carbon pools after harvesting and fire in relation to time since disturbance (TSD, years). A: Living biomass; B: deadwood; C: forest floor; and D: mineral soil (0–100 cm). Symbols represent stand means for each age class (±1 SE; $n = 3$). Left/Lower: Living biomass carbon pools after harvesting and fire in relation to time since disturbance (TSD, years) for A: aboveground biomass; B: coarse roots; C: medium roots; and D: fine roots. Symbols represent stand means for each age class (±1 SE; $n = 3$). Right/Upper: Deadwood carbon pools after harvesting and fire in relation to time since disturbance (TSD, years) for A: snag and stump aboveground biomass; B: snag and stump coarse roots; C: downed woody debris (DWD); and D: dead medium and fine roots. Symbols represent stand means for each age class (±1 SE; $n = 3$). Right/Lower: Total ecosystem carbon after harvesting and fire in relation to time since disturbance (TSD, years). Symbols represent means for each age class (±1 SE; $n = 3$)

(iv) make the conditions of surface extreme with higher temperatures and lower soil moisture. This leads to decreasing water infiltration, increasing erosion, and destruction of soil biota. High soil heating (>120 °C), as in high-intensity forest fires, is detrimental to the soil conditions, especially its physical and biological properties. This implies that the successful use of prescribed burning for management necessitates limited fuel loads to control soil heating (Agbeshie et al. 2022).

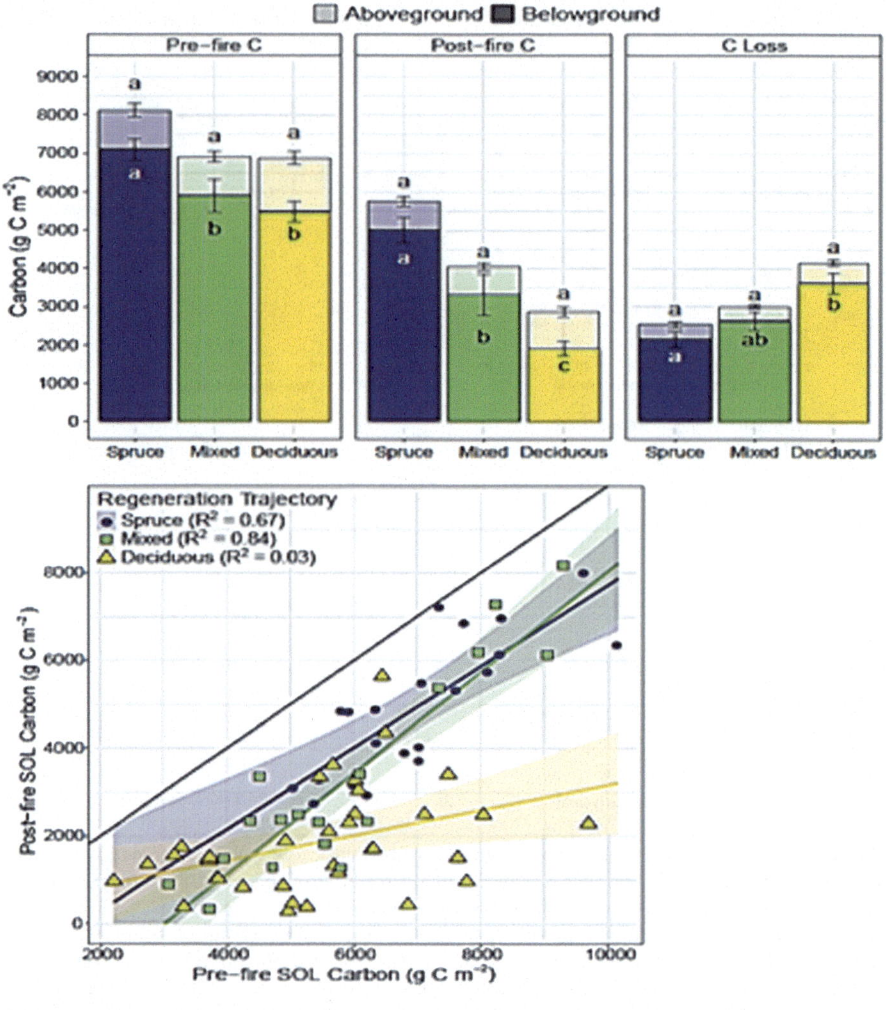


**Fig. 7.5** Wildfire impacts on tree species composition and carbon loss in boreal forests (Mack et al. 2021). Open Access. Courtesy of American Association for the Advancement of Science. Upper: Ecosystem carbon pools and wildfire losses across successional trajectories, with carbon above and below ground. Trajectories include stands that returned to spruce dominance after fire (blue), transitioned to dominance by deciduous trees (yellow), or transitioned to a mixture of black spruce and deciduous species (green). Letters represent significant differences ($P < 0.05$) between trajectories within the above- and belowground pools. Lower: Wildfires affect soil organic layer carbon pools (SOLC) and successional trajectories. The relationship of between- and post-fire SOLC pools was estimated across three successional trajectories of tree regeneration

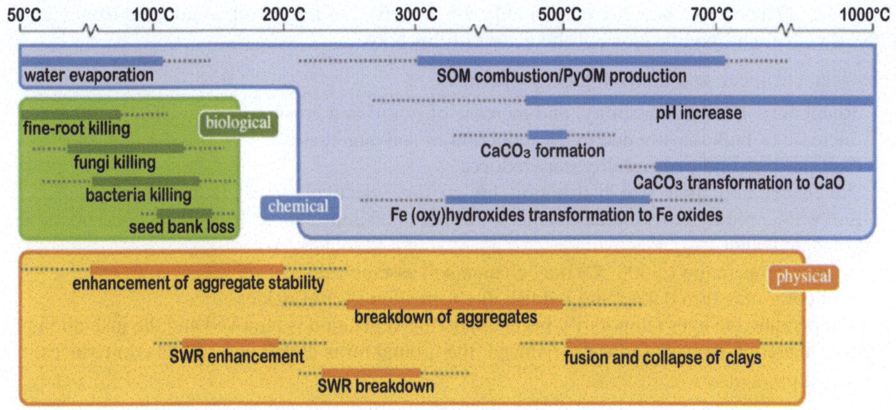


**Fig. 7.6** Impacts of fire-induced temperatures on soil properties compiled by Agbeshie et al. (2022) based on Santin et al. (2018). Open Access. Courtesy of Springer Nature

### *7.3.3 Impacts of Prescribed Burning on Soil Properties*

In forest management, controlled fire or prescribed burning is used to improve soil conditions for natural and artificial regeneration and to make nutrients available for growth later in succession. On the other hand, many plant and animal species are dependent on the specific environments that wildfires or prescribed burning create. Such sites are characterized by partly burnt wood and long-term effects on the physical, chemical, and biological properties of the soil. The effects of fire on forest ecosystems are related to the burning severity, which is indicated by the temperature and duration of the fire. In reforestation, fires with low-to-moderate severity are sufficient for eliminating undesired species and making more nutrients available for trees seeded/planted in a site (Table 7.2).

According to Certini (2005), fires with high severity may also have following harmful effects on the soil: "significant removal of organic matter, deterioration of both structure and porosity, considerable loss of nutrients through volatilization, ash entrapment in smoke columns, leaching and erosion, and marked alteration of both quantity and specific composition of microbial and soil-dwelling invertebrate communities." However, the pre-fire levels of most site properties can be recovered and even exceeded if plants recolonize the site rapidly.

### *7.3.4 Impacts of Prescribed Burning in Soil Management in Boreal Forestry*

In boreal conditions, fire is a major disturbance that naturally affects the long-term ecosystem dynamics. In prescribed burning, fire affects the soil, and especially soil

**Table 7.2** Overview of how fire may modify the properties of forest soil, as adopted from Certini (2005), with slight modifications of the author of this book

| Physical, chemical, and mineralogical properties |
|---|
| • Reduction of soil permeability, and increase of runoff and erosion.<br>• Increase of bulk density due to the reduction of soil aggregates.<br>• Soil porosity and permeability may reduce.<br>• No change in particle size distribution, but erosion selectively transfers fine fractions.<br>• Soil acidity reduces (pH increases) due to release of alkaline cations (Ca, Mg, K, Na) bound in soil organic matter.<br>• Higher temperature (>500 °C) may change soil assemblages.<br>• Soil color may turn darker and reddish due formation of iron oxides.<br>• Temperature changes temporarily because sharing vegetation is removed and the insulation of organic matter on the soil surface is reduced; the ground turns darker and soil albedo reduces. |
| Chemical properties |
| • Quantity of soil organic matter, especially in litter and humus above the mineral soil, reduces with burning but recovers to pre-fire levels over time.<br>• Quality of soil organic matter changes for the enrichment of the fraction more recalcitrant to biochemical attack. This is due to both selective burning of fresh residues (leaves, twigs, etc.), and neoformation of aromatic and highly polymerized (humid-like) compounds. Burnt material, representing the products of incomplete burning, may have a residence time of centuries or even millennia.<br>• Availability of nutrients tends to increase in the short term.<br>• Organic nitrogen bound in organic matter, but unavailable, is partly volatilized and partly mineralized to ammonium in a form available to biota. Ammonium is biochemically transformed to nitrate, which tends to rapid leach if not bound in vegetation. Nitrogen availability returns to pre-fire levels in a few years. Organic phosphorus mineralizes to orthophosphate but the loss through volatilization and leaching out of the soil is low. Orthophosphate precipitates in a slightly available mineral form, if not rapidly taken up. Calcium, magnesium, and potassium increase with burning in the short term.<br>• Exchange capacity decreases proportionally to the loss of organic matter.<br>• Base saturation increases due to the prevailing release of bases from burning organic matter. |
| Biological properties |
| • Microbial biomass decreases but recovers to pre-fire levels based on the rate of plant recolonization of the site.<br>• Composition of the microbial community changes as a consequence of the selective effects of fire on some groups of microorganisms and the modification in vegetation. In general, fungi decline more than bacteria.<br>• Biomass of soil-dwelling invertebrates decreases, but less than that of micro-organisms.<br>• Composition of soil-dwelling invertebrate communities changes, but the time to recovery of the pre-fire assemblage varies. |

temperature, by (i) reducing the insulating litter and humus layers on mineral soil; (ii) increasing the heat flow to mineral soil by increasing heat conductance (reducing the thickness and packing litter and humus layers); and (iii) increasing the capacity of the soil surface to absorb heat (black soil surface). In this regard, Fig. 7.7 demonstrates the fire-induced changes in a site measured by Viro (1969) for prescribed burning in southern boreal conditions. Surface temperature at the litter/humus layer increased compared to no burning. Burning also increased the temperature below the humus layer through the soil profile, where the majority of tree roots

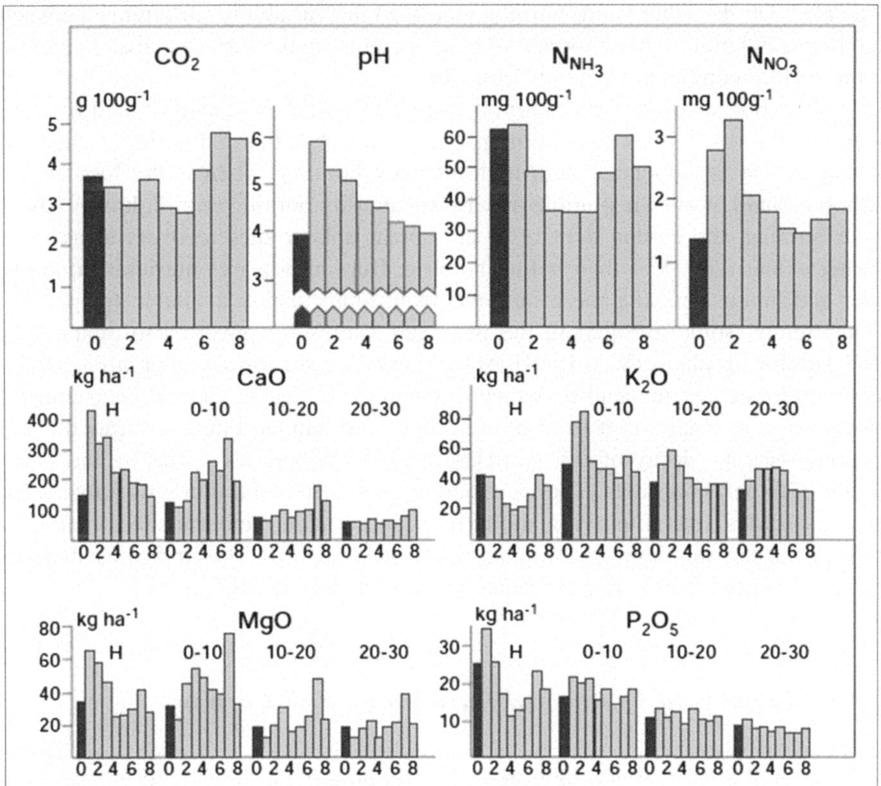


| Measuring place | Temperature, °C | | |
|---|---|---|---|
| | Open area, burnt | Tree stand, not burnt | Open area, not burnt |
| Air, 5 cm above soil surface | 24 | 21 | 25 |
| Humus surface | 31 | 18 | 24 |
| In mineral soil surface | 16 | 9 | 13 |
| Mineral soil, 10 cm below surface | 13 | 9 | 11 |
| Mineral soil, 20 cm below surface | 12 | 8 | 10 |

**Fig. 7.7** Effects of prescribed burning on chemical and physical properties of moraine soil in southern boreal forests (Viro 1969). Courtesy of Natural Resources Institute Finland. Upper: Chemical properties in a burnt forest site in the southern boreal zone. Legend: H = humus, black column = not yet burnt before burning, 1 = less than a year since burning, and time since burning of 2 = 2–3, 3 = 4–6, 4 = 7–9, 5 = 10–12, 6 = 13–19, 7 = 20–32 and 8 = 33–50 years. Below: Effects of prescribed burning on temperature in air close to the soil surface, on the surface of humus, on the surface of mineral soil below the humus, and in the mineral soil profile

is located. On the other hand, burning increased the variability (difference between maximum and minimum temperature) of temperature in the surface soil and air close to the surface compared to the unburnt site.

Viro (1969) also found that burning reduced the carbon density in soil by 25% over a period of 20 years after burning compared to that before burning. Reduction of soil carbon is associated with high-intensity burning, whereas the loss of soil carbon is small or even negligible under low-intensity burning (e.g., Johnson 1992). After burning, the carbon density in the boreal upland sites recovers to an even greater extent (20–25%) than before burning. Burning releases nutrients bound in litter and humus, making these nutrients even more readily available for growth. Concurrently, some nitrogen in the litter and humus is converted to ammonium ($NH_4$) usable for plants (Viro 1969). At the same time, the formation of nitrate ($NO_3$) increases for decreasing acidity, which is due to the release of calcium, phosphorus, potassium, and magnesium from burning litter and humus. Fire also heats the soil and increases the decay of soil particles, stones, and bedrock, affecting the availability of mineral nutrients. The increased content of mineral nutrients remains for several decades, thus improving site fertility. In burnt and unburnt sites, the chemistry in deeper soil layers is similar (Neff et al. 2005). This indicates that the immediate impacts on boreal moraines are restricted to the surface soil.

### *7.3.5 Impacts of Prescribed Burning on Soil Carbon*

Kulmala et al. (2014) found that prescribed burning of slash tends to increase soil moisture and temperature, reduce soil acidity, and increase ammonium ($NH_4$-N) and nitrate ($NO_3$-N) compounds. The experiment was conducted in a mature Norway spruce forest (*Picea abies*) in middle boreal conditions (62° N). The forest was located on a moderately fertile site (*Myrtillus* site type, MT), where the measures were conducted over three years following burning (Fig. 7.8). The field layer of the ground vegetation was dominated by blueberry (*Vaccinium myrtillus*), wavy hair grass (*Deschampsia flexuosa*), and squirrel berry (*Maianthemum bifolium*). The vegetation in the bottom layer was dominated by mosses like red-stemmed feather moss (*Pleurozium schreberi*), wind-blown moss (*Dicranum polysetum*), and layer moss (*Hylocomium splendens*).

Before the experiment, the stem wood stocking in the study area was 400 $m^3$ $ha^{-1}$ (Kulmala et al. 2014). For treatment and measurements, the area was divided into (i) the control area (no cutting, no burning); (ii) the clear-cutting area with no burning; and (iii) the clear-cutting area with burning. This burning experiment was characterized as follows:

- After the clear-cutting, the biomass of the aboveground slash was 47,000 kg $ha^{-1}$.
- The total amount of burnt organic matter in the experiment was 44,900 kg (SE ± 10,900 kg), which comprised 28,900 kg $ha^{-1}$ of slash (SE ± 6900 kg),

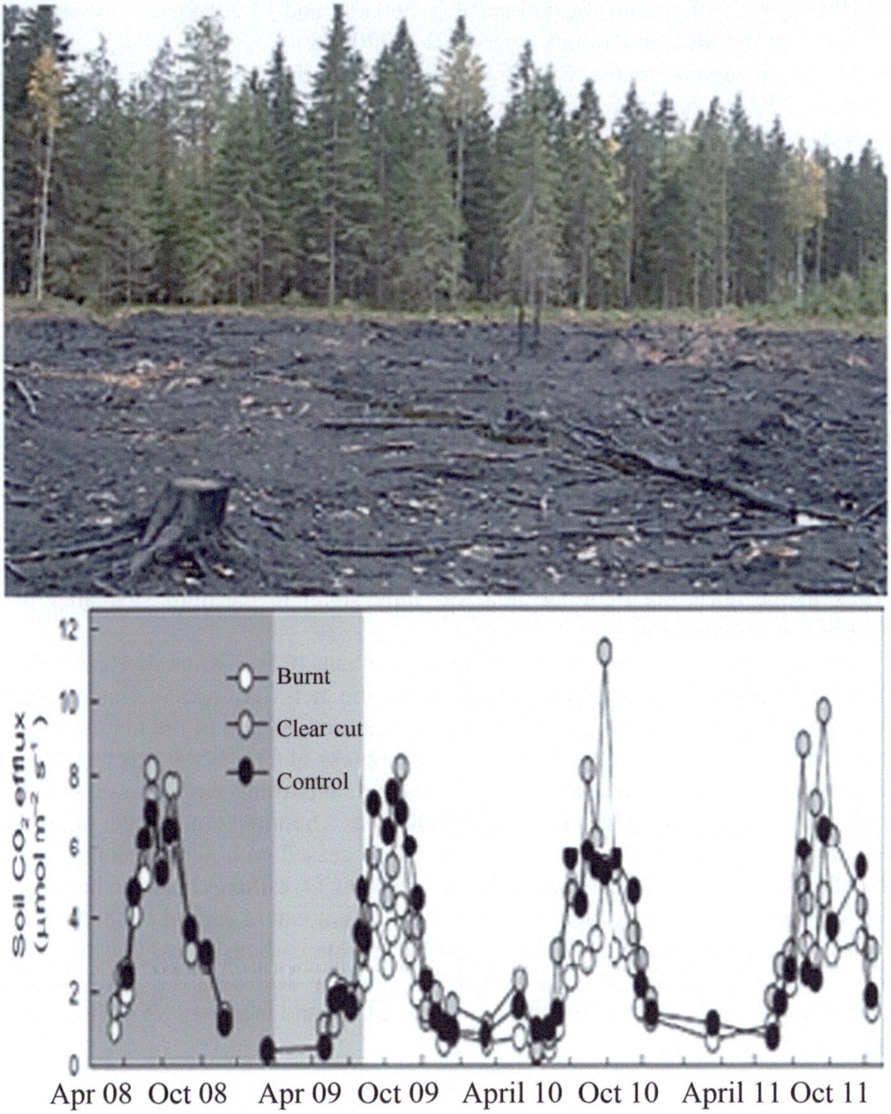


**Fig. 7.8** An experiment for investigating how prescribed burning affects soil properties in southern boreal forests. Upper: Burnt area representing Norway spruce and Scots pine dominated upland site on medium fertile site (*Myrtillus* site type) before harvest and burning in the background. The burnt area is not that used in measurements. Courtesy of Tapio Finland. Below: Means of soil $CO_2$ fluxes for the control, clear-cutting, and burnt clear-cutting area (dark grey shading), after clear-cutting but before burning (light grey shading), and after burning (no-shading) (Kulmala et al. 2014). Permission of Elsevier

1900 kg $ha^{-1}$ of ground vegetation (SE ± 360 kg), and 14,200 kg $ha^{-1}$ of organic matter in the litter and humus layers (SE ± 7200 kg).

- The burnt organic matter was 55,800 kg $ha^{-1}$, with 35,800 kg $ha^{-1}$ (78%) of slash, 2300 kg $ha^{-1}$ (100%) of ground vegetation, and 17,600 kg $ha^{-1}$ (22%) of soil organic matter.
- The mean depths of the litter layer were 2.8 ± 0.26 cm in the uncut forest and 2.9 ± 0.29 cm after the clear-cutting and the prescribed burning.
- The change in the depth of the humus layer in the burnt clear-cutting area was statistically significant: from 3.7 ± 0.41 to 2.5 ± 0.32 cm. However, the spatial variation was high, the standard deviation being about 50% of the mean values.

According to Kulmala et al. (2014), clear-cutting increased the moisture, temperature, pH, and the nitrogen in ammonium and ($NH_4$) and nitrate ($NO_3$) compounds in the soil compared to those to in the uncut control site. Concurrently, the root respiration tended to decline in the first year after clear-cutting, with a small reduction in soil $CO_2$ efflux. In the next two years, the $CO_2$ efflux was higher than in the clear-cutting area. The increase was related to increased decay due to elevated soil moisture and temperature. In this two-year period, the combined clear-cutting and burning reduced the $CO_2$ efflux. Consequently, the temperature response to $CO_2$ efflux in the burnt clear-cutting area was only small in the third year after the experiment initialization.

Regardless of treatment, soil was a sink for methane ($CH_4$) (Kulmala et al. 2014). After the clear-cutting and burning of slash, the net $CH_4$ uptake immediately decreased, but one year later, the uptake was the same as that in the mature forest used as the control. Following clear-cutting, the uptake of $CH_4$ in soil was like that in the control site. Parallel measurements in the site with burning slash showed that the emission of volatile organic compounds (VOC) was a hundred times larger than that before burning. Nevertheless, the VOC emissions ceased within three months after the burning. In two and a half years, the subsequent $CO_2$ effluxes from the burnt site were lower than that at the clear-cut site. The total amount of carbon emitted due to prescribed burning was higher due to the immediate carbon losses. On the other hand, the clear-cutting and slash burning temporarily decreased the ability of the boreal Norway spruce forest to bind atmospheric $CH_4$ and increased VOC emissions (Kulmala et al. 2014).

### *7.3.6 Impacts of Prescribed Burning on Soil Carbon and Nitrogen*

Based on a meta-analysis, Johnson, and Curtis (2001) found that harvesting, on average, had significant or no effect on soil carbon and nitrogen, with significant impacts depending on harvesting mode and tree species (Fig. 7.9). In this regard, the harvesting of only coniferous sawlogs increased (18%) soil carbon and nitrogen, while whole-tree harvesting decreased (6%) both soil carbon and nitrogen. Johnson

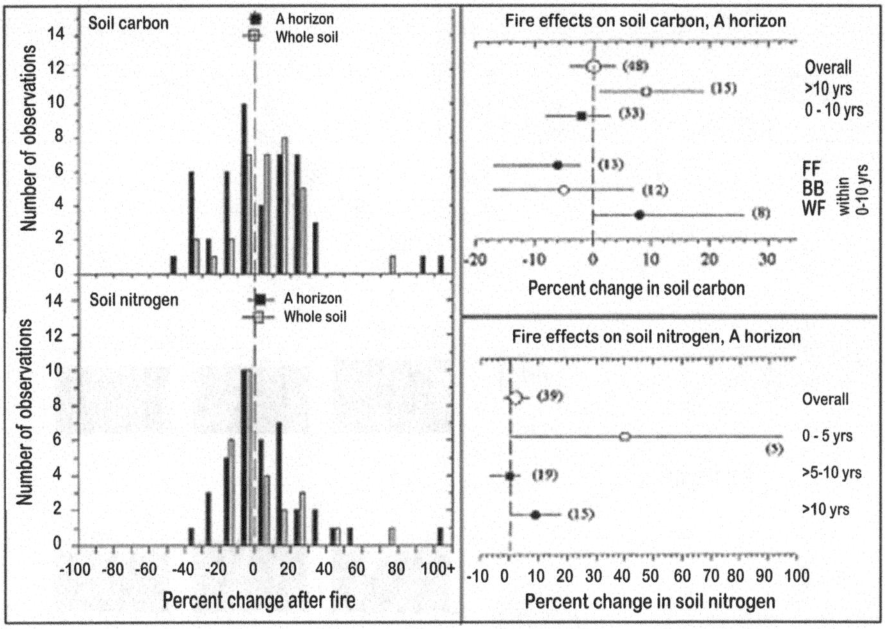


**Fig. 7.9** Effects of fire on soil carbon and nitrogen based on a meta-analysis (Johnson and Curtis 2001). Permission of Elsevier. Left: Distribution of the number of studies showing the effect of fire on soil carbon (upper panel) and nitrogen (lower panel). Right: Results for the A horizon non-parametric meta-analysis of soil carbon (upper panel) and nitrogen (lower panel). Numbers 0–5, 6–10, and +10 years refer to time since treatment, with FF = prescribed fire, BB = prescribed burning of slash after harvest, and WF = wildfire

and Curtis (2001) also found that soil carbon was lower after prescribed burning than after wildfire over a 10-year period compared to changes in the control sites. This difference between prescribed burning and wildfires was attributed to the differences of carbon bound in charcoal and recalcitrant, hydrophobic organic matter and to the effects of naturally invading post-fire vegetation capable to fix nitrogen at different colonizing rate into the study areas.

## 7.4 Mechanical Soil Preparation

### *7.4.1 Impacts on Soil Properties*

Mechanical soil management affects soil carbon and nitrogen through changes in soil biological, physical, and chemical properties (Table 7.1). There is a wide variety of ploughs, excavators, and other devices available for mechanical soil management. In mechanical management, the ground vegetation is partly removed from the surface of mineral soil (litter, and soil detritus), with mixing soil organic matter

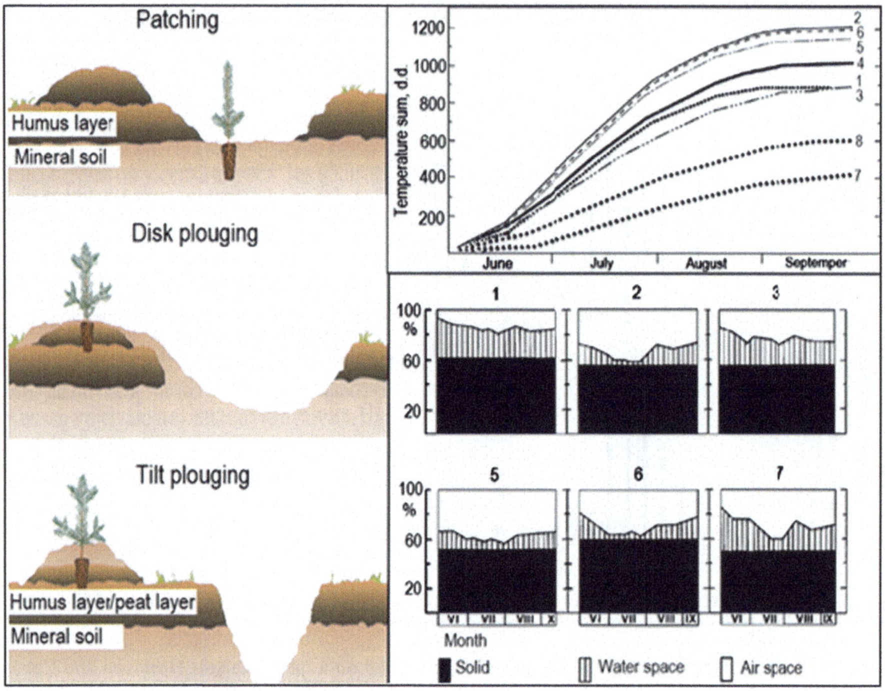


**Fig. 7.10** Schematic depicting how various soil management methods affect the properties of planting sites, with effects on soil temperature and moisture content in boreal podzol on the thermal conditions in soil (Upper) and the share of air and water filling soil (Below) in the northern boreal zone over a growing season (66° N) (Kauppila and Lähde 1975). Courtesy of Natural Resources Institute Finland. Left: Examples of various methods for soil preparation for providing various microsites for seedlings. Right/Upper: Temperature sum related to soil management: 1 = patching; 2 = tilt ploughing; 3 = disk tilt ploughing; 4 = mixing organic and inorganic surface layers; 5 = 40 cm high mounding; 6 = 50 cm high mounding; 7 = uncut no soil management; 8 = no soil management on open area. Right/Lower: Volume percent of solid particles, water, and air space in soil profile of 0–10 cm in managed site: 1 = scarifying; 2 = tilt ploughing; 3 = shoulder ploughing; 4 = rotary tilting; 5 = 40 cm high mounding; 6 = 50 cm high mounding: 7 = uncut with no soil management

with the surface of mineral soil. Consequent formation of mounds and pits provides seedbeds/sites for seeding and planting. Wherever waterlogging is no problem, the removal of litter and humus and opening of the mineral soil through scarification is enough to improve the establishment, survival, and growth of seedlings (Fig. 7.10). In waterlogged upland soils, more intensive measures like mounds are needed to increase soil aeration to improve root growth for enhancing the establishment and growth of seedlings. In many cases, these microsites enhance reforestation on waterlogged soils, including those with a high share of fine soil fractions and low soil aeration. Increasing soil temperature and nutrient supply from decaying organic matter in mounds make these microsites preferable for seeding and planting.

Figure 7.10 demonstrates the effects of soil management on soil temperature in northern boreal conditions. This field experiment by Kauppila and Lähde (1975) showed that ploughing with intensive disturbance substantially increased the accumulation of the temperature sum in soil. Similarly, mounds and pits increased the temperature sum in the surface soil. However, the temperature increase remained lower than that in ploughed soils, showing that the increase of temperature sum was related to the intensive soil preparation for long-term improving soil temperature and aeration. This is emphasized by the slow accumulation of temperature sum on soil treated when using patches with low mounds and shallow pits. Even in this case, the soil temperature was higher than that in untreated sites, with an undisturbed litter and humus layer on mineral soil. However, the variability of temperature in mechanically disturbed soil is lower than that on prescribed-burning soils (Leikola 1974; Pohtila 1977). Only on the sites treated with rotary tilling, humus content deeper in soil increases, due to mixing of organic layers and the mineral soil below.

In general, soil management increases the evaporation of soil water due to higher soil temperature. The increase is especially large after prescribed burning, where humus and mineral soil below the humus dry more rapidly in black soil than on unburnt sites. This also holds for mechanical soil management, except the effects on soil moisture, which are smaller than on burnt sites. On the other hand, the reduction of ground vegetation, litter, and humus allows more water to flow into the soil profile due to reduced water interception. This also holds for the mechanically treated sites, except ploughing with increasing the surface flow. However, the effects of soil management on soil moisture seem to remain only local despite the increase of soil aeration and decrease of water content (Fig. 7.10). Under high ground water levels, soil management like ploughing may affect soil moisture even between furrows if the ground water level is lowered throughout the whole site.

### 7.4.2 Soil $CO_2$ Efflux in Sites with and Without Logging Residues

In even-aged management, clear-cutting is widely followed by site preparation and subsequent planting/seeding to establish a new tree generation for the next production cycle. In this context, carbon accumulated in the previous production cycle (legacy carbon) still decays, but litter from growing seedlings gradually compensates the decayed soil carbon (de novo litter and humus). Decay may also increase due to the soil preparation in regeneration areas with seedlings in their early growing and developing phase. Over time, the carbon in litterfall from seedlings and carbon from recovering ground vegetation exceeds the decay in soil carbon (litter and humus), and the accumulation of carbon in soil increases gradually with accelerating rate over time.

Pumpanen et al. (2004) investigated how clear-cutting, and subsequent soil preparation, affects the soil $CO_2$ efflux in the mid-boreal zone (61° N). The study site was dominated by a mixture of mature Scots pine and Norway spruce at the age of 130 years. The soil parent material represented a glaciofluvial deposit, with a texture varying from coarse to fine sand. The effect of harvesting and site preparation on soil $CO_2$ efflux was studied on a clear-cut site with four soil treatments, including mounding, exposed E and C horizons (Fig. 7.11), and undisturbed soil with and without logging residues left on the site. The site was moderately fertile (*Myrtillus* site type, MT) representing a dry subarea for Scots pine and a mesic subarea for Norway spruce. In the mesic subarea, the ground vegetation was dominated by the dwarf shrubs (e.g., *Vaccinium vitis-idaea, Vaccinium myrtillus*), with scattered herbs and grasses. The forest floor was covered with mosses (e.g., *Dicranum polysetum, Pleurozium schreberi, Hylocomium splendens*). The aboveground biomass in the ground vegetation ranged from 13.7 to 16.3 kg $m^{-2}$ for the dry and mesic subsites.

According to Pumpanen et al. (2004), the site preparation accounted for over 75% of the total variance in soil $CO_2$ effluxes. Clear-cutting and removal of logging residues also decreased instantaneous $CO_2$ effluxes from the soil surface by 40%. However, the highest instantaneous effluxes were measured from mounds and from undisturbed soil containing logging residues. These sampling points showed effluxes twice as high as those in the control forest or in places, where the logging residue was removed. The lowest instantaneous effluxes, about 18% of those in the control forest, were measured from the exposed C horizon. In more details:

- Annual effluxes in the forest were 1900 g $CO_2$ $m^{-2}$ before harvesting, and 3242, 2845, and 2926 g $m^{-2}$ in the three successive years after harvesting in sites with logging residues. If logging residues were removed, the annual efflux decreased to that prior to clear-cutting.
- Annual $CO_2$ efflux from the logging residue was 1423 g $m^{-2}$ during the first year after clear-cutting, equal to 388 g $m^{-2}$ of carbon assuming even distribution of logging residue in the site. Thus, some 23% of the total carbon pool in the aboveground logging residues was released during the first year after clear-cutting.
- Annual carbon emissions from the O, E, and B horizons were 352 g $m^{-2}$ during the first year after clear-cutting. Obviously, about 20% of the emissions was originating from decomposing roots. The decomposition rate was most rapid during this period, slowing in subsequent years.

Based on (i) the measured $CO_2$ emission rate, (ii) the reduction observed in the decomposition rate, and (iii) the aging of decomposing material, Pumpanen et al. (2004) concluded that the decomposition of logging residues is slow. Consequently, large amounts of non-decomposed material still remained at the time point at which the carbon uptake exceeded carbon losses, with the site turning from source to sink.

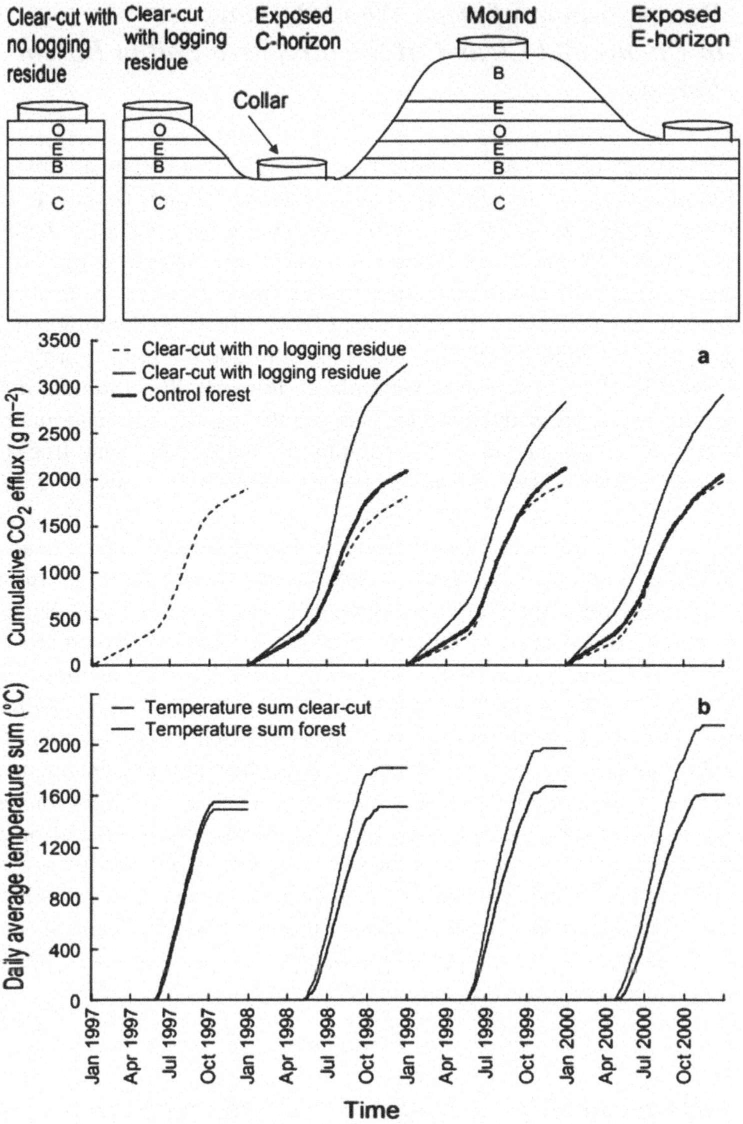


**Fig. 7.11** Schematic depiction of the site preparations used in the experiment, and annual soil $CO_2$ effluxes and soil temperature sum (Pumpanen et al. 2004). Courtesy of Natural Resources Institute Finland. Upper: For the experimental layout, eight blocks were established in the clear-cut area for subsequent soil treatment with four methods: (i) in each block, the soil treatment included clear-cutting with logging residues left on the site; (ii) in the site with an exposed C horizon, the O horizon and surface of the mineral soil were removed; (iii) in site with an exposed E horizon, the O horizon was removed; and (iv) for the mound, mineral and organic soil were mixed. Middle: Cumulative soil $CO_2$ efflux from the respective sampling points in the period 1997–2000. Lower: Cumulative temperature sum in the respective sampling points in the period 1997–2000

### 7.4.3 Decomposition of Dead Wood Residues in Uncut- and Clear-Cut Site Prepared Soil in Boreal Forests

Coarse dead woody debris has a significant role in retaining carbon and nutrients in soil, and in providing habitats for invertebrates and vertebrates dependent on dead wood. In this respect, Finér et al. (2016) conducted experiments to identify how clear-cutting and mechanical soil preparation affect decomposition of dead coarse woody debris compared to the decomposition in uncut forest sites. In the experiments, standard wooden stakes of Scots pine (*Pinus sylvestris*), loblolly pine (*Pinus taeda*), and aspen (*Populus tremula*) were placed in uncut and in nearby clear-cut sites (62° N) (Fig. 7.12). The stakes were placed horizontally: (i) on the top of the surface organic layer; (ii) on the interface between the organic layer and mineral soil; and (iii) vertically in the mineral soil to a depth of 20 cm. Every year, five stakes of each tree species were collected from each stake location for analyses (Finér et al. 2016).

Finér et al. (2016) found that clear-cutting and soil preparation tend to accelerate the initial decomposition of wood, thus affecting the loss/retention of carbon and nutrients in forest sites depending on the intensity of harvesting and management. The mean mass loss in the clear-cut area was 59.8% of the original mass of each stake. This was larger than in the uncut area, where the percent mass loss was $19.8 \pm 3.0$ for both pines and $43.3 \pm 5.1$ for aspen (Fig. 7.12). On the other hand, the mass loss for aspen was faster than for Scots pine and loblolly pine in the uncut area over the whole monitoring period. Finér et al. (2016) further found that the mass loss on the surface of the organic layer was 6–10% faster than that at the mineral soil interface or in the mineral soil. On the top of the organic layer in the clear-cut site, mass loss was 32–35% slower than at the deeper layers, due to the low moisture at the soil surface. Furthermore, mass loss correlated positively with the soil temperature ($r > 0.94$). Finér et al. (2016) also found that mass loss in the uncut forest was positively correlated with the nitrogen content of the wooden stakes, which increased the decomposition of dead woody debris.

### 7.4.4 Management Impacts on Soil Carbon and Nitrogen Stocks

Management such as clear-cutting and site preparation change the carbon and nitrogen stocks in boreal soils, as found by Piirainen et al. (2015) in boreal conditions (63° N). They monitored the carbon and nitrogen stocks in the forest floor and mineral soil over a 10-year period after clear-cutting and soil preparation in sites where logging residues were left after harvesting. The experiment was established on a moderately fertile site (*Myrtillus* site type, MT) dominated by mature Norway spruce forest mixed with scattered Scots pines, Pendula birch (*Betula pendula*),

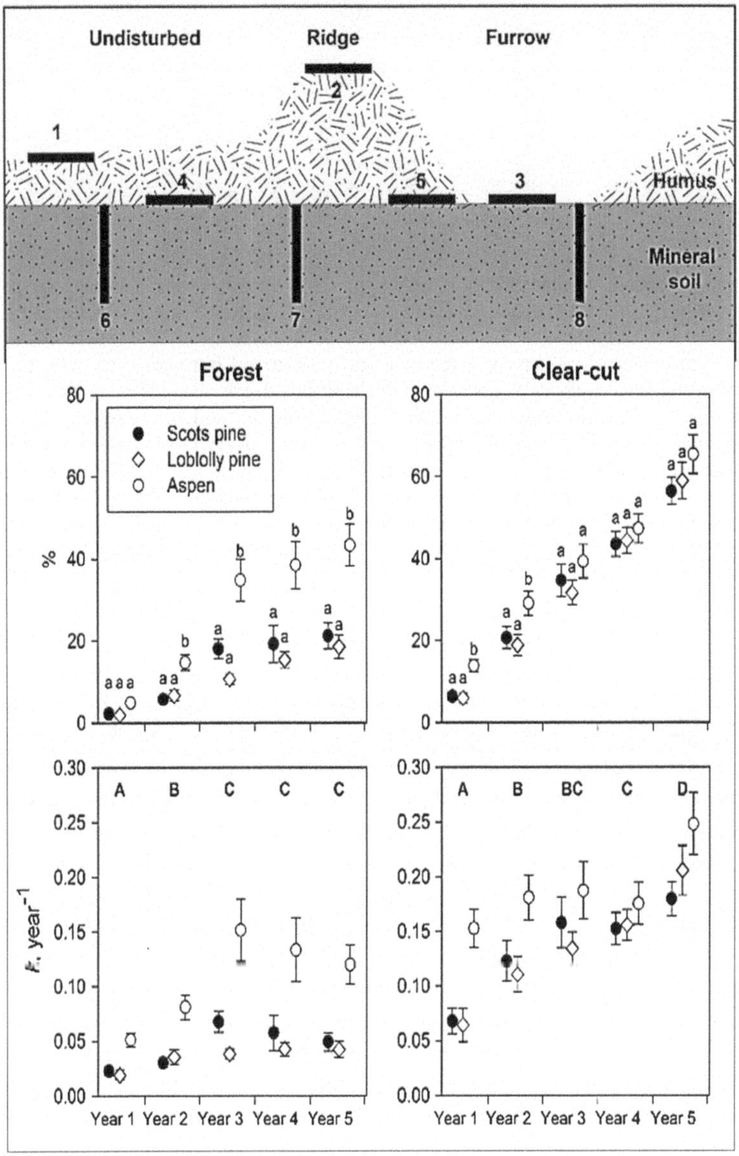


**Fig. 7.12** Initial decomposition of wood in boreal forest sites as a function of time since clear-cutting and site preparation (Finér et al. 2016). Permission of Elsevier. Upper: Experimental layout with wooden stakes used to monitor the decay of wood in different subsites. The numbers 1–3 are stakes located on the soil surface; 3–5 are stakes located at the organic layer/mineral soil interface; and 6–8 are stakes in the mineral soil. A similar layout of stakes was also used in the uncut forest site. Middle: Mass loss of stakes as a function of their tree species and location in the experimental sites in the uncut and clear-cut forests as a function of treatment time. Lower: Mass loss in decay rate (*k*) as a function of the experimental period and tree species of stakes in the uncut and clearcut forests as function of treatment time

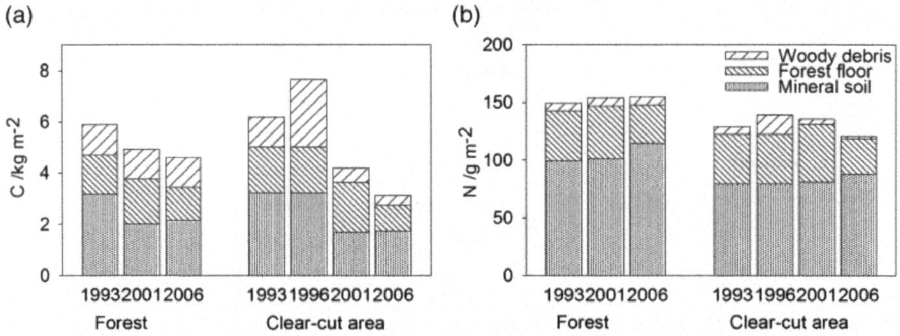


**Fig. 7.13** Soil-management-driven changes in soil carbon and nitrogen stocks. In the analysis, carbon and nitrogen were divided into those in the (i) woody debris above the forest floor and coarse debris (diameter > 10 mm) in the forest floor (excluding stumps and coarse roots); (ii) forest floor (including litter, fine roots); and (iii) mineral soil 0–20 cm deep. Observations occurred before (1993), immediately after (1996), and 5 years (2001) and 10 years (2006) after clear-cutting. Soil harrowing was conducted in 1998 (Piirainen et al. 2015). Permission of Wiley. Left: Carbon in the untreated (years 1993, 2002, and 2006) and treated plots (years 1993, 1996, 2001, and 2006). Right: Nitrogen in the untreated (years 1993, 2002, and 2006) and treated plots (years 1993, 1996, 2001, and 2006)

Pubescent birch (*Betula pubescens*), and aspen (*Populus tremula*). The forest was uneven-aged, with the oldest trees being up to 170 years old, and the soil was a haplic podzol. The mean depth of the forest floor and the eluvial and illuvial horizons was 3, 12, and 15 cm, respectively (Piirainen et al. 2015). The clear-cutting in the treated plots was performed in August 1996. The clear-cut plots were harrowed with a disk-plow in September 1998 and planted in June 1999 with one-year-old potted Scots pine seedlings at a density of 2300 seedlings per hectare (Fig. 7.13).

According to Piirainen et al. (2015), the soil carbon stock above the mineral soil after 5 years was lower (>20%) in the treated area than in the uncut forest, and after 10 years < 50% lower (Fig. 7.13). The corresponding nitrogen stock was marginally larger (< 5%) after 5 years, but lower (< 20%) after 10 years. In the mineral soil, there were no changes. Carbon and nitrogen were only lost from the furrows and not from the other microsites. Carbon and nitrogen were not lost when compared with the carbon and nitrogen stocks in the untreated site. Nitrogen utilized by the regenerating vegetation or immobilized by the stumps at the same site indicated that nitrogen stock remained after the clear-cutting, thus being retained at the site.

## 7.5 Impacts of Fire and Dead Wood on the Biodiversity of Temperate and Boreal Forests

In boreal forests, decaying coarse wood is important in supporting biodiversity by maintaining a variety of habitats and hosting epixylic bryophytes, lichens, polypore, and other decomposer fungi and invertebrates (e.g., Essen et al. 1997). Coarse woody debris exposed to fire provides specific site conditions due to its reduced decay rate. Fire affects the structure, properties, and turnover of coarse woody debris. Such changes make an important contribution to the structure and properties of habitats in natural boreal forests. In this regard, coarse roots, stumps, and woody logging residues not harvested are important in managing and conserving the forest biodiversity.

Shorohova et al. (2008) investigated the impacts of fire on the decay of coarse woody debris with comparison to coarse woody debris not exposed to fire, as used in prescribed burning. The study concerned Scots pine (*Pinus sylvestris*), Norway spruce (*Picea abies*), and birch (*Betula pendula* and *B. pubescens*) in southern Finland (61° N). The focus was on the stumps representing two 40-year phases of chronosequences, with and without prescribed burning after clear-cutting. Firstly, Shorohova et al. (2008) assumed that the decomposition rate and the consequent mass loss in stumps was lower by the presence of bark with higher decay-resistance to fire than the wood. Secondly, the decomposition rate and the mass loss in stumps were positively affected by the mechanical and biological fragmentation of bark (Fig. 7.14).

Shorohova et al. (2008) used an exponential model to estimate the annual rate ($k$) of decomposition for aboveground stump wood (Olson 1963): $\rho = \rho_o\, x\, e^{-k_w x\, t}$, where $\rho$ is the wood bulk density (Mg m$^{-3}$), $\rho_o$ is the initial bulk density, and $k_w$ is the annual decay rate (yr$^{-1}$). To assess the temporal variability, the value of $k$ for each stump was calculated as $k_w = \frac{\ln(\rho_o) - \ln(\rho)}{t}$, with the percentage mass loss being $\delta_w = (\rho_o\text{-}\rho)\, x\, \rho_o^{-1})$ x 100. Concurrently, the specific mass of bark ($M_b$, Mg m$^{-2}$) with the mass ($m$, Mg) per surface area ($s$, m$^2$) was calculated with $M_b = {}^{m}\!/_{s}$.

For aboveground stumps, Shorohova et al. (2008) estimated the decay rate ($k$) of wood to be 0.048, 0.052, and 0.068 yr.$^{-1}$ for Scots pine, Norway spruce, and birches, respectively. The study further showed that bark decomposed faster than wood, and the fragmentation of bark increased the decomposition rate of wood. Burning reduced the decay of wood and bark in Scots pine and the bark in Norway spruce. Concurrently, burning had no effect on Norway spruce wood or birch bark. Furthermore, the decomposition of bark of all the tree species was slower in stumps with larger diameter, but only slightly slower in birch wood.

Shorohova et al. (2008) concluded that (i) the decomposition rates of wood and bark are species-specific; (ii) the fragmentation of bark increases the loss of burnt woody debris; and (iii) decay rate ($k$) is species-, bark-, and wood-specific, with effects on carbon loss and dynamics in woody debris on burnt sites. Such refinements regarding the impacts of fire on the decay rate of wood and bark, and the fragmentation of bark, might be necessary for optimizing the management of ecosystem carbon stocks and the conservation of biodiversity. Hence, proper prescribed burning for restoration decreases the decomposition rate and consequently ensures the longer persistence of coarse woody debris, such as in stumps, for maintenance of biodiversity in managed forests (Shorohova et al. 2008).

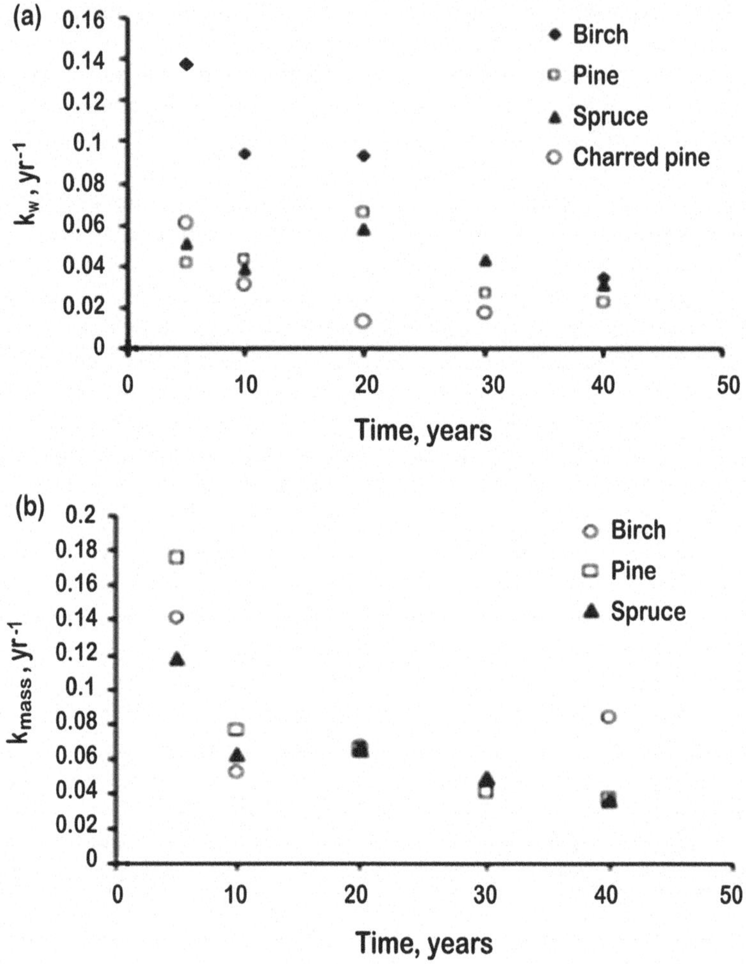


**Fig. 7.14** Annual decomposition rate of charred wood ($k_w$) (Upper) and the decomposition rate of whole mass of stump ($k_{mass}$) (Lower) for Scots pine, Norway spruce, and birches. The values are estimated based on the equation $\rho = \rho_o\, x\, e^{-kx\, t}$, where the decomposition rate k is $k_w$ or $k_{mass}$ for uncharred wood or whole stump mass (Shorohova et al. 2008). Permission of Elsevier

## 7.6 Concluding Remarks

In forests, ecosystem processes cycle carbon, nutrients, and water within/through the ecosystem in maintaining carbon stocks and modifying the structure and functioning of soils, including the stocks of water, nutrients, and organic matter. They affect nutrient fluxes and availability for uptake in physiological processes driving autotrophic and heterotrophic production in the ecosystem. The properties of soil carbon and the soil environment control ecosystem processes such as nutrient uptake, decomposition of soil organic matter with cycling of nutrients, and availability of water.

Boreal forest soil includes large carbon pools, with a significant role in the global carbon cycle. In upland sites, soil is like a profile of mineral material (parent material, basic soil components) with litter and the humus layer on mineral soil. Concurrently, vegetation alters soil physical and chemical properties, with impacts on soil biology and chemistry and on the growth and success of plants. Soil represents long-term interactions between mineral particles, climate, trees, ground vegetation, soil animals, and microbial communities. These interactions are of primary importance in management of boreal forests for various ecosystem services like timber and biomass, with further effects on the absorption of atmospheric carbon to mitigate climatic warming.

Management has various effects on soil through succession in different phases of stand- and landscape-scale chronosequences. Especially, forest regeneration with soil preparation has important and long-term effects on soil process for sustaining the productivity of forest site. Soil preparation tends to enhance the decomposition of soil carbon and translocate carbon down in the soil profile. The rate of decomposition of surface organic matter is poorly understood in terms of the depth of mineral soil and depending on the site preparation practice, site fertility, and soil type. However, the effects of soil disturbance on soil carbon stocks are overridden by rapid reforestation and increasing tree growth, which increases carbon input to soil. This reduces the long-term effects on soil carbon due to site preparation and harvest of timber and biomass in forestry.

## References

Agbeshie AA, Abugre S, Atta-Darkwa T, Awuah R (2022) A review of the effects of forest fire on soil properties. J For Res 33:1419–1441

Amiro BD, Stocks BJ, Alexander ME, Flannigan MD, Wotton BM (2001) Fire, climate change, carbon, and fuel management in Canadian boreal forest. Int J Wildland Fire 10:405–413

Certini G (2005) Effects of fire on properties of forest soil: a review. Oecologia 1434:1–10

DeLuca TH, Boisevenue C (2012) Boreal forest soil carbon: distribution, function, and modelling. Forestry 85(2):162–184

Essen PA, Ehnström B, Ericson L, Sjöberg K (1997) Boreal forests. Ecol Bullet 46:16–47

Finér L, Jurgensen M, Palviainen M, Piirainen S, Page-Dumroese D (2016) Does clear-cut harvesting accelerate initial wood decomposition? A five-year study with standard wood material. For Ecol Manag 5:10–18

James J, Harrison R (2016) The effect of harvest on forest soil carbon: a meta-analysis. Forests 7(12):308

Johnson DW (1992) Effects of forest management on soil carbon storage. Water Air Soil Pollut 64: 83–120

Johnson DW, Curtis PS (2001) Effects of forest management on soil C and N storage: meta-analysis. For Ecol Manag 140:227–238

Kauppila A, Lähde E (1975) Koetuloksia maan käsittelyn vaikutuksista metsämaan ominaisuuksiin Pohjois-Suomessa. Summary: on the effects of soil treatments on forest soil properties in North-Finland. Folia Forestalia 230:1–29

Kolström T, Kellomäki S (1993) Tree survival in wildfire. Silva Fennica 27(4):277–281
Kulmala L, Aaltonen H, Berninger F, Kieloaho A-J, Levula J, Bäck J, Hari P, Kolari P, Korhonen JFJ, Kulmala M, Nikinmaa E, Pihlatie M, Vesala T, Pumpanen J (2014) Changes in biogeochemistry and carbon fluxes in a boreal forest after the clear-cutting and partial burning of slash. Agric For Meteorol 188:33–44
Leikola M (1974) Muokkauksen vaikutus metsämaan lämpösuhteisiin Pohjois-Suomessa. Summary: effect of soil preparation on soil temperature conditions of forest regeneration areas in northern Finland. Communicationes Instituti Forestalis Fenniae 84(2):1–64
Luyssaert S, Schulze ED, Borner A, Knohl A, Hessenmoller D, Law BE, Ciais P, Grace J (2008) Old-growth forests as global carbon sinks. Nature 455(7210):213–215
Lyford WH, MacLean DW (1966) Mound and pit microrelief in relation to soil disturbance and tree distribution in New Brunswick, Canada. Harvard Forest Paper No. 15, Petersham
Mack MC, Walker XJ, Johnstone JF, Alexander HD, Melvin AM, Jean M, Miller SN (2021) Carbon loss from boreal forest wildfires offset by increased dominance of deciduous trees. Science 372:280–283
Neff JC, Harden JW, Gleixner G (2005) Fire effects on soil organic matter content, composition, and nutrients in boreal interior Alaska. Can J For Res 35:2178–2187
Olson JS (1963) Energy storage and the balance of producers and decomposers in ecological systems. Ecology 44(2):322–331
Oris F, Asselin H, Ali AA, Finsinger W, Bergeron Y (2014) Effect of increased fire activity on global warming in the boreal forest. Can J For Res 22:206–219
Piirainen S, Finér L, Starr M (2015) Changes in forest floor and mineral soil carbon and nitrogen stocks in a boreal forest after clear-cutting and mechanical site preparation. Eur J Soil Sci 66: 735–743
Pohtila E (1977) Reforestation of ploughed sites in Finnish Lapland. Seloste: aurattujen alueiden metsänviljely Lapissa. Communicationes Instituti Forestalis Fenniae 91(4):1–98
Powlson DS, Stirling CM, Jat ML, Gerard BG, Palm CA, Sanchez PA, Cassman KG (2014) Limited potential of no-till agriculture for climate change mitigation. Nat Clim Chang 4:678–683
Pumpanen J, Westman CJ, Ilvesniemi H (2004) Soil $CO_2$ efflux from a podzolic forest soil before and after forest clear-cutting and site preparation. Boreal Environ Res 9:199–212
Santin C, Otero XL, Doerr SH, Chafer CJ (2018) Impact of a moderate/high-severity prescribed eucalypt forest fire on soil phosphorous stocks and partitioning. Sci Total Environ 621:1103–1114
Seedre M, Taylor AR, Brassard BW, Chen HY, Jõgiste K (2014) Recovery of ecosystem carbon stocks in young boreal forests: a comparison of harvesting and wildfire disturbance. Ecosystems 17:851–863
Shorohova E, Kapitsa E, Vanha-Majamaa I (2008) Decomposition of stumps in a chronosequence after clear-felling vs. clear-felling with prescribed burning in a southern boreal forest in Finland. For Ecol Manag 255(10):3606–3612
Ulanova N (2000) The effect of windthrow on forests at different spatial scales: a review. For Ecol Manag 135:155–167
Vanha-Majamaa I, Tuittila E-S, Tonteri T, Suominen R (1996) Seedling establishment after prescribed burning of clear-cut and partially cut mesic boreal forest in southern Finland. Silva Fennica 30(1):31–35
Viro PJ (1969) Prescribed burning in forestry. Communicationes Instituti Forestalis Fenniae 67(7): 1–49
Zackrisson O (1977) Influence of forest fire on the north Swedish boreal forest. Oikos 29(1):22–32

# Chapter 8
# Reforestation with Impacts on Carbon Stocks in Boreal Forest Ecosystem

**Contents**




**Abstract** Clear-cutting ceases the formation of foliage and root litter in trees with the consequent decline of soil carbon. Clear-cutting reduces soil carbon stocks most in the forest floor (litter, humus) but also deeper in the soil profile. The preexisting litter and humus decline until carbon uptake recovers in ground vegetation (shrubs, herbs, grasses). Consequent reforestation with enhancing growth of seedlings further increases the recovery of soil carbon stocks and slows down the loss of soil carbon. Tree species selection and reforestation practices such as planting density determine





S. Kellomäki, *Forest Management for Timber Production and Climate Change Mitigation*, Managing Forest Ecosystems 44,
https://doi.org/10.1007/978-3-031-71575-4_8

the rate of carbon sequestration and stocking in soil after clear-cutting. Tree species also affect the sequestration and stocking of soil carbon depending on the amount and quality of litter. Litterfall and harvesting residues further affect the carbon stocks in the litter layer and mineral soil, thus maintaining the soil detritus or humus layer below the surface litter layer. Deeper in the soil profile, the impacts on carbon density are smaller and occur over the long term.

## 8.1 Carbon Management Through Reforestation

### *8.1.1 Carbon Uptake and Losses in Relation to Management and Harvest*

Any disturbance, such as a management operation, changes the structure and functioning of boreal forest ecosystems. In even-aged management, the change is profound: from clear-cutting through reforestation, precommercial and commercial management, to the clear-cutting and initiation of the next production cycle. The rate of carbon and nutrient cycles and the consequent dynamics of site properties are inherent to those induced by the previous production cycle. For uneven-aged management, the carbon stocks in trees and soil are less variable compared to those under even-aged management. However, the mean carbon stocks (trees, soil) tend to be equal under even-aged management than that under uneven-aged management, if the management recommendations are followed (Kellomäki et al. 2019).

During secondary succession, the temporal amount of carbon in soil detritus is linked to the amount before the stand-replacing disturbance event, such as clear-cutting *with only stem wood removed*. As expected, legacy carbon in woody detritus decreases during succession, while de novo carbon increases following reforestation (Fig. 8.1; Box 8.1). Over time, the amount of total carbon is dependent on the amount of carbon removed in the harvest. According to Harmon (2009), no removal of legacy carbon yields a reverse shape, with the consequence that the amount of woody carbon in the middle phase of succession falls below that of the later phase. The lower the intercept, the slower the reestablishment of de novo carbon is compared to that occurring under faster reestablishment.

The effects of clear-cutting with *both stem wood and logging residues* deviate substantially from those under the stem-wood-only harvest. In the harvest of stem wood and logging residues, legacy carbon is virtually missing at the beginning of the new production cycle; that is in reforestation after clear-cutting (Fig. 8.1; Box 8.1). De novo carbon sequestered in the new tree generation with litterfall partly compensates the loss of legacy carbon, but the recovery period is longer than under stem-wood-only harvests. The removal of logging residues for energy production reduces the soil carbon over a longer period compared to that under the harvest of only stem wood. Furthermore, carbon in the harvest of stem wood and logging residues is

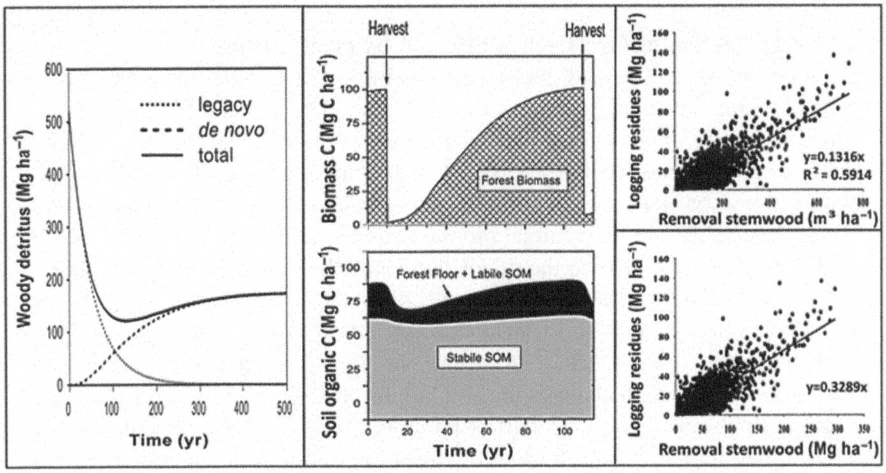


**Fig. 8.1** Carbon stocks in boreal forests over time in relation to harvest pattern. Left: overview of a model of woody detritus mass through succession, where legacy carbon mass represents that left in the site by the disturbance, such as clear-cutting with stem wood harvest, following a negative exponential function in decline. De novo carbon is related to the litterfall and mortality of trees in the established forest. The total woody detritus is based on de novo carbon if the stand age exceeds 250 years, after which no previous legacy carbon is present (Harmon 2002, 2009). Permission of Oxford University Press. Middle: simulation of carbon dynamics in the aboveground biomass and soil induced by harvesting for temperate Norway spruce forest under a 100-year rotation. The share of labile soil organic matter (carbon in forest floor and stable soil organic matter) is 25% of the total soil carbon (Jandl et al. 2007a, b). Permission of Elsevier. Right: logging residues on thinned and clear-cut areas just after treatment as a function of the removal of stem wood in volume. Logging residues include the branches with foliage, coarse roots, upper parts of stems, and stumps of Scots pine, Norway spruce, and pendula and pubescent birches. Dots represent simulated mean values for Finnish National Forest Inventory sample plots under the current climate (Kilpeläinen and Baker 2013). Courtesy of Springer Nature

about 30 40% larger than that in the stem wood only harvest (Kilpeläinen and Baker 2013).

In forestry, the share of timber and energy biomass depends on the age structure of forest resources exposed to harvesting and management. In both cases, the harvest gain per area is higher in the final cut than in thinning, but there is a strong correlation between the harvest of timber and energy biomass (Fig. 8.1). The harvest for energy biomass is further increased if stumps and coarse roots are harvested for energy biomass in clear-cut areas. In Finland, the harvestable biomass potential in energy production is of the magnitude of 16 million m$^3$ yr.$^{-1}$ (Helynen et al. 2007), with 45% originating from young stands in precommercial thinning and stands in the first commercial thinning phase. The remaining 55% represents harvesting of logging residue from final cuts representing the upper parts of the stem, branches, stumps, and coarse roots. Harvest of coarse roots mainly occurs from sites dominated Norway spruce.

**Box 8.1 Carbon in forest soil is affected by clear-cutting**
In clear-cutting, the flow of photosynthates through tree roots into soil ceases, thus inducing substantial changes in the decay processes and fungal communities in soil (Kohout et al. 2018). The soil below trees can be divided into three separate zones: (i) fine roots, (ii) rhizosphere, and (iii) bulk soil mixed with organic particles, with carbon content declining down the soil profile. These zones differ in their fungal biomass content and enzyme activity, being higher in the root zone over the rhizosphere, and the bulk soil. In clear-cutting, the enzyme activity of the rhizosphere declines and even disappears, gradually resembling that of bulk soil (Prescott 2005, 2010). Thus, soil organic matter represents the dying roots of trees and vascular ground cover species and fungal mycelia, which changes the soil habitat. In 2 years of clear-cutting, the mass of fine roots declines to half of the initial mass. In Kohout et al. (2018), this implied a mass loss of 5 Mg $ha^{-1}$ from fine roots. Root endophytic fungi may have a significant role in the early phases of this process. In soil, activity of extracellular enzymes also decreases in the long term following the end of rhizodeposition by tree roots (Kohout et al. 2018).

## *8.1.2 Carbon Fluxes in Clear-Cut Areas*

Clear-cutting alone and in combination with soil scarification and reforestation increases the emission of carbon dioxide from soil, with a concurrent absence of carbon input from litterfall. In this context, Vestin et al. (2020) measured the greenhouse gas emissions (GHGs, carbon dioxide ($CO_2$), methane ($CH_4$), nitrous oxide ($N_2O$)) related to harvesting. The experimental sites were in central Sweden (60° N), with a mean annual air temperature of 6.5 °C and mean annual precipitation of 575 mm. Measurements began a year after clear-cutting, soil scarification, and planting. The soil was a sandy-loamy glacial till covered with an organic layer 3–10 cm thick. The preharvest forest stand was a mixture of Scots pine (*Pinus sylvestris*, 58%) and Norway spruce (*Picea abies*, 40.5%), with a small fraction of birch (*Betula* spp., 1.5%). Prior to harvesting, the ground vegetation was dominated by bilberry (*Vaccinium myrtillus*) and feather mosses (*Hylocomium splendens* and *Pleurozium schreberi*). Following the harvest, soil was mounded or harrowed for planting in May 2010. In total, 2500 seedlings $ha^{-1}$ were planted (50% each of Scots pine and Norway spruce). Following harvesting, grasses and shrubs became dominant, thus having a detrimental impact on the seedling survival. Therefore, an additional 2200 seedlings $ha^{-1}$ (100% Norway spruce) were planted in May 2012.

Vestin et al. (2020) found that soil moisture affected the emissions of each greenhouse gas (GHG), with a contribution to the total GHG budget. Rapid return of ground vegetation at the moist plot reduced the total emissions of $CO_2$ compared to that at the dry plot. This was partially offset by high $CH_4$ emissions. Waterlogging following clear-cutting switched the plots from sinks to sources of $CH_4$. In addition,

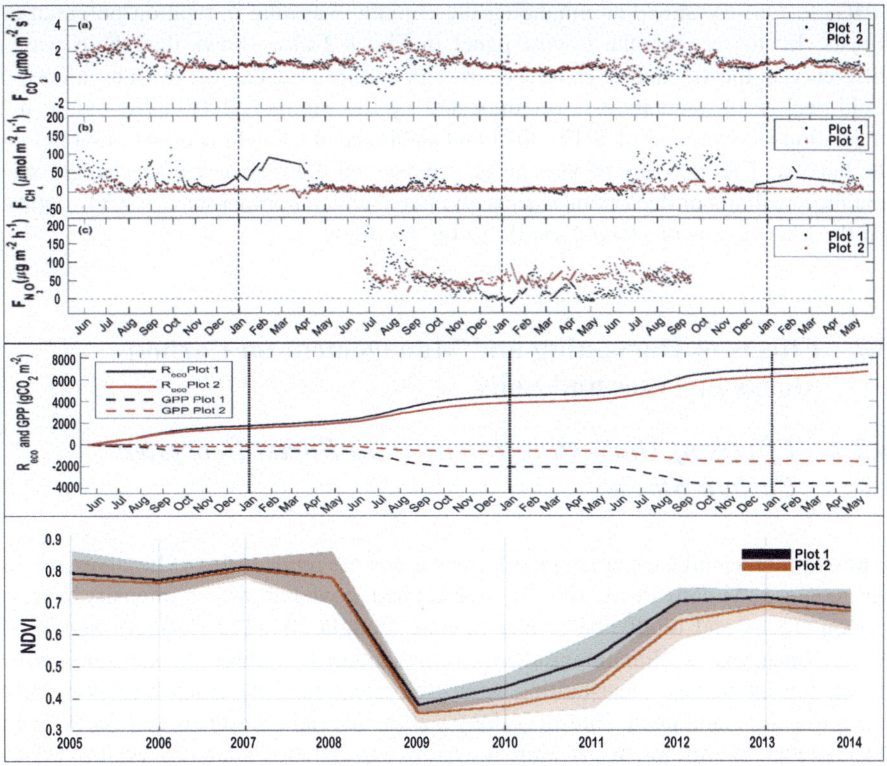


**Fig. 8.2** Impacts of clear-cutting and harvesting on greenhouse gas fluxes in a boreal forest (Vestin et al. 2020). Open Access. Courtesy of MDPI. Upper: daily average greenhouse gas (GHG) fluxes at plot 1 (black dots) and plot 2 (red dots): (a) $CO_2$ fluxes from May 2010 to May 2013, (b) $CH_4$ fluxes from May 2010 to May 2013, and (c) $N_2O$ fluxes from June 2011 to September 2012. Middle: ecosystem respiration ($R_{eco}$) and gross primary productivity (*GPP*) in g $CO_2$ m$^{-2}$ from May 2010 through May 2013 on plot 1 ($R_{eco}$: black line; GPP: black dashed line) and plot 2 ($R_{eco}$: red line; *GPP*: red dashed line). Below: Landsat-derived summertime normalized difference vegetation index (NDVI) from 2005 to 2014, covering both pre- and postharvest conditions in both plots. The shaded bands represent the standard deviation of the spatially and temporally averaged data used to create an annual summertime NDVI

there were periods of $N_2O$ uptake at the moist plot, but both plots were net sources of $N_2O$ on an annual basis (Fig. 8.2). The total 3-year carbon budgets were 4107 g $CO_2$-equivalent m$^{-2}$ and 5274 g $CO_2$-equivalent m$^{-2}$ at the moist and dry plot, respectively. The role of $CO_2$ was overwhelming, contributing 91.8% to the total carbon budget at the moist plot and 98.2% at the dry plot.

Regarding only the full year with $N_2O$ measurements, the total GHG budgets were 1070 g $CO_2$-eqvivalents m$^{-2}$ and 1696 g $CO_2$-eqvivalents m$^{-2}$ at the moist and dry plot. At the moist plot, $CH_4$ contributed 3.7% to the total GHG budget, while the $N_2O$ contribution was 7.3%. At the dry plot, $CH_4$ and $N_2O$ contributed up to 1.5% and 7.6%. These findings emphasize the importance of the consideration of the three

GHGs in forestry aimed at enhancing the climatic-warming mitigation potential of forests. In this regard, the lowest panel in Fig. 8.2 also shows that there was a difference in the vegetation cover in early 2009 before the harvest, which induced a rapid decline of vegetation cover indicated by the summer NDVI. Thereafter, the vegetation recovered until 2012–2013 and stabilized at a level lower than that before the harvest. The recovery of vegetation represented the recovery of herbs, grasses, and the seedlings of deciduous shrubs and trees, which compensated in $CO_2$ uptake for the low success of planted seedlings on the plots.

## 8.2 Effects of Harvesting and Management on Carbon Stocks in Trees and Soils

### *8.2.1 Effects of Harvesting Intensity on Forest Ecosystem Carbon Stocks*

Forests provide sinks for atmospheric carbon and are repositories of biodiversity. In this context, Simard et al. (2020) investigated how harvesting intensity affects carbon stocks and biodiversity in temperate Douglas fir (*Pseudotsuga menziesii*) forests spanning a climatic gradient from semiarid to humid conditions. Five harvesting intensities (clear-cutting, seed-tree retention, 30% patch retention, 60% patch retention, and uncut controls) along a climatic gradient were used (Fig. 8.3). In this context, increasing harvesting intensity reduced the total ecosystem carbon stock. The aboveground carbon stock and carbon stock in trees decreased within a year of harvesting. The losses were negatively correlated with climatic aridity. In humid forests, the decline in total ecosystem carbon stock was up to 50% for harvesting through clear-cutting, while the carbon loss in large-patch retention harvesting was 30%. In arid forests, the loss for the same harvesting regimes ranged from 60% to 8%. In lower-retention harvests, small-patch retention protected both carbon stocks and biodiversity in the arid forests, while seed-tree cutting performed similarly or even better in the humid forests.

According to Simard et al. (2020), the belowground carbon stocks declined by an average of 29% after harvesting. The loss took place from the forest floor, while there was no loss from the mineral soil. Regarding the standing and coarse deadwood in soil (secondary carbon pools), carbon stocks declined regardless of harvesting regime, cutting intensity, or site aridity (Fig. 8.3). Carbon stocks in logging residues and stumps increased after harvesting. In understory plants, the carbon pool declined across all harvesting intensities in the humid forests but increased in arid forests. In harvesting, the highest retention yielded the lowest carbon losses and highest increase in biodiversity, with the opposite impacts occurring in clear-cutting. Partial retention of canopy trees can reduce losses in carbon, biodiversity, and other ecosystem services.

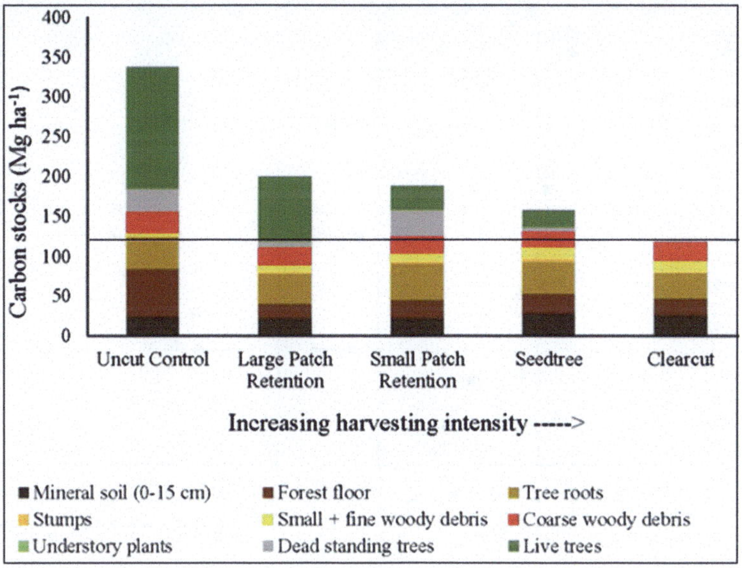


**Fig. 8.3** Mean carbon pools for each harvesting regime along a harvesting intensity gradient (Simard et al. 2020). Open Access. Courtesy of Frontiers Media. Horizontal black line added by the author of this book for scaling the level of carbon stock vs. increasing harvest intensity. Further explanations of the author of this book for harvest treatments included in the study. Clear-cut (no tree retention). Single tree retention (retention of 25 large Douglas fir stems $ha^{-1}$ at 25 m spacing), which resembles traditional tree-seed cutting with single parent trees representing wide location (density 20–100 parent trees $ha^{-1}$) in natural regeneration like that of Scots pine in northern Europe. Small patch retention (retention of 30% of the stand area in 30-m-wide unconnected patches, with all trees cut in the remaining 70% of the stand), which resembles shelterwood cutting with densely located patches of parent trees (density 200–400 parent trees $ha^{-1}$) like that for Norway spruce in natural regeneration in northern Europe. Large-patch retention (retention of 60% of the stand area with all trees cut in the remaining 40% of the stand) resembling the patch-scale selective cutting under uneven-aged management in regeneration of Norway spruce or Scots pine in northern Europe. Uncut control (full retention)

### 8.2.2 *Effects of Forest Management on Soil Carbon and Nitrogen Stocks*

In northern Europe, thinning and clear-cutting is widely used in the production of timber and biomass. Clear-cutting for harvesting, followed by planting, initializes the next production cycle in even-aged forestry. Harvesting tends to reduce soil carbon up to 10% in the soil profile, with the greatest loss of carbon occurring from the forest floor in temperate and boreal forests (Johnson 1992). Clear-cutting reduces soil carbon stocks most in the forest floor (litter, humus) but also deeper within the soil profile, as found by Mayer et al. (2020). The total carbon loss in temperate forests was up to 30% in the forest floor, while the effect of clear-cutting on carbon

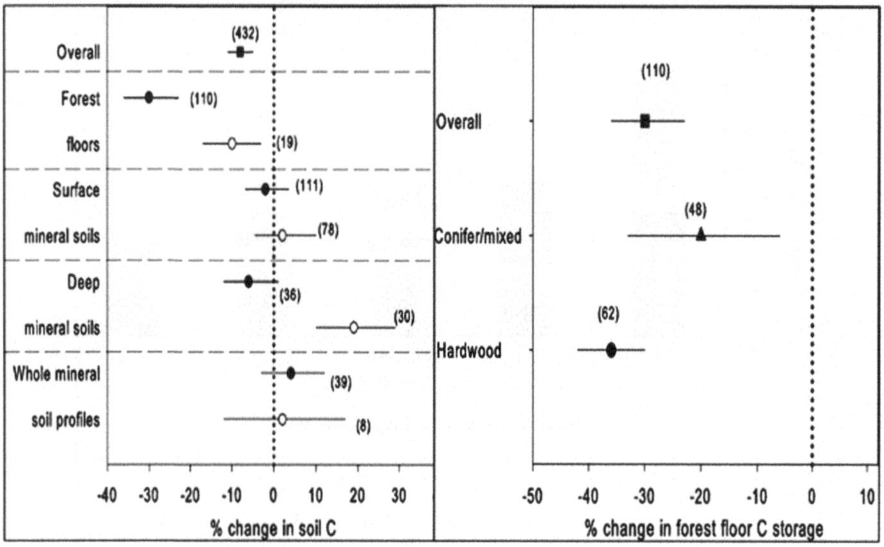


**Fig. 8.4** Impacts of harvesting on soil carbon in temperate forests (Nave et al. 2010). Permission of Elsevier. Meta-analysis shows the magnitude of change in a parameter (i.e., the "effect size") in response to an experimental treatment across a wide range of experimental systems and conditions. The effect of treatment is indicated by the response ratio $\ln(R) = \ln\left(X^{-E}/X^{-C}\right)$, where $X^{-E}$ is the carbon value in the site exposed to harvesting and $X^{-C}$ is the carbon value in the site not exposed to harvesting. The number of response ratios ($k$) from a given publication depends on the number of experimental conditions, e.g., soil carbon in sites with four thinning levels, plus one site with no thinning representing the control. Left: changes in soil carbon from the forest floor to deep mineral soil (20–100 cm) due to harvesting. The numbers in parentheses indicate the number of studies used in the meta-analysis to estimate the mean value and confidence of changes in carbon in different parts of the soil profile. Right: changes in carbon in forest floor related to the mixture of tree species

was smaller deeper in the soil. Consequently, the carbon loss in the whole soil profile due to clear-cutting was <10%.

A similar distribution of soil carbon loss related to harvesting was found by Nave et al. (2010) in the meta-analysis of temperate forests, based on more than 400 studies throughout the world (Fig. 8.4). This summary showed that harvesting (thinning, clear-cutting) reduced soil carbon by up to 8 ± 3% on average. According to Nave et al. (2010), the soil carbon stock responded differently depending on harvesting pattern and mineral-soil type. The forest floor was likely to lose more carbon than the mineral soil. At the same time, the carbon stock in the forest floor reduced 30 ± 6%. Nave et al. (2010) further found that carbon losses from the forest floor were lower in coniferous/mixed stands (20%) than in deciduous stands (36%) (Fig. 8.4). According to Nave et al. (2010), the recovery of carbon stock in the forest floor takes a period of 50–70 years or more depending on the soil conditions.

James and Harrison (2016) also found that different harvesting (clear-cutting, thinning) combined with harvesting intensity (stem-wood-only versus whole-tree harvesting) affected soil carbon stocks. The analysis was based on 945 databases for coniferous and deciduous forests, or mixes thereof, throughout temperate and boreal biomes. The mean loss of soil carbon was 11.2%. However, the responses varied in different soil depths through the soil profile, with the highest losses (30.2%) being in the O horizon (top layer of soil composed of decaying organic matter). The carbon loss was also clear (3.3%) deeper within the upper soil carbon pools (0–15 cm depth). Even in deep soil (60–100 cm), a loss of soil carbon (17.7%) was induced by the removal of tree biomass in harvesting. James and Harrison (2016) found that carbon loss was dependent on soil type, with the highest losses in podzols and clays. The study encompassed multiple chronosequences, showing that carbon stocks in the forest floor and mineral soil started to recover in one to five decades following harvesting. For example, the recovery of soil carbon in podzols requires several decades (Nave et al. 2010; James and Harrison 2016).

Johnson and Curtis (2001) also used a meta-analysis to define how different management/harvesting techniques affect soil carbon and nitrogen stocks in temperate/boreal conditions. Harvesting induced losses of soil organic matter and nutrient contents related to site preparation. In this regard, Fig. 8.5 shows how the observations of soil carbon and nitrogen are distributed in relation to the percentage change induced by harvesting. On average, harvesting had a small effect on soil carbon and nitrogen. Nevertheless, harvesting of saw logs increased (18%) soil carbon and nitrogen stocks. This was because harvest residues were left on the site in stem-based harvest, while whole-tree harvesting reduced (6%) carbon and nitrogen due to removal of carbon- and nitrogen-rich mass as shoots and foliage. The increase of soil carbon and nitrogen occurred only after harvesting of coniferous saw logs.

## 8.3 Tree Species Selection Impacts on Soil Carbon

### *8.3.1 Effects of Single Tree Species on Soil Carbon*

In forestry, the selection of tree species in reforestation is widely related to the productivity of species and the need to produce timber for a given purpose. Tree species selection also affects the carbon sequestration and stocks in forest ecosystems. For combined production of timber and carbon services (carbon sequestration and carbon stocks in forest ecosystems), the selection of tree species is among the key decisions for optimized management in the scale of forest landscape. According to Jandl et al. (2007a, b), the proper selection of tree species for terrestrial carbon stocks is important for the following reasons:

- The growth of wood over time and wood density are species-dependent, with species-specific effects on carbon sequestration capacity.
- Population dynamics with species-specific spacing affect the stocking of tree stands, with subsequent effects on carbon sequestration.
- Rooting pattern and depth are species-specific, affecting carbon sequestration in soil carbon stocks.

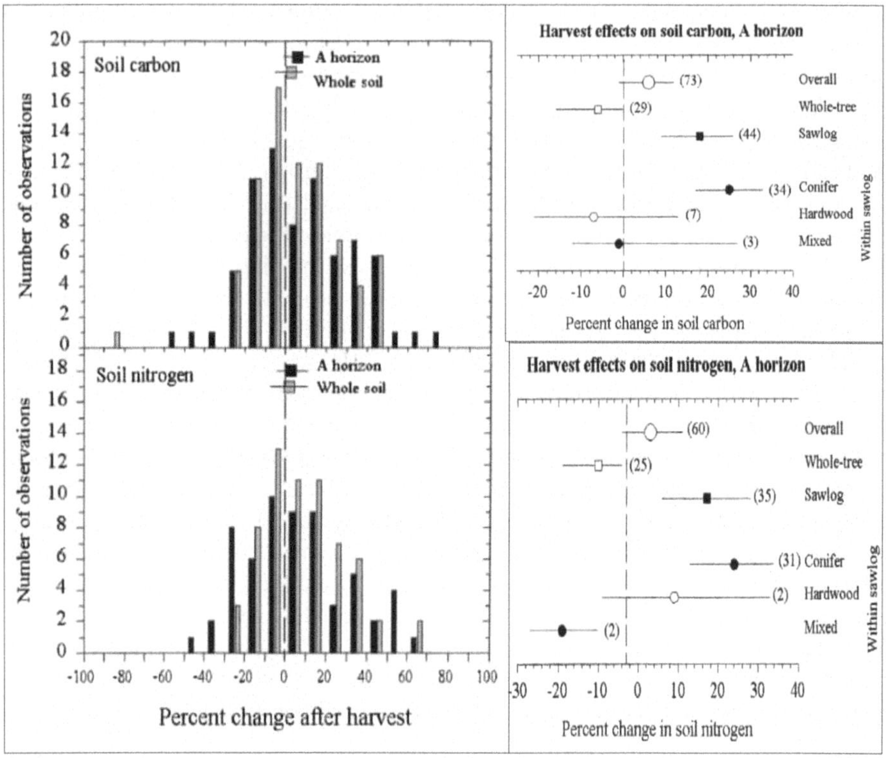


**Fig. 8.5** Effect of management and harvesting on soil carbon and nitrogen, based on Johnson and Curtis (2001). Permission of Elsevier. Left: distribution of the number of studies showing the effect of harvesting on soil carbon (upper panel) and nitrogen (lower panel) in A horizons and whole soils in deciles of percent change from the control. Right: harvesting effect on soil carbon (upper panel) and soil nitrogen in A horizons (lower panel)

- The life span of trees is species-specific, with effects on lifelong carbon sequestration in above- and belowground carbon stocks.
- Vulnerability to disturbances is species-specific, with effects on the life span of trees and lifelong carbon sequestration in ecosystems.

Trees affect soil carbon depending on the amount and quality of litter produced and on species. Litterfall and harvesting residues form the litter layer on mineral soil and maintain the soil detritus or humus layer below the litter layer (e.g., Vesterdal et al. (2013)). Decomposing litter also feeds humus particles deeper in the soil profile, helping water infiltrate down to the parent material and groundwater and further outside the site. This is common on humus-rich sites, where rainfall exceeds the outflow of water outside the roots of trees. In this regard, trees supply a large amount of carbon to the soil profile through root growth and mortality, which renews and maintains the humus mixed among soil particles in the soil profile.

Regarding the long-term impacts of tree species on soil carbon, Hansson et al. (2011) investigated the role of common boreal tree species in driving the carbon dynamics in Scots pine (*Pinus sylvestris*), Norway spruce (*Picea abies*), and Pendula birch (*Betula pendula*) 50 years after planting. The site was on a heather (*Calluna*-dominated) moorland in a hemiboreal area (56° N) in southwestern Sweden. The experimental stands had been thinned four to five times. The measured basal area was 15.4 $m^2$ $ha^{-1}$ for Scots pine, 29.3 $m^2$ $ha^{-1}$ for Norway spruce, and 15.5 $m^2$ $ha^{-1}$ for birch. The study plots were characterized as pure (containing only a single species) or slightly mixed stands, such as those with Scots pine and birch or Scots pine and Norway spruce. The climate, parent material, land-use history, and soil type were similar in the study plots. This implies that the differences in soil properties were related to the dominating tree species used in reforestation of the given plots (Fig. 8.6).

Hansson et al. (2011) found that the soil properties were similar regardless of main tree species and the species mixture in each study plot. The effects of tree species were most apparent in the humus layer and decreased deeper in the soil profile (Fig. 8.6). Down to a depth of 20–30 cm, the differences in soil properties related to the dominant tree species were small. In the Norway spruce stands, the carbon stored in the mineral soil down to 30 cm was 7.3 kg C $m^{-2}$, in birch stands 4.1 kg C $m^{-2}$, and in Scots pine stands 4.9 kg C $m^{-2}$. The variability in soil carbon was related to the between-species differences in the type and amount of litterfall. The fine litterfall was 2.3 Mg $ha^{-1}$ $yr.^{-1}$ for Scots pine, 1.2 Mg $ha^{-1}$ $yr.^{-1}$ for birch, and 2.0 Mg $ha^{-1}$ $yr.^{-1}$ for Norway spruce. Including coarse litter, the litterfall was 2.6 and 2.5 Mg $ha^{-1}$ $yr.^{-1}$ for Scots pine and Norway spruce. In birch, the combined amount of fine and coarse litter was 1.2 Mg $ha^{-1}$ $yr.^{-1}$.

Hansson et al. (2011) further found that the thickness of the humus layer was 6.7 cm in Norway spruce stands. In Scots pine and birch stands, the thickness of the humus layer was smaller with 4.7 cm and 2.1 cm, respectively. The total soil carbon stock, including the humus layer and mineral soil of 0–30 cm in depth, was significantly higher in Norway spruce stands (7270 g C $m^{-2}$) than in Scots pine stands (4922 g C $m^{-2}$), and birch stands (4084 g C $m^{-2}$). The same order remains in the findings of Stendahl et al. (2010): the carbon stock in organic soils tends to be higher in Norway spruce (9.2 kg C $m^{-2}$) than in Scots pine (5.7 kg C $m^{-2}$), increasing along with rising temperatures. The same distribution pattern held for soil nitrogen. The total amount of nitrogen was higher in Norway spruce stands (349 g N $m^{-2}$) than in birch (240 g N $m^{-2}$) and Scots pine (269 g N $m^{-2}$) stands (Hansson et al. 2011). The ratio between the total amount of carbon and nitrogen (C/N) in the profile was clearly lower for birch (17) and Scots pine stands (18) than for Norway spruce stands (20), with a similar pattern for the humus layer (Fig. 8.6).

Hansson et al. (2011) concluded that less than one rotation period is enough for clear differences to emerge in many soil properties. This holds especially for the humus layer between birch, Scots pine, and Norway spruce in the stand adjacent to each other. Pure Norway spruce stands sequestered more carbon than pure Scots pine and birch stands as further found by Stendahl et al. (2010). The simulations over several rotations showed that SOC was 22% higher in Norway spruce than in Scots

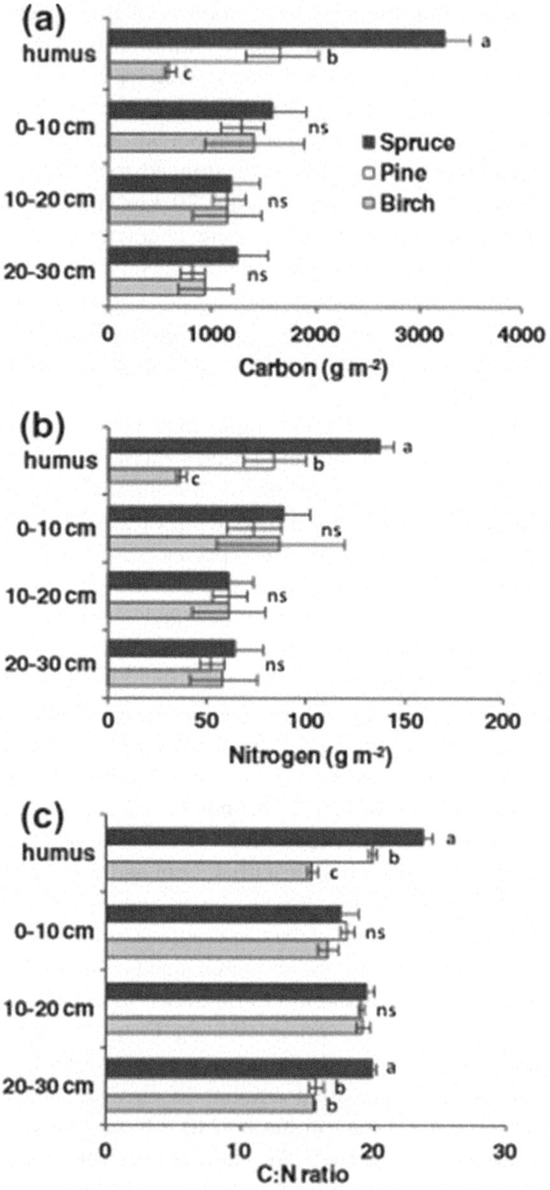


**Fig. 8.6** Differences in (**a**) the amount of carbon (g C $m^{-2}$), (**b**) the amount of nitrogen (g N $m^{-2}$), and (**c**) the carbon/nitrogen ratio at different depths for the stands of Scots pine, Norway spruce, and birch. Different letters indicate significant differences between tree species ($p < 0.05$), and "ns" indicates nonsignificance (Hansson et al. 2011). Permission of Elsevier

pine stands, which may be the difference in the number of harvesting residues. The soil organic matter related to ground cover species was higher under Scots pine than under Norway spruce due to differences in the quality and amount of litter production in ground vegetation (Stendahl et al. 2010).

### 8.3.2 Carbon Sequestration in Mixed Forests

In boreal conditions, forests often comprise mixtures of coniferous and deciduous species. This is especially common on fertile sites (e.g., Payne et al. 2019). Depending on tree species, carbon is taken up at different rates, thus affecting the carbon accumulation of trees, soil detritus, and the soil profile more than in forests occupied by a single species, which may occur on poor or moderately fertile sites. However, appropriate selection of tree species for growing in pure stands or in a mixture of several species may enhance the carbon sink capacity of forest ecosystems, as found by Blaško et al. (2020).

Blaško et al. (2020) investigated the capacity of carbon sequestration in pure Scots pine and Norway spruce, and their mixture, in southern boreal conditions (60° N) on sites of moderate fertility. The study was based on a 60-year-old experiment for monocultures and a mixture of Norway spruce and Scots pine on the sites equally suitable for the growth of both species. The focus was to assess the importance of tree species selection in the carbon management of boreal forest ecosystems both above and below the soil surface, with the aim of influencing the carbon cycle and thus mitigating climatic change. The more specific aim was to evaluate the properties of both species regarding single species and species mixture effects on ecosystem carbon and nitrogen stocks. Total standing volume, aboveground tree biomass, fine-root biomass, carbon and nitrogen pools in tree biomass and soil, litterfall inputs, and soil $CO_2$ emissions were measured in the stand scales.

Blaško et al. (2020) found that the stocking and growth in Scots pine was double compared to that of Norway spruce stands (Fig. 8.7). The difference in litterfall was related to the stocking difference, thus showing higher litterfall for Scots pine than for Norway spruce, with the amount of litterfall being dependent on the stocking of tree species occupying the site. However, foliage mass in Norway spruce was higher than in Scots pine stands, while the accumulation of carbon in humus was lower than in the Scots pine-dominated stand. The carbon sinks in the aboveground tree biomass and in soil were larger in Scots pine than in Norway spruce stands. Blaško et al. (2020) further found that the carbon dioxide effluxes from soil and fine-root mass in the soil were larger in the sites occupied by Norway spruce than that of Scots pine.

On the other hand, Blaško et al. (2020) found that there was "no significant mixing effect in the mixed stands, given the levels of the measured variables did not exceed levels of the most productive monoculture with the exception of higher SOC stocks (soil organic carbon stocks) in the deeper (10–20 cm) mineral-soil layer in the mixed stands." This implies that Scots pine and Norway spruce are not complementary to each other in terms of carbon sequestration and management. The mixed stands may perform similarly to monocultures. This finding indicates that management for multispecies stands may not reduce the carbon uptake and accumulation of carbon in boreal forest ecosystems, as also emphasized by Mäkipää et al. (2023).

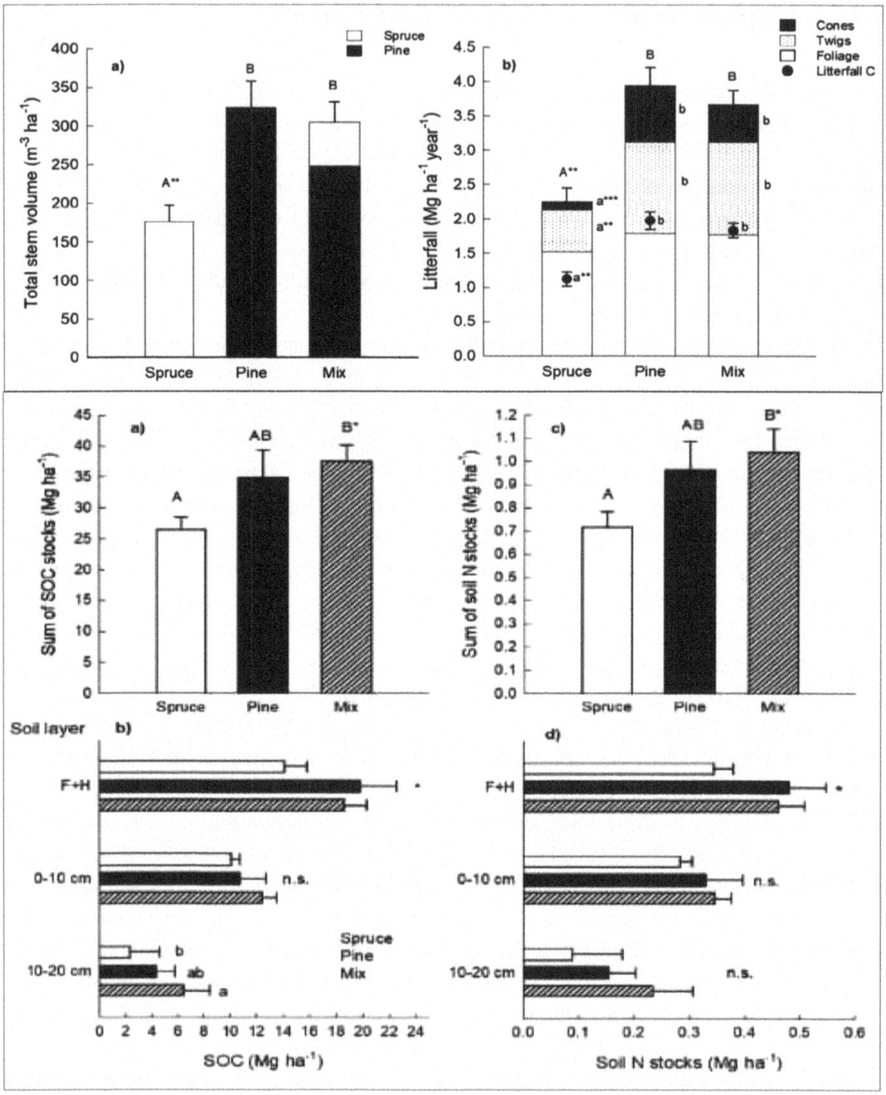


**Fig. 8.7** Effects of tree species and their combinations on carbon sequestration and stocks for pure Scots pine (*Pinus sylvestris*) and Norway spruce stands (*Picea abies*) and a mixture of both species (Blaško et al. 2020). Permission of Elsevier. Upper: total stocking (m$^3$ ha$^{-1}$) and annual total litterfall (Mg ha$^{-1}$ yr.$^{-1}$). Middle: sum of soil organic carbon stocks (Mg ha$^{-1}$) and sum of soil nitrogen stocks (Mg ha$^{-1}$). Lower: organic soil carbon (Mg ha$^{-1}$) and nitrogen in soil profile in different layers

## 8.4 Impacts of Planting Density on Forest Ecosystem Carbon Stocks

### *8.4.1 Stand Density for Increased Soil Carbon*

Clear-cutting leads to the loss of soil carbon due to the termination of foliage and root litter from harvested trees. Loss of soil carbon is further driven by the decay of preexisting litter and humus in the soil profile until the new tree generation is established through reforestation. In this context, Na et al. (2021) investigated how the stand density affects the sequestration and storage of carbon in the forest ecosystem (trees and soil) in 8-year-old pure hybrid larch (*Larix* spp.) plantations established at five densities: 833, 1000, 2000, 3300, and 4400 seedlings per hectare. The sites represented hybrid larch in boreal mountainous forests in northeastern China (45° N).

In the analysis, the focus was on experimental plots with densities of 2000, 3300, and 4400 seedlings per hectare, representing low (LD), medium (MD), and high (HD) density, respectively (Table 8.1). Na et al. (2021) found that soil carbon stock increased along with increasing stand density. The amount of soil carbon in forest floor and roots correlated with the increases of dissolved organic carbon and litter and carbon input from roots (Table 8.1). Concurrently, soil respiration decreased most in the stands of higher tree density under warm and most conditions representing the highest productivity in the growing period. The reduced decay of soil organic matter indicated by lowered respiration was related to a reduced levels of decay of litter. The stimulated inputs and reduced exports of carbon from the forest floor resulted in a 40% higher soil carbon stock in the high-density stand compared to that in the low-density stand in 8 years after planting.

### *8.4.2 Combining Effects of Stand Density and Tree Species for Increased Soil Carbon*

In soil carbon stocks, the input of litter above and below the soil has an important role in maintaining carbon sequestration in the forest ecosystem. The interaction between tree species and spacing affects the amount of litter contributing carbon to the soil surface and, through decay, into soil detritus, with impacts on the deposition of carbon at different depths in the soil profile. This is affected by the growth and development of trees through chronosequences and the consequent forest management. However, stand density tends to have only small or nonexistent effects on soil carbon, as was in young stands found by Menyailo et al. (2022). They assumed the absence of an effect of stand density is related to a process in which the old soil carbon is replaced by fresh soil carbon, resulting in zero net carbon sequestration, with a change in the quality of the soil organic matter.

**Table 8.1** Carbon experiment sites occupied by larch (*Larix* spp.) in boreal mountainous forests in northeastern China (45° N) (Na et al. 2021)

| Density of trees per hectare | Plant biomass and carbon stocks in forest floor and roots | | | | | | |
|---|---|---|---|---|---|---|---|
| | Biomass, g m$^{-2}$ | | Carbon stock, g m$^{-2}$ | | Annual litter fall, g m$^{-2}$ | | Loss, % yr.$^{-1}$ |
| | Forest floor | Roots | Forest floor | Roots | First year | Second year | |
| 2000 (LD) | 544 ± 51.1a | 485 ± 141a | 169 ± 14.5a | 178 ± 74.0a | 250 ± 87.8 | 325 ± 15.4 | 59.1 ± 7.5 |
| 3300 (MD) | 612 ± 74.8a | 843 ± 241.5ab | 239 ± 22.0b | 337 ± 79.3a | 337 ± 47.2 | 348 ± 44.5 | 57.7 ± 5.9 |
| 4400 (HD) | 647 ± 47.6a | 12,161 ± 446.5b | 207 ± 38.4ab | 445 ± 85.5a | 346 ± 9.7 | 369 ± 11.9 | 46.6 ± 8.6 |

| Density, trees ha$^{-1}$ | Thickness of soil layers from the soil surface down the soil profile, cm | | | | |
|---|---|---|---|---|---|
| | 0–10 | 10–20 | 20–40 | 40–60 | Soil profile, 0–60 |
| 2000 (LD) | 4.6 ± 0.8a (53) | 2.1 ± 0.3ab (23) | 1.2 ± 0.2a (13) | 0.9 ± 0.1a (11) | 8.8 ± 0.5a (100) |
| 3300 (MD | 5.8 ± 1.6a (59) | 1.6 ± 0.1a (17) | 1.3 ± 0.4a (13) | 1.1 ± 0.0a (11) | 9.8 ± 1.6ab (100) |
| 4400 (HD) | 6.3 ± 1.5a (51) | 2.7 ± 0.4b (22) | 1.7 ± 0.9a (13) | 1.6 ± 0.6a (13) | 12.3 ± 1.3b (100) |

Explanation for the table. Left/upper: plant biomass and carbon storage of the forest floor and roots in a larch plantation with varying densities of trees. Right/upper: annual litterfall mass and rate of litter mass loss with varying densities of trees. Lower: soil carbon storage at different layers with different densities of trees, with the numbers in parentheses indicating the distribution of carbon through the soil profile (0–60 cm)

Legend for planting densities trees per hectare—LD (low density) = 2000, MD (middle density) = 3300, and HD (high density) = 4400—with the densities in seedlings ha$^{-1}$. Values are given as mean ± SE ($n = 3$), and the figures in brackets are percentages of soil carbon storage. Different lowercase letters indicate significant differences by density ($P < 0.05$) based on Tukey's test. Open Access. Courtesy of Springer Nature

Menyailo et al. (2022) further investigated whether selected tree species at varying stand densities affect the dynamics of soil carbon in an afforestation experiment established 30 years prior for Norway spruce (*Picea abies*), Siberian larch (*Larix sibirica*), and Scots pine (*Pinus sylvestris*). The experimental plots represented 18 levels of stand density, originally ranging from 500 to 125,000 seedlings per hectare (Fig. 8.8). Plots representing five density levels were selected for the study of the content of carbon and nitrogen in the mineral soil from the soil surface (0–5 cm depth). Based on the soil samples, the content and stability of soil organic matter was defined using (i) carbon decomposability (mineralization of carbon per unit of soil carbon) and (ii) primability (susceptibility of soil organic matter to microbial priming). The main findings are collated in Figs. 8.8 and 8.9 based on Menyailo et al. (2022).

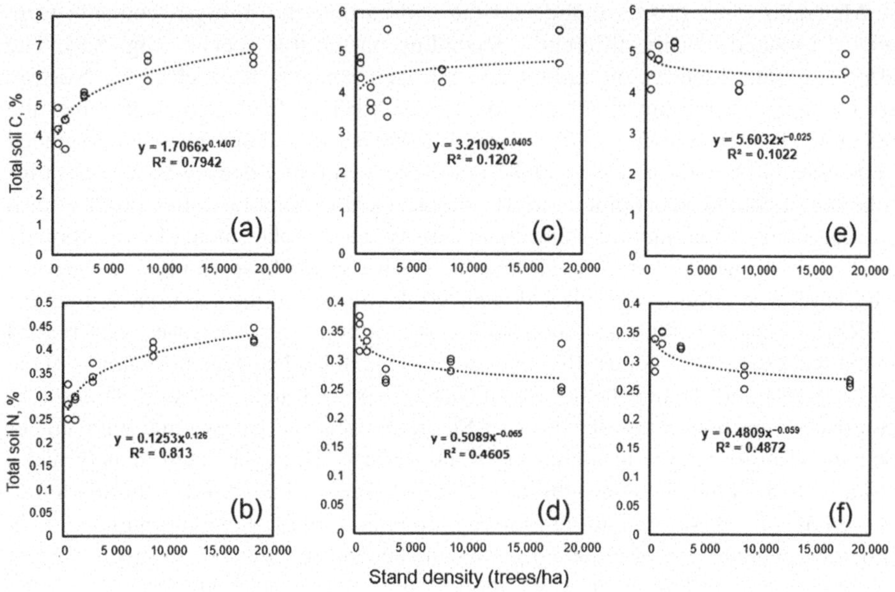


**Fig. 8.8** Effects of tree species and stand density on the total soil organic matter and nitrogen content (Menyailo et al. 2022). Points represent the mean values ± SE, $n = 3$. Open Access. Courtesy MDPI. Upper: soil carbon. Lower: soil nitrogen for Norway spruce (**a**, **b**), Siberian larch (**c**, **d**), and Scots pine (**e**, **f**) as a function of five levels of stand density

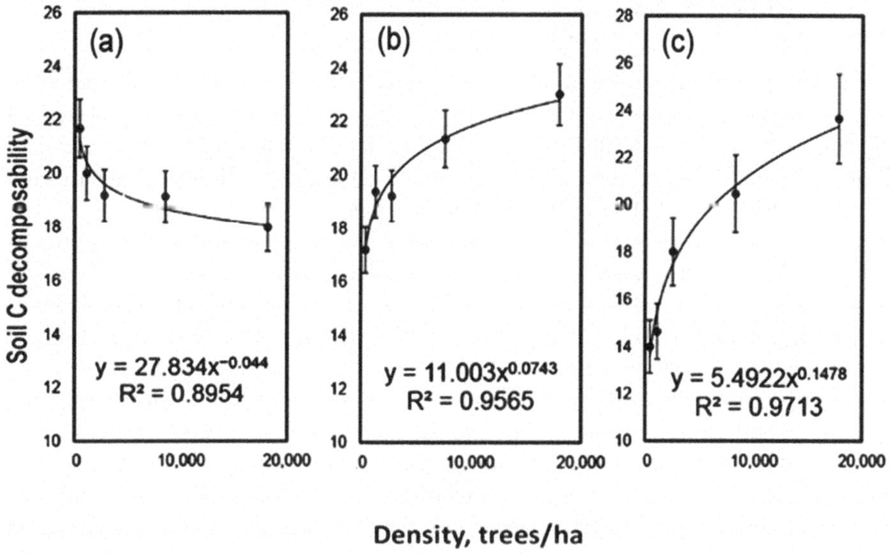


**Fig. 8.9** Effects of tree species and stand density on the decomposability of soil organic matter (Menyailo et al. 2022). Points with bars represent the mean values ± SE, $n = 3$. Open Access. Courtesy MDPI. Left: Norway spruce (**a**). Middle: Siberian larch (**b**). Right: Scots pine (**c**). Figure extracted from the original Fig. 8.8 for layout reasons

Menyailo et al. (2022) found that the soil carbon and nitrogen contents were related to stand density differently depending on the tree species (Fig. 8.8). The effects of stand density on carbon and nitrogen contents were clear for Norway spruce under increasing stand density. Under Siberian larch and Scots pine, the effect on nitrogen content was negligible or nonexistent, and nitrogen content tended to decline along with increasing stand density. The missing density-based effect for Siberian larch and Scots pine is due to replacement of the old soil carbon by fresh soil carbon representing increasing stand density. Such replacement of old carbon by fresh carbon under Siberian larch and Scots pine could have resulted in zero net carbon sequestration in the soil and changed the quality of the soil organic matter.

Regarding soil carbon decomposability, the carbon losses/increases were related to the stand density of dominating tree species (Fig. 8.9). The decomposability of the soil carbon reduced when the density of Norway spruce stands increased. This was in contrast to the increasing densities of Siberian larch and Scots pine, with higher decomposability related to increasing stand density. Thus, the stand density had a strong effect on the decomposability of the soil organic matter and accumulation of carbon in soil stocks. According to Menyailo et al. (2022), these correlations likely provide optional ways of organizing the management of carbon stocks over a forest landscape by considering how to allocate the appropriate combination of trees species in seedling stands of varying densities to different forest sites.

### *8.4.3 Tree Growth and Carbon Stocks Related to Soil Management*

The effects of soil management on carbon sequestration in trees, ground vegetation, and soil in boreal forests were summarized in Fig. 8.10 by Saksa (2020) based on Mjöfors et al. (2017) and Sikström et al. (2020) for common Nordic tree species like Scots pine and Norway spruce. This summary shows that reforestation based on planting/seeding is improved if an upland site is prepared mechanically by mixing organic matter and mineral soil. The increasing decay of organic matter increases nutrient release. Consequently, the site preparation increases the survival and growth of seedlings but also carbon emissions, with impacts on subsequent carbon balance and carbon stock in soil. On the other hand, Mjöfors et al. (2017) found that mechanical site preparation increased ecosystem carbon stocks compared to those in the unprepared sites (Fig. 8.10). The effect of site preparation (disc trenching, mounding, and ploughing) was related to the increased carbon stock in trees. Over 25 years, this experiment showed a similar response for Scots pine, Norway spruce, and lodgepole pine (*Pinus contorta*). According to Mjöfors et al. (2017), the tree carbon stock was highest when mounded or disc trenched and lowest in unprepared plots. The effects of site preparation on soil carbon extended no deeper than 30 cm in the soil profile.

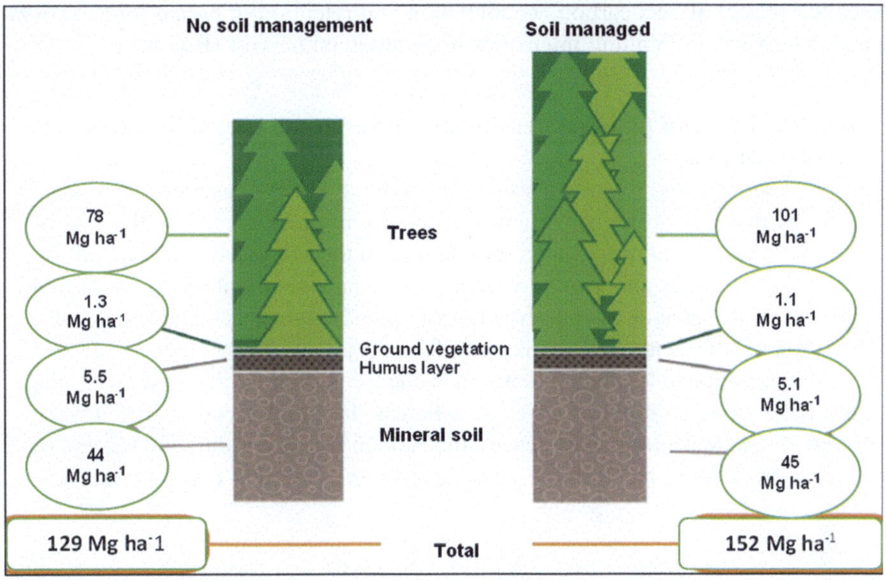


**Fig. 8.10** Effect of soil management on carbon sequestration in trees, ground vegetation, and soil in boreal forests, summarized by Saksa (2020) based on Mjöfors et al. (2017) and Sikström et al. (2020). Courtesy of Natural Resources Institute Finland

### *8.4.4 Planting Density, Carbon Retention, and Accumulation in Managed Forest Ecosystems*

In forestry, carbon sequestration refers to the process of capturing and storing atmospheric carbon dioxide in forest ecosystems. From the growth and yield studies, carbon sequestration indicates the uptake rate of de novo carbon for temporary retaining in ecosystem depending on the removal of carbon outside the ecosystem in forestry or the decay of soil organic matter linked to ecosystem carbon dynamics. From the management point of view, de novo carbon compensates the loss of the old carbon (legacy carbon), thus maintaining/increasing the soil carbon. This makes the carbon management sensitive to the forest management and timber harvest through the successional dynamics of chronosequences.

From the management point of view, the spacing (stand density) in reforestation is a key factor affecting carbon uptake in the precommercial phases. Over the following phases of chronosequence, these impacts gradually disappear, but the initial density affects the early phase of commercial management. The narrow spacing tends to allow early commercial thinning, while the wide spacing allows delaying thinning or reducing thinning intensity and frequency for maintaining/conserving carbon stocks in commercial cuttings, e.g., Kellomäki et al. (2013). In these simulations, the focus was on how the spacing in the precommercial phase (seedling phase) affects carbon accumulation and retention in boreal forest ecosystems under varying thinning intensities in commercial harvest (Box 8.2).

**Box 8.2 Effect of planting density on carbon in soil over time under regular thinning**

Carbon sequestered in an ecosystem may remain in the ecosystem over years or even decades through rotation. Carbon retention is related to the rate of carbon uptake in biomass and carbon loss from the ecosystem and depends on the growth of trees and ground vegetation, management (enhancement of tree growth, soil management), harvesting (frequency, intensity), and the decomposition of litter and organic matter in the ecosystem.

The duration of carbon retention in the ecosystem is referred to as the residence time ($\tau$) in the ecosystem, whereas the turnover rate is the inverse value of residence time. Residence time is obtained by dividing the storage of carbon (e.g., kg C $ha^{-1}$) in the ecosystem by the emission rate ($q$, e.g., kg C $ha^{-1}$ $yr.^{-1}$):

$$\tau = \frac{\text{Capacity ofa system to hold carbon}}{\text{Rate of carbon flow through the system}} = \frac{\text{Storage}}{q}. \tag{8.1}$$

The residence time is assumed to represent steady state conditions. The residence time begins when carbon enters the system and ends when carbon leaves the system. Under management, the forest is unbalanced, representing varying amounts of carbon in the ecosystem depending on its uptake rate and loss rate, which are partly based on the phase of ecosystem development.

Kellomäki et al. (2013) carried out the model-based simulations, in which carbon stocks, emissions, and carbon residence time were related to varying thinning intensity. The focus was on the stands of Norway spruce, with the initial density of 1800, 2700, and 3600 seedlings $ha^{-1}$. The simulations concerned sites of moderate fertility (*Myrtillus* site type, MT) in central Finland (62° N), applying a 100-year rotation. Thinning included (i) no thinning (NoThin), (ii) conventional thinning (CU), and (iii) thinning with reduced intensity (CU + 30%BA). In no thinning (NoThin), thinning was excluded but even in this case the rotation ended in clear-cutting. In conventional thinning (CU), the rules of practical management were used with the removal of trees from below. Under thinning with reduced intensity (CU + 30% BA), the thinning intensity in terms of basal area (BA) was increased by 30% after thinning compared to that used under conventional thinning. The effects of initial density in reforestation and of varying thinning regimes on carbon dynamics were assessed based on (i) carbon stock at the end of rotation, (ii) mean carbon stock throughout the rotation, (iii) carbon emissions, and (iv) carbon residence time, summarized in Fig. 8.11 and Table 8.2.

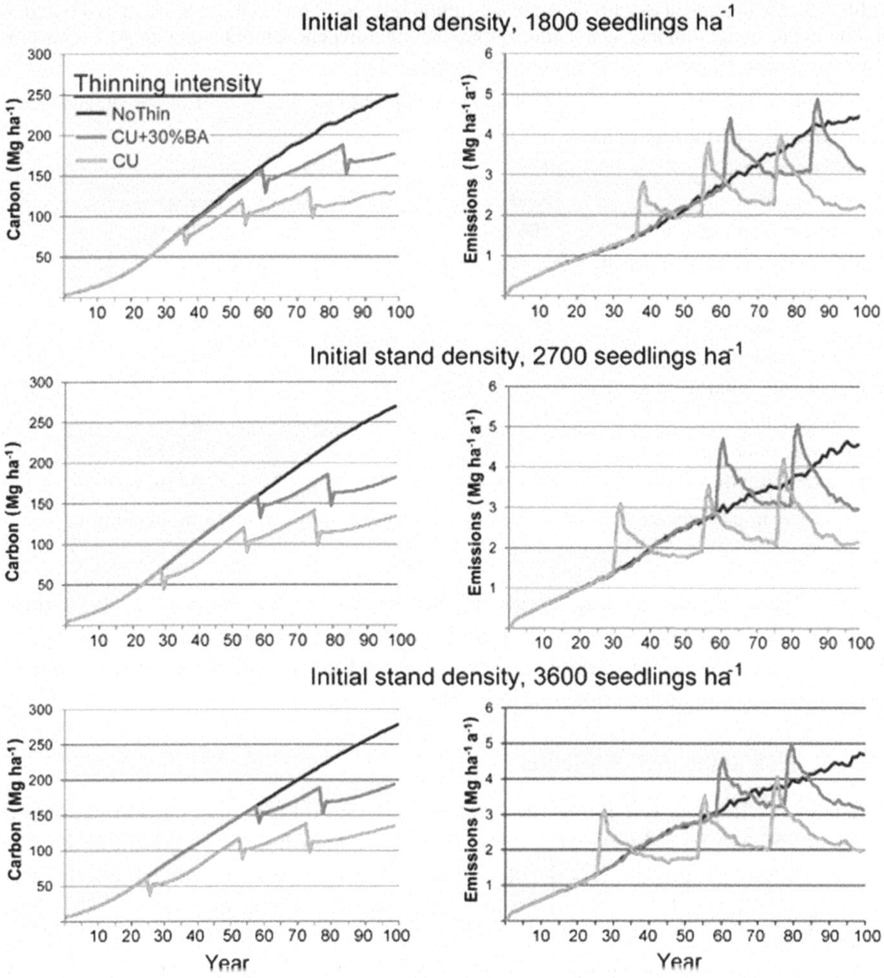


**Fig. 8.11** Carbon stocks, emissions in a boreal forest ecosystem based on the initial density of tree stands (1800, 2700, and 3600 seedlings per hectare) and thinning intensity (Kellomäki et al. 2013). Courtesy of Springer Nature. Left: stocking of carbon (de novo carbon in trees and soil) in the ecosystem through a chronosequence as a function of initial stand density and varying thinning intensity. Right: carbon emissions as a function of initial stand density and varying thinning intensity through the chronosequence

In general, the simulations showed that thinning reduces carbon stocks through a chronosequence, but the carbon stocks through rotation can be increased by reducing the thinning intensity or thinning frequency (Table 8.2). In this context, there was no clear correlation between the carbon stocking and initial density in afforestation; that is, management through the maturing chronosequence tends to eliminate the effects of the initial density. Nevertheless, the effects of initial density were clear in the precommercial phase, with more rapid carbon accumulation than under wide

**Table 8.2** Mean amount of carbon and mean carbon residence time in the ecosystem as a function of thinning using no thinning, conventional thinning, or conventional thinning with 30% reduction of thinning intensity based on the simulations in Fig. 8.11

| Management regime | Carbon in rotation, Mg ha$^{-1}$ | Carbon retains, years |
|---|---|---|
| Initial density 1800 seedlings ha$^{-1}$ | | |
| • No thinning | 127– | 51– |
| • Current thinning | 83 (65) | 45 (88) |
| • Current thinning +30% | 106 (84 | 49 (96) |
| Initial density 2700 seedlings ha$^{-1}$ | | |
| • No thinning | 135– | 52– |
| • Current thinning | 86 (64) | 45 (87) |
| • Current thinning +30% | 113 (84) | 47 (90) |
| Initial density 3600 seedlings ha$^{-1}$ | | |
| • No thinning | 138– | 51– |
| • Current thinning | 86 (61) | 45 (88) |
| • Current thinning +30% | 118 (85) | 47 (92) |

The numbers in parentheses indicate the percentage in relation to that under no thinning

spacing. Thus, dense seedling stands in afforestation likely compensate the carbon loss after clear-cutting more rapidly than is possible under wide planting spacing. In the following, details of carbon sequestration and storing in ecosystem related to varying management are collected in Table 8.2.

Carbon stock at the end of rotation

- Under no thinning (NoThin), the stock of new carbon (de novo carbon) extended up to 250 Mg ha$^{-1}$ by the end of the 100-year rotation if the stand initial density was 1800 seedlings ha$^{-1}$. If the stand initial density was increased up to 2700 and 3600 seedlings ha$^{-1}$, the carbon stock increased slightly: up to 270 and 280 Mg ha$^{-1}$.
- Under conventional thinning (CU), the stocking at the end of the rotation was 130 Mg ha$^{-1}$ regardless of initial stand density.
- Under reducing thinning intensity (CU + 30% BA), the stocking at the end of the rotation was 170, 180, and 190 Mg ha$^{-1}$ for increasing initial stand density from 1800 to 3600 seedings ha$^{-1}$.

Mean carbon stock throughout the rotation

- Under no thinning (NoThin), the mean stock over the rotation was 127, 135, and 138 Mg ha$^{-1}$ with increasing initial stand density from 1800 to 3600 seedlings ha$^{-1}$.
- Under conventional thinning (CU), the mean carbon stock over the rotation was only slightly increased from 83 to 85 Mg ha$^{-1}$ in relation to management option.
- Under increased thinning intensity (CU + 30% BA), the mean carbon stock over the rotation increased from 107 to 118 Mg ha$^{-1}$ in relation to management option.

Carbon emissions

- Regardless of initial stand density, carbon emissions were similar under no management. Under thinning, the emissions increased immediately after cutting due to the decay of logging residues. This fluctuation temporarily exceeded the emissions under no management, but in the long term, the emissions under thinning stabilized at a lower rate than those under no thinning due to the decay of logging residues, with a reduction of emissions.

Carbon residence time

- Regarding the carbon residence time, there were differences related to the initial stand density. Firstly, the increase of residence time in the precommercial phase increased along with increasing initial stand density, with the highest residence time in the stand of the highest density (3600 seedlings $ha^{-1}$). Secondly, the carbon residence time increased along with increasing initial stand density, especially in the simulation using the highest initial density. The simulation further indicated that in very dense seedling stands may reduce the carbon residence time due to the excess mortality of trees.

## 8.5 Tree Breeding for Increasing Carbon Sequestration Through Reforestation

### *8.5.1 Tree Provenances for Carbon Management*

In reforestation, the productivity of tree species in relation to site properties is among the main factors affecting carbon sequestration in even-aged management. Rapid growth is an important trait in increasing the yield in a given period, thus shortening the production cycle, and enhancing carbon accumulation in management. In a geographic context, genetic variation is typical for any tree species, splitting species into subpopulations or provenances locally adapted to the prevailing conditions. The differences between provenances are inherent, and they are recognizable from generation to generation, but different provenances can still interbreed.

In boreal conditions, reforestation with trees from lower boreal latitudes tends to increase productivity in higher boreal latitudes. In Nordic countries, such provenance transfer may increase the growth of Norway spruce by up to 40% compared to the growth of local provenances used in plantations in similar conditions. The growth increase is related to the longer duration of yearly growth compared to the local provenance. However, the northward transfer may increase mortality and winter damage as in Norway spruce (e.g., Napola 2014). Such damage is not common for the southward transfer of northerly provenances, but the productivity remains lower than that for the local provenance.

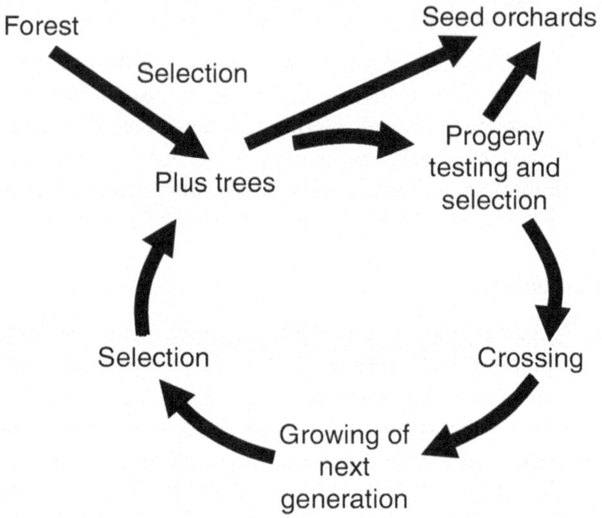


**Fig. 8.12** The breeding cycle is initiated with the phenotypic selection of plus trees in natural stands followed by the establishment of seed orchard and progeny tests (Ruotsalainen 2010, 2014). Courtesy of Sorbifolia

In tree breeding, selection is the chief method of enriching genes for improved performance of trees and their populations or provenances. Selection is based on a comparison in which the trees belonging to the accepted unit are better than average or better than some other reference used to assess the success of selection. In the first breeding phase or in phenotypic selection, a few trees per hectare with desirable properties or plus trees (i.e., the very best trees) are identified from wild populations (Fig. 8.12). In the second breeding phase, the basic population includes trees (progeny) that are the offspring of plus trees. In the third cycle and onward, the offspring of selected trees from the previous cycle are used in selection. In breeding, the genes controlling given traits are enriched. Nevertheless, breeding does not deplete genetic diversity, because the initial breeding population consists of thousands of plus trees, and the clonal composition varies among seed orchards. The genetic variation is also maintained by open pollination. When advanced generations of seed orchards are used, the genetic gain in terms of volume growth may increase by 10% for the breeding cycle.

## 8.5.2 *Genetic Gains in Breeding for Growth and Yield*

Experiments based on the first phenotypic selection have shown that breeding may increase growth only to a limited extent (Table 8.3). In many cases, the effect of variability in growing conditions is larger than the effect of genotype selection. When using seeds from first-generation seed orchards, the volume growth of Scots pine (*Pinus sylvestris*) may increase by up to 10–30%, as presented by Rytter et al. (2013) and Jansson (2017), among others. A growth increase of a similar magnitude

**Table 8.3** Genetic gains for various growth traits in selected tree species, based on Swedish and Finnish studies modified slightly from Jansson et al. (2017, p. 275)

| Species, trait, and alphabetic (a–j) footnote to short explanation of experiments and origin of data for defining genetic gain (1) | Genetic gain, % | Reference |
|---|---|---|
| Scots pine | | |
| • Volume/unit area over rotation (a) | 10 | Rosvall et al. 2001 |
| • Volume/unit area over rotation (a) | 25 | Rosvall et al. 2001 |
| • Volume/unit area over 30 years (b) | 12 | Jansson 2007 |
| • Volume/unit area over 30 years (b) | 25 | Jansson 2007 |
| • Volume/unit area over rotation (b) | 8.3–10.7 | Andersson et al. 2007 |
| • Stem volume growth over 27 years (b) | 18.9 | Andersson et al. 2007 |
| • Mean annual yield/area over rotation (d) | 11.5 | Haapanen et al. 2018 |
| • Mean annual yield/area over rotation (e) | 23.9 | Haapanen et al. 2018 |
| Total mean | 18 | |
| Norway spruce | | |
| • Volume/unit area over rotation (f) | 10 | Rosvall et al. 2001 |
| • Volume/unit area over rotation (c) | 25 | Rosvall et al. 2001 |
| • First-generation seed orchard (d) | 10–25 | Kvaalen et al. 2008 |
| Total mean | 18 | |
| Birch | | |
| • Volume/unit over rotation (i) | 5–20 | Rosvall et al. 2001 |
| • Stem volume growth over 8–12 years (j) | 26–29 | Hagqvist and Hahl 1998 |
| Total mean | 20 | |

(1) Explanation for experiments and data for different tree species with notes a–j
• Scots pine: (a) first-round seed orchard, (b) third-round seed orchard, (c) controlled-pollinated progenies of selected plus tree, (d) first-generation seed orchard, (e) 1.5-generation seed orchard.
• Norway spruce: (f) first-round seed orchard, (g) third-round seed orchard, (h) first-generation seed orchard.
• Birch: (i) seed orchard with selected plus trees, (j) seed orchard progenies of second generation, selected plus trees.

seems to hold for Norway spruce (*Picea abies*) and birch (*Betula pendula*), but the variability, including larger values, is higher for these species than for Scots pine. In the future, improved seeds will be nearly exclusively used in nurseries and even to some extent in direct seeding. However, the realized gain will depend substantially on the intensity of cultivation, maintenance, and management of forest plantations (e.g., Kellomäki et al. 2009; Kilpeläinen et al. 2010; Rytter et al. 2013). For example, Norway spruce originating from Estonia (i.e., 100–200 km to the south of Finland) is successful in reforestation in southern Finland up to 100 km north of the southern coastal area, up to 61° N (Napola 2014).

### *8.5.3 Genetic Gains in Improving Growth in Forestry*

Haapanen et al. (2018) used a growth and yield model to (i) examine genetic gains in early growth and timber quality traits with two types of Scots pine (*Pinus sylvestris*) seed orchard stock related to the response to two phases of plus-tree selection; (ii) simulate the likely changes in yield at the stand level throughout a production cycle representing the rotations widely used in moderately poor sites (*Vaccinium* site type, VT) and fertile sites (*Myrtillus* site type, MT) for Scots pine; and (iii) assess the realized field performance of 1.5-generation orchard seed lots. Three types of improved Scots pine seed lots were compared to unimproved seed lots using stand-based simulations. Seed lots represented (i) five seed lots from wild Scots pine stands ("unimproved"), (ii) two seed lots from first-generation seed orchards ("SO1.0g"), and (iii) five 1.5-generation orchard seed lots.

Haapanen et al. (2018) conducted model simulations for southern Finland (61° N, temperature sum 1200 d.d.) using an initial spacing of 2500–4444 seedlings per hectare. Figure 8.13 shows the development of basal area of different genetic stocks of Scots pine grown using the current forest management rules for private forestry in Finland. Seed lots used in the experiments behind the simulations are given in the insert, with further explanation in the text. Genetic gain in mean annual yield was (i) 10.6–12.1% (mean: 11.5%) for mature SO1.0 g, (ii) 15.8–16.0% (mean: 15.9%) for juvenile SO1.5 g, and (iii) 22.9–24.8% (mean: 23.9%) for mature SO1.5 g. Furthermore, there was a 1.3–2.4% (mean: 1.9%) additional gain from genetic thinning.

Haapanen et al. (2018) used the basal area weighted mean diameter to indicate the timing of the final cut. The increased growth rate reduced the rotation length by 4–17 years depending on orchard generation, level of pollen contamination, and genetic thinning (Fig. 8.13; Box 8.3). Based on the seedlings from mature first-generation orchards, the rotation length reduced by 5.5 years. When using genetic thinning, the rotation length reduced by 6.5 years compared to that for reforestation with unimproved seedlings. For the seedlings representing mature 1.5-generation orchards, the rotation length was reduced by 13.5 years, while the gain based on thinning was 15 years. The mean stem wood yield was 541 $m^3$ $ha^{-1}$. When the 3% discount rate was used, the increase of bare-land value for improved stock varied from 766 to 1836 € $ha^{-1}$ (25.2–60.5%). The proportional genetic gains in bare-land values were greater than the gains in annual yield.

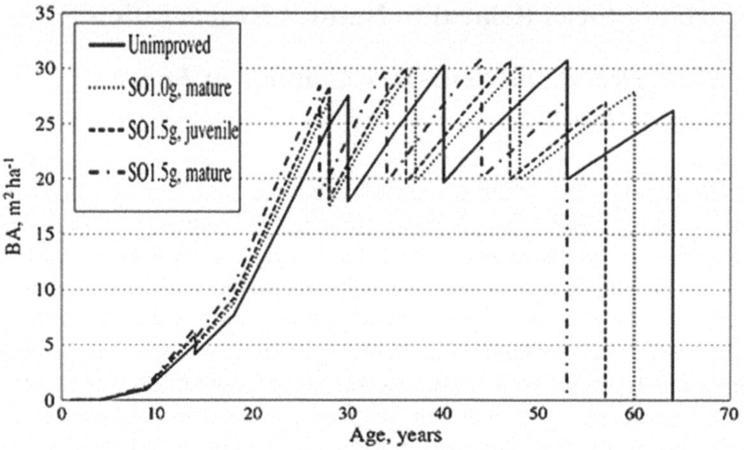


**Fig. 8.13** Development of basal area (BA) of different genetic stocks of Scots pine (*Pinus sylvestris*) grown sites of moderately poor (*Vaccinium* site type, VT) and fertile (*Myrtillus* site type, MT) sites in southern Finland (61° N) under a temperature sum of 1200 d.d. using the current forest management rules for private forestry (Haapanen et al. 2018). Seed lots used in the experiments behind the simulations are given in the insert, with further explanation in the text. Courtesy of Natural Resources Institute Finland

**Box 8.3 Productivity of improved genotypes under intensive management**

Under intensive even-aged management, primary production can be increased substantially if using high-productivity species and/or provenances, proper spacing and cleaning in plantations, regular thinning, optimized stocking and nitrogen fertilization, and proper rotation in relation to the life cycle of species. Routa et al. (2011), for example, found that the most productive clones of Norway spruce could potentially increase biomass yield up to 6 Mg ha$^{-1}$ yr.$^{-1}$ in early rotation when using a planting density of 2500 seedlings ha$^{-1}$. This was 70% more than the mean over several clones included in the simulations for the moderately fertile sites (*Myrtillus* site type, MT) in the mid-boreal zone. Routa et al. (2011, 2013a, b) further found that intensive management, consisting of high precommercial density and appropriate intensity in commercial thinning combined with fertilizing, enhanced the production of timber and energy biomass. The effects of fertilization were especially pronounced with moderate or low site fertility. In general, reduced thinning intensity with increased stocking through rotation seemed to increase the productivity of Scots pine and Norway spruce in managed boreal forests (e.g., Briceño-Elizondo et al. 2006).

## 8.6 Carbon Stocks Related to Natural Regeneration

### *8.6.1 Shelterwood and Seed-Tree Cuttings in Forest Regeneration*

In northern Europe, the shelterwood and seed-tree cuttings are widely used to regenerate naturally Scots pine, Norway spruce, and birches (e.g., Leinonen et al. 1989). Regarding chronosequence, the natural regeneration is likely successful in both cases if seedlings are established naturally under mature tree stand in the density of 400–500 trees ha$^{-1}$ before final cutting for regeneration. In this phase, the establishment of seedlings is common in stands occupied by mature trees growing in small groups, thus allowing open areas (gaps) to exist between groups. Gaps break the canopy cover of dominating trees (parent trees) and provide seedbeds with increasing soil temperature and moisture for enhancing germination of seeds and establishment of seedlings. This is demonstrated in Fig. 8.14 for natural regeneration of Scots pine, with (i) seed-tree cutting on the left with small seedlings few years after the cutting and (ii) on the right with shelterwood cutting with seedlings of varying size born in canopy gaps during 10–20 years after the cutting.

According to Cajander (1934), the establishment and the growth of seedlings is slow if the canopy gap size is $<100$ m$^2$ in the 10-year period after regenerative cutting. In a gap size $>100$ m$^2$, the growth increased along with increasing gap size over time. In shelterwood cutting, 100–300 parent trees per hectare are left representing the basal area in the range of 14–16 m$^2$ ha$^{-1}$ on fertile sites. It is evident that the effect of shelterwood cutting on soil carbon is likely of the magnitude as that due to intensive thinning. However, regeneration success varies depending on the quantity and quality of the seed crop and the seedbed properties that affect the germination of seeds and establishment of seedlings. In the shelterwood and seed-tree cuttings, the establishment of seedlings involves several seed crops following each other from year to year, producing a seedling stand with varying size and age distribution regardless of tree species (Fig. 8.15).

In *shelterwood cutting*, 100–300 parent trees are left per hectare for seeding the site and sheltering the soil for excess temperature variability on soil surface, stabilize moisture conditions in seedbeds, and curb excess growth of ground vegetation (Fig. 8.15). Shelterwood cutting falls between the last commercial thinning of the mature tree stand and clear-cutting aiming at planting/seeding. Shelterwood cutting is mainly used in reforestation of Norway spruce on moraine soils not prepared or only slightly prepared for avoiding damages in parent trees.

In *seed-tree cutting*, 20–50 up to 100 parent trees per hectare are retained for natural seeding of reforestation areas. These parent trees give no or only marginal shelter opposite to shelterwood cutting. In reforestation, seed-tree cutting falls between the shelterwood cutting and clear-cutting with few retention trees per hectare left for conservation. In using seed-tree cutting, proper soil preparation enhances the regeneration success for improving the seedbed and reduces the competing ground vegetation. In using seed-tree cutting, soil preparation is

**Fig. 8.14** Natural regeneration of Scots pine. Left: seed-tree cutting with small seedlings. Right: shelterwood cutting with seedlings born in canopy gaps. Photos by Heli Peltola

acceptable because the damages in parent trees remain marginal for wide spacing between parent trees (Fig. 8.15).

### *8.6.2 Impacts of Shelterwood Cutting on Ecosystem Carbon*

The impacts of cutting for regeneration with shelterwood and seed-tree cuttings on ecosystem carbon are poorly understood. In general, shelterwood mainly effects on tree carbon, while the impacts on soil carbon are marginally if soil preparation is excluded. Impacts resemble those induced by the final, commercial thinning before the end of the rotation in clear-cutting or shelterwood cutting. In seed-tree cutting, the effects on the ecosystem carbon dynamics in trees are nearly the same as to clear-cutting for planting/seeding. This holds also for soil carbon when preparing soil for enhancing the establishment and growth of seedlings.

The study of Scott et al. (2004) is among the few that have explored the changes in carbon stocks and net carbon exchange over 1 year after shelterwood cutting (Fig. 8.16). The measurements dealt with a temperate forest (45° N) dominated by coniferous trees (e.g., red spruce (*Picea rubens*), balsam fir (*Abies balsamea*), white pine (*Pinus strobus*)). Prior to shelterwood cutting, the basal area of trees was 30 $m^2$ $ha^{-1}$, and 76 Mg C $ha^{-1}$ was stored in the above- and belowground biomass. After

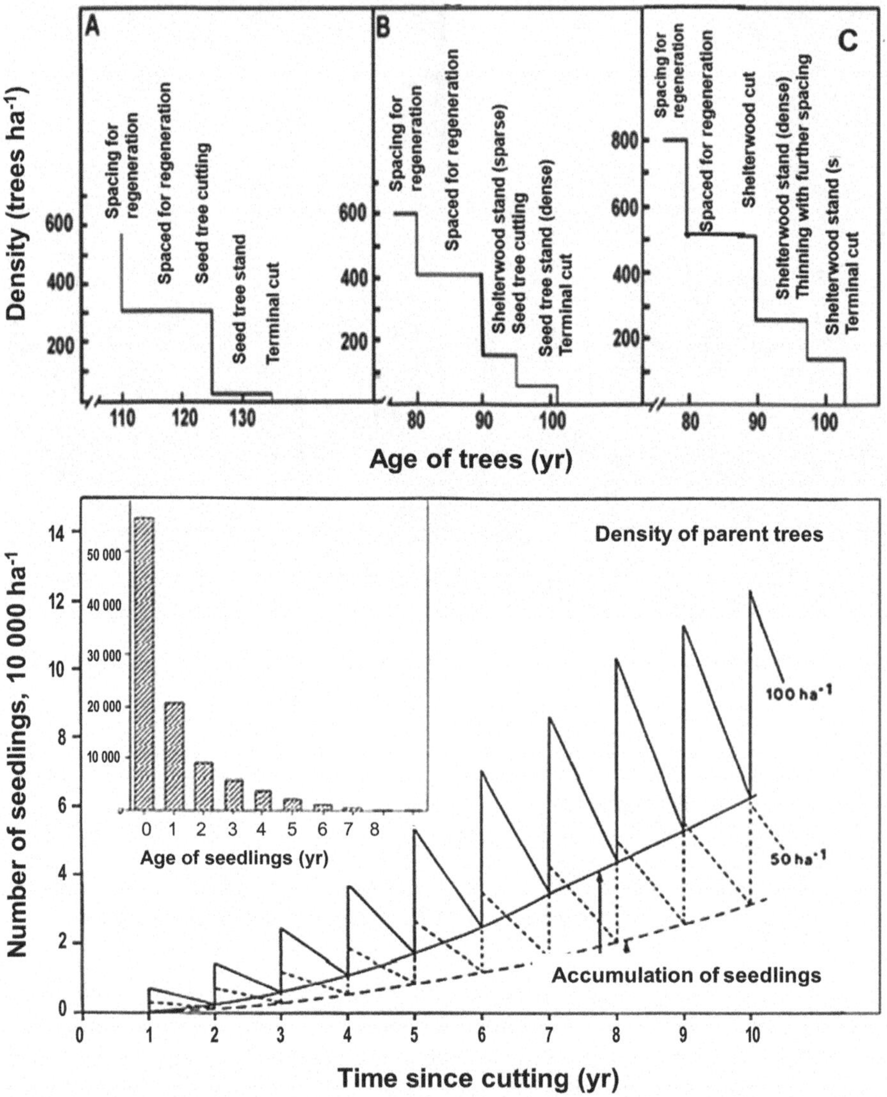


**Fig. 8.15** Outlines of some methods used in natural regeneration with some requirements needed for successful results. Upper: outline presentation of how seed-tree and shelterwood cutting are used in the classic form in natural regeneration. Courtesy of Natural Resources Institute Finland. Lower: cumulative number of seedlings in a naturally regenerated site using seed-tree cutting with 50 and 100 parent trees per hectare in 8 years since the regenerative cutting. The insert demonstrates the age distribution of seedlings over the regeneration period of 8 years (Kellomäki et al. 1987). Courtesy of Finnish Society of Forest Science

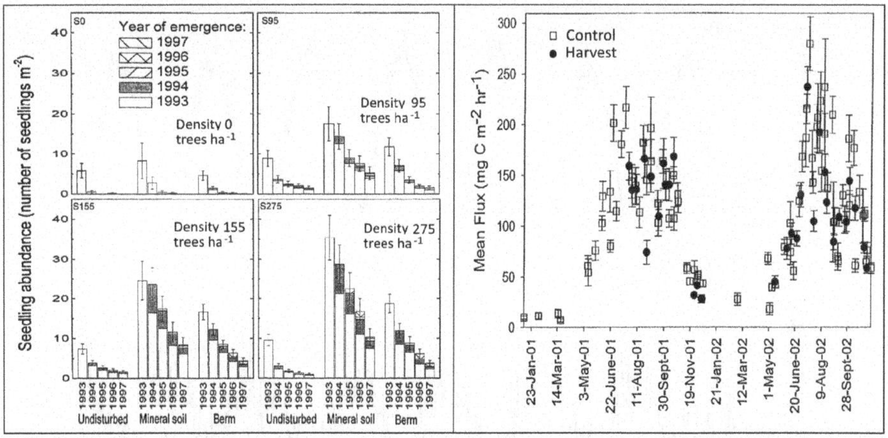


**Fig. 8.16** Natural regeneration and impacts on soil carbon efflux in coniferous forests reforested through shelterwood cutting. Left: an example of seedling establishment in Norway spruce (*Picea abies*) stands of varying density of parent trees over 5 years after shelterwood cutting and soil management, indicated by emergence year and survival of seedlings to the end of each year. Soil was undisturbed and disturbed, disturbance providing sites with open mineral soil and sites with a mixture of upturned humus and mineral soil (berm). The parent tree density was: S0 = clear-cut area with no parent trees; S95 = density of 95 parent trees ha$^{-1}$; S155 = 155 parent trees ha$^{-1}$; and S275 = 275 parent trees ha$^{-1}$ (Nilsson et al. 2002). Permission of Elsevier. Right: carbon fluxes of soil respiration in the control stand and in the stand exposed to shelterwood cutting (2001–2002) in a plantation, with selected coniferous species representing a temperate forest, detailed in the text. Each point represents the mean of measurements based on 16 monitoring chambers (Scott et al. 2004). Courtesy of Springer Nature

harvesting, the basal area reduced by 30%, while carbon loss was 15 Mg carbon ha$^{-1}$. The harvest yielded 5.3 Mg carbon ha$^{-1}$ in aboveground and 5.2 Mg C ha$^{-1}$ in root/stump biomass. At the same time, the leaf area index (LAI) and litterfall declined by 40% after the harvest. Nearby eddy covariance measurements showed that the net ecosystem exchange (*NEE*) prior to harvesting was about 1.8 Mg ha$^{-1}$ yr.$^{-1}$. Following harvesting, *NEE* declined to about 18%, which was less than expected based on the changes in basal area and LAI. Furthermore, soil respiration declined slightly after harvesting, indicating no major soil carbon loss (Scott et al. 2004).

Thorne et al. (2020) also examined how shelterwood cutting may affect carbon dynamics in forest ecosystems (Fig. 8.17). The experimental forest was planted in 1939 on a site of sandy soil using eastern white pine (*Pinus strobus*). In 2014, the mean height of pines was 21.8 ± 1.7 m, and the stand density was 421 ± 166 trees ha$^{-1}$. The forest further included balsam fir (*Abies balsamea*) and native hardwood trees like oak (*Quercus velutina*, *Q. alba*), red maple (*Acer rubrum*), and some wild black cherry trees (*Prunus serotina*). The forest was first thinned in 1983, randomly removing 30% of trees. In 2012, a second cutting was conducted, and 30% of the dominant overstory trees were selectively removed. This improved the availability of light and water in the site/soil, subsequently increasing the growth of the remaining

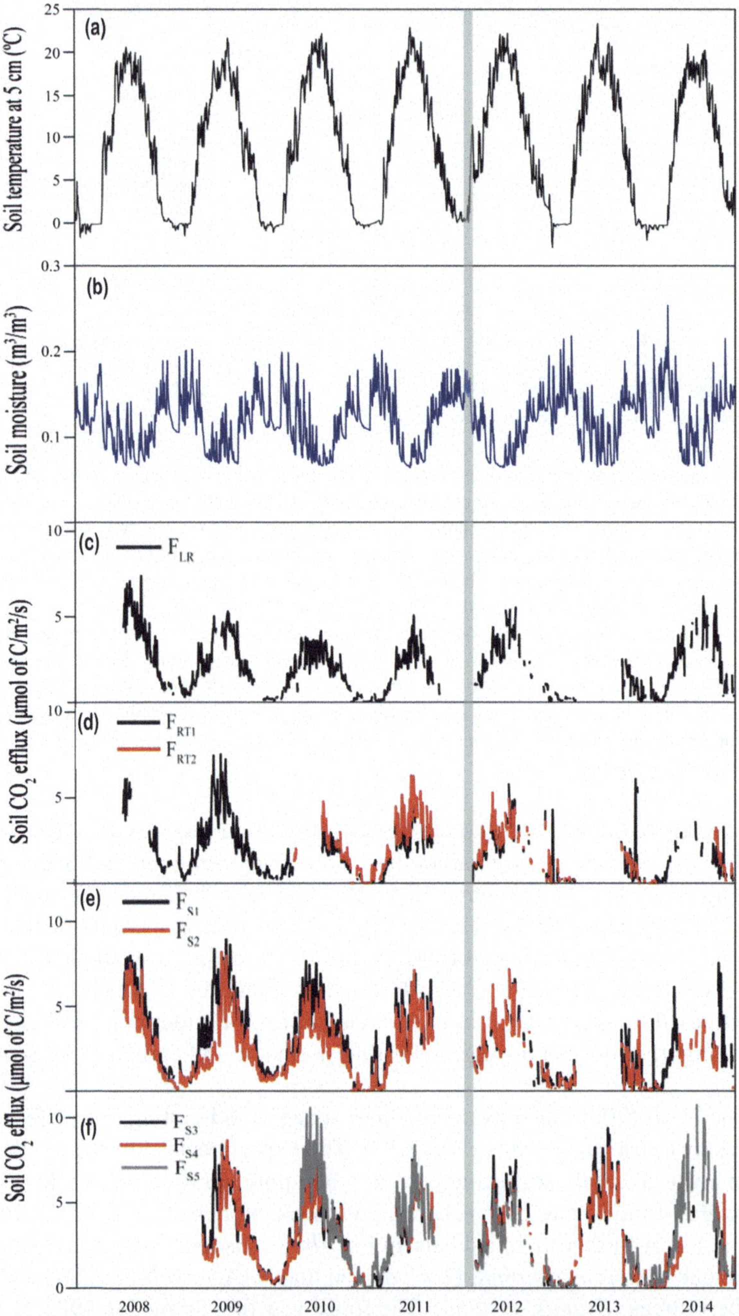


**Fig. 8.17** Effects of shelterwood cutting on carbon dynamics in a forest ecosystem, showing the response of total soil $CO_2$ efflux and its components in a mature forest dominated by white pine (*Pinus strobus*) (> 82%) located in Southern Ontario, Canada (Thorne et al. 2020). Open Access. Courtesy of MDPI. The gray line denotes the timing of the shelterwood cutting event. The subfigures indicate (**a**) daily values of soil temperature at 5 cm depth, (**b**) soil moisture averaged

trees. In the shelterwood cutting, the mean basal area was reduced by 13% and the peak leaf area index by 35%. Consequently, 43 Mg ha$^{-1}$ of foliage, 200 Mg ha$^{-1}$ of living branches, and 129 Mg ha$^{-1}$of dead branches were added on the forest floor in the plots. At the same time as logging residues, 515 Mg ha$^{-1}$ roots died and were transferred to litter. These biomass components would subsequently decompose, thus contributing to ecosystem respiration.

Warren and Ashton (2014) also investigated how shelterwood cutting changes carbon in the soil and forest floor of oak (*Quercus* spp.) hardwood forests in the northeastern (42° N) USA. In the experimental area, the tree species mixture included oaks (*Quercus* spp.), hickories (*Carya* spp.), and beech (*Fagus spp.*). In shelterwood cutting, the basal area was reduced from 37 m$^3$ ha$^{-1}$ down to 11 m$^2$ ha$^{-1}$ while maintaining tree species diversity. Seven to 8 years later, half of the standing trees (5 m$^2$ ha$^{-1}$ of basal area) were cut. The rest of the trees remained in the final cut, letting the trees grow in irregular patterns with the seedling stand initiated by shelterwood cutting. The impact of shelterwood cutting on soil carbon was compared to the soil carbon stocks in unharvested sites. According to Warren and Ashton (2014) (Fig. 8.18),

- The amount of carbon in the soil surface (the upper 30 cm) in harvested sites was significantly lower (6.01 kg C m$^{-2}$) than in the control sites (8.36 kg C m$^{-2}$).
- Harvesting increased the carbon in coarse woody debris up to 3.16 kg C m$^{-2}$ compared to that in uncut control sites, with 0.62 C kg m$^{-2}$. There were differences between sites and treatment, with an interaction between both.
- When combining carbon in surface soil with carbon in litter and coarse woody debris, there was no significant difference between the sites exposed to shelterwood cutting and unharvested sites.
- Over the 20-year time span, the carbon content of sites after harvesting was slightly reduced regardless of carbon pool. The largest decline of carbon occurred in coarse woody debris, from 4.5 kg C m$^{-2}$ down to 0.40 kg C m$^{-2}$. In this period, the carbon in coarse woody debris was close to that in the control sites.

**Fig. 8.17** (continued) between 0 and 30 cm depths, (**c**) soil $CO_2$ efflux from mineral-soil efflux ($F_{LR}$), (**d**) root trenched efflux ($F_{RT1}$ and $F_{RT2}$), (**e**) soil $CO_2$ efflux from the initial group of control ($F_{S1}$ and $F_{S2}$) chambers, and (**f**) soil $CO_2$ efflux from the latest group of control ($F_{S3}$, $F_{S4}$, and $F_{S5}$) chambers, from 2008 to 2014. According to Thorne et al. (2020), there were no significant differences in both soil temperature and soil moisture between the preharvesting (2008–2011) and postharvesting (2012–2014) periods (Fig. 8.17). However, shelterwood cutting reduced the total soil $CO_2$ effluxes by 37%. From heterotrophic sources, the postharvesting soil $CO_2$ effluxes also reduced significantly, by 27%. For the horizon sources in mineral soil, no significant difference was found between the pre- and postharvesting measurements. An analysis of ecosystem respiration ($R_E$) showed postharvesting increase. However, the overall net ecosystem carbon exchange (*NEE*) was similar pre- and postharvesting. Thus, there was an increase in the gross ecosystem productivity following harvesting, which compensated for the increased carbon losses from increasing ecosystem respiration. After harvesting, carbon dynamics in the forest ecosystem will also change rapidly due to the establishment of offspring during initiation of reforestation and the growth of new tree generations

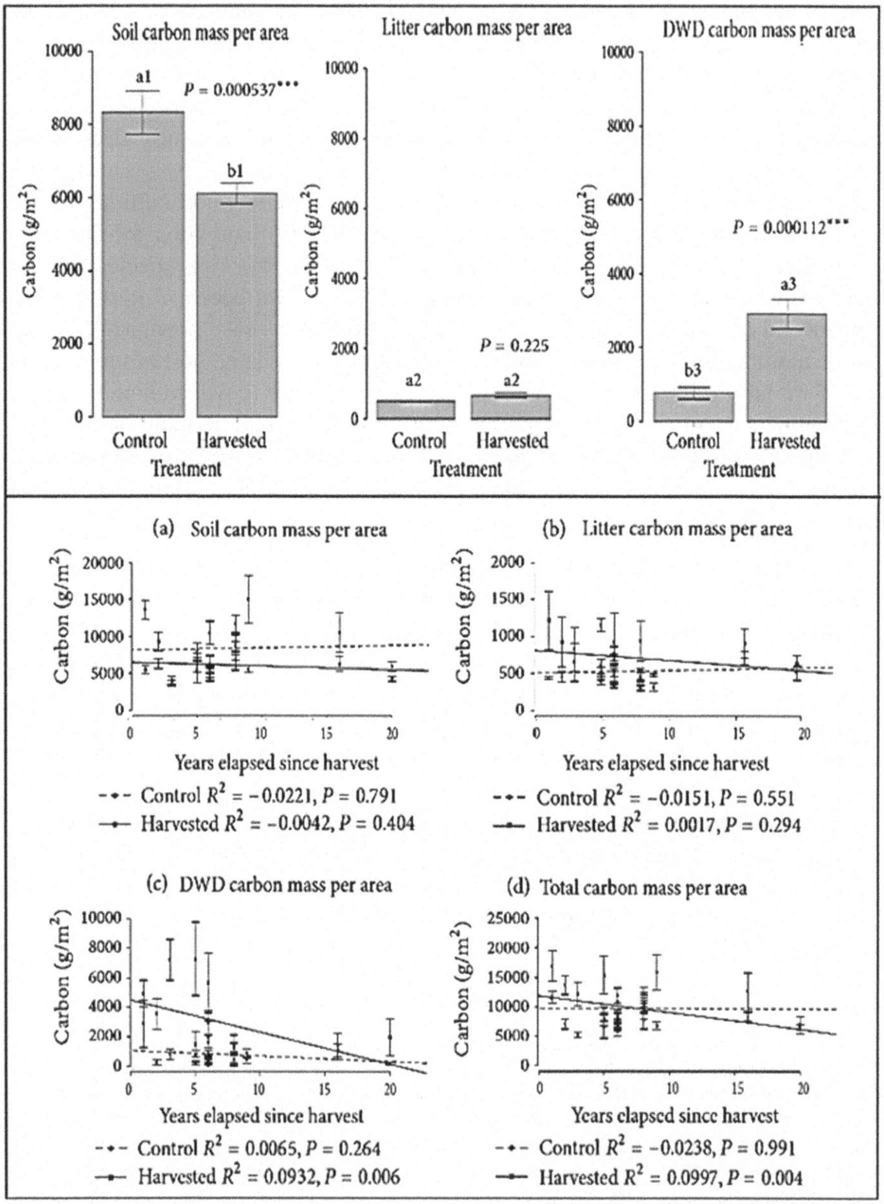


**Fig. 8.18** Impacts of shelterwood cutting on soil carbon, litter on soil, and dead woody debris in temperate deciduous forests (Warren and Ashton 2014). Open Access. Courtesy of Hindawi. Upper: carbon in each pool (carbon in soil, carbon in litter, and carbon in dead wood) between treated and untreated sites (control sites). Lower: regression between carbon stock and carbon efflux in each carbon pool over a 20-year period in untreated sites and sites exposed to shelterwood cutting

Warren and Ashton (2014) further found the soil carbon stocks decreasing in the mineral soil, but no effects in the litter layer occurred in shelterwood cutting. They concluded that soil carbon tends to decline after shelterwood cutting but the total soil carbon stock per the top 10 cm did not deviate from that in the uncut sites.

Puhlick et al. (2016) also studied how regeneration after shelterwood affects ecosystem carbon. They further compared the impacts on carbon stocks to similar stocks in an untreated ecosystem and an ecosystem treated with selective harvesting. The specific goals of this study were to evaluate the carbon stocks in the total ecosystem and in wood products related to wood harvested in cuttings. The impacts of various management treatments were assessed, including (i) selection cutting, (ii) shelterwood cutting, (iii) commercial clear-cutting, (iv) and no management. The experiment was established in mixed-species stands representing temperate forest with the treatments being done in the 1950s and the measurements of the ecosystem carbon pools in 2012. When compared across managed treatments, the main impacts were as follows (Fig. 8.19):

- Commercial clear-cutting had the lowest total ecosystem carbon stocks (reduction by an average of 21%) than other treatments ($P < 0.05$), while the selection and shelterwood treatments had similar total ecosystem carbon stocks.
- Inclusion of the carbon stored in wood products did not influence the differences in carbon storage among treatments.
- In the reference stand, the total carbon stock was $247.0 \pm 17.7$ Mg ha$^{-1}$ (mean ± SD), compared to $161.7 \pm 31.3$ Mg ha$^{-1}$ in the managed stands and $171.2 \pm 31.7$ Mg ha$^{-1}$ in wood-based products.

Puhlick et al. (2016) concluded that timing of harvests, tree species, and the amount of timber removed in harvesting of different tree species influenced carbon storage in wood products. This emphasized the combined impacts of forest management on carbon stocks and balance for sequestering carbon in the forest ecosystem and in wood-based products. Subsequently, the mean total carbon stock in the ecosystem was highest under regular selection and shelterwood cuts and lowest under commercial clear cutting after 60 years under management. This assessment held even though the wood used in wood-based products was excluded.

### 8.6.3 *Impacts of Shelterwood Cutting on Ecosystem Carbon and Nitrogen*

According to Marozas et al. (2018), shelterwood cutting is an option for clear-cutting with planting in reforestation of boreal forests. In this regard, Marozas et al. (2018) monitored the changes in organic carbon, total nitrogen, nitrate and ammonium nitrogen, available phosphorus, and potassium in mineral soil 1, 3, 5, and 7 years after shelterwood cutting for reforestation of Scots pine (*Pinus sylvestris*) (Fig. 8.20). The mean concentration of organic carbon in the soil mineral layer in 1- to 7-year-old

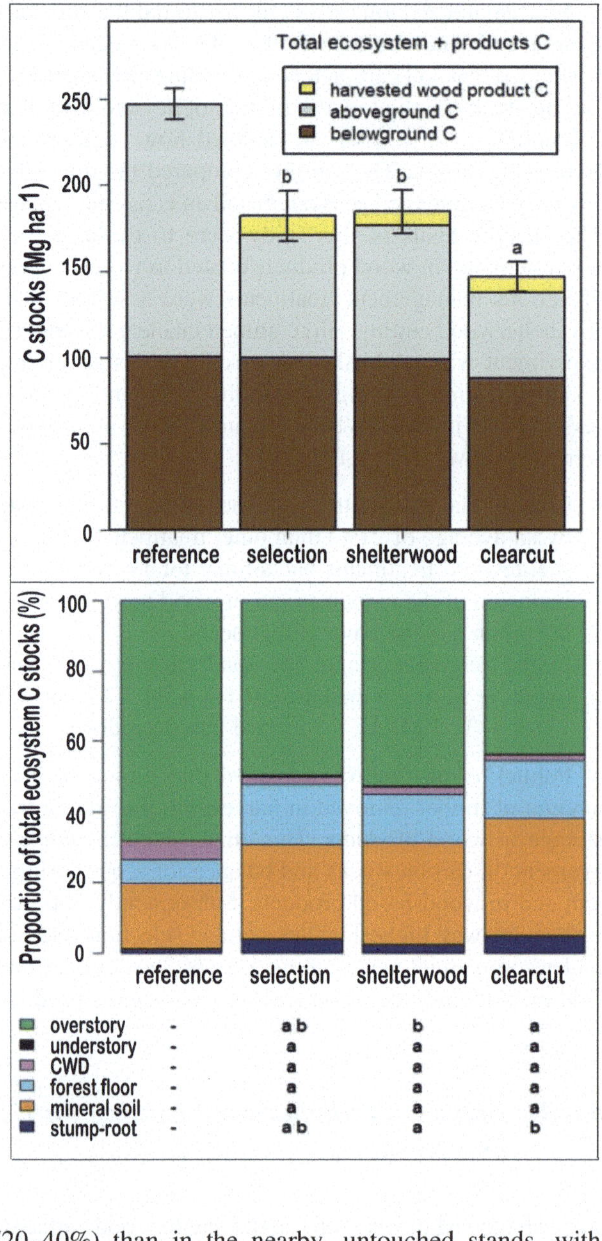


**Fig. 8.19** Impacts of uneven-aged selective cutting, shelterwood cutting, and clear-cutting on a temperate forest ecosystem, distributed among different carbon pools for trees, ground vegetation, and soil (Puhlick et al. 2016). Courtesy of Canadian Science Publishing. Upper: mean and standard error of total ecosystem carbon plus carbon stored in harvested wood products (Mg ha$^{-1}$). Different letters indicate significantly different least-squares means at the average relative volume of coarse fragments in the mineral soil (33.6%). The reference was not included in pairwise comparisons. Lower: carbon storage in ecosystem pools expressed as a share of total ecosystem carbon stock following nearly 60 years of management. Different letters indicate significantly different least-squares means. The reference was not included in pairwise comparisons, and the overstory carbon pool includes the aboveground and belowground portions of living trees and shrubs expected for fine roots. CWD refers to coarse woody debris

cutting sites was higher (20–40%) than in the nearby, untouched stands, with culmination 3–5 years since the cutting. The nitrogen contents increased immediately after the cutting, with an increase of 20% higher than in uncut sites. Thereafter, nitrogen content declined gradually slightly below that in uncut sites. At the same time, the concentration of carbon in relation to nitrogen (C/N) was quite same as uncut sites in 3 years since the cutting. Thereafter, the value C/N increased up to 40%

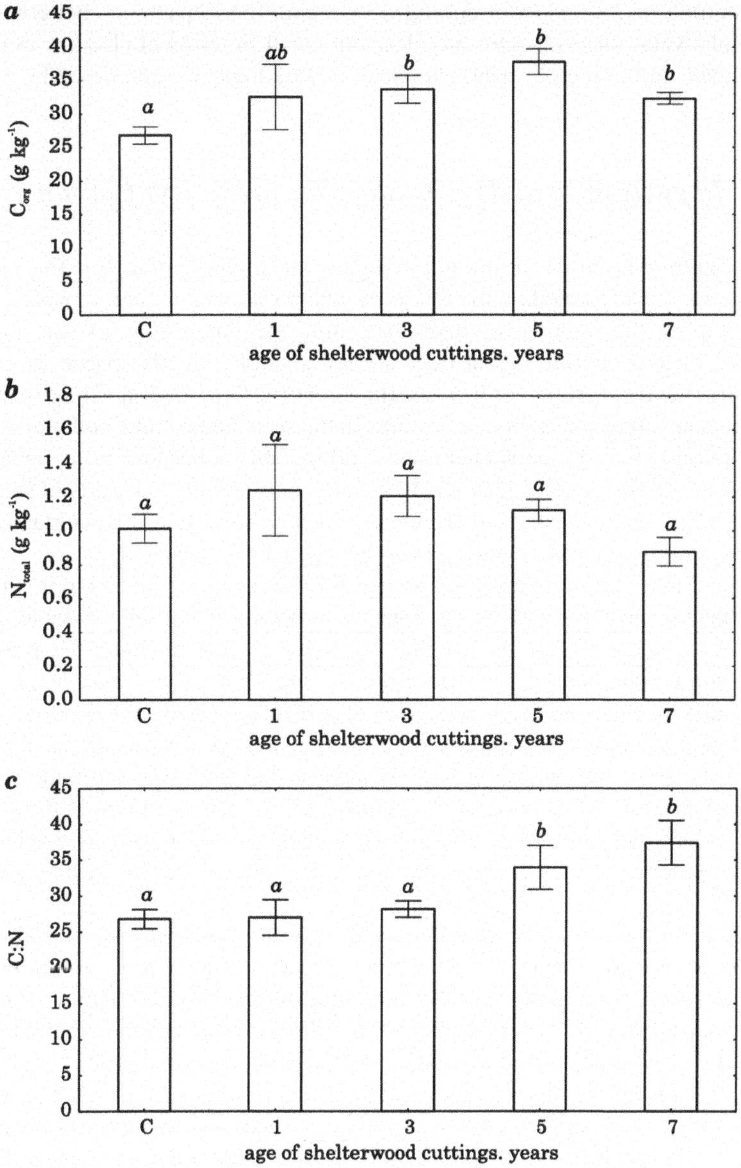


**Fig. 8.20** Effect of shelterwood cutting for Scots pine on soil carbon (upper), nitrogen (middle), and the carbon/nitrogen ratio (lower) as a function of time in years after cutting (Marozas et al. 2018). Courtesy of Polish Society for Magnesium Research. The superscript letters in the rows indicate significant ($P \leq 0.05$) differences between means (Marozas et al. 2018)

for the cutting in 7 years from cutting. In general, the impacts of shelterwood are small. Relatively, these changes are clear but small in terms of changes in the soil microclimate and in the physical, chemical, and biological properties (Fig. 8.20).

### 8.6.4 Impacts of Seed-Tree Cutting on Ecosystem Carbon

Seed-tree cutting is a final cutting following the last thinning of mature trees capable of producing seeds. Excluding the phase-by-phase removal of parent trees, seed-tree cutting implies the removal of trees exceeding the dimensions of saw logs and pulpwood. In this cutting, parent trees are left standing widely spaced as a source of seeds for the regeneration of new seedling cohorts. Seed-trees are distributed over the site in single trees or in groups of trees. In the seed-tree cutting, soil management is more feasible than in the shelterwood cutting. This is because in the latter case, mechanized soil disturbance may easily disturb roots and the butt parts of stems and result in a risk of root and stem wood rot. Such damage is more avoidable in the seed-tree cutting due to the wide spacing between seed-trees.

The seed-tree cutting is most successful for the regeneration of birch if the soil management is intensive with a large fraction of open mineral soil available. In such sites, it takes several years for ground vegetation to recover like after prescribed burning for preparation of the reforestation area (Vanha-Majamaa et al. 1996). Seeding may occur from the parent trees at a distance of several hundred meters. Raulo (1981) recommends that only 10–20 parent trees per hectare are needed in the center of the regeneration area if birch is also seeded from the surrounding forests (see also Sarvas 1948; Raulo and Mälkönen 1976). The seed-tree cutting is also successful for Scots pine, especially in poor sites on sand-rich soils, where intensive soil management provides a large amount of open mineral soil for seedbeds (Sarvas 1948).

In regenerating Scots pine in mineral sites, 50–100 dominant and/or codominant trees per hectare are required. However, soil management reduces the required density of parent trees, with only 20–40 parent trees per hectare required in intensively managed sites. In the seed-tree cutting, the share of remaining parent trees is 5–10% of that available in mature tree stands ready before reforestation. This implies that 90–95% of carbon in potential parent trees will be removed, with the same share of reduction of litter input to soil. The impact of seed-tree cutting on soil carbon is like that in clear-cutting. Therefore, in successful regeneration, it takes likely more than10–20 years to offset the carbon losses in soil organic matter and to recover carbon inputs including litterfall from parent trees, growth of seedlings, and ground vegetation. In rough comparison to finding of Simard et al. (2020) for Douglas fir, the amount of carbon in ecosystem after seed-tree cutting is 20–40% of that in uncut area in trees and soils, which is likely less than in Scots pine than Norway spruced dominated forest.

## 8.7 Concluding Remarks

In northern Europe, thinning and clear-cutting is widely used in the production of timber and biomass. Clear-cutting with planting/seeding initializes the next production cycle in even-aged forestry. Harvesting tends to reduce soil carbon up to 10% in the soil profile, with the highest loss in the forest floor in temperate and boreal forests. Clear-cutting reduces soil carbon stocks most in the forest floor (litter, humus), but also deeper in the soil profile. The selection of tree species in reforestation is widely based on the productivity of tree species and the need to produce timber for a given purpose. Tree species also affect soil carbon depending on the amount and quality of litter. Litterfall and harvesting residues affect the carbon stocks in the litter layer and mineral soil, thus maintaining the soil detritus or humus layer below the litter layer. Decomposing litter further feeds humus particles deeper into the soil, carried by infiltrating water to the parent material and groundwater. In combining the production of timber and carbon services (carbon sequestration and carbon stocks in forest ecosystems), proper tree species selection is among the key factors in forest management in a forest landscape.

## References

Andersson B, Elfving B, Persson T, Ericsson T, Kroon J (2007) Characteristics and development of improved *Pinus sylvestris* in northern Sweden. Can J For Res 37:84–92

Blaško R, Forsmark B, Gundale M, Lundmark T, Nordin A (2020) Impacts of tree species identity and species mixing on ecosystem carbon and nitrogen stocks in a boreal forest. For Ecol Manag 458:117783

Briceño-Elizondo E, Garcia-Gonzalo J, Peltola H, Kellomäki S (2006) Carbon stocks and timber yield in two boreal forest ecosystems under current and changing climatic conditions subjected to varying management regimes. Environ Sci Policy 9(3):237–252

Cajander EK (1934) Kuusen taimistojen vapauttamisen jälkeisestä pituuskasvusta. Deutsches Referat: über den Höhenzuwachs der Fichten Phlanzenbestände nach der Befreiung. Communicationes Instituti Forestalis Fenniae 19.5:1–59

Haapanen M, Hynynen J, Ruotsalainen S, Siipilehto J, Kilpeläinen M (2018) Realized and projected gains in growth, quality, and simulated yield of genetically improved Scots pine in southern Finland. Eur J For Res 135:997–1009. https://doi.org/10.1007/s10342-016-0989-0

Hagqvist R, Hahl J (1998) Rauduskoivun siemenviljelysten jalostushyöty Etelä- ja Keski-Suomessa. Reports from the Foundation for Forest Tree Breeding 13:1–32

Hansson K, Olsson B, Olsson M, Johansson U, Kleja DB (2011) Difference in soil properties in adjacent stands of Scots pine, Norway spruce and silver birch in SW Sweden. For Ecol Manag 262:522–530

Harmon ME (2002) Carbon sequestration in forests: addressing the scale question. J For 99(4): 24–29

Harmon ME (2009) Woody detritus mass and its contribution to carbon dynamics of old-growth forest: the temporal context. In: Wirth C, Gleixner G, Heimann M (eds) Old-growth forests: function, fate, and value. Ecological studies 207:159–190

Helynen S, Flyktman M, Asikainen A, Laitila J (2007) Metsätalouteen ja metsäteollisuuteen perustuvan energialiiketoiminnan mahdollisuudet. Scenarios of wood-based energy business in Finland for forestry and forest industry. VTT Research Notes 2397

James J, Harrison R (2016) The effect of harvest on forest soil carbon: a meta-analysis. Forests 7(12):308

Jandl R, Vesterdal L, Olsson M, Bens O, Badeck F, Rock J (2007a) Carbon sequestration and forest management. CABI Rev 2(17):1–15

Jandl R, Lindner M, Vesterdal L, Bauwens B, Baritz R, Hagedorn F, Johnson DW, Minkkinen K, Byrne KA (2007b) How strongly can forest management influence soil carbon sequestration? Geoderma 137:253–268

Jansson G (2007) Gains from selecting *Pinus sylvestris* in southern Sweden for volume per hectare. Scand J For Res 22:185–192

Jansson G, Hansen JK, Haapanen M, Kvaalen H (2017) The genetic and economic gains from forest tree breeding programmes in Scandinavia and Finland. Scand J For Res 32(4):273–286

Johnson DD, Curtis PS (2001) Effects of forest management on soil C and N storage: meta-analysis. For Ecol Manag 140:227–238

Kellomäki S, Hänninen H, Kolström T, Kotisaari A, Pukkala T (1987) A tentative model for describing the effects of some regenerative processes on the properties of natural seedling stand. Silva Fennica 21(1):1–10

Kellomäki S, Koski V, Niemelä P, Peltola H, Pulkkinen P (2009) Management of forest ecosystems. In: Kellomäki S (ed) Forest resources and sustainable management, 2nd edn. Paper Engineers' Association/Paperi ja Puu Oy, Jyväskylä, pp 252–373

Kellomäki S, Kilpeläinen A, Alam A (2013) Effects of bioenergy production on carbon sequestration in forest ecosystems. In: Kellomäki S, Kilpeläinen A, Alam A (eds) Forest bioenergy production: management, carbon sequestration and adaptation. Springer Science + Business Media, New York, pp 125–158

Kellomäki S, Strandman H, Peltola H (2019) Effects of even-aged and uneven-aged management on carbon dynamics and timber yield in boreal Norway spruce stands: a forest ecosystem model approach. Forestry 92:635–647

Kilpeläinen A, Baker M (2013) Integration of management measures for bioenergy production from spatial and temporal perspectives in a forest region – the case of Finland. In: Kellomäki S, Kilpeläinen A, Alam A (eds) Forest bioenergy production. Springer Science+Business Media, New York, pp 111–122

Kilpeläinen A, Routa J, Peltola H, Zubizarreta-Gerendiain A, Pulkkinen P, Kellomäki S (2010) Effects of genetic entry and competition on above ground biomass production of Norway spruce in southern Finland. For Ecol Manag 259:2327–2332

Kohout P, Charvátová M, Štursová M, Mašínová T, Tomšovský M, Baldrian P (2018) Clearcutting alters decomposition processes and initiates complex restructuring of fungal communities in soil and tree roots. ISME J 12:692–703

Kvaalen H, Steffenrem A, Johnsen Ø, Edvarsen ØM, Johnsås OR, Øyen HB (2008) Foredla plantemateriale, skjøtsel for vekst og kvalitet. Norsk Skogsbruket 6:26–28

Leinonen K, Leikola M, Peltonen A, Räsänen PK (1989) Kuusen luontainen uudistuminen Pirkka-Hämeen metsälautakunnassa. Summary: natural regeneration of Norway spruce in Pirkka-Häme forestry board district, Southern Finland. Acta Forestalia Fennica 209:1–53

Mäkipää R, Abramoff R, Adamczyk B, Baldy V, Biryol C, Bosela M, Casals P, Dondini M, Filipek C, Garcia-Pausas J, Gros R, Göröryová E, Hashimoto S, Hassegawa M, Immonen P, Laiho R, Li H, Li Q, Luyssaert S, Menival C, Mori T, Naudts K, Santonja M, Smolander A, Toriyama J, Tupek B, Ubeda X, Verkerk PJ, Lehtonen A (2023) How does management affect soil C sequestration and greenhouse gas fluxes in boreal and temperate forests? A review. For Ecol Manag 529:120637

Marozas V, Sasnauskienė J, Dautartė A, Gavenauska SA, Sabienė N, Armolaitis K (2018) Effect of shelterwood cuttings on soil chemical properties in Scots pine (*Pinus sylvestris* L.) forests in Europe's hemi boreal zone, in Lithuania. J Elem 23(1):353–367

Mayer M, Prescott CE, Abaker WEA, Augusto L, Cécillon L, Ferreira GWD, James J, Jandl R, Katzensteiner K, Laclau J-P, Laganière J, Nouvellon Y, Paré D, Stanturf JA, Vanguelova EI, Vesterdal L (2020) Tamm review: influence of forest management activities on soil organic carbon stocks: a knowledge synthesis. For Ecol Manag 466:118127

Menyailo OV, Sobachkin RS, Makarov MI, Cheng C-H (2022) Tree species and stand density: the effects on soil organic matter contents, decomposability, and susceptibility to microbial priming. Forests 13(2):284

Mjöfors K, Strömgren M, Nohrstedt H-Ö, Johansson M-B, Gärnenäs AI (2017) Indications that site preparation increases forest ecosystem carbon stocks in the long term. Scand J For Res 32(8): 717–725

Na M, Sun X, Zhang Y, Sun Z, Rousk J (2021) Higher stand densities can promote soil carbon storage after conversion of temperate mixed natural forests to larch. Eur J For Res 140:373–386

Napola M-L (2014) Puuston kehityksen vaikutus kuusikon pintakasvillisuuteen. Metsänhoitotieteen laudaturtyö. MSc thesis. Helsingin Yliopiston metsänhoitotieteen laitos

Nave LE, Vance ED, Swanston CW, Curtis PS (2010) Harvest impact on soil carbon storage in temperate forest. For Ecol Manag 259:857–866

Nilsson U, Gemmel P, Johansson U, Karlsson M, Welander TU (2002) Natural regeneration of Norway spruce, Scots pine and birch under Norway spruce shelterwoods of varying densities on a mesic-dry site in southern Sweden. For Ecol Manag 161:133–145

Payne NJ, Cameron DA, Leblanc J-D, Morrison IK (2019) Carbon storage and net primary productivity in Canadian boreal mixed wood stands. J For Res 30(5):1667–1678

Prescott CE (2005) Do rates of litter decomposition tell us anything we really need to know? For Ecol Manag 220:66–74

Prescott CE (2010) Litter decomposition: what controls it and how can we alter it to sequester more carbon in forest soils? Biogeochemistry 101:133–149

Puhlick JJ, Weiskittel AR, Fernandez IJ, Fraver S, Kenefic LS, Seymour RS, Kolka RK, Rustad LE, Brissette L (2016) Long-term influence of alternative forest management treatments on total ecosystem and wood product carbon. Can J For Res 46:1404–1412

Raulo J (1981) Koivukirja. Gummerus Oy, Jyväskylä

Raulo J, Mälkönen E (1976) Koivun luontainen uudistaminen muokatulla kangasmaalla. Summary: natural regeneration of birch (*Betula verrucosa* Ehrh. and *B. pubescens* Ehrh.) on tilled mineral soil. Folia Forestalia 252:1–14

Rosvall O, Jansson G, Andersson B, Ericsson T, Karlsson B, Sonesson J, Stener LG (2001) Genetiska vaster i nuvarande och framtida fröplantager och klonbelandningar. Redugörelse nr 1, Uppsala. Skogsforsk

Routa J, Kellomäki S, Peltola H, Asikainen A (2011) Impacts of thinning and fertilization on timber and energy wood production in Norway spruce and Scots pine: scenario analyses based on ecosystem model simulations. Forestry 84(2):159–175

Routa J, Pyörälä P, Alam A, Kilpeläinen A, Peltola H, Kellomäki S (2013a) Integrated production of timber and energy biomass in forestry. In: Kellomäki S, Kilpeläinen A, Alam A (eds) Forest bioenergy production: management, carbon sequestration and adaptation. Springer Science + Business Media, New York, pp 57–79

Routa J, Kellomäki S, Strandman H, Bergh J, Pulkkinen P, Peltola H (2013b) The timber and energy biomass potential of intensively managed cloned Norway spruce stands. Glob Chang Biol 5:43–52

Ruotsalainen S (2010) Alkuperän vaikutus puulajin menestymiseen. Summary: the importance of provenance for the performance of tree species. Sorbifolia 41(4):149–172

Ruotsalainen S (2014) Increased production through forests tree breeding. Scand J For Res 294: 333–344

Rytter L, Johansson K, Karlsson B, Stener LG (2013) Tree species, genetics and regeneration for bioenergy feedstocks in Northern Europe. In: Kellomäki S, Kilpeläinen A, Alam A (eds) Forest bioenergy production: management, carbon sequestration and adaptation. Springer Science + Business Media, New York, pp 7–37

Saksa T (2020) Metsänhoidon keinot vastata muuttuvaan ilmastoon. PowerPoint presentation. Luke

Sarvas R (1948) Tutkimuksia koivun uudistumisesta Etelä-Suomessa. Summary: research on the regeneration of birch in South-Finland. Communicationes Instituti Forestalia Fenniae 35(4): 1–91

Scott NA, Rodrigues CA, Hughes H, Lee JT, Davidson EA, Dail DB, Malerba P (2004) Changes in carbon storage and net carbon exchange one year after an initial shelterwood harvest at Howland Forest, ME. Environ Manag 33(Supplement):S9–S22

Sikström U, Hjelm K, Holt Hanssen K, Saksa T, Wallertz K (2020) Influence of mechanical site preparation on regeneration success of planted conifers in clearcuts in Fennoscandia – a review. Silva Fennica 54(2):10172

Simard SW, Roach WJ, Defrenne CE, Pickles BJ, Snyder EN, Robinson A, Lavkulich LM (2020) Harvest intensity effects on carbon stocks and biodiversity are dependent on regional climate in Douglas-Fir forests in British Columbia. Front For Global Change 3:88

Stendahl J, Johansson M-B, Eriksson E, Nilsson Å, Langvall O (2010) Soil organic carbon in Swedish spruce and pine forests – differences in stock levels and regional patterns. Silva Fennica 44(1):5–21

Thorne R, Khomik M, Hayman E, Arain A (2020) Response of soil $CO_2$ efflux to shelterwood harvesting in a mature temperate pine forest. Forests 11:304

Vanha-Majamaa I, Tuittila ES, Tonteri T, Suominen R (1996) Seedling establishment after prescribed burning of clear-cut and partially cut mesic boreal forest in southern Finland. Silva Fennica 30(1):31–35

Vesterdal L, Clarke N, Sigurdsson BD, Gundersen P (2013) Do tree species influence soil carbon stocks in temperate and boreal forests? For Ecol Manag 309:4–18

Vestin P, Mölder M, Kljun N, Cai Z, Hasan A, Holst J, Klemedtsson L, Lindroth A (2020) Impacts of clear-cutting of a boreal forest on carbon dioxide, methane, and nitrous oxide fluxes. Forests 11:961

Warren K, Ashton MS (2014) Change in soil and forest floor carbon after shelterwood harvests in a New England oak-hardwood forest, USA. Int J For Res 2014:527236

# Chapter 9
# Impacts of Precommercial Management on Forest Carbon Dynamics

## Contents



**Abstract** In even-aged forestry, boreal forest sites are carbon sources for up to 10–30 years following the final cutting and regeneration (clear-cutting with planting/seeding or cutting for natural regeneration with shelterwood/seed-tree cutting), before shifting to carbon sinks. This period covers the seedling phase with increase of growth and accumulation of carbon in trees and soil, affecting the rate of the shift from a carbon source to a carbon sink. In this context, precommercial management refers to the management of seedling stands prior to commercial management. In even-aged forestry, management includes thinning and harvest of trees for timber in the period before final cutting. Precommercial management may also include cleaning, that is, the removal of trees or tree species likely to disrupt the growth of seedlings selected for future management. In precommercial management, energy biomass may be harvested, but the main objective is to control the growth and development of seedling stands for producing timber and biomass in commercial management through pole and middle-aged phases in the chronosequence.

 
S. Kellomäki, *Forest Management for Timber Production and Climate Change Mitigation*, Managing Forest Ecosystems 44,
https://doi.org/10.1007/978-3-031-71575-4_9

## 9.1 Dynamics of Carbon Sources and Sinks in Early Succession in Managed Forests

### 9.1.1 Coexistence of Different Tree Species and Precommercial Management

In northern Europe, Scots pine (*Pinus sylvestris*) and Norway spruce (*Picea abies*) are commonly grown in pure or mixed-species stands. Both species also grow in mixtures with birch (*Betula pendula* and *B. pubescens*) in moderate fertile and fertile sites (*Myrtillus* site type, MT, or higher fertility) (Saksa and Kankaanhuhta 2007). In such mixtures, different species share the same environment, which leads to interspecific and intraspecific competition for resources. Through succession, competition tends to reduce resource availability, reducing the growth of varied tree species depending on their sensitivity to resource supply (Huston 1979).

Over a chronosequence, the competing capacity of different tree species alone or in mixtures affects carbon sequestration in tree stands (Fig. 9.1). The consequent changes in carbon dynamics are related to the species-specific response to changing functioning and structure trees/stands over successional phases. This holds for precommercial management (early cleaning, precommercial thinning with spacing), which has strong impacts later on the growth and development of tree stands in commercial phase. In this regard, precommercial management refers to the management of seedling stand with producing energy biomass but excluding timber before transition from seedling phase to pole phase under further maturation. In early commercial phase, management refers to thinning producing mainly timber, but energy biomass is still available for harvest after precommercial phase.

### 9.1.2 Impacts of Management and Climatic Warming in Early Succession

In a mixture of Scots pine and birch, competition involves a gradual decline of growth in the shade-intolerant Scots pine under the shading created by the rapid growth of birch. In precommercial management, the proper control of thinning and cleaning of dominant birch makes it possible to maintain the tree species mixture of Scots pine and birch from precommercial through to commercial phases (Hari et al. 1982; Kellomäki et al. 1988). On the other hand, birch benefits from climatic warming, while Scots pine and Norway spruce benefit less. Under warming, birch could outcompete Scots pine by increasing shading, as shown in a simulation experiment in Fig. 9.1.

In this simulation experiment, Scots pine and birch grew in a moderately fertile site (fertility $\geq$ *Myrtillus* site type, MT) originally planted with Scots pine. Birch

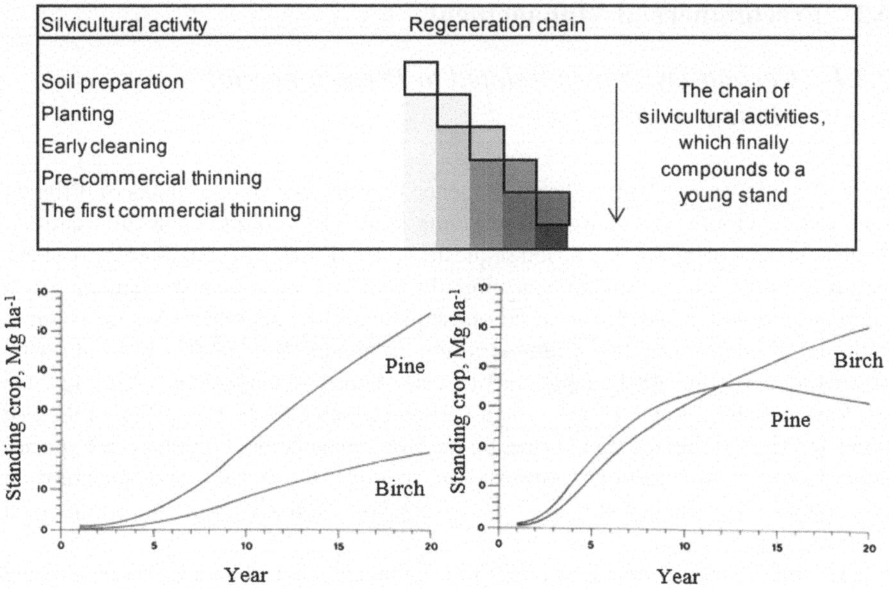


**Fig. 9.1** Forest regeneration, growth, and development of seedlings in a mixture and precommercial management with accumulation of biomass in seedlings. Upper: schematic depiction of a sequence of interlinked management operations for planted Norway spruce, from soil management through reforestation to the maturity of the tree stand ready for the first commercial thinning. The increasing grey indicates the increasing impacts of a management operation on the structure and functioning of a tree stand ready for commercial thinning (Uotila 2017). Courtesy of Finnish Society of Forest Science. Lower/left: effect of the current climate on the growth of Scots pine and birch in a mixed stand in a moderately fertile site (*Myrtillus* site type, MT) on moraine soil in southern Finland (62° N) (Kellomäki et al. 1988). Courtesy of Finnish Society of Forest Science. Lower/right: effects of warming climate on the growth of Scots pine and birch in a mixed stand in a moderately fertile site (*Myrtillus* site type, MT) on moraine soil in southern Finland (62° N) (Kellomäki et al. 1988). Courtesy of Finnish Society of Forest Science

invaded the site naturally, as is typical for Scots pine plantations in moderately fertile sites without cleaning. Under the current climate, the coexistence of these tree species is possible if birch invades several years after (5–7 years' delay) Scots pine is planted (Fig. 9.1) (Hari et al. 1982; Kellomäki et al. 1988). The initial stand included 2000 Scots pine and 1000 birch seedlings per hectare. The climate was assumed to be 2 °C higher over the year than under the current climate. Concurrently, the temperature sum increased from the current level of 1300 d.d. to 1700 d.d. However, the changes in precipitation and atmospheric $CO_2$ were excluded from the simulations. Whenever Scots pine is preferred, proper cleaning, density, and spacing are required in early management.

## 9.2 Precommercial Management

### 9.2.1 Carbon Dynamics Related to Precommercial Management

In even-aged management, a chronosequence is composed of tree stands of different ages and developmental phases, from young seedlings to old trees in full maturity. From a forestry perspective, carbon sequestration and stocks in ecosystems provide a means of mitigating climatic warming through appropriate management. Such mitigation would proceed over a chronosequence, the structure of which changes with progression from the young-seedling to mature-tree phase, thus allowing regeneration for the next management cycle. Forest management cycles involve succession from phase to phase, which varies according to the properties of the tree stand and site. At the tree stand scale, phase-based management involves tree species preferencing and selection, reforestation, control of density and cleaning in precommercial management, and harvesting of biomass prior to commercial management.

Precommercial management refers to the management of stands in the seedling phase, in which tree spacing is controlled and stands are cleaned, but no commercial cuttings are performed. In *spacing*, the stand density is controlled in line with future management objectives preferred in commercial management. Spacing aims at improving the growth and development of the phase-by-phase sequence of tree stands forming the chronosequence. In cases where the initial density of a Scots pine seedling stand, for example, is narrow (density $>$ 2500 seedlings $ha^{-1}$), the spacing needs to be performed before the mean height of seedlings exceeds 3–4 m (Äijälä et al. 2014). Concurrently, *cleaning* aims at removing the seedlings of trees/tree species that will likely disrupt the growth of seedlings targeted for future management. This is the case for Scots pine planted in fertile sites (e.g., *Myrtillus* site type, MT) that facilitate the natural regeneration of birches growing outside the planting site. In this situation, the rapidly growing birch seedlings will likely cast shade on the planted Scots pine seedlings, reduce their growth, and induce excess mortality.

While yields of energy biomass and even pulp wood can be increased through the appropriate spacing in precommercial phase, the greatest timber yield increases are achieved through future thinning in commercial management. In the mature seedling phase prior to commercial phase, Scots pine stand density is 1600–2000 trees $ha^{-1}$ after thinning and spacing. This allows the first commercial thinning to be conducted in the perspective of 10–15 years after regeneration in southern and boreal forests in Finland (Äijälä et al. 2014). On the other hand, the removal of competing trees reduces the stand-based carbon uptake. This reduction also occurs in the carbon accumulation in soil, thus slowing the recovery of ecosystem carbon to prior levels (e.g., Lindgren and Sullivan 2013; Dempster 2022). However, the proper afforestation and management of seedling stands through early and mature seedling phases increase carbon sequestration over subsequent phases in the chronosequence.

### 9.2.2 Cleaning

In precommercial management, cleaning refers to the removal of conspecific trees or tree species that may disturb the growth of the preferred trees or tree species through the different phases of the chronosequence toward full maturity. In management prioritizing coniferous trees (Scots pine, Norway spruce), the focus of cleaning is on removing deciduous seedlings (e.g., birch, aspen) to improve the survival and growth of conifers. Cleaning may occur during the early development of seedling stands and is necessary on fertile sites (*Myrtillus* and *Oxalis-Myrtillus* site types, MT and OMT) with fine soil texture and peat-rich soils that have been well prepared (e.g., disk-trenching, patching) for reforestation. This is because these favorable growing conditions allow deciduous species growing outside a stand to invade the site through natural regeneration.

In southern and mid-boreal conditions, early cleaning in Scots pine and Norway spruce plantations should occur 3–5 years after seeding/planting. This increases, for example, the diameter growth of Norway spruce by 20–30% compared to that under no early cleaning (Uotila et al. 2012; Uotila and Saksa 2014). Cleaned trees such as silver birch, downy birch, aspen, and gray alder tend to recover through stump and root sprouts, but the growth of such plants is lower than the growth of seedlings before cleaning, thus reducing the need for additional cleaning.

Cleaning also implies the selection for preferred trees in pure or mixed stands, (i) wherever the site fertility allows the requisite growth or (ii) where several tree species can coexist in a mixture (Jakkila and Pohtila 1978). Although cleaning involves the removal of trees disturbing the regular growth and development of trees selected for future management, some of the disturbing trees may be retained for restocking the stand. In such partial cleaning, the trees to be cleaned may be removed from above or from below, as shown in Table 9.1. In the former case, the growth-suppressing birches above Scots pines are removed. In the latter case, birches smaller than Scots pines are removed, while some birches taller than Scots pines are retained. In both cases, cleaning increases the growth and survival of Scots pines; that is, the radial and even height growth of Scots pines in cleaned plots is higher than in uncleaned plots. However, excessively heavy cleaning may reduce the quality of stem wood of Scots pines because of a coincident increase of branch growth at the lower pole in cleaned plots (Ikäheimo and Norokorpi 1986.

### 9.2.3 Thinning and Spacing

In precommercial management, thinning refers to the control of the density and spacing of mature seedling stands not yet yielding timber. The timing of precommercial thinning affects the reforestation method used, such as seeding in patches on mounds. For example, patches of Scots pine seedlings should be thinned and cleaned as soon as the mean height of seedlings exceeds 1–2 m (Äijälä et al.

**Table 9.1** Example of the effects of cleaning of deciduous species on the growth and development of Scots pine in northern boreal conditions (Ikäheimo and Norokorpi 1986

| Variable | No treatment | Full cleaning | Cleaning from above | Cleaning from below |
|---|---|---|---|---|
| • Mortality, % | 72 | 33 | 44 | 56 |
| • Mean height, m | 4.2 | 3.2 | 3.4 | 3.3 |
| • Basal area, $m^2$ $ha^{-1}$ | 1.1 | 1.3 | 0.7 | 0.9 |
| • Mean growth, $m^3$ $ha^{-1}$ $yr.^{-1}$ | 0.42 | 1.1 | 0.95 | 0.71 |
| • Stem wood volume, $m^3$ $ha^{-1}$ | 1.0 | 3.6 | 2.3 | 2.9 |
| • Slenderness (h/d0), m $dm^{-1}$ | 5.4 | 4.2 | 4.6 | 4.5 |
| • Thickness of living branch, mm | 14 | 20 | 17 | 18 |
| • Height of crown bottom, dm | 9.9 | 5.5 | 6.8 | 6.1 |

Legend: d0 = stem diameter at the stem butt; h = stem length

2014). In planted stands with a single seedling on each patch, thinning should be conducted when the mean height of seedlings exceeds 3–4 m. Thinning may be delayed, but not later than when the mean height is 5–7 m. However, the timing of tending depends on the management goals, like the quality of timber used, for example, in the joinery industry. In this context, early thinning enhances the growth of branches on the lower stem of Scots pine, thus reducing the quality of future timber. In contrast, delayed thinning reduces the growth of branches on the lower stem, thus increasing the early dying and pruning-off of dead branches and improving the quality of timber obtainable later from the lower stem of mature trees (Fig. 9.2).

Figure 9.3 shows the relationship between stand density and the accumulation of mass in Scots pine in the seedling/pole phase. Even during the initial growth and development, thinning with wide spacing allows a higher accumulation of mass in the stem and branches than that allowed by narrow spacing. Over time, these differences increase, and wide spacing yields a threefold amount of mass and carbon in stem wood per tree in 20 years compared to that in narrow spacing. Wide spacing increases the branch growth and delays the death and pruning-off of lower branches. However, the amount of branch mass in relation to stem mass declines over time even in a widely spaced stand. Over time, trees are differentiated into size classes based on stand density. The size distribution becomes skewed, especially when the trees are growing in narrow spacing. The differentiation of the canopy into canopy layers also implies the death of most suppressed trees related to the initial stand density.

**Fig. 9.2** Precommercial management of a tree stand in the seedling phase. Courtesy of Tapio Finland. Left: spacing and cleaning of a mixture of coniferous and deciduous species in a stand in the seedling phase. Right: precommercial thinning of a naturally seeded Scots pine stand in the seedling phase

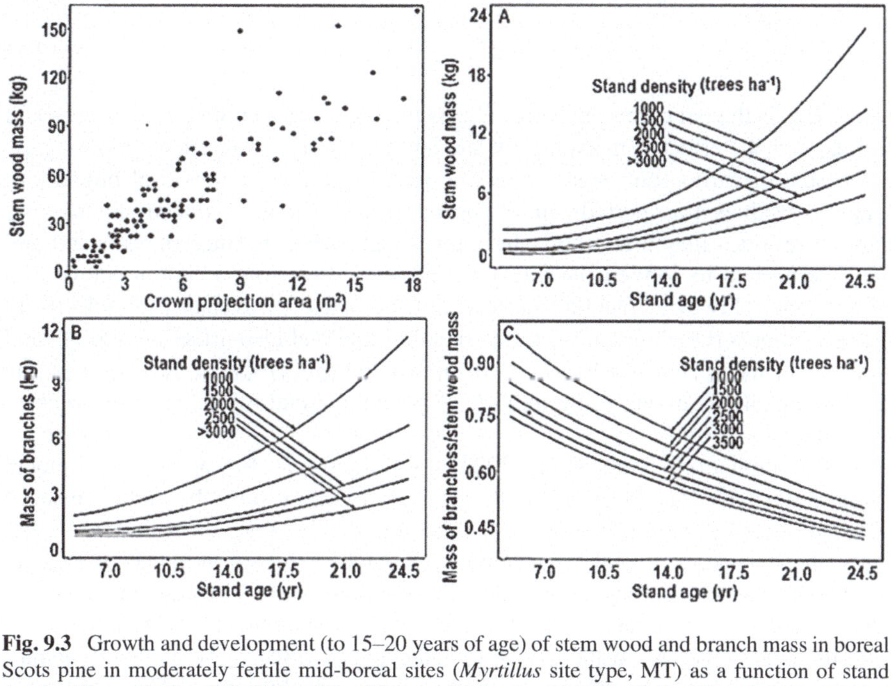


**Fig. 9.3** Growth and development (to 15–20 years of age) of stem wood and branch mass in boreal Scots pine in moderately fertile mid-boreal sites (*Myrtillus* site type, MT) as a function of stand density (Seppo Kellomäki, University of Eastern Finland, unpublished). Left/upper: accumulation of stem mass in relation to crown projection. Right/upper (**a**): accumulation of stem mass in relation to spacing. Left/lower (**b**): accumulation of branch mass in relation to spacing. Right (**c**): ratio between branch and stem wood mass, indicating the branchiness of trees

## 9.3 Effects of Precommercial Management on Carbon Sequestration

### *9.3.1 Outline of Impacts of Precommercial Management on Ecosystem Carbon*

In practice, the biomass of trees cleaned or removed in precommercial management is harvested as energy biomass. This reduces the carbon in trees and litter, allowing a negligible amount of carbon accumulation on the soil in logging residues. On the other hand, cleaned and thinned trees may be left on the site, thus increasing the carbon in soil as well as the carbon emissions from litter and soil organic matter, as demonstrated by Petty et al. (2024). The authors conducted model simulations dealing with the carbon balance in young Norway spruce (*Picea abies*) forest from before the developmental phase (from planting) to spacing in the mature seedling phase. In the context of ecosystem carbon, the net carbon balance through the precommercial phase is affected by both the carbon sequestration and emissions. The carbon balance is also affected by emissions induced by harvesting and transportation of biomass outside the site:

$$C_{net} = C_{seq} + C_{man} + C_{decomp} + C_{harv} \tag{9.1}$$

where $C_{\mathrm{net}}$ is the net carbon balance; $C_{\mathrm{seq}}$ is the sequestration of carbon in trees and soil; $C_{\mathrm{man}}$ is the effect of management and harvesting on ecosystem carbon; $C_{\mathrm{decomp}}$ is the sum of carbon emissions from soil; and $C_{\mathrm{harv}}$ is the removal of biomass in logging residues transported out of the ecosystem. In this context, removal of logging residues may affect both the growth and carbon balance of the stand and the time needed to convert the forest regeneration area from a carbon source to a carbon sink. However, planting density and the origin (provenance or selectively breed seedlings (breeding gain)) of the planting material also affect the growth and carbon balance of tree stands, as was demonstrated by Petty et al. (2024) using the model simulations in analyzing impacts of precommercial management on ecosystem carbon.

In the simulations, Petty et al. (2024) used a gap-type forest ecosystem model (Sima model, Kellomäki et al. 2008; Fig. 9.4) to study how the carbon balance performs after a clear-cut in relation to precommercial management. In the simulations, the management options combined (i) removal or non-removal of logging residues in the final felling before planting, (ii) planting of Norway spruce (*Picea abies*) seedlings with and without breeding improvement, and (iii) planting at various stand densities. At the tree, soil, and ecosystem scale, the carbon dynamics were monitored from the planting to the first commercial thinning. The simulations were conducted on a moderately fertile site (*Myrtillus* site type, MT) on podzol soil in the mid-boreal forest zone in Central Finland (62° N) under the current climate. The temperature sum was 1100 d.d., and mean annual precipitation was 500–600 mm.

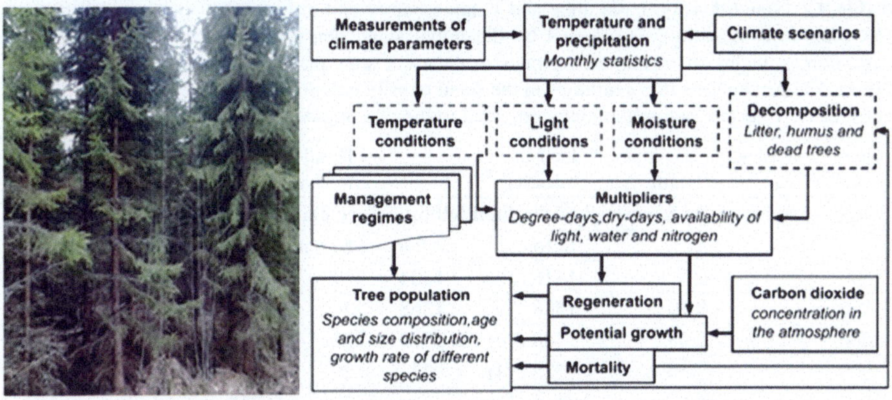


**Fig. 9.4** Overview of the forest ecosystem model used to simulate the dynamics of biomass and carbon in Norway spruce from the seeding phase to the pole phase (Tables 9.2 and 9.3; Fig. 9.5). Left: a typical Norway spruce stand at the first commercial thinning. Right: overview of the ecosystem simulation model (Sima) (Kellomäki et al. 2008) used in generating data for the analysis. Open Access. Courtesy of PMC

The simulation options included several management regimes combining the planting density, the impacts of breeding on growth, and the planting in sites with/without logging residues from the previous management cycle's clear-cutting. The specifications of the possible management scenarios were as follows:

- Initial soil conditions in the simulations were the same regardless of the management option.
- Planting density was 1500, 2000, 2500, or 3000 seedlings per hectare.
- Seedlings were either of local origin with no breeding gain in growth or nonlocal seedlings with a breeding gain of 10%, 20%, or 30% compared to that in the seedlings of local origin.
- Planting was conducted on the sites with or without logging residues.

The simulations extended through the precommercial phase for 33 years, representing the time to the first commercial thinning at the end of the precommercial phase in the chronosequence. Regardless of management scenario, the thinning was performed from below, with a target basal area after thinning of 18 $m^2$ $ha^{-1}$ for all simulation cases. The year of thinning was based on the dominant height (12 m) and basal area (22 $m^2$ $ha^{-1}$) equal to the values of the baseline simulation using a planting density of 2000 seedlings $ha^{-1}$. Thinning followed the management recommendations widely used in private forestry in Finland (Äijälä et al. 2014).

The main findings of Petty et al. (2024) concerning how different management scenarios affected the structure of Norway spruce stands are listed in Table 9.2. The findings represent management scenarios on growth and development from the initial seedling phase to trees in the mature pole phase before the first commercial thinning (at the age of 33 years).

**Table 9.2** Stand density, basal area, and volume before and after the first commercial thinning (33 years after planting), as affected by the initial stand density, genetic gain in growth, and exclusion or inclusion of harvest residues in clear-cut sites planted in simulations (Petty et al. 2024). The percentages in parentheses in the stand density column indicate the natural mortality of seedlings per hectare before thinning compared to the planting density

| Planting density, trees $ha^{-1}$ | Genetic gain, % | Density, trees $ha^{-1}$ | | Basal area, $m^2$ $ha^{-1}$ | | Volume, $m^3$ $ha^{-1}$ | |
|---|---|---|---|---|---|---|---|
| **Logging residues not harvested in the final felling before planting** | | | | | | | |
| | | Before | After | Before | After | Before | After |
| 1500 | 0 | 1418 (5) | 1304 | 19 | 18 | 111 | 112 |
| 1500 | 30 | 1388 (7) | 1368 | 26 | 18 | 177 | 128 |
| 2000 | 0 | 1788 (10) | 1370 | 22 | 18 | 127 | 108 |
| 2000 | 30 | 1722 (14) | 974 | 29 | 18 | 194 | 123 |
| 2500 | 0 | 2116 (15) | 1430 | 24 | 18 | 138 | 104 |
| 2500 | 30 | 2040 (18) | 1066 | 32 | 18 | 205 | 120 |
| 3000 | 0 | 2508 (16) | 1546 | 27 | 18 | 151 | 104 |
| 3000 | 30 | 2346 (22) | 1136 | 34 | 18 | 211 | 115 |
| **Logging residues harvested in the final felling before planting** | | | | | | | |
| | | Before | After | Before | After | Before | After |
| 1500 | 0 | 1400 (7) | 1388 | 17 | 18 | 101 | 109 |
| 1500 | 30 | 1376 (8) | 944 | 24 | 18 | 163 | 124 |
| 2000 | 0 | 1790 (10) | 1452 | 21 | 18 | 117 | 105 |
| 2000 | 30 | 1744 (13) | 1016 | 28 | 18 | 182 | 118 |
| 2500 | 0 | 2148 (14) | 1542 | 23 | 18 | 128 | 102 |
| 2500 | 30 | 2116 (15) | 1116 | 31 | 18 | 196 | 116 |
| 3000 | 0 | 2512 (16) | 1446 | 26 | 18 | 137 | 100 |
| 3000 | 30 | 2320 (23) | 1200 | 32 | 18 | 194 | 113 |

(i) Before the first commercial thinning, the density of seedling stands declined 5–23% from that used in the initialization of simulations, with the increasing mortality along with increasing initial stand density. Mortality tended to be higher for improved seedlings than for those of local origin at the same initial stand density. The dense seedling stands with enhanced growth showed strong competition between seedlings, with an increasing risk of mortality as the seedlings matured. Harvesting/non-harvesting of logging residues seemed to have no effect on the mortality.

(ii) After the first commercial thinning, the same basal area (18 $m^2$ $ha^{-1}$) remained regardless of the management scenarios. As a result, the stem volume varied from 100 to 128 $m^3$ $ha^{-1}$, and the stand density from 944 to 1546 trees $ha^{-1}$, in simulations.

(iii) Stem volume before the first commercial thinning ranged from 111 to 151 $m^3$ $ha^{-1}$ for stands planted with seedlings of local origin, regardless of the initial stand density, where logging residues were not harvested. Where improved seedlings (30% higher growth rate) were used, the stem volume ranged from 177 to 211 $m^3$ $ha^{-1}$.

(iv) When logging residues inherent from the previous clear-cutting were harvested before planting, the stem volume before the first commercial thinning was lower than when residues remained on the site. The stem volume ranged from 101 to 137 $m^3$ $ha^{-1}$ for seedlings of local origin and from 163 to 194 $m^3$ $ha^{-1}$ for improved seedlings.
(v) Dominant height remained similar before and after the precommercial thinning, while the mean diameter increased because thinning was performed from below.

### 9.3.2 Effect of Management on Carbon Stock in Trees and Soil

According to Petty et al. (2024), the mean carbon stock in trees depended on (i) the initial stand density, (ii) the quality of seedlings (local/improved with a 30% increase in growth), and (iii) the presence or absence of logging residues on the site. In this context, Table 9.3 shows the carbon stocks and dynamics in the trees, soil, and ecosystem depending on the management scenario, from the initial seedling phase to young trees in the pole phase before the first commercial thinning at the age of 33 years. The key findings were as follows:

(i) Mean carbon stocks in the trees, soil, and ecosystem increased along with increasing initial stand density and the quality of seedlings (local/improved with a 30% increase in growth) regardless of the presence or absence of logging residues on the site.
(ii) When logging residues remained in the sites, the mean carbon stock in planted trees ranged from 15 to 22 Mg C $ha^{-1}$ for the seedlings of local origin and from 22 to 28 Mg C $ha^{-1}$ for the improved seedlings used in the simulations.
(iii) When logging residues were removed from the sites before planting, the mean carbon stock in trees ranged from 13 to 21 Mg C $ha^{-1}$ for the simulations with local seedlings and from 20 to 28 Mg C $ha^{-1}$ for the simulations with improved seedlings.
(iv) When logging residues were left on the planted sites, the mean soil carbon ranged from 40 to 44 Mg C $ha^{-1}$ and from 42 to 45 Mg C $ha^{-1}$ for the seedlings of local origin and the improved seedlings, respectively.
(v) When logging residues were harvested from the sites before planting, the mean soil carbon stock decreased, ranging from 35 to 37 Mg C $ha^{-1}$ and from 34 to 39 Mg C $ha^{-1}$ when using the seedlings of local origin or improved seedlings, respectively.
(vi) Higher planting density and improved seedlings increased the mean carbon stock in trees and soil, as did the presence of logging residues left on site in the previous final felling before regeneration.
(vii) Carbon stock in the ecosystem was more sensitive to the carbon accumulation in trees than that in soil, regardless of management scenario.

**Table 9.3** Mean carbon stocks in trees and soil, with the carbon balance over a 33-year period from initialization of the simulations to the first commercial thinning for different management scenarios. The years in which carbon balance was <0 indicate higher carbon uptake than emissions in the ecosystem (Petty et al. 2024)

| Planting density, trees $ha^{-1}$ | Genetic gain, % | Mean tree C, Mg $ha^{-1}$ | Mean soil C, Mg $ha^{-1}$ | Mean soil $CO_2$ emissions, $g\ m^{-2}\ yr^{-1}$ | Mean $CO_2$ uptake of trees, $g\ m^{-2}\ yr^{-1}$ | Mean C balance (NEE), $g\ m^{-2}\ yr^{-1}$ |
|---|---|---|---|---|---|---|
| **Logging residues not harvested in the final felling before planting** | | | | | | |
| 1500 | 0 | 15 | 40 | 1208 | 987 | 221 |
| 1500 | 30 | 22 | 42 | 1243 | 1314 | −71 |
| 2000 | 0 | 17 | 40 | 1217 | 1094 | 123 |
| 2000 | 30 | 25 | 44 | 1250 | 1445 | −195 |
| 2500 | 0 | 20 | 41 | 1230 | 1185 | 45 |
| 2500 | 30 | 28 | 43 | 1232 | 1546 | −314 |
| 3000 | 0 | 22 | 44 | 1251 | 1262 | −11 |
| 3000 | 30 | 31 | 45 | 1242 | 1630 | −388 |
| **Logging residues harvested in the final felling before planting** | | | | | | |
| 1500 | 0 | 13 | 35 | 1011 | 932 | 79 |
| 1500 | 30 | 20 | 34 | 987 | 1235 | −248 |
| 2000 | 0 | 16 | 36 | 1011 | 1033 | −22 |
| 2000 | 30 | 24 | 39 | 1038 | 1372 | −334 |
| 2500 | 0 | 18 | 36 | 1014 | 1116 | −102 |
| 2500 | 30 | 27 | 38 | 1025 | 1473 | −448 |
| 3000 | 0 | 21 | 37 | 1014 | 1192 | −177 |
| 3000 | 30 | 28 | 39 | 1024 | 1536 | −512 |

### 9.3.3 *Effect of Management on Carbon Uptake, Emissions, and Balance*

Petty et al. (2024) further analyzed the carbon dynamics of young Norway spruce stands over 33 years—from planting, through stands at the seedling phase, and up to the first commercial thinning in the pole phase. Carbon dynamics were investigated in terms of the carbon uptake in trees, the carbon emissions from soil, and the carbon balance (*NEE*) between the carbon uptake and emissions. As with the mean carbon stocks in the trees, soil, and ecosystem (Table 9.3), the impacts of management on the carbon uptake, emissions, and balance were related to (i) the initial stand density, (ii) the presence or absence of logging residues on the site, and (iii) the quality of seedlings (local/improved with a 30% increase in growth).

When logging residues remained on the site, the mean $CO_2$ uptake in trees ranged from 987 to 1262 g C $m^{-2}$ yr.$^{-1}$ with seedlings of local origin and 1314–1630 g C $m^{-2}$ yr.$^{-1}$ with improved seedlings (Table 9.3). On the other hand, when logging residues were harvested, the carbon uptake was lower, ranging from 932 to 1192 g C m$^{-2}$ yr.$^{-1}$ with the seedlings of local origin and from 1235 to 1536 g C m$^{-2}$ yr.$^{-1}$ with improved seedlings. Use of higher planting density and improved seedlings increased the mean $CO_2$ uptake of trees. Concurrently, the mean carbon emission from soil ranged from 1208 to 1251 g C m$^{-2}$ yr.$^{-1}$ regardless of seedling origin and initial stand density. Furthermore, when logging residues were harvested, the mean carbon emission from soil was lower, ranging from 987 to 1038 g C m$^{-2}$ yr.$^{-1}$ depending on seedling origin and initial stand density (Table 9.3).

The mean ecosystem carbon balance (trees and soil) over the 33-year simulation period ranged from a carbon source (positive carbon balance values, soil $CO_2$ emissions > $CO_2$ uptake of trees) to carbon sink (negative carbon balance values, soil $CO_2$ emissions < $CO_2$ uptake of trees), depending on management scenario (Table 9.3). When logging residues remained on the sites before planting, the mean carbon balance ranged from 221 to −11 g C m$^{-2}$ yr.$^{-1}$ with the seedlings of local origin and from −71 to −388 g C m$^{-2}$ yr.$^{-1}$ with the improved seedlings. This held for initial stand densities of 1500–3000 seedlings ha$^{-1}$ (Table 9.3). The mean carbon balance was positive at initial stand densities of 1500–2500 seedlings ha$^{-1}$ with the seedlings of local origin. In contrast, the use of improved seedlings induced a systematically negative carbon balance regardless of initial stand density, thus increasing ecosystem carbon stocks.

When logging residues were harvested from the regeneration site, the mean carbon balance over the 33-year period ranged from 79 to −177 g C m$^{-2}$ yr.$^{-1}$ with the seedlings of local origin and from −248 to −512 g C m$^{-2}$ yr.$^{-1}$ with improved seedlings, for initial stand densities of 1500–3000 seedlings ha$^{-1}$ (Table 9.3). The mean carbon balance was negative in most of the simulation scenarios, excluding the scenario with an initial stand density of 1500 seedlings ha$^{-1}$ and seedlings of local origin. Carbon uptake exceeded carbon loss and improved the carbon balance most when the seedlings with improved growth were used together with higher planting density (>2000 seedlings ha$^{-1}$) than that typically used in forest regeneration. The removal of logging residues also improved the carbon balance.

Figure 9.5 shows a sequence of the forest carbon dynamics over the simulated 33-year management period (Petty et al. 2024). The carbon loss from soil increased rapidly over the first 5 years regardless of the management scenario. Thereafter, the loss of soil carbon reduced gradually regardless of whether logging residues had been harvested in the final felling before forest regeneration with planting. At the same time, the carbon accumulation in trees increased gradually during the first 10 years. Thereafter, the accumulation increased over the precommercial phase (from year 10 to year 33 from the seedling phase to the pole phase) when the initial stand density was less than 2500 seedlings ha$^{-1}$ and the growth rate was moderate. On the other hand, the period of rapid carbon accumulation was reduced, and the total carbon accumulation leveled off when high initial spacing and improved seedlings with high growth rates were used in the simulation. This implied that the narrow spacing and high growth increased mortality due to strong competition in the stand, thus reducing the gains from rapid carbon accumulation in trees and soil (Table 9.3).

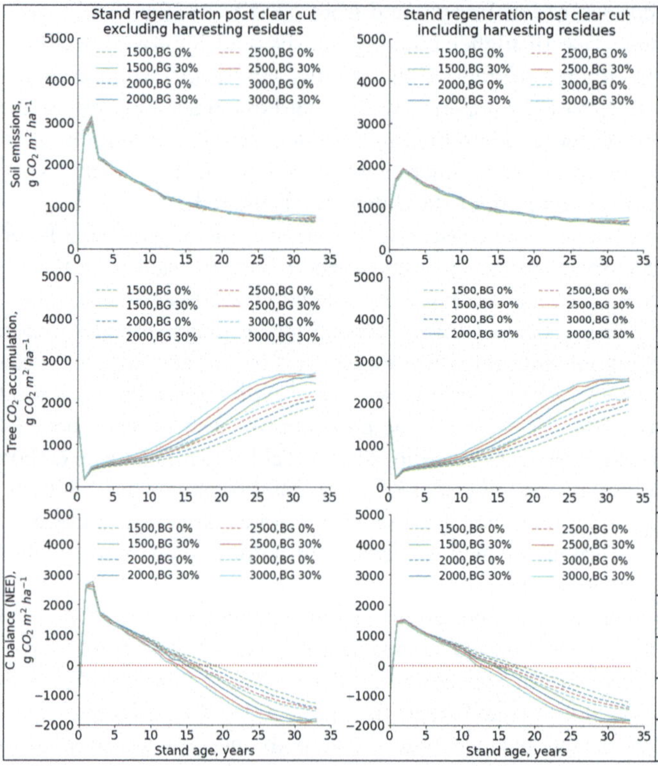


| Planting density, trees $ha^{-1}$ | Genetic gain, % | $CO_2$ balance (NEE) < 0, years | |
|---|---|---|---|
| | | Excluding harvesting | Harvesting residuals |
| 1,500 | 0 | 19 | 18 |
| 1,500 | 10 | 18 | 17 |
| 1,500 | 20 | 18 | 17 |
| 1,500 | 30 | 17 | 15 |
| 2,000 | 0 | 18 | 17 |
| 2,000 | 10 | 17 | 16 |
| 2,000 | 20 | 16 | 15 |
| 2,000 | 30 | 15 | 15 |
| 2,500 | 0 | 17 | 16 |
| 2,500 | 10 | 16 | 15 |
| 2,500 | 20 | 15 | 14 |
| 2,500 | 30 | 14 | 13 |
| 3,000 | 0 | 16 | 15 |
| 3,000 | 10 | 15 | 14 |
| 3,000 | 20 | 15 | 14 |
| 3,000 | 30 | 14 | 12 |

**Fig. 9.5** Carbon dynamics and stocks of Norway spruce (*Picea abies*) stands with different properties under precommercial management (Petty et al. 2024). Left: effect of precommercial management on the mean carbon emission from soil, carbon accumulation in trees, and ecosystem carbon balance related to (i) planting density and use of seedlings of local origin (BG 0%) and enhanced growth (BG 30%, breeding gain) and (ii) logging residues from the clear-cutting excluded (left) or included (right) in the site before planting. Right: time needed from post-clear-cutting planting for the recovery of the carbon balance from carbon source to sink ($NEE < 0$), depending on planting density and growth rate (breeding or genetic gain, %) of seedlings with and without the removal of logging residues

Figure 9.5 shows that a negative carbon balance (carbon sink with $NEE < 0$) was achieved 14–19 and 12–18 years after post-clear-cutting planting. This was dependent on planting density and growth rate (genetic gain, %) of seedlings with or without the removal of logging residues before planting. The shortest duration was achieved with a narrow spacing of 3000 seedlings $ha^{-1}$ and a 30% increase in growth due to breeding. In contrast, the longest recovery period occurred with wide spacing and local-origin seedlings. The rapid growth of seedlings partly compensated for the wide spacing, which tended to increase the time needed to balance the carbon loss and uptake. Consequently, the balancing of carbon loss and uptake was 5–6 years more rapid for narrow spacing using improved seedlings compared to wide spacing with unimproved seedlings regardless of whether logging residues remained on the site. Removal of logging residues reduced the duration; it took to transform the forest into a carbon sink from a source by a few years.

Based on Fig. 9.5, the time to achieve a $CO_2$ balance (*NEE*) < 0 (Recovery time) in years is linearly related both to planting density (Planting density), breeding gain (Breeding gain), and harvesting of logging residues (Harvest = 1, no Harvest = 0) before planting of seedlings, as per the following:

$$\begin{aligned} \textit{Recovery time} = 22.063 - 0.002 \times \textit{Planting density} - 0.084 \\ \times \textit{Breeding gain} - 1.062 \times \textit{Harvest} \end{aligned} \tag{9.2}$$

When logging residues were removed, the recovery time for the planting density of 1500 local-origin seedlings $ha^{-1}$ was 18 years. At the same planting density, the recovery time was reduced by 3–15 years if planting with improved seedlings was conducted without removing logging residues. Similarly, for the planting density of 3000 seedlings $ha^{-1}$, excluding the breeding gain, the recovery time was 15 years, with a reduction of 3 years. The same planting density of seedlings with a 30% breeding gain reduced the recovery time to 12 years, yielding a reduction in recovery time of 6 years.

The simulation findings of Petty et al. (2024) were close to those of Grelle et al. (2023), who estimated the recovery of carbon sink/source balance based on eddy covariance measurements in southern Sweden for Norway spruce (*Picea abies*), Scots pine (*Pinus sylvestris*), and larch (*Larix* × *eurolepis*). The annual carbon fluxes returned from positive (sources) to negative (sinks) in 8–13 years after forest regeneration depending on the site productivity and management. The return to a carbon sink took place in a period equal to 15% of the rotation length at these sites. In general, the stands reached a carbon balance under conventional reforestation in a period equal to 30% of the rotation period in these sites. Grelle et al. (2023) summarized that the managed forest in the southern boreal zone of Sweden is a carbon source during a small part of the rotation period. The length of the source period is shorter in fertile sites with high productivity than in poor sites with low productivity. On the other hand, a reduced woody debris left on the site reduces the length of the source period. Achat et al. (2015) found that, for a set of boreal forests in North America and Europe, the relative reduction rate of soil carbon culminated 5–10 years after cutting. However, the full recovery was likely much longer, exceeding 20 years, as found by Grelle et al. (2023) for boreal forests.

## 9.4 Effects of Precommercial Management on Carbon Sequestration throughout Rotation

The simulation of Petty et al. (2024) dealt mainly with how precommercial management may affect the growth and development of seedling stands from early to mature phases, prior to the pole phase and thinning. The study of Shaedel et al.

(2017) had a similar focus, investigating the impacts of precommercial management on the functioning and structure of tree stands in mature phases in a chronosequence. This latter study addressed the total aboveground carbon stocks in a 54-year-old western larch (*Larix occidentalis*) stand. The measurements concerned the long-term effects of precommercial thinning on aboveground carbon stock and its distribution among four aboveground carbon pools: (i) living overstory, (ii) living understory/mid-story, (iii) woody detritus, and (iv) forest floor. The carbon in these pools was further subdivided into the carbon accumulated prior to the initiation of the growth and development of the current stand (legacy carbon) and the carbon accumulated by the current stand after the initiation (non-legacy carbon). The densities of trees in the plots were 494, 890, and 1640 trees per hectare, which were originally chosen to determine the ideal spacing for the growth of western larch. Additionally, the comparative analysis included measurements concerning unthinned stands (Shaedel et al. 2017).

Shaedel et al. (2017) found no influence of precommercial thinning on the total non-legacy carbon over 54 years after treatment (Fig. 9.6). In living trees, carbon was nearly identical across the assessed densities, showing that trees had grown much larger at low density than at high density. Under low density, the carbon in the understory trees, mid-story trees, and ground vegetation was much higher than under no thinning, where there was substantially more non-legacy carbon in woody detritus than that in thinned stands. On the other hand, pools of legacy carbon did not vary significantly with density but made up a substantial proportion of aboveground carbon stocks. In non-thinned sites, legacy carbon in dead wood formed the long-term carbon stock in 62 years after initialization of these stands, representing 20–25% of aboveground carbon stocks. There were no trade-offs between managed stands in relation to the growth and development of trees and understory vegetation. Consequently, this maximized the aboveground carbon stocks for a long period following experiment initialization. Shaedel et al. (2017) concluded that precommercial thinning before canopy closure and strong between-tree competition could likely be used to enhance the adaptation of trees to climatic warming, with the capacity to mitigate climate change.

## 9.5 Concluding Remarks

In precommercial management, energy biomass may be harvested, but the main objective is to control the growth and development of seedling stands for producing timber and biomass in commercial management through pole and middle-aged phases in the chronosequence. In this regard, the soil carbon in a regeneration site includes the logging residues and soil carbon inherent before cutting. Over time, this carbon (legacy carbon) decreases but is compensated for by carbon (de novo carbon) sequestered in seedlings established in reforestation (seeding/planting in artificial and seeding in natural regeneration). Over several years following the cutting, the ecosystem carbon tends to decline until reaching the turning point at which carbon

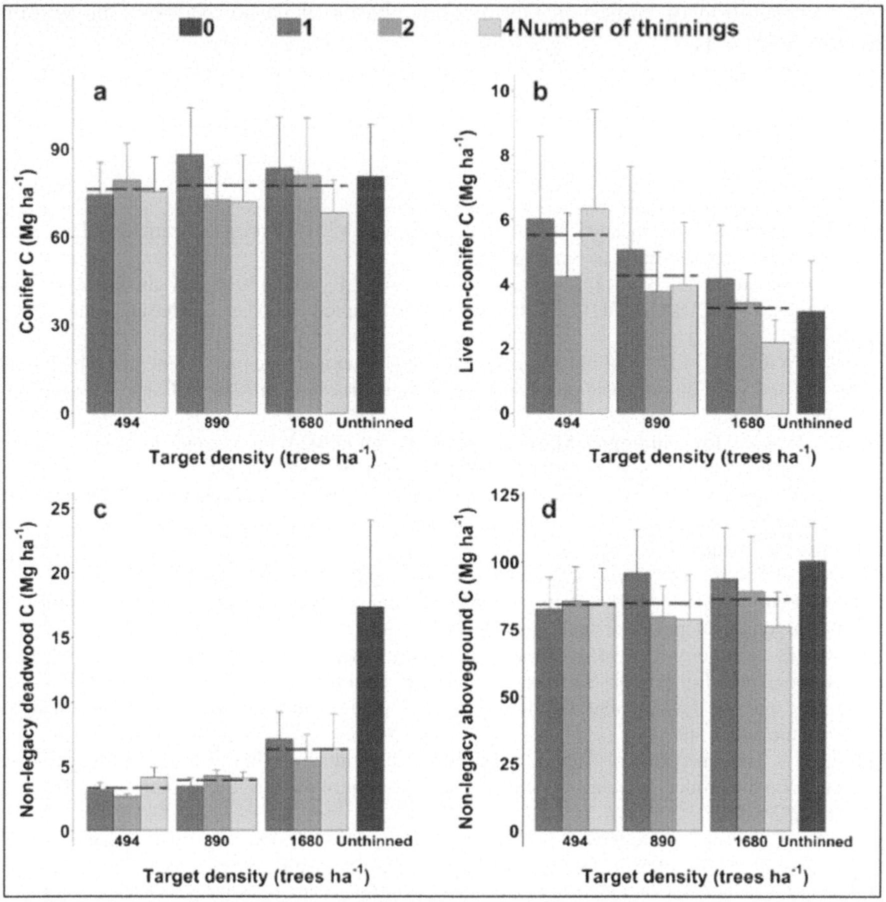


**Fig. 9.6** Effects of thinning on three carbon pools: (**a**) living conifers, (**b**) living non-conifers, and (**c**) non legacy deadwood. Panel (**d**) is the sum of the three previous pools. The bars are grouped by target density, shown on the x-axis. Within each density level, three different thinning regimes (one, two, or four entries) were used to achieve the target stand density. The dashed lines across the three grouped bars are the average of all number-of-entry treatments within density and represent the main effect of density on carbon stocks. Error bars represent one standard error of the mean (Shaedel et al. 2017). Permission of Elsevier

sequestration in the seedlings and ground vegetation exceeds carbon loss from the ecosystem. Thereafter, the ecosystem carbon increases along with the growth of seedlings, which sequester carbon. This compensates for the carbon lost through decay since the regenerative cutting, with the shift rate from a carbon source to a carbon sink depending on the management regime. In boreal conditions, the compensation of carbon loss may take 10–30 years depending on management, as indicated in some experimental and model-based studies. There is currently no evidence that precommercial management affects the total carbon sequestration and accumulation of carbon in commercial phases of carbon stocks representing mature-tree stands.

## References

Achat DL, Fortin M, Landmann G, Ringeval B, Augusto L (2015) Forest soil carbon is threatened by intensive biomass harvest. Sci Rep 5:15991

Äijälä O, Koistinen A, Sved J, Vanhatalo K, Väisänen P (eds) (2014) Hyvän metsänhoidon suositukset—METSÄNHOITO. Kehittämiskeskus Tapion julkaisuja, Metsäkustannus Oy, Helsinki

Dempster WR (2022) Effects of planting, vegetation management, and pre-commercial thinning on growth and yield of lodgepole pine regenerated after harvesting in Alberta, Canada. Forests 13: 1–15

Grelle A, Hedwall P-O, Strömgren M, Håkansson C, Bergh J (2023) From source to sink—recovery of the carbon balance in young forests. Agric For Meteorol 330:109290

Hari P, Kellomäki S, Mäkelä A, Ilonen P, Kanninen M, Korpilahti E, Nygren M (1982) Metsikön varhaiskehityksen dynamiikka. Summary: dynamics of early development of tree stands. Acta Forestalia Fennica 1977:1–42

Huston M (1979) A general hypothesis of species diversity. Am Nat 113(1):81–101

Jakkila J, Pohtila E (1978) Perkauksen vaikutus taimiston kehitykseen Lapissa. Summary: effect of cleaning on development of sapling stands in Lapland. Folia Forestalia 360:1–27

Ikäheimo E, Norokorpi Y (1986) Perkauksen vaikutus männyn istutustaimikoiden kehitykseen, laatuun ja tuhoihin Pohjois-Suomessa. Summary: the effect of cleaning on the incidence of damage and the development and quality of scots pine plantations in northern Finland. Folia Forestalia 647:1–49

Kellomäki S, Hänninen H, Kolström T (1988) Model computations on the impacts of the climatic change on the productivity and silvicultural management of the forest ecosystem. Silva Fennica 22(1):293–305

Kellomäki S, Peltola H, Nuutinen T, Korhonen KT, Harri SH (2008) Sensitivity of managed boreal forests in Finland to climate change, with implications for adaptive management. Philos Trans R Soc Lond B Biol Sci 363(1501):2341–2351. https://doi.org/10.1098/rstb.2007.2204.

Lindgren PMF, Sullivan TP (2013) Long-term responses of tree and stand growth of young lodgepole pine to pre-commercial thinning and repeated fertilization. For Ecol Manag 307: 155–164

Petty A, Pikkarainen L, Strandman H, Kellomäki S, Solonen H, Kilpeläinen A, Peltola H (2024) Influence of planting density and breeding gain on carbon balance in young boreal Norway spruce stands. Manuscript under preparation. University of Eastern Finland, School of Forest Sciences

Saksa T, Kankaanhuhta V (2007) Metsänuudistamisen laatu ja keskeisimmät kehittämiskohteet Etelä-Suomessa. Metsänuudistamisen laadun hallintahankkeen loppuraportti. Metsäntutkimuslaitos. Gummerus kirjapaino Oy, Jyväskylä

Shaedel MS, Larson AJ, Affleck DLR, Belote RT, Goodburn JM, Page-Dumroese DS (2017) Early forest thinning changes aboveground carbon distribution among pools, but not total amount. For Ecol Manag 389:187–198

Uotila K (2017) Optimization of early cleaning and precommercial thinning methods in juvenile stand management of Norway spruce stands. Doctoral Dissertation,. University of Helsinki

Uotila K, Rantala J, Saksa T (2012) Estimating the need for early cleaning in Norway spruce plantations in Finland. Silva Fennica 46(5):683–693

Uotila K, Saksa T (2014) Effects of early cleaning on young *Picea abies* stands. Scand J For Res 29(2):111–119

# Chapter 10
# Commercial Thinning and Selective Cutting with Impacts on Forest Carbon

## Contents




**Abstract** In even-aged management, thinning refers to the harvesting of timber and biomass through forest chronosequences extending from pre-commercial management to the final cutting and subsequent reforestation for the next rotation. Thinning includes the harvesting of trees in the pole and middle-aged phases. This increases tree spacing to enhance (i) the growth and development of the remaining trees and (ii) the provision of timber and biomass. Throughout a rotation, thinning involves the





S. Kellomäki, *Forest Management for Timber Production and Climate Change Mitigation*, Managing Forest Ecosystems 44,
https://doi.org/10.1007/978-3-031-71575-4_10

periodic removal of trees from below or above to achieve a given spacing of the remaining trees in patches, with the consequent growth and development faster than before thinning. Thinning removal includes (i) the proactive removal of timber otherwise lost in natural mortality and (ii) increased allocation of productivity to a smaller number of trees to enhance their growth and development from the thinning phase to tree maturity. In uneven-aged management, selective cutting focuses on the removal of larger individual trees, which create space for the natural regeneration and ingrowth of seedlings through the canopy of dominating large trees. In uneven-aged management, rotation is excluded, with the aim of maintaining a skewed forest structure. This implies that seedlings, young, and pole-phase trees are dominant among middle-aged and mature trees. This is achieved by harvesting maturing/mature trees for the manufacturing and energy industries facilitating the establishment of seedlings in canopy gaps and increasing growth of remaining larger trees.

## 10.1 Thinning/Selective Cutting, Growth, and Yield

### *10.1.1 Harvesting and Logging in Management*

*Harvesting* is a general term used to describe the cutting and further preparation of timber/biomass for transport to wood-based production outside the forest ecosystem. In even-aged management, thinning refers to the periodic removal of trees from stands in patches/sites to increase the amount of space for the remaining trees to grow and develop based on the removal of trees from below (suppressed trees) or from above (dominating trees) (Fig. 10.1). This also holds for uneven-aged management, which combines the selective cutting and harvesting of timber with concurrent facilitation of natural regeneration. In selective cuttings, the focus is on the removal of dominant trees of varying size occupying a site without a patch-based distribution. In both even- and uneven-aged management, harvesting is the process of cutting timber and moving it to a particular location for subsequent transport. In this context, logging may include skidding as well as on-site processing such as the division of the tree biomass into timber, including separation of stem wood into saw logs, pulpwood, and energy biomass. The latter refers to stem wood unsuitable for timber and logging residues such as slash and stumps usable for energy biomass.

In even-aged management, carbon management is based on the carbon taken up in growth and on the carbon removed in harvesting throughout rotation from the seedling phase to full maturity (e.g., Vuokila 1980). In the seedling phase of trees in dense stands, growth will rapidly compensate the carbon lost after clear-cutting/soil management, thus enhancing the recovery of ecosystem carbon stocks (Fig. 10.1). In this developmental phase, dense seedling spacing also allows for an increase of carbon stocks in the pre-commercial and commercial phases. These further increases and retains carbon stocks in the forest ecosystem even under regular thinning. Concurrently, higher ecosystem carbon stocks in maturing tree stands may be retained and obtained in the phase of thinning, as shown by Garcia-Gonzalo et al.

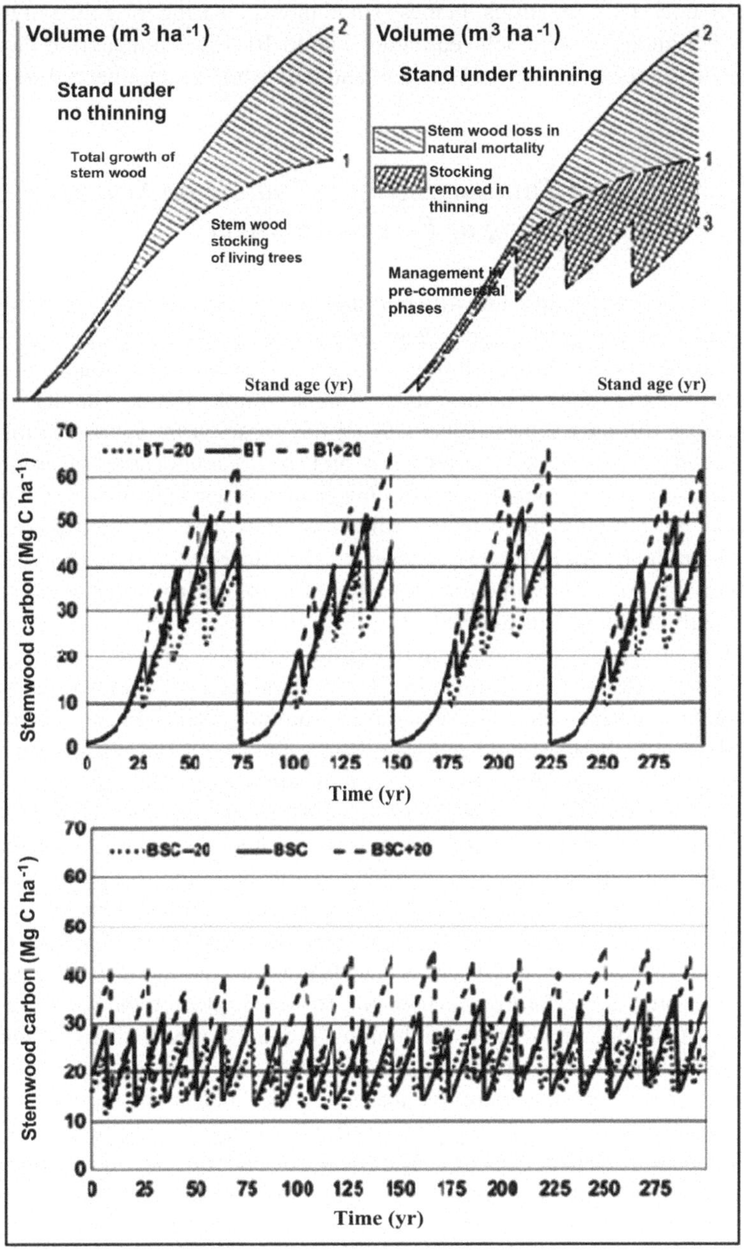


**Fig. 10.1** Schematic depictions of how harvesting impacts on the growth and development of tree stands/trees under even-aged and uneven-aged management. Compiled the author of this book. Left/upper: growth and development of a tree stand under no management from reforestation over time. Right/upper: growth and development of a tree stand under repeated thinning under even-aged management, from reforestation to final cutting. Middle: several cycles in even-aged management with harvesting through thinning and clear-cutting/reforestation over a rotation period. Legend: BT ± 20 indicates baseline thinning (BT) with a 20% increase/reduction of basal area for the tree

(2007a, b) in model simulations. In these simulations, the triggering time of thinning and thinning intensity were delayed/reduced up to 10–15% compared to the timing and intensity under ordinary management and harvesting in commercial forestry.

### *10.1.2 Impacts of Thinning/Selective Cutting on Structure and Functioning of Forest Ecosystems*

Thinning is used to modify the structure and functioning of forest ecosystems to produce selected ecosystem goods and services such as energy biomass, timber, amenity values, and so forth. Both in even-aged and uneven-aged management, the removal of trees involves the harvesting of trees in pole and middle-aged phases. Consequently, thinning removal includes (i) the proactive removal of timber and biomass otherwise lost in natural mortality and (ii) the reallocation of productivity due to increased spacing, which enhances the growth of a smaller number of trees. In both cases, the growing space of the remaining trees increases, and biomass and timber are produced for different purposes, as demonstrated under even-aged management in Fig. 10.1. The same principles hold for uneven-aged management using selective cutting removal of trees.

Under even-aged management, thinning interventions after the seedling phase can be divided into (i) early commercial (early thinning), (ii) commercial, (iii) and rehabilitation interventions. Early commercial thinning refers to the first thinning that yields energy biomass and pulpwood for commercial purposes. Commercial thinning refers to any thinning intervention after early thinning, the harvest representing mainly pulpwood and saw logs. The last thinning for wide spacing prior to regenerative cutting mainly increases saw logs and may also increase the potential for natural regeneration. In rehabilitation thinning, the focus is also on spacing, but mainly on the removal of trees in poor conditions with low technical value. This is expected to improve the amount and value of future yields and to avoid additional attacks by fungi and insects. Low phenotypic conditions may also indicate low genetic potential for future management. However, phenotypic variability tends to cover genetic difference; that is, thinning provides only a small potential for genetic improvement in tree populations.

Under even-aged management, the frequency and intensity of thinning interventions are dependent on the site fertility/growth rate of trees, management objectives, harvesting methods, and tree species. The frequency and intensity of thinning should be limited to such a degree that thinning interventions are profitable while allowing the stocking to be maintained at a sufficiently high level to sustain productivity

**Fig. 10.1** (continued) carbon remaining after thinning per hectares. Lower: several cycles of uneven-aged management cycles involving the selective cuttings for the natural regeneration and creation of space for maturation of ingrowing trees over time. Legend: BSC ± 20 indicates baseline selection (BSC) with a 20% increase/reduction of basal area for the tree carbon remaining after selection per hectare through the production cycle. In Nordic countries, such as Finland, thinning is typically done twice over a rotation of 80 years on fertile sites. The consequent reduction of stocking reduces the total growth but increases the growth of the remaining single trees (e.g., Hynynen et al. 2019). Rapidly increasing growth of the remaining trees does not compensate for the effect of reduction of stocking on the total growth compared to tree growing in stands not thinned. However, the enhanced growth of single trees increases the value of the remaining trees for use as pulpwood to saw logs depending on the timing and intensity of thinning.

In uneven-aged management, selective cutting focuses on the removal of larger individual trees mainly destined for saw log production. Such cutting allows space for (i) regeneration and (ii) the ingrowth of seedlings and pole-phase trees through shading upper canopy layers. This later replaces the dominating mature trees removed in cutting. In uneven-aged management, rotation is excluded, thus maintaining a skewed forest structure, with the dominance of seedlings, young, and pole-phase trees in density. Concurrently, the density of mature trees is wide allowing establishment of seedlings in gaps induced in removal of dominant mature trees. In uneven-aged management, the harvest of both saw logs and pulpwood is based on stem-based logging, but the biomass harvest is also based on the whole-tree selection. This concerns especially trees growing at excessively high density in the pre-commercial phase, which delays regeneration and ingrowth of seedlings through upper canopy layers. In uneven-aged management, the selection interval for the combined reforestation and harvesting of timber is short (e.g., 10–25 years) depending on tree species, site fertility/growth rate, and selection frequency and intensity (Fig. 10.2).

### *10.1.3 Thinning and Selective Cutting in Different Developmental Phases over Time*

For even-aged management, Vuokila (1980) divided stand-based chronosequences into the phases with a focus on (i) management and (ii) utilization in relation to the phase of growth and development from seedlings to mature trees ready for harvesting and regeneration. The phases are separated by the culmination of the annual growth rate over the production cycle, or rotation, under even-aged management (Fig. 10.3).

Under even-aged management, the growth of trees increases rapidly during the *management phase*. In this phase, pre-commercial management increases the growth rapidly toward the stocking of trees that fulfil the desired dimensions of timber (pulpwood, saw logs). Concurrently, the stocking of trees follows the upper and lower limits of thinning removal, which determine the spacing of the remaining trees. This also holds for uneven-aged management, with increasing growth of the remaining trees and reforestation and ingrowth of seedlings over time (Fig. 10.3). In the *utilization phase*, the annual growth declines while the biomass stocking may

**Fig. 10.2** Example of Norway spruce stands mature for thinning under even-aged management. Courtesy of Tapio Finland. Left: pole-phase Norway spruce in the phase of early commercial thinning. Right: mature Norway spruce for commercial thinning before the regeneration phase

still increase substantially, as occurs in the late management phase of even-aged management. In the utilization phase, the thinning mode and intensity are less important than in the management phase. The consequent economic return increases due to the thinning of maturing trees. In the utilization phase, maintaining stocking through late rotation is important for maintaining productivity through the whole production cycle. Thinning in the utilization phase also makes it possible to modify the tree stand by selecting parent trees for use in seed-tree or shelterwood cuttings for natural regeneration (Fig. 10.3).

Under even-aged management, the basal area is reduced to the lower dominant height/basal area trajectory (Fig. 10.3). The lower trajectory may also include the effects of harvesting operations, which may reduce the stocking to a level lower than optimal from the perspective of growth and yield. Further reduction of stocking may be risky for the sustainability of harvesting and management. This is indicated by the ultimate trajectory, below which the stocking (minimum stocking) should not be reduced in thinning. In general, the basal area may be reduced by up to 30% in the first commercial thinning. Subsequent thinning interventions, the intensity is 20–30% of the basal area for each intervention, with the basal area defined based on dominant height/basal area trajectory. On the other hand, the upper and lower limits control the timing and intensity of thinning. This allows maintaining necessary or higher carbon stocks in trees and in soil, with the consequent impacts on litterfall and the further accumulation of carbon in litter and humus in soil.

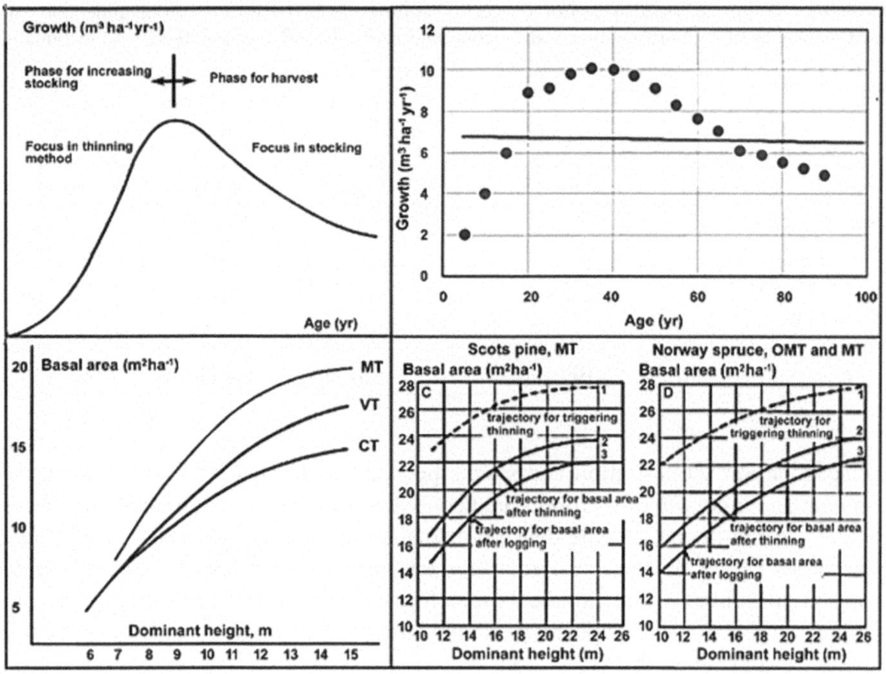


**Fig. 10.3** Schematic depicting how thinning rules relate to the age/dominant height of a tree stand and stem wood growth/basal area of trees in a tree stand under even-aged management based on Vuokila (1980). Left/upper: schematic graph showing the thinning focus throughout a rotation (stand age) based on the annual growth in a tree stand under even-aged management (Vuokila 1980). Courtesy of WSOY. Right/upper: growth of managed boreal Scots pine on a fertile site (*Myrtillus* site type, MT) over a production cycle under even-aged management (Koivisto 1962). Courtesy of Natural Resources Institute Finland. The horizontal line indicates the mean growth (7.1 $m^3$ $ha^{-1}$ $yr.^{-1}$) throughout the production cycle in a managed forest based on the growth and yield tables. Left/lower: basal area of Scots pine in southern boreal forests as a function of the dominant height of trees and related to site fertility (site type), extending from poor (*Calluna* site type, CT) through to moderately poor (*Vaccinium* site type, VT) and fertile (*Myrtillus* site type, MT) (Mäkinen 1959). Courtesy of Natural Resources Institute Finland. Right/lower: examples of thinning rules using the dominant height/basal area trajectory for thinning of Scots pine in a moderately fertile site (*Myrtillus* site type, MT) and the same for thinning of Norway spruce in a fertile site (and *Oxalis-Myrtillus* site type, OMT, and *Myrtillus* site type, MT) (Vuokila 1980). Courtesy of WSOY. Line 1: trajectory for triggering thinning; Line 2: trajectory for basal area after thinning; Line 3: trajectory for basal area after logging

Under uneven-aged management, the timing and intensity of cutting is defined only with the prevailing basal area. This is because there is no stand-based chronosequence for specifying the thinning pattern to meet the stand-based growth and development unlike under even-aged management. Under uneven-aged management, the increase of upper and lower limits affects regeneration by reducing the establishment of naturally born seedlings that drive the dynamics of trees (Table 10.1). Under uneven-aged management, there are limited possibilities to

**Table 10.1** Suggested basal area prior to selection cutting and after selection cutting under uneven-aged management of Norway spruce and Scots pine across boreal forests (60–70° N) in Finland (Äijälä et al. 2014)

| Dominance of tree species and site | Basal area prior to selection cutting, $m^2\ ha^{-1}$ | Basal area after selection cutting, $m^2\ ha^{-1}$ | | |
|---|---|---|---|---|
| | | Southern boreal | Middle boreal | Northern boreal |
| **Norway spruce** | | | | |
| • Fertile (MT) | ≥20 | About 12 | About 11 | About 10 |
| • Very fertile (OMT) | ≥22 | About 13 | About 12 | About 11 |
| **Scots pine** | | | | |
| • Very poor (CT) | ≥16 | About 11 | About 10 | About 8 |
| • Poor (VT) | ≥19 | About 11 | About 10 | About 8 |

increase carbon sequestration and stocks per land unit. This is because the establishment and ingrowth of seedlings are keenly controlled by the dominant large trees with the strong impacts on the density, growth, and uneven-aged structure of forest over time. Under uneven-aged management, selection occurs at 10- to 25-year intervals in southern and middle boreal forests, and at even longer intervals in northern boreal forests, depending on the growth rate of trees and the selection intensity (Äijälä et al. 2014).

## 10.2 Effects of Harvesting of Timber and Biomass on Soil Carbon and Nitrogen

### *10.2.1 Effects of Thinning on Net Ecosystem Carbon Exchange and Carbon Sequestration*

Growth and yield studies show that thinning reduces the total growth in boreal tree stands depends on the timing and intensity of thinning but increases the growth of individual trees. The stand-based reduction of the growth compared to the growth under no thinning is linked with the removal of foliage area, which reduces the stand-based gross primary productivity (*GPP*). Concurrently, thinning also affects the autotrophic ($R_a$) and heterotrophic ($R_h$) components of ecosystem respiration ($R_{eco}$). This involves a reduction in maintenance respiration, decomposition of harvesting residues, and changing foliage litterfall, root litter, and root exudates. Furthermore, changes in nutrient availability and input after thinning may modify the contribution of both *GPP* and $R_{eco}$ to net ecosystem carbon exchange (*NEE*). Increasing growth of the remaining trees does not compensate for the effect of the reduction of stocking on the total growth. Nevertheless, the stand-based growth is allocated among a reduced number of surviving trees. Further over years, the stand-

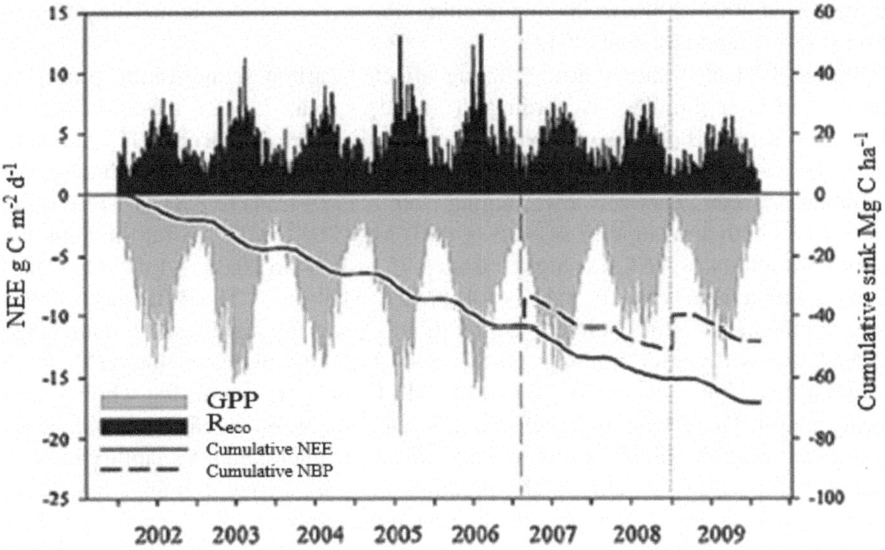


**Fig. 10.4** Net ecosystem carbon exchange at a Sitka spruce (*Picea sitchensis*) plantation from 2002 to 2009 through the pre-commercial and commercial management phases (Saunders et al. 2012). Permission of Elsevier. Black shading denotes the daily loss of carbon through ecosystem respiration ($R_{eco}$), whereas gray shading denotes carbon uptake through *GPP*. The solid black line indicates the cumulative *NEE*-based carbon sink, while the dashed line indicates the cumulative-based carbon sink and post-thinning net biome production (*NBP*). The dashed vertical line indicates the timing of the first thinning, and the dotted vertical line the timing of the second thinning

based growth may recover to the level before thinning, but the mean stem wood stocking over the rotation tends to remain lower than that under no thinning.

At the stand scale, processes controlling carbon uptake in photosynthesis and carbon losses in respiration affect both the short- and long-term carbon sequestration potential, which emphasizes the importance of management operations such as thinning. In this context, Saunders et al. (2012) investigated the effects of thinning on the *NEE* of a Sitka spruce (*Picea sitchensis*) stand in a hemiboreal region (52° N) under maritime climatic conditions in the Irish midlands. The stand was established in 1988 at a density of 2300 seedlings per hectare. The first thinning intervention took place in December–January 2006/2007. The harvesting was selective, with harvesting residues left in rows to reduce soil disturbance by the harvesting machines. In the initial thinning, trees were removed from each size class, while in the second thinning (in the autumn of 2008), mostly larger trees were removed. In the second thinning, the same thinning corridors were used as in the first thinning.

According to Fig. 10.4, the stand density was 2367 trees per hectare in the period of 2002–2006. In the first thinning, stand density was reduced by 25% to 1768 trees per hectare. Before the first thinning, the basal area increased from 31 to 48 $m^2$ $ha^{-1}$, dropping by 17% to 40 $m^2$ $ha^{-1}$ in thinning. The reduction of stand density and basal area indicated the reduction of leaf area index, which increased from 4 to 6 $m^2$ $m^{-2}$

in the period 2002–2006. In the first thinning, the leaf area index reduced by 17% to 4.9 $m^2$ $m^{-2}$ (Saunders et al. 2012).

Figure 10.4 also shows how thinning affected carbon sequestration processes before and after thinning. According to Saunders et al. (2012), the experimental forest was a net carbon sink with *NEE* sink strength in the range from 6.75 (±1.19) to 10.33 (±1.41) Mg C $ha^{-1}$ $yr.^{-1}$ over the monitoring period. Before thinning, the variation between years was low, ranging from 8.44 (±1.34) to 8.81 (±1.09) Mg C $ha^{-1}$ $yr.^{-1}$, with a mean *NEE* of 8.66 (±1.43) Mg C $ha^{-1}$ $yr.^{-1}$. After thinning, the annual variability of *NEE* was higher, from 6.75 (±1.19) to 10.33 (±1.41) Mg C $ha^{-1}$ $yr.^{-1}$, with a mean value of 8.41 (±1.42) Mg C $ha^{-1}$ $yr.^{-1}$. The difference of *NEE* between the mean values before and after thinning was not significant, but the range of *NEE* was wider after thinning. Furthermore, Fig. 10.4 indicates that *NPP* varied between 13.22 (±3.72) and 18.94 (±4.88) Mg C $ha^{-1}$ $yr.^{-1}$ over the whole monitoring period. There were no significant differences between the mean annual values before and after thinning. However, there was clear interannual variability in *NPP* and $R_{net}$ impacting on *NEE* in the prethinning years. After thinning, both dead biomass and litterfall declined, while fine root turnover (*FRT*) increased after thinning.

Saunders et al. (2012) further found that the seasonality of *NEE* was similar over the monitoring period. Figure 10.4 shows that the highest values for carbon uptake were between 14 and 19 g C $m^{-2}$ $d^{-1}$, coinciding with maximum mean daily temperatures and the highest irradiances. The highest carbon losses ranged between 7 and 13 g C $m^{-2}$ $d^{-1}$, occurring at higher temperatures and in the periods of high plant productivity. The annual net carbon gains were pronounced during the summers/growing seasons, while carbon losses dominated in winter periods. Thinning removal reduced the strength of the carbon sink from 68.51 to 48.1 Mg C $ha^{-1}$ by 30% over the study period.

### 10.2.2 Sensitivity of Soil Carbon Stocks and Dynamics to Commercial Thinning

Commercial thinning tends to decrease soil carbon stocks by reducing the aboveground litterfall and belowground litter forming in soil (Vesterdal et al. 1995; Jandl et al. 2007). The impacts of thinning are most evident on soil surface layers (organic matter including litter and humus), while its effects are small in the mineral soil below the organic layers. This holds even deeper in the soil profile down to the parent material, as in boreal podzols. In this context, Zhang et al. (2018) conducted a meta-analysis of how thinning changes the growing conditions (the increase of soil temperature and moisture and the availability of nitrogen) and the soil organic carbon, litter production, and soil respiration based on thinning intensity (Fig. 10.5). The response of carbon dynamics to thinning was indicated by the response ratio (RR) (Equation 10.1 below Fig. 10.5):

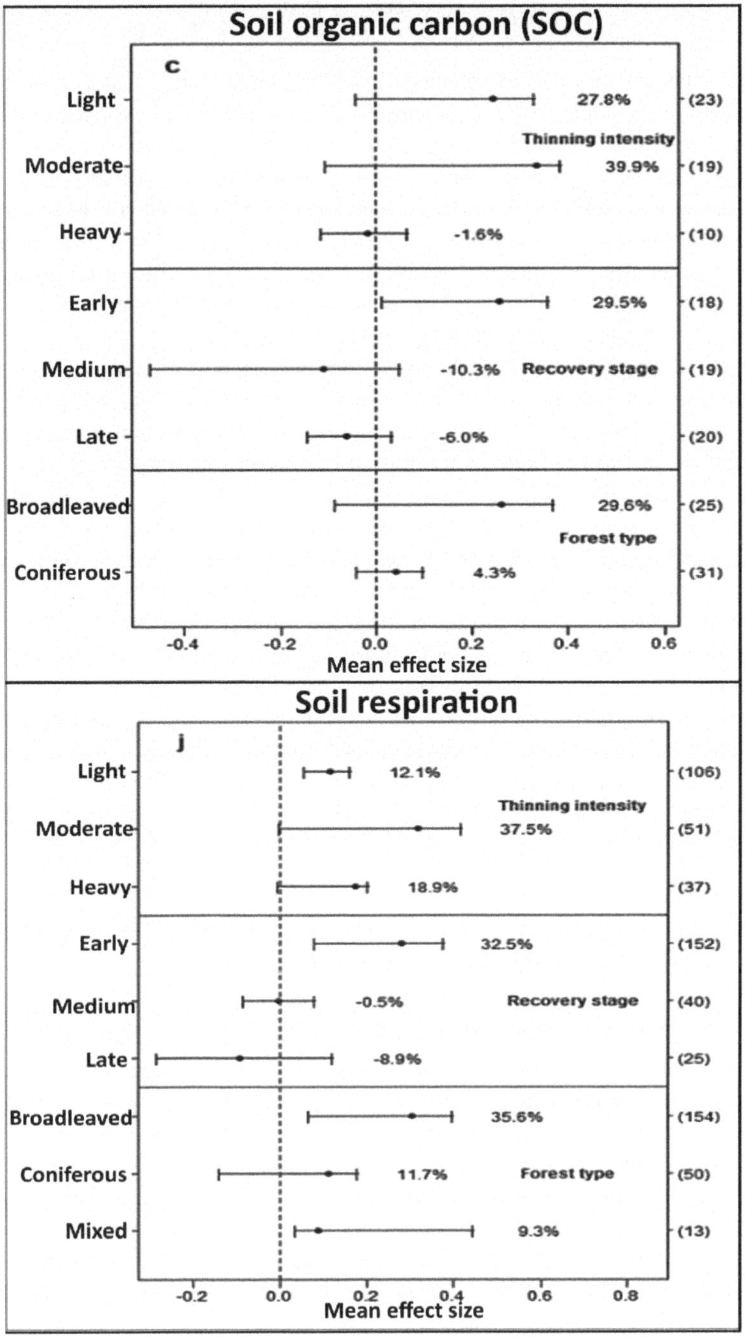


**Fig. 10.5** Mean effect of thinning on changes in soil carbon stocks (organic carbon and respiration) for boreal forests (Zhang et al. 2018). Permission of Elsevier. The sites were located at latitudes from 27 to 57° N, with mean annual temperature from −3 °C to +24 °C and precipitation from 280 to 2500 mm. The numbers in parentheses indicate the number of plots used in the analysis for the given parameters in the subfigures

$$\ln RR = \ln\left(\overline{X_t}/\overline{X_c}\right) = \ln\overline{X_t} - \ln\overline{X_c} \tag{10.1}$$

where $\overline{X_t}$ is the mean value of selected variables in tree stands exposed to thinning and $\overline{X_c}$ is the mean value of the same variables in the stands not exposed to thinning. In the analysis, three degrees of thinning intensity were distinguished based on the percentage removal of volume, basal area, or stand density (Zhang et al. 2018): (i) light thinning (<33%), (ii) moderate thinning (33–67%), and (iii) heavy thinning (>67%). The thinning impacts were identified in three stages after thinning: (i) early stage: < 2 years after thinning; (ii) medium stage: 2–5 years after thinning; and (iii) late stage: >5 years after thinning. The recovery time extended over 40 years, as also found in the comprehensive review by Mäkipää et al. (2023).

Zhang et al. (2018) found that the loss of soil carbon was highest in the first years after thinning. Increasing decay was related to increasing soil temperature and moisture due to decreasing canopy shading allowing more radiation and precipitation to fall on the soil surface. Thinning elevated soil mean temperature (+8.7%), with a subsequent increase in soil respiration (+29.4%) (Fig. 10.5). Under light (<33%) and moderate (33–67%) thinning, the soil carbon efflux (soil respiration) increased. Changes in growing conditions had no pronounced change on fine root biomass or soil carbon stocks, including total soil carbon, soil organic carbon, and carbon in microbial biomass. However, thinning reduced litterfall (23.7%) from the early stage to the late stage in both coniferous and deciduous forests. Thinning significantly increased soil respiration in both deciduous (35.6%) and mixed forests (9.3%), but not markedly in coniferous forests. This indicated the quality difference in litterfall based on variability in tree species dominance. In conclusion, Zhang et al. (2018) recommended the use of light thinning, because this thinning intensity had small impacts on carbon dynamics and stocks and a short recovery period to the prethinning state.

### *10.2.3 Effects of Harvesting on Soil $CO_2$ Efflux*

In forest ecosystems, soil $CO_2$ efflux is a critical component of ecosystem respiration ($R_E$), which is generally driven by variability in soil temperature and soil moisture. According to Schlesinger and Andrews (2000), the efflux of carbon dioxide from forest soil comprises 30–80% of the total ecosystem respiration depending on the growth rate of trees, with the concurrent addition of litter on the soil surface and humus in soil profile. Soil respiration is related to litterfall, and there is a direct relationship between soil respiration and net primary production (*NPP*) across the main forest domains. Soil respiration further has close links to forest growth and management operations that improve site fertility and increase foliage and root litter. Harvesting of timber and biomass also changes soil properties, including by improving soil temperature and moisture for microbial activity, thus affecting the decay of litter and soil organic matter, with impacts on soil $CO_2$ efflux.

In regular forestry, even-aged management extends through different development phases of tree stands, from the establishment phase with seedlings to a fully mature stand ready for the next production cycle. In harvesting, changes in long-term carbon dynamics are linked with logging in the pole, old, and mature phases over a chronosequence, as demonstrated by Chen et al. (2020) for managed boreal forests. They used a meta-analysis to summarize findings from 77 field studies to investigate how forest logging affects soil $CO_2$ efflux. In their data search, Chen et al. (2020) used the following combinations of terms relevant for the study problem: (i) forest logging/harvesting/thinning/cutting/clear-cutting/selective cutting and (ii) soil respiration/$CO_2$/carbon emission. The studies were accepted for the meta-analysis if (i) they included a reference treatment, i.e., an unharvested forest plot, and (ii) they were field-based experiments established in forest ecosystems. Raw data were either extracted from published tables or obtained by digitizing published graphs. The soil $CO_2$ efflux, $Q_{10}$, soil temperature (ST), soil moisture (SM), litter production (LP), fine root biomass (FRB), and associated soil properties (e.g., soil organic carbon (SOC), total nitrogen (TN), soil pH, soil C/N ratios, and soil microbial biomass carbon (MBC)) in the control and treatment groups were also recorded. The response ratio of $Q_{10}$ indicates the temperature sensitivity of $CO_2$ emission based on the chemical reactions related to forest logging (Fig. 10.6).

According to Chen et al. (2020), forest logging potentially reduces carbon sequestration in forest ecosystems, inducing the following main findings:

- Forest logging increased soil $CO_2$ efflux by 5.0% over the growing season, but the mean effect was nonextant across all studies included in the analysis. The variability of the soil $CO_2$ between studies was high related to forest type, logging type, and time since logging.
- Forest logging increased the soil $CO_2$ efflux by 4.4% in coniferous forests. In mixed and hardwood forests, logging induced no change in the soil $CO_2$ efflux.
- Logging type had a significant effect on soil $CO_2$ efflux: thinning increased soil $CO_2$ efflux by 12.1%, but clear-cutting decreased the efflux by 8.6%.
- Time since logging increased the soil $CO_2$ efflux for 2 years after logging. Thereafter, the efflux reduced over 3–6 years after logging, but the efflux again increased for 6 years from logging.
- Forest logging reduced fine root biomass and increased dead organic matter in the soil profile, with a consequent increase of soil $CO_2$ efflux. This was due to the accelerated decomposition of organic matter related to elevated soil temperature and moisture and to changes in the quality of the organic substrate.

Chen et al. (2020) also analyzed the values of the response ratio of soil $CO_2$ efflux in relation to biophysical factors (Fig. 10.7). These factors included soil temperature (ST), soil moisture (SM), soil organic carbon (SOC), soil total nitrogen (TN), soil pH, soil C/N ratios, soil microbial biomass carbon (MBC), litter production (LP), and fine root biomass (FRB). According to Fig. 10.7, logging increased ST and SM and reduced soil acidity (pH) (up to 9.0%, 6.3%, and 1.6%, respectively) with increasing soil $CO_2$ efflux. On the other hand, logging reduced the values of soil C/N ratio, litter production (LP), and fine root production (FRB) by 4.5%, 20.3%,

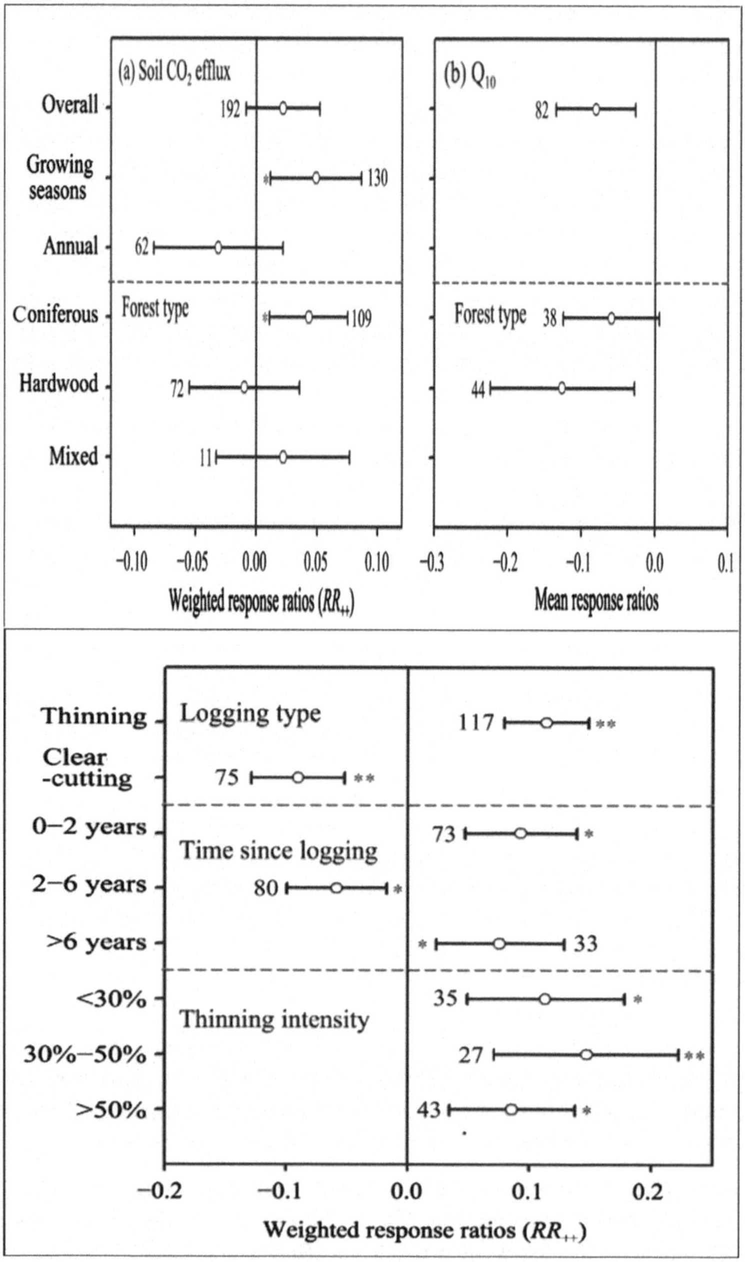


**Fig. 10.6** Impacts of forest logging on soil $CO_2$ efflux (Chen et al. 2020). Open Access. Courtesy of Sciences in Cold and Arid Regions. Left/upper: weighted response ratio (*RR*++) of soil $CO_2$ efflux to forest logging. Right/upper: mean response ratios of $Q_{10}$ indicating temperature sensitivity based on the chemical reactions to forest logging. Lower: weighted response ratio (*RR*++) of soil $CO_2$ efflux categorized according to logging type, time since logging, and thinning intensity. Bars represent *RR*++ in the 95% confidence interval (CI). If the CI did not overlap with zero, the response was considered significant (*$P < 0.05$, **$P < 0.01$, ***$P < 0.001$). Vertical lines are drawn at *RR* + + = 0. The sample size for each variable is shown next to the bar

and 23.4%, respectively. There were no clear changes in soil organic carbon (SOC), total nitrogen (TN), or carbon in microbial mass (MBC). Concurrently, the relationship between the response ratio of soil $CO_2$ efflux and soil moisture (SM) was of a parabola opening downward, while the correlation between response ratios in relation to soil temperature (ST), fine root biomass (FRB), and soil C/N ratio was negative.

Chen et al. (2020) concluded that soil $CO_2$ efflux increases in boreal forests because of forest logging, thus reducing carbon sequestration in forest ecosystems. This holds for both thinning and clear-cutting. However, the variability in logging impacts between the experiments included in the meta-analysis was high, explained by forest type, logging type, and time since logging. In logging, fine root biomass declined, thus reducing autotrophic respiration in soil. The subsequent increase of heterotrophic respiration in litter and soil organic matter increased the overall soil $CO_2$ efflux, induced by elevated soil temperature. Chen et al. (2020) further showed that changes in soil organic matter increase the decay of litter and soil detritus because of logging of timber and biomass.

### *10.2.4 Effects of Harvesting on Carbon Stocks in Forests*

James and Harrison (2016) also carried out a meta-analysis to identify how harvesting likely affects carbon in forest ecosystems. The database included a total of 945 responses to harvesting found in 112 publications from around the world, representing NE North America, NW North America, SE North America, SW North America, Europe, Asia, and the Pacific (Australia, New Zealand), with pure coniferous and deciduous forests and mixed forests of both. Harvesting included clear-cutting and thinning, using stem-only and whole-tree harvesting.

According to James and Harrison (2016), harvesting reduced soil carbon stocks by 11.2% on average, with a 95% confidence interval (CI) [14.1%, 8.5%] (Fig. 10.8). The variation in responses occurred between the layers at different depths in the soil profile. Carbon losses were highest in the O-horizon (30.2%). Deeper in the soil, the losses were smaller but still significant (3.3%), such as for carbon stocks in the surface soil (0–15 cm depth). Carbon loss through harvesting was detectable even in deep soil (60–100+ cm), at 17.7%. In this regard, spreading of harvest residues over the site enhanced the loss of soil carbon (10.9%). The loss was most increased in the mineral soil (17.5%), while the removal of harvest residues indicated no significant addition of losses to soil carbon. On the other hand, soil preparing for reforestation, like prescribing burning and mechanical soil treatment, reduced soil carbon. In the burned soils, the carbon loss was 15.2% more than that in soil not burned. This effect was especially pronounced in the O-horizon, i.e., in the topsoil with mixture of mineral particles and organic matter to a depth of 13–25 cm. Similarly, mechanical soil treatment for seeding/planting reduced carbon in topsoil but less or nonexistent than occurring in prescribed burning. James and Harrison (2016) summarize that the response of soil carbon varied between soils with varying properties, with recovery

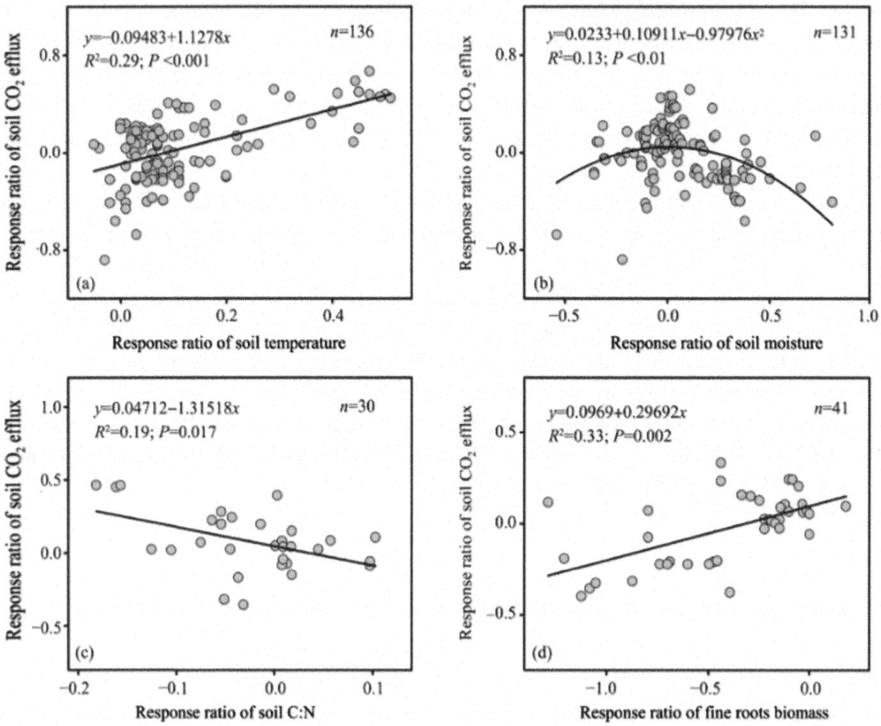


**Fig. 10.7** Impacts of forest logging on soil properties, including soil $CO_2$ efflux (Chen et al. 2020). Open Access. Courtesy of Sciences in Cold and Arid Regions. Relationships between the response ratio of soil $CO_2$ efflux and the response ratio of soil temperature, soil moisture, soil C/N ratios, and fine root biomass

occurring within several decades following the harvesting and soil management for reforestation.

## *10.2.5 Effects of Management and Biomass Removal on Soil Carbon*

In forestry, management and intensity and frequency of harvesting affect the production of timber and biomass and forest residues and logging slash as well. Concurrently, harvesting of timber and biomass affect the stocks of carbon and nitrogen in the ecosystem, with further growth and carbon uptake and carbon emission into the atmosphere. It also affects the formation of carbon stocks at varying depths in the soil profile due to water flow. In this context, James et al. (2021) conducted a meta-analysis to determine (i) how different methods of biomass removal and conventional harvesting affect soil carbon and nitrogen stocks in forest

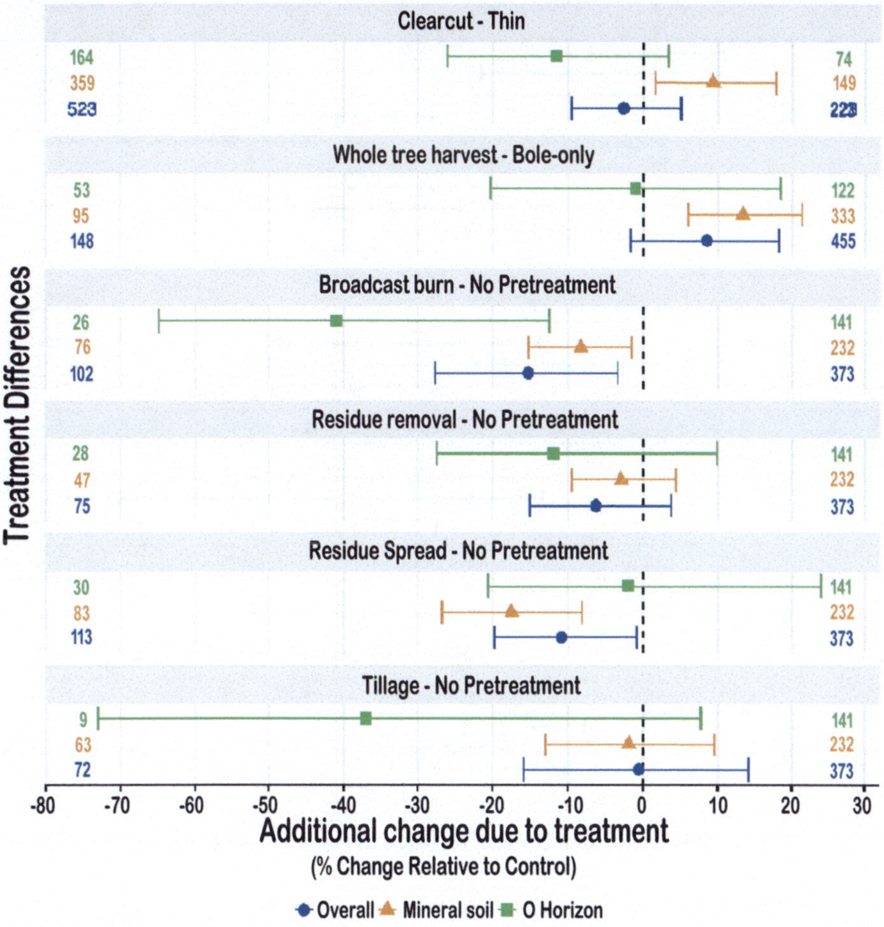


**Fig. 10.8** Differences in response of soil C to harvesting between treatment strategies. (James and Harrison 2016). Open Access. Courtesy of MDPI. Differences with positive differences represent reduced loss of C due to more intensive treatment, and negative differences represent increased losses of C due to more intensive treatment. Point estimates indicate differences between mean effect sizes ±95% confidence intervals. Mean effect differences with confidence intervals overlapping the dashed line (0%) show no significant difference between the harvesting, residual management, or pretreatment strategies. Number of response ratios (k) for the intensive treatment in each comparison are on the left and for the less intensive treatment on the right

soil; (ii) how these effects differ with soil depth; and (iii) how long the effects on soil carbon and nitrogen persist or how long they take to be reversed.

James et al. (2021) compiled soil carbon and nitrogen data with key statistics (e.g., standard deviation) from 142 scientific articles published since 1979. In the analysis, the studies dealing with forest harvesting and biomass removal were divided into different categories represented by the below terms and abbreviations,

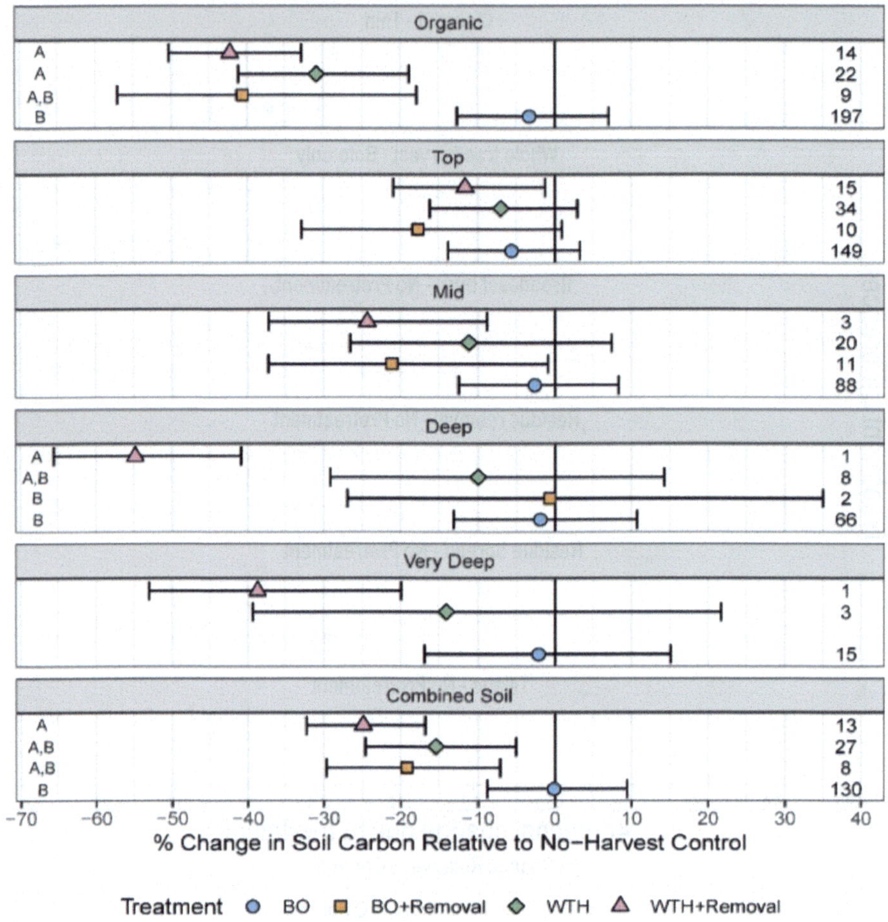


**Fig. 10.9** Effect of bole-only harvesting (BO), whole-tree harvesting (WTH), and biomass removal (+removal) on soil carbon in organic horizon and at different mineral soil depths based on meta-analysis. Estimates are shown with a 95% confidence interval. On the left, different capital letters are significantly different from each other ($p < 0.05$). The number of observations from literature included in each estimate is listed on the right (James et al. 2021). Permission of Elsevier

further explained in the caption of Fig. 10.9: (i) no harvesting control, (ii) bole-only harvesting (BO, partial harvesting, or clear-cutting) in regular harvesting, (iii) BO with partial removal of logging slash and/or O-horizon (BO removal), (iv) whole-tree harvesting (WTH), and (v) WTH with slash and O-horizon removal (WTH + removal). James et al. (2021) used the response ratio (*RR*) to determine how each treatment affected the stocks of soil carbon and nitrogen:

$$RR = \ln\left(\frac{\overline{X_1}}{\overline{X_2}}\right) \tag{10.2}$$

where $\overline{X_1}$ is the mean carbon or nitrogen stock at a specific soil depth for an individual treatment and $\overline{X_2}$ is the mean carbon or nitrogen stock at the corresponding depth in the control (unharvested treatment). The SE of *RR* was calculated as.

$$SE[RR] = \sqrt{\frac{SE\left[\overline{X_1}\right]^2}{\overline{X_1}^2} + \frac{SE\left[\overline{X_2}\right]^2}{\overline{X_2}^2}} \tag{10.3}$$

According to James et al. (2021), the studies that did not report measures of variance were not included in the meta-analysis. The response ratio (and variance) can be back-transformed to be expressed as percentage change relative to the control:

$$\%Change = \left(\frac{\overline{X}_t - \overline{X}_c}{\overline{X}_c}\right) \times 100 = \left(\frac{\overline{X}_t}{\overline{X}_c} - 1\right) \times 100 = \left(e^{(RR)} - 1\right) \times 100 \tag{10.4}$$

According to James et al. (2021), bole-only harvesting with partial removal of logging slash and/or O-horizon (BO + removal), and whole-tree harvesting (WTH), reduced the soil carbon by 19.2% and 15.4% (Fig. 10.9). The reduction was less than that for whole-tree harvesting with the harvesting of slash and O-horizon (WTH + removal), which was 24.9%, thus involving significantly less soil carbon than in non-harvesting controls across combined soil depths. Only regular harvesting through partial or clear-cutting (BO) had no effect on soil carbon. In different mineral soil horizons, only the BO, BO + removal, and WTH + removal treatments had significantly less carbon than that in controls. However, there was a high degree of heterogeneity in treatment response between the studies used in the analysis.

## 10.3 Effects of Thinning and Selective Cutting on Soil Carbon and Nitrogen Stocks

### *10.3.1 Effects of Conventional and Intensive Biomass Harvesting on Forest Carbon*

In managed boreal forest ecosystems, carbon is largely contained in soil stocks (>60% of the total ecosystem carbon). Consequently, management and repeated harvesting reduce the stocks and change the dynamics of soil carbon, as shown by Achat et al. (2015). Their meta-analysis focused on temperate and boreal forests in North America and Europe. The data were sourced from experiments on (i) carbon in unharvested forests, (ii) carbon in forests grown under conventional harvesting, and (iii) carbon in forests grown under intensive harvesting.

Under conventional harvesting, only stem wood was harvested, whereas under intensive harvesting, stem wood and logging residues (stumps, branches, and foliage) were collected for energy. The impacts of different management and harvesting modes on soil carbon stocks were analyzed as a function of soil depth and along the entire soil profile, with soil carbon being distributed among four soil layers (Fig. 10.10): forest floor (F: organic soil layer above the mineral soil profile), top mineral soil (T: mean sampling depth $\leq$ 10 cm), mid soil (M: 11–20 cm), and deep soil (D: >20 cm). Consequently, the stocks of carbon (in Mg C $ha^{-1}$) were calculated in each soil layer: in the mineral soil profile (e.g., TMD = T + M + D) and in the organic plus mineral soil profile (e.g., FT = F + T, FTMD = F + T + M + D) (Achat et al. 2015).

*Conventional harvesting* reduced soil organic carbon in the forest floor (F layer) by 22% compared to no harvesting (Fig. 10.10). Carbon loss was detectable over a decade after the harvesting, and the loss was fully compensated more than half a century after the harvesting. There were only minor differences in carbon loss from the forest floor between thinning and clear-cutting, excluding the 10-year period after harvesting. The response of carbon in the upper mineral layer (T layer) remained stable. However, carbon was lost if the topsoil layer was disturbed in clear-cutting using heavy machinery or by pre-reforestation soil preparation. Under conventional harvesting, the combined carbon loss from the forest floor and topsoil (FT) was related to the loss from the forest floor. This implies that the carbon loss from the upper part of the soil profile was compensated by carbon accumulation deeper in layer D (depth > 20 cm) and above all medium layers (M). Carbon losses and carbon gains were in balance, that is, not significantly different from zero (the mean value for the complete FTMD soil profile = −6% SOC).

*Intensive harvesting* strongly reduced carbon stocks in woody debris (WD) and in the F layer relative to stem-only harvesting (Fig. 10.10). There was a general reduction of soil carbon stock due to a clear decrease of carbon in the organic soil layers (WD and F) and the mineral soil (TMD). After a decade, the loss of soil organic carbon was still detectable in the forest floor (F layer of organic soil layer above the mineral soil profile) but not in the topsoil (T). However, there were negative relationships between carbon losses in the F layer and in mineral soils, indicating that the carbon lost from the forest floor accumulated in the mineral layers. The possible vertical fluxes were low, and they could not compensate carbon losses from mineral soil layers over a decade following intensive harvesting. When compared with unharvested forests, conventional harvesting had a moderate positive impact on soil carbon stocks, while intensive harvesting had a negative impact on soil carbon stocks. Thus, intensive harvesting induced high carbon losses in comparison with untouched forests, with carbon losses of 37% in the forest floor and 7% in deeper soil layers (Achat et al. 2015).

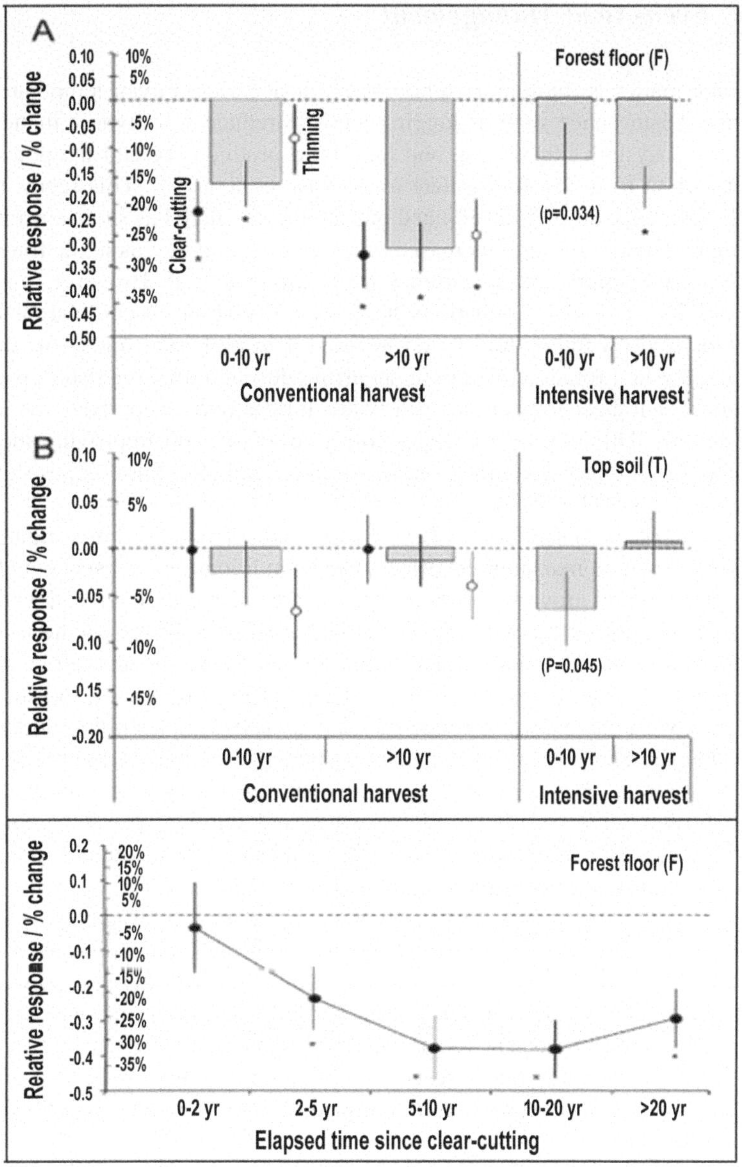


**Fig. 10.10** Effects of conventional and intensive harvesting on soil organic carbon (SOC) stocks in soil as a function of soil depth (individual soil layers) and in the entire soil profile (cumulated soil layers) (Achat et al. 2015). Open Access. Courtesy of Springer Nature. Upper: effects of conventional and intensive harvesting on SOC stocks in the forest floor (F) and top mineral soil (T) in relation to time elapsed since harvesting. (A) Forest floor. (B) Topsoil. Effects were assessed over two periods (0–10 years and > 10 years since harvesting, with means ± standard errors). Lower: effects of conventional clear-cutting harvesting on SOC stocks in forest floor (F) in relation to time elapsed since harvesting. Effects were assessed for five periods (0–2, 2–5, 5–10, 10–20, and > 20 years since harvesting, with means ± standard errors)

### 10.3.2 Even-Aged Management

Soil organic matter is the major carbon stock in boreal and even temperate forest ecosystems. Disturbance such as logging tends to reduce soil organic matter, with impacts on ecosystem functioning and long-term productivity and the potential to sequester carbon from the atmosphere (e.g., Nave et al. 2010). This seems to hold especially for clear-cutting in even-aged management, which has short-term impacts on soil organic matter through increased decay of litter and humus in the forest floor and ceased litter inputs and accumulation of carbon in soil. However, long-term studies and forest model simulations have shown that the accumulation of soil organic matter and carbon stocks tend to recover to preharvesting levels on most sites during the first rotation. Thinning occurring during a rotation also impacts soil carbon stocks, but these impacts are clearly less understood compared to the impacts of clear-cutting. Thinning also changes forest structure with improving the availability of soil water and nutrients for trees left in a stand after thinning, as shown by Briceno-Elizondo et al. (2006).

Using model-based calculations, Briceno-Elizondo et al. (2006) investigated how thinning in even-aged management affected site conditions and subsequent changes in photosynthesis and growth of trees in relation to available nitrogen and water. The model simulations focused on Scots pine (*Pinus sylvestris*) in the southern (62° N) and northern (66° N) sites. In both sites, thinning increased the amount of water in precipitation per remaining trees, with a consequent increase of soil moisture with enhanced carbon uptake, as also found by Ge et al. (2011) in a similar model-based exercise. The increase of water per remaining trees was greater in the north than in the south, whereas the increase in available nitrogen was larger in the south. Thinning further increased the available water and nitrogen per remaining trees along with increasing thinning intensity and increasing photosynthetic production and the growth of mass per remaining trees (Fig. 10.11).

Based on field experiments, Jurgensen et al. (2012) also investigated the impacts of periodic thinning in even-aged management on soil carbon and nitrogen stocks in a 134-year-old red pine (*Pinus resinosa*) forest in Minnesota (47° N), USA, and a 104-year-old hardwood forest in Wisconsin (45° N), USA. In the red pine forest, five thinning regimes were used. The remaining basal area was 13.8, 18.4, 22.7, 27.6, and 32.1 $m^2\ ha^{-1}$. In the hardwood forests, the optional remaining basal area of stands was 13.8, 17.2, and 20.6 $m^2\ ha^{-1}$ in thinning conducted five times over 50 years, as shown in Table 10.2.

In the red pine forest, the most intensive thinning (remaining basal area of 13.8 $m^2\ ha^{-1}$) and uncut control yielded higher carbon and nitrogen stocks than the other thinning regimes. On the other hand, multiple thinning had no effect on carbon and nitrogen pools in the forest floor and on the surface of the mineral soil (30-cm depth). However, the variability in basal area correlated positively with carbon and nitrogen stocks in the forest floor under light thinning (remaining basal area of 32.1 $m^2\ ha^{-1}$), whereas the correlation with the carbon and nitrogen stocks in the A horizon was negative. Jurgensen et al. (2012) concluded that stem-only

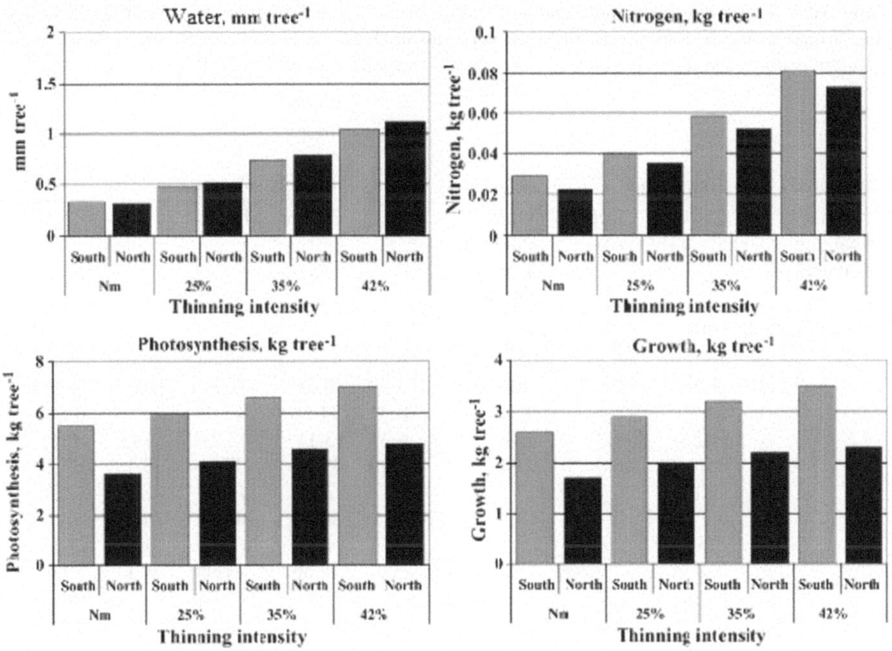


**Fig. 10.11** Effect of thinning on water and nitrogen availability and on photosynthesis and growth of Scots pine (*Pinus sylvestris*) in mid- and northern boreal forests (Briceno-Elizondo et al. 2006). Permission of Elsevier. Effect of thinning on the mean precipitation and nitrogen per remaining trees in a Scots pine stand over 30 years of simulation, with consequent changes in photosynthetic production and growth in the southern (62° N) and northern (66° N) stand. Thinning options were (i) no thinning (Nm) and (ii) the reduction of basal area by 25%, 35%, and 42% per thinning from below. In each case, the initial density was 1900 trees ha$^{-1}$, the mean diameter 14 cm, and mean height 10 m

removal and bioenergy production would have small impacts on total ecosystem soil carbon and nitrogen stocks.

### 10.3.3 Uneven-Aged Management

Even-aged forests usually function as carbon sinks during most of their rotation. However, after clear-cutting, they become sources of carbon for a period of several years. Applying uneven-aged management with selective cutting maintains tree cover and reduces the environmental impact on the forest floor. Unfortunately, no large-scale experimental studies are available to make exhaustive comparisons of growth and yield and impacts on carbon sequestration and stocks between even-aged and uneven-aged management. However, Vuokila (1970) found that thinning from above (resembling selective) cutting under even-aged management increased the

**Table 10.2** Remaining basal area and soil carbon after different thinning treatments in (A) a red pine (*Pinus resinosa*) experiment in Minnesota and in (B) a hardwood experiment in Wisconsin. Standard error of the mean is given in parentheses (Jurgensen et al. 2012). Permission of Wiley

| Basal area | | Forest floor | | A horizon | | Forest floor + A horizon | B horizon | | Total mineral soil |
|---|---|---|---|---|---|---|---|---|---|
| Treatment, remaining basal area, $m^2$ $ha^{-1}$ | | Carbon, Mg $ha^{-1}$ | Depth, cm | Mineral soil bulk density, Mg $m^{-3}$ | % | Carbon, Mg $ha^{-1}$ | | | |
| A: Red pine experiment | | | | | | | | | |
| 13.8 | 23.9a | 10.6 | 6.5 | 0.96 | 8.69 | 47.2a | 57.8a | 25.6 | 72.8 |
| 18.4 | 27.8b | (1.53) | 6.4 | 1.06 | 5.60 | (0.8) | (2.3) | (0.9) | (1.6) |
| 23.0 | 27.1b | 12.3 | 6.1 | 0.86 | 8.31 | 34.6b | 47.0b | 29.1 | 64.0 |
| 27.5 | 31.1c | (1.26) | 6.2 | 0.98 | 6.33 | (3.1) | (3.2) | (5.1) | (6.2) |
| 32.1 | 35.0d | 8.0 | 6.0 | 0.91 | 7.48 | 39.3b | 47.3b | 23.1 | 62.5 |
| | | (1.15) | | | | (2.6) | (3.7) | (1.0) | (2.3) |
| | | 8.7 | | | | 36.3b | 45.0b | 24.8 | 61.1 |
| | | (0.93) | | | | (1.1) | (1.8) | (0.4) | (1.1) |
| | | 9.4 | | | | 35.5b | 44.9b | 24.1 | 59.6 |
| | | (0.74) | | | | (3.8) | (3.1) | (3.2) | (6.4) |
| Clear-cut | 38.5e | 9.9 (0.58) | 7.8 | 1.08 | 6.74 | 50.7a (5.4) | 60.6a (5.9) | 24.3 (1.7) | 74.9 (6.8) |
| B: Hardwood experiment | | | | | | | | | |
| 13.8 | 16.1a | 4.4 | 6.8 | 0.97 | 6.64 | 38.4 | 42.8 | 63.8 | 102.2 |
| 17.2 | 19.0b | (1.01) | 6.8 | 0.97 | 6.78 | (5.4) | (5.7) | (8.2) | (3.6) |
| 20.7 | 22.4c | 4.3 | 6.8 | 0.92 | 8.39 | 41.2 | 45.5 | 58.7 | 99.9 |
| | | (0.80) | | | | (8.3) | (4.4) | (6.1) | (2.2) |
| | | 4.9 | | | | 43.2 | 48.1 | 61.2 | 104.4 |
| | | (1.27) | | | | (14.2) | (13.2) | (13.2) | (1.2) |
| Clear-cut | 34.2d | 6.1 (1.89) | 6.3 | 0.93 | 8.17 | 42.0 (8.4) | 48.2 (6.9) | 63.8 (2.6) | 105.8 (7.8) |

Values with different letters are statistically different at $p < 0.05$

timber yield compared to that obtained under thinning from below. This was possible, if thinning from above using the same basal area had been done from below in the early phases of chronosequence (management phase). This sustained high stocking throughout the rotation for the utilization phase, when thinning from above was used before the terminal cut (Vuokila 1970, 1977).

Regarding the effects of different management strategy on boreal soil carbon dynamics, Kumpu et al. (2018) measured soil $CO_2$ efflux in even- and uneven-aged Norway spruce stands (*Picea abies*) in southern boreal conditions (60° N) in Finland using a closed chamber method with fixed measuring points. The study included two uneven-aged stands and two even-aged stands (a clear-cut site and a mature even-aged stand). Soil moisture and soil temperature were measured at the same time as soil $CO_2$ efflux. Soil cores were collected from the topsoil of each study plot to determine soil carbon and nitrogen concentrations. The main findings were as follows:

- Mean soil $CO_2$ efflux through the summer was highest in the clear-cut plot (0.367 mg $m^{-2}$ $s^{-1}$) while slightly lower in the uneven-aged stands (0.298 and 0.257 mg $m^{-2}$ $s^{-1}$) and lowest in the mature even-aged stand (0.224 mg $m^{-2}$ $s^{-1}$).
- There was no statistically significant difference in soil $CO_2$ efflux between even- and uneven-aged stands of the same site fertility, soil moisture, and soil temperature. Soil $CO_2$ efflux increased steadily with soil temperature, whereas increasing soil moisture increased soil $CO_2$ efflux at lower moisture levels but only moderately at higher soil moisture levels.
- Soil carbon and nitrogen concentration did not differ between study plots of the same fertility.
- Uneven-aged forest structure did not prevent an increase in soil $CO_2$ efflux after cutting. However, the high variation in soil $CO_2$ efflux rates within the uneven-aged stands suggests that the $CO_2$ efflux can be controlled with the intensity of the selective cutting.

### 10.3.4 *Effects of Even-Aged Vs. Uneven-Aged Management on Ecosystem Carbon*

The impacts of even-aged management on timber yield and carbon dynamics have been widely studied experimentally, unlike those of uneven-aged management. A long-term study reported by Nilsen and Strand (2013), which compared the carbon storage and fluxes in boreal Norway spruce under uneven-aged and even-aged management, is unique in this respect (Table 10.3). Under uneven-aged management, selection cuttings were done through an 81-year period, while the stand exposed to even-aged management was established 81 years ago after clear-cutting. In the uneven-aged stand, the timber production was measured for 81 years while in the even-aged stand for 52 years. Carbon stocks were estimated based on inventory data, tree-based biomass functions, soil samples, and carbon analyses from trees and soil.

The present carbon stock in trees (including roots) was 210 Mg C $ha^{-1}$ in the even-aged stand and 76 Mg C $ha^{-1}$ in the uneven-aged stand (Table 10.3). Concurrently, the soil carbon stock was 178 C Mg $ha^{-1}$ and 199 Mg C $ha^{-1}$ for the even-aged and for uneven-aged stands. The difference was due to a 20% higher carbon content in the humus layer in the uneven-aged stand compared to that in the even-aged stand. The total ecosystem carbon stock was 388 Mg C $ha^{-1}$ for even-aged and 275 Mg C $ha^{-1}$ for uneven-aged management. The ecosystem carbon stock for the even-aged management was about 30% larger than that for under uneven-aged management.

The timber production in the uneven-aged stand was 95% of that in the even-aged stand, thus allowing less carbon sequestration than in even-aged management

**Table 10.3** Long-term growth and biomass and carbon stocks in boreal Norway spruce (*Picea abies*) stands representing uneven-aged and even-aged management in the experiment reported by Nilsen and Strand (2013) for analyzing how management regime affects carbon stocks and fluxes in the ecosystem scale. Courtesy of Finnish Society of Forest Science

| | Tree stands in 2003 | | | |
|---|---|---|---|---|
| Management regime | Start of experiment | Density, trees $ha^{-1} > 2.5$ cm dbh | Stem wood stocking, $m^3$ $ha^{-1}$ | Dominant height, m |
| Uneven-aged | 1923 | 1266 | 222 | 27.9 |
| Even-aged, light thinning | 1951 | 1043 | 710 | 28.1 |
| Even-aged, medium thinning | 1951 | 1120 | 649 | 27.4 |
| Management regime | **Total biomass and carbon in standing trees and ground vegetation** | | | |
| | Total biomass in trees, Mg $ha^{-1}$ | Stem wood biomass, Mg $ha^{-1}$ | Carbon in total tree biomass, Mg $ha^{-1}$ | Carbon in ground vegetation, Mg $ha^{-1}$ |
| Uneven-aged | 155 | 79 | 76 | 0.39 |
| Even-aged, light thinning | 450 | 266 | 220 | 0.70 |
| Even-aged, medium thinning | 407 | 211 | 199 | 0.52 |
| Management regime | **Growth and carbon in trees** | | | |
| | Mean annual growth, $m^3$ $ha^{-1}$ $yr.^{-1}$ in the past 81 years from the start | Total carbon Accumulation in trees In 81 years, Mg $ha^{-1}$ | Current stem Volume growth, $m^3$ $ha^{-1}$ $yr.^{-1}$ | Current carbon accumulation, kg $ha^{-1}$ $yr.^{-1}$ |
| Uneven-aged | 10.0 | 225 | 11.3 | 4130 |
| Even-aged, light thinning | 11.0 | 276 | 23.0 | 5590 |
| Even-aged, medium thinning | 10.0 | 248 | 24.2 | 5270 |
| Management regime | **Carbon on/in soil surface and profile** | | | |
| | Thickness of O-horizon, cm | Carbon in O-horizon, Mg $ha^{-1}$ | Carbon in soil (30 cm), Mg $ha^{-1}$ | Carbon in soil (100 cm), Mg $ha^{-1}$ |
| Uneven-aged | 8.30 (2.47) | 53.7 (8.3) | 103.2 (30.5) | 198.8 (50.4) |
| Even-aged | 7.25 (2.26) | 43.6 (4.8) | 100.8 (15.8) | 178.1 (16.0) |

Explanation of the table

- Upper part: Outlines of tree stands in the experimental sites.
- Upper/middle: Total biomass in standing trees and carbon in tree biomass and ground vegetation.
- Lower/middle: Mean annual growth of stem wood including natural mortality in an 81-year period under uneven-aged management and the mean annual growth in the 52-year period under even-aged management and the current stem wood growth and current accumulation of carbon.
- Lower: Thickness of organic layer on mineral soil with carbon stock and the carbon stock in soil profile.

(Table 10.3). The total net carbon sequestration in trees for 81 years, excluding the present soil carbon stock, amounted to 37 Mg C $ha^{-1}$ in the even-aged stands. The average O-horizon thickness in the uneven-aged stand was 1 cm larger than in the even-aged stand, but this difference was not statistically significant. The soil carbon efflux was higher under even-aged than uneven-aged management over growing season (Nilsen and Strand 2013).

Kellomäki et al. (2019) also found that the timber harvest was nearly the same as in Norway spruce stands under both even-aged and uneven-aged management regimes in the simulation-based analysis. This also held for saw logs, whereas the pulp wood yield was larger under even-aged than under uneven-aged management, as previously found by Pukkala et al. (2011a, b). However, the NPV with a 3% interest rate was clearly smaller under even-aged than uneven-aged management when the even-aged production cycle was initiated by planting on bare land. This has also been found in many previous studies (e.g., Tahvonen et al. 2010; Pukkala et al. 2010, 2011a, b). When using young or mature stands in initializing the calculation, however, NPV for even-aged management tended to be larger than that for uneven-aged management. Pukkala (2016) suggested that even-aged management combining clear-cutting and planting on bare land might be more profitable than optimal continuous-cover forestry if the planted Norway spruce seedlings have 20% faster diameter and height growth than naturally regenerated seedlings.

## 10.4 Carbon Stocks during Rotation under Different Thinning Regimes and Tree Species

### *10.4.1 Regimes for Exploring the Sensitivity of Ecosystem Carbon to Thinning*

Throughout a rotation, the timber yield is closely related to tree species and thinning intensity and frequency and growth rate/productivity. These affect the stocking and growth of stem wood through the managed chronosequence. In even-aged management with thinning, the managed chronosequence includes different developmental phases, from young stands in the pole phase to old mature stands in the regenerative phase. In any developmental phase, the growth is highest under no thinning, while thinning reduces stocking, with a concurrent reduction of total growth. Nevertheless, the thinning pattern affects the thinning intensity and the timing of thinning depending on the target stocking through the rotation. The remaining stem wood stocking in different thinning interventions is demonstrated in Fig. 10.12 as a function of the basal area defining the thinning event in relation to the dominant height of trees.

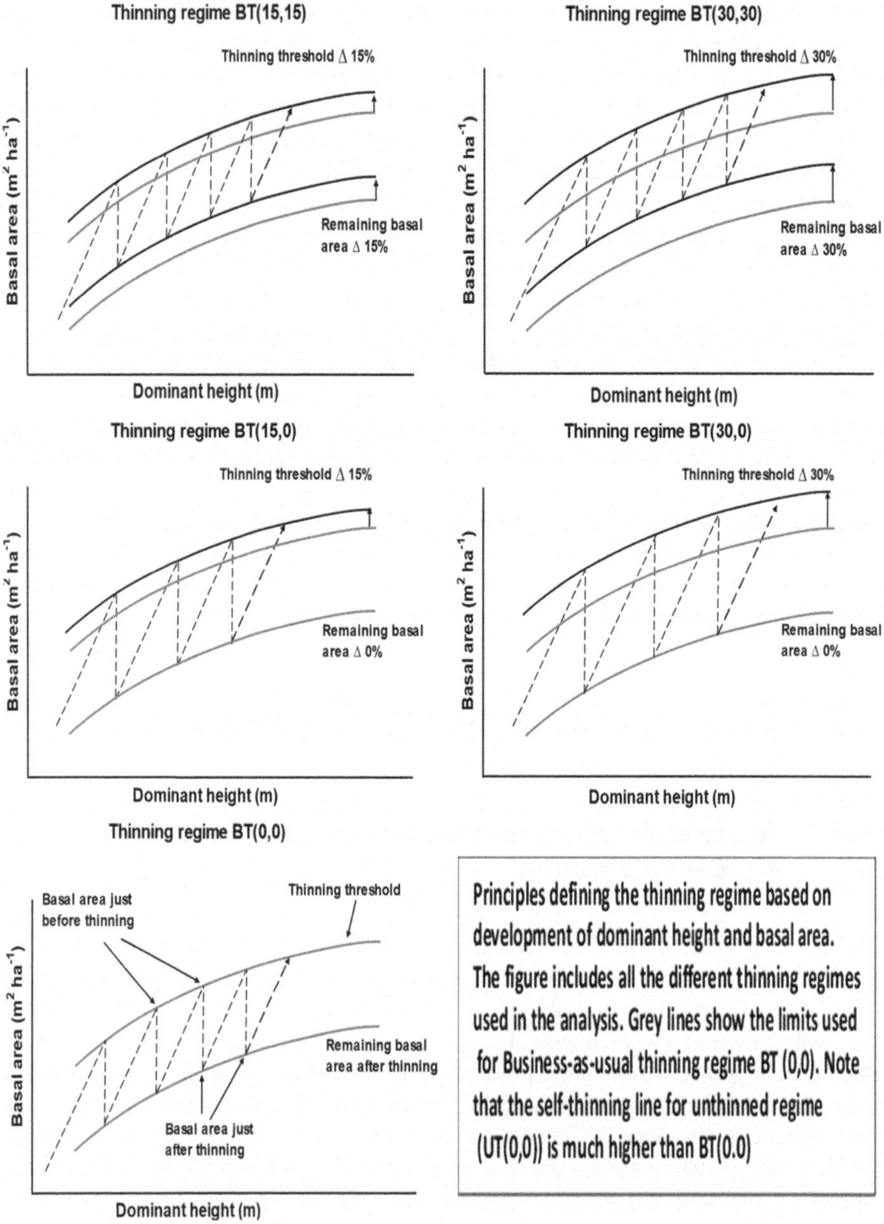


**Fig. 10.12** Modified thinning regimes used by Garcia-Gonzalo (2007) in simulations of the effects of different thinning regimes on carbon in managed boreal forest ecosystems. The thinning regime (intensity and timing of thinning) were given as a function of basal area defining the thinning relative to the dominant height of trees. Courtesy of Finnish Society Forest Science. Legend for the thinning patterns: BT (0,0) = thinning following the current management recommendations. BT (x, y) = the change of the upper and lower limits of thinning intensity, where the numbers refer to the percentage changes relative to the thinning intensity (e.g., BT (15,15), 15% changes in both limits) under the current management recommendations (BT (0,0)) for thinning

### 10.4.2 Overview of Simulations of Forest Structure

Garcia-Gonzalo et al. (2007a, b) investigated how the different thinning regimes presented in Fig. 10.12 may affect carbon stocks in boreal forest ecosystems dominated by Scots pine (*Pinus sylvestris*), Norway spruce (*Picea abies*), or birch (*Betula pendula*) relative to the thinning rules applied in practical forestry. The simulations used an ecophysiological model to assess how management affects carbon stocks over a rotation in southern boreal forests (Box 10.1). The focus of the assessment was on the effects of thinning on the total carbon stock, with the carbon in aboveground and belowground biomass and carbon in timber yield.

**Box 10.1 Overview of the Model Used in Simulations**
In the assessment, Garcia-Gonzalo et al. (2007a, b) used an ecophysiological forest ecosystem model to simulate the implications of management on the growth and carbon dynamics of tree trees. Forest growth is related to the structure and functioning of tree populations/stands under edaphic and climatic factors (environment) and ecophysiological properties of trees affecting the canopy, photosynthesis, and consequent total tree growth (Kellomäki and Väisänen 1997; Ge et al. 2011):

$$G_T = f(P_{nc}) = f(R_n) \times f(L) \times f(g_{cs}) \times f(W_a) \times f(N_{up}) \times f(T, C_a, P) \times f(M) \quad (10.5)$$

where $G_T$ is the total growth (foliage, branches, stem, coarse roots, fine roots), $P_{nc}$ is canopy net photosynthesis, $R_n$ is net radiation interception, $L$ is leaf area, $g_{cs}$ is canopy conductance, $W_a$ is available soil water, $N_{up}$ is uptake of nitrogen, $T$ is air temperature, $C_a$ is atmospheric $CO_2$, $P$ is precipitation, and $M$ is management.

The management exercise utilizes the study of Garcia-Gonzalo et al. (2007a, b). The forest area located in eastern Finland (63° N), with a mean altitude of 94 m above sea level, with the temperature sum 1,0.200 d.d. and annual precipitation 500–700 mm. Several representative stands were selected for the simulation to make an assessment over the whole study area (Table 10.4). Most of the stands were located on moderately fertile and fertile sites (*Myrtillus* and *Oxalis-Myrtillus* site types, MT and OMT) dominated by Norway spruce, while Scots pine was common on moderately poor sites (*Vaccinium* site type, VT), and silver birch existed in a mixture with the conifers. In even-aged management, thinning was conducted from below, targeting suppressed trees. The stem wood of the trees removed in thinning and in final cutting was converted to saw logs (with a minimum diameter of 15 cm at the top) and pulpwood (with minimum top diameter 6 cm). The remainder of the stem and logging residues included branches, needles, and roots.

**Table 10.4** Total mean carbon stock in forest ecosystems (carbon in trees and soil) under the current climate over a period of 100 years and effects of changing management on the growth change (%) of total carbon sequestered compared to that under the reference management (BT (0,0)) (Garcia-Gonzalo et al. 2007b). Permission of Springer Nature

| | Total C in forest ecosystem | | Total C in trees (above+ below) | | Total C aboveground | | Total C belowground | | Total C in soil | |
|---|---|---|---|---|---|---|---|---|---|---|
| Management regime | Mg C ha$^{-1}$ | % | Mg C ha$^{-1}$ | % | Mg C ha$^{-1}$ | % | Mg C ha$^{-1}$ | % | Mg C ha$^{-1}$ | % |
| Scots pine | | | | | | | | | | |
| BT (0,0) | 68 | - | 31 | - | 27 | - | 3.5 | - | 37 | - |
| BT (15,0) | 70 | 3 | 32 | 6 | 28 | 6 | 3.7 | 5 | 38 | 2 |
| BT (30,0) | 73 | 7 | 34 | 12 | 30 | 12 | 3.9 | 10 | 39 | 3 |
| BT15,15) | 73 | 7 | 34 | 12 | 30 | 12 | 3.9 | 10 | 39 | 3 |
| BT (30,30) | 78 | 14 | 37 | 22 | 33 | 23 | 4.3 | 20 | 31 | 7 |
| UT (0,0) | 95 | 38 | 45 | 47 | 40 | 47 | 5.1 | 44 | 50 | 31 |
| Norway spruce | | | | | | | | | | |
| BT (0,0) | 128 | - | 46 | - | 35 | - | 11.4 | - | 82 | - |
| BT (15,0) | 131 | 2 | 48 | 4 | 36 | 4 | 11.8 | 4 | 83 | 1 |
| BT (30,0) | 136 | 6 | 52 | 12 | 39 | 13 | 12.5 | 10 | 84 | 2 |
| BT15,15) | 135 | 6 | 51 | 10 | 39 | 11 | 12.3 | 8 | 84 | 3 |
| BT (30,30) | 142 | 11 | 56 | 20 | 43 | 21 | 13.2 | 16 | 86 | 5 |
| UT (0,0) | 188 | 47 | 90 | 94 | 69 | 96 | 21.3 | 87 | 98 | 20 |
| Silver birch | | | | | | | | | | |
| BT (0,0) | 56 | - | 22 | - | 20 | - | 2.3 | - | 34 | - |
| BT (15,0) | 58 | 3 | 23 | 6 | 21 | 6 | 2.5 | 6 | 35 | 2 |
| BT (30,0) | 60 | 7 | 25 | 12 | 22 | 12 | 2.6 | 11 | 35 | 3 |
| BT15,15) | 61 | 8 | 25 | 13 | 22 | 13 | 2.6 | 13 | 36 | 5 |
| BT (30,30) | 64 | 15 | 27 | 25 | 24 | 25 | 2.9 | 25 | 37 | 9 |
| UT (0,0) | 80 | 44 | 35 | 62 | 31 | 62 | 3.8 | 61 | 45 | 32 |
| Management unit (average) | | | | | | | | | | |
| BT (0,0) | 106 | - | 40 | - | 32 | - | 8.5 | - | 66 | - |
| BT (15,0) | 109 | 3 | 42 | 5 | 33 | 5 | 8.8 | 4 | 67 | 1 |
| BT (30,0) | 112 | 6 | 45 | 12 | 36 | 13 | 9.3 | 10 | 68 | 3 |
| BT15,15) | 112 | 6 | 44 | 11 | 35′ | 11 | 9.2 | 9 | 68 | 3 |
| BT (30,30) | 118 | 11 | 48 | 21 | 38 | 22 | 9.9 | 17 | 70 | 6 |
| UT (0,0) | 154 | 45 | 73 | 83 | 58 | 83 | 15.4 | 81 | 81 | 23 |

Legend for the thinning patterns

- BT (0,0) = thinning following the current management recommendations.
- UT (0,0) = no thinning but clear-cutting at the end of rotation.
- BT (x, y) = the change of the upper and lower limits of thinning intensity, where the numbers refer to the percentage changes relative to the thinning intensity under the current management recommendations (BT (0,0)) for thinning.

## *10.4.3 Responses of Carbon in Trees and Ecosystem*

The carbon stock in trees followed the same pattern as the total carbon stock in the ecosystem, but thinning affected the carbon stock in trees more than that in the ecosystem (Table 10.4). An increase in the thinning threshold increased the mean carbon stock in trees over a period of 100 years, especially when the remaining basal area was increased. The same pattern also held for soil carbon, although the relative change was lower than in trees (Garcia-Gonzalo et al. 2007a, b). The main results are summarized below for the management options described in Table 10.4, involving baseline thinning (BT (0,0)) and thinning with 30% reduced intensity (BT (30,30)). Varying combinations of both the triggering and remaining basal area in thinning were also used in the simulation.

- In trees, the carbon stock under no thinning (UT (0,0)) was 83% higher than that under baseline thinning (BT (0,0)). This pattern was similar for both above- and belowground carbon for all tree species. The corresponding increase of carbon in soil ranged from 1% for BT (15,0) to 6% for BT (30,30) compared to that of BT (0,0), which yielded 66 Mg C ha$^{-1}$ over a period of 100 years.
- In soil, the carbon stock under no thinning (UT (0,0)) was 23% higher than that under BT (0,0). This pattern held for all tree species. Under BT (30,30), carbon in soil increased up to 7% for Scots pine and up to 9% for Silver birch. In Norway spruce, the increase was lower, with a maximum of 5% for BT (30,30). Under no thinning, the increase of soil carbon compared to BT (0,0) was 31% for Scots pine, 32% for Silver birch, and 20% for Norway spruce.

Table 10.4 also shows the total carbon stock in forest ecosystems under different thinning options as a percentage change compared to the total carbon stock under the reference management (BT (0,0)). The management effects were tree species-specific, but the effects of different management options on the total carbon stock were similar regardless of tree species. In Scots pine, the maximum increase of the total carbon stocks for the management option BT (30,30) was 14% higher than that under BT (0,0). This also held for birch, with an increase of maximum ecosystem carbon by 15% under BT (30,30). In Norway spruce, the increase in the total ecosystem carbon stock was higher than that for the other tree species, with the maximum increase being 11% compared to that under BT (30,30).

Garcia-Gonzalo (2007) concluded that carbon stock in forest ecosystems and carbon in timber yields can be increased concurrently over a period of 100 years (Table 10.4; Fig. 10.13). This implies that appropriate management should allow for a higher stocking of trees to be sustained throughout the rotation. Baseline thinning regime (BT (0,0)) aimed at rapid growth and short rotations might not be optimal for climatic change mitigation via increasing carbon sequestration under the current climate. This held for the total ecosystem carbon stock under the management option BT (30,30), which sustained high forest stocking, with high carbon sequestration compared to that under the current thinning rules used in forestry.

There are opportunities to change thinning rules and tree species used in cultivated reforestation to increase carbon sequestration in situ and concurrently to increase timber production. In Fig. 10.13, the tree species-based fitting implies that the total mean timber harvest (yield) for Scots pine was about 100 Mg ha$^{-1}$, with mean total ecosystem carbon stock of about 70 Mg C ha$^{-1}$. For Norway spruce, the mean timber yield of 150 Mg ha$^{-1}$ implied total ecosystem carbon stock of about

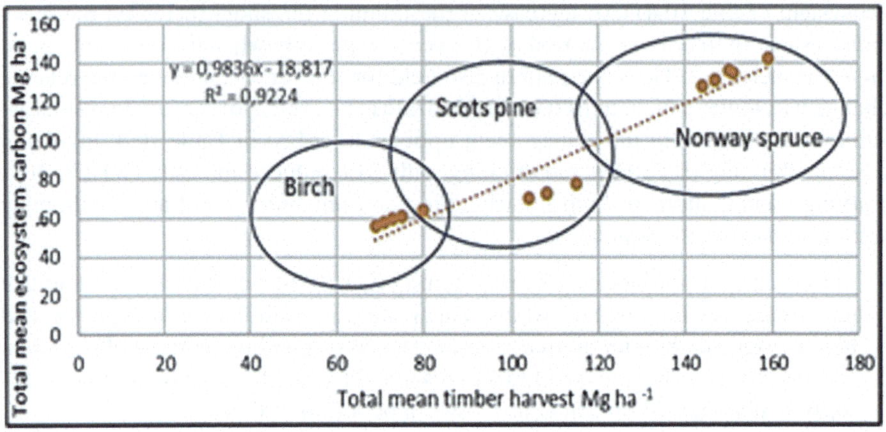


**Fig. 10.13** Total mean ecosystem carbon relative to total mean carbon in harvested timber as a function of tree species and management type (given in Table 10.4) over a 100-year rotation for three tree species based on Garcia-Gonzalo et al. (2007b). Complied by the author of this book

130 Mg C ha$^{-1}$. For birch, the mean timber yield was about 70 Mg ha$^{-1}$ and mean carbon ecosystem carbon was 60 Mg C ha$^{-1}$. These differences indicate that Norway spruce is more efficient at sequestering carbon in an ecosystem than Scots pine and birch, where stem wood production used a larger share of NPP than other organs producing most of the litter. The litter production in Norway spruce using NPP exceeded that in Scots pine per unit of timber yield.

## *10.4.4 Carbon in Managed Forest Landscapes*

According to Garcia-Gonzalo et al. (2007a, b), the structure of a forest landscape (e.g., species and age class distribution) is among the key factors affecting carbon stocks and yields of timber and biomass. In seedling stands, carbon is lost for several years until its increasing uptake in tree growth exceeds carbon loss from the soil. In maturing trees, carbon uptake decreases, with a concurrent decline of carbon accumulation in soil due to reducing litterfall. Concurrently, the amount and value of timber in maturing trees increases due to increasing share of sawn wood.

At the scale of forest landscapes, the distribution of young and mature tree stands has a strong effect on sequestrating carbon in producing timber. Garcia-Gonzalo et al. (2007a) conducted simulations of how forest age class distribution affects timber production and carbon sequestration in a forest landscape. In the simulations, the tree stands were divided per tree species into age classes: 0–20, 21–40, 41–70, and > 70 yr. Four age class distributions were created: (A) distribution dominated by intermediate age classes (normal distribution), (B) distribution not dominated by any age class (equal distribution), (C) distribution dominated by early age classes (left-skewed distribution), and (D) distribution dominated by old age classes (right-skewed distribution). Stem wood growth, timber yield (saw logs and pulpwood), and carbon sequestration were calculated for a 100-year simulation period over the area, applying the different age class distributions, forest management regimes, and climates. The cost of carbon sequestration was calculated based on indirect pricing, that is, the cost of reducing timber production when increasing carbon sequestration (Garcia-Gonzalo et al. 2007a).

According to Garcia-Gonzalo et al. (2007a), the timber yield was affected by both the initial age class distribution and forest management regime, but the effects of age class distribution on carbon stocks were small under the simulation over 100 years (Fig. 10.14). Assuming the current climate and baseline thinning (BT (0,0)), the highest values of timber and carbon stock (687 $m^3$ $ha^{-1}$ and 106 Mg C $ha^{-1}$, respectively) were obtained when the initial landscape was dominated by mature stands. On the other hand, the domination of young stands gave the lowest values (573 $m^3$ $ha^{-1}$ of timber and 103 Mg C $ha^{-1}$ of carbon). Garcia-Gonzalo et al. (2007a) also found that climatic warming produced the same pattern, but climatic warming increased the timber yield up to 18% and carbon stocks up to 6% depending on management and on the landscape structure. Regardless of the initial age class distribution, any increase in tree stocking tended to increase the timber yield, the net present value (NPV), and the carbon stocks. The timber yield and NPV were lowest under no thinning (UT (0,0)). Carbon stocks were highest under no thinning (UT (0,0)), that is, 45% greater than under baseline thinning (BT (0,0)) (Garcia-Gonzalo et al. 2007a).

Figure 10.14 shows that the timber yield and the amount of carbon increased when the stocking of trees was increased compared to that under baseline thinning (BT (0,0)). According to Garcia-Gonzalo et al. (2007a), the potential cost (potMC) under the current climate per additional Mg of carbon was from 32 to 41 € $Mg^{-1}$ depending on the age class distribution, when maximized carbon stock (UT (0,0)) and maximized timber yield (BT (30,30)) were obtained. Thus, an additional 32 Mg $ha^{-1}$ (normal distribution) to 36 Mg C $ha^{-1}$ (left-skewed distribution) would be stored in the forest ecosystem at a cost of 32–41 € $Mg^{-1}$ depending on the landscape structure. Under the changing climate, the additional amount of carbon that could be stored if using no thinning (UT (0,0)) ranged from 39 to 42 Mg $ha^{-1}$. In this case, the cost varied from 31 to 41 € $Mg^{-1}$. The increasing discount rate reduced the marginal cost of carbon sequestration (Garcia-Gonzalo 2007; Garcia-Gonzalo et al. 2007a, b).

Under the current marginal cost of carbon sequestration (curMC), the shift from baseline thinning (BT (0,0)) to no thinning (UT (0,0)) under the current climate increased carbon stock by 44–47 Mg $ha^{-1}$ depending on the initial age class distribution (Fig. 10.14). Under the changing climate, the increase was 50–53 Mg $ha^{-1}$. Concurrently, the cost increased 27 € $Mg^{-1}$ under the current climate and 28 € $Mg^{-1}$ at most under climatic warming. The real option approach (roMC) showed that carbon stock was increased when shifting from management regime BT (0,0) to BT (30,30), with no loss in NPV regardless of the climate scenario. In this case, the thinning regime maximizing NPV clearly increased carbon

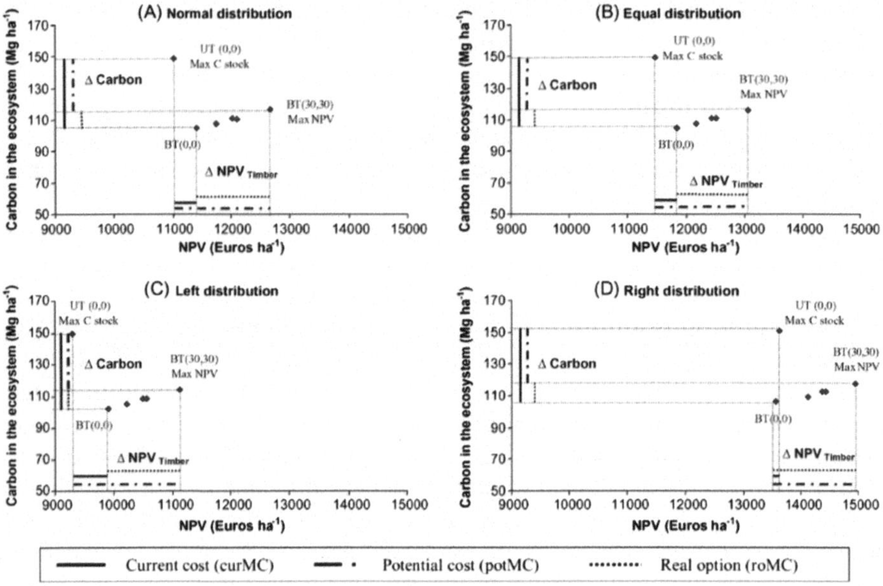


**Fig. 10.14** Relationship between the carbon in the forest ecosystem and the net present value (NPV) of timber (discount rate of 3%, with simulation initialization on bare ground) when using different thinning regimes (Garcia-Gonzalo et al. 2007a). Permission of Elsevier. For thinning regimes, see Table 10.4. The explanations for the concepts attached below Fig. 10.14 are as follows. The potential marginal cost of carbon sequestration (potMC) involves the differences in carbon stock and in the NPV of timber, representing management regimes maximizing carbon stock and NPV. The current marginal cost of carbon sequestration (curMC) indicates the differences in carbon stock and NPV of timber production, when shifted from the current management to one maximizing carbon stock. The real optional marginal cost of carbon sequestration (roMC) involves the differences in carbon stock and in NPV when shifted from the current management to management increasing both carbon stock and the NPV of timber production compared to current management

stock compared to baseline thinning (BT (0,0)). This choice would increase carbon stock by 11–12 Mg C ha$^{-1}$ in the forest ecosystem depending on the initial landscape structure (Garcia-Gonzalo 2007; Garcia-Gonzalo et al. 2007a, b).

## 10.5 Thinning, Rotation Length, and Regeneration Cutting of Maturing Forest

### *10.5.1 Carbon Balance in Relation to Rotation Length in Even-Aged Management*

In even-aged management, the rotation length defines the production cycle of timber and biomass from the previous regeneration cutting with reforestation to next regeneration cutting with regeneration for initiation of the next production cycle. Over the production cycle, carbon accumulation in the ecosystem increases through the maturation of chronosequence of trees, with a gradual increase of carbon stock in stem wood and soil carbon. However, growth and yield studies indicate that the increase of stem wood stock declines through maturation beyond the intersection of current annual growth and mean annual growth of stem wood. Concurrently, there is a subsequent decline in the growth and carbon sequestration in the crown (branches, foliage) and roots (coarse and fine roots), which follows an allometric relationship between different tree organs. This implies a gradual decline in litter production and a concurrent decline in carbon on the soil and deeper in the soil profile. This was also found by Kaipainen et al. (2004) in model simulations of how the increase of rotation length affects the reduction of greenhouse gases at the European scale. These simulations also indicated that the growth rate reduced, but accumulation of carbon stock in trees increased when rotation length was increased.

Lundmark et al. (2018) also assumed that extended rotation length reduces growth and yield under even-aged management while increasing the ecosystem carbon (Table 10.5). They simulated the growth and development of carbon stocks

**Table 10.5** Simulation options used to identify how rotation length affects the productivity and carbon stocks in an ecosystem occupied by Norway spruce (*Picea abies*) and Scots pine (*Pinus sylvestris*) in poor and fertile sites. Costs (€ per Mg of carbon) for improving carbon balance were calculated for the marginal effect on the improved balance by increasing the rotation length by 10, 20, or 30 years for Norway spruce and Scots pine occupying high- and low-productivity sites based on Lundmark et al. (2018). Courtesy of NRC Research Press

| Site index and extension of rotation in years vs. reference rotation | Costs in net present value (NPV), € $ha^{-1}$ | Climatic benefits vs. extended to reference rotation, € $ha^{-1}$ | Cost in € per Mg carbon for improved carbon balance |
|---|---|---|---|
| Norway spruce | | | |
| G18; +10 | 37 | 4.0 | 8 |
| G18; +20 | 339 | 7.2 | 47 |
| G18; +30 | 713 | 10.1 | 71 |
| G30; +10 | 946 | 12.4 | 76 |
| G30; +20 | 3389 | 19.4 | 175 |
| G30; +30 | 6077 | 22.8 | 266 |
| Scots pine | | | |
| T16; +10 | 285 | 2.4 | 120 |
| T16; +20 | 813 | 4.8 | 168 |
| T16; +30 | 1383 | 6.8 | 202 |
| T24; +10 | 1168 | 5.0 | 235 |
| T24; +20 | 2624 | 8.5 | 308 |
| T24; +30 | 4199 | 10.6 | 396 |

• Site index refers to dominant height at the age of 100 years indicating the site fertility with the consequent productivity
• High productivity represented a site index of 30 m for Norway spruce (G30) and 24 m for Scots pine (T24), while the figures for low-productivity sites were 18 m (G18) and 16 m (T16)
• Costs indicate the reduction of net present value (NPV) relative to reference rotation
• NPV includes the net revenue from final cut and the NPV for future generations if future generations would continue to be managed using extended rotation

in four even-aged Norway spruce stands of high and four low productivities. Productivity was indicated with the site index. For Norway spruce, the site index was G18 and G30 and, for Scots pine, T16 and T24. In the simulations, the current rotation length was further extended by +10, +20, and + 30 years to identify how the extension of current rotation length affects ecosystem carbon stocks and the mean timber yield. The reference rotation length was 110 and 75 years for Norway spruce and 110 and 85 years for Scots pine depending on site index. The stem wood growth was 3.0 and 10.0 $m^3$ $ha^{-1}$ $yr.^{-1}$ for Norway spruce and 3.2 and 6.1 $m^3$ $ha^{-1}$ $yr.^{-1}$ for Scots pine depending on site index.

According to Lundmark et al. (2018), a moderate increase of rotation length (10 years) increased both the mean carbon stock and the mean timber yield in Norway spruce (Table 10.5). Regardless of duration (20- and 30-year increase to the current rotation), extended rotation length increased the mean carbon stock but reduced the mean timber yield. Concurrently, this further reduced carbon storage in forest products and decreased substitution effects. On the other hand, Lundmark et al. (2018) found that the reduction of rotation length (10 years) decreased both the mean carbon stocks and timber yield, whereas a moderate increase of rotation length may slightly increase climate benefits. However, there was a trade-off between the temporary gain in increasing carbon stocks and the permanent loss of productivity, reducing the substitution potential of forest products in mitigating climatic warming.

### *10.5.2 Clear-Cutting and Site Preparation Effects on Soil Carbon and Nitrogen*

In even-aged management, clear-cutting is widely used to terminate the current production cycle for the next production cycle. Prior to reforestation, the site may be prepared for natural regeneration or for planting/seeding through plantation-type regeneration. The combination of clear-cutting with soil preparation directly affects soil carbon by ceasing the litterfall to the soil organic matter and increasing the decay of soil legacy carbon. In this context, Olsson et al. (1996) investigated how clear-cutting affected soil carbon when using varying harvest intensity: (i) conventional stem wood harvesting, leaving harvesting residues on site; (ii) harvesting all above-ground tree parts except needles; or (iii) whole-tree aboveground harvesting, leaving no harvesting residues. Soil carbon and nitrogen contents were determined at four coniferous forest sites across southern and northern Sweden, which were Scots pine and Norway spruce dominated sites with 15–16 years elapsed since cutting. The impacts of harvesting intensity on the carbon in the forest floor and the upper mineral layer (0–20 cm) were also examined, excluding coarse woody detritus, dead branches, and roots.

Regardless of tree species and cutting intensity, Olsson et al. (1996) found that the content of carbon and nitrogen in the humus layer decreased both in southern (56° N) and northern (64° N) sites (Fig. 10.15). At the same time, carbon and nitrogen

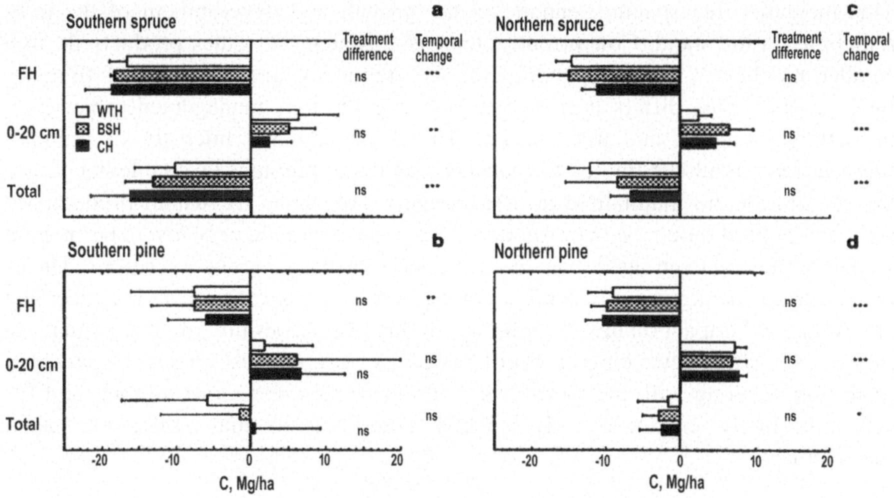


**Fig. 10.15** Changes in the amounts of soil carbon during the period of 15–16 years after harvesting in the southern (left) and northern (right) sites depending on tree species, the depth of the soil profile, and harvesting intensity. Legend for the abbreviations: *FH* humus layer; 0–20 cm = mineral soil below the humus layer; *WTH* whole-tree harvesting; *CH* conventional harvesting with logging residues left on site; *BSH* branch and stem harvesting, but with needles left on site. Furthermore, *ns* effect not statistically significant; *effect statistically significant with the risk <0.05, ** with the risk <0.01, and *** with the risk <0.001 (Olsson et al. 1996). Permission of Elsevier

content increased in the mineral soil. Total carbon content from the soil surface down to 20 cm decreased for Norway spruce by 22% in the north and 17% in the south. This implied a reduction of 22% and 13% of nitrogen, respectively. For Scots pine, there were no differences in carbon pools between thinning intensities in the south, while in the north, soil carbon decreased by 7% and nitrogen increased by 7%. No general effect of harvesting intensity on soil carbon or nitrogen pools was evident over the sites. However, there was a site and treatment interaction for nitrogen, indicating site-specific effects of harvesting intensity on nitrogen. In the northern Norway spruce site, whole-tree harvesting and the removal of woody residues reduced nitrogen pools in the humus more than under conventional harvesting. The opposite trend occurred for Norway spruce in the south, where conventional harvesting reduced the total nitrogen more than under the other harvesting intensities.

## 10.6 Concluding Remarks

In even-aged management, thinning is used to modify the structure and functioning of forest ecosystems to produce selected ecosystem services such as timber and carbon. Thinning is performed in the pole and middle-aged phases to harvest trees.

This increases tree spacing, enhancing the growth and development of the trees remaining in the stand. Concurrently, increased spacing allocates productivity to a smaller number of trees depending on the frequency and intensity of thinning interventions. This differs according to the site fertility, management objectives, harvesting methods, and tree species. The frequency and intensity of thinning interventions should be limited to such a degree that thinning is profitable but allows the stocking to be maintained at a sufficiently high level to sustain productivity through the production cycle or rotation. Until now, there are only few experimental studies on how uneven-aged management affects the dynamics of soil carbon. Under uneven-aged management, selective cutting maintains tree cover and reduces the environmental impact on the forest floor, but this likely does not prevent an increase in soil $CO_2$ efflux after cutting. Soil $CO_2$ efflux may increase under uneven-aged management along with the increasing soil temperature and moisture, but the $CO_2$ efflux is likely enhanced only slightly compared to that under even-aged management.

## References

Achat DL, Fortin M, Landmann G, Ringeval B, Augusto L (2015) Forest soil carbon is threatened by intensive biomass harvest. Sci Rep 5:15991

Äijälä O, Koistinen A, Sved J, Vanhatalo K, Väisänen P (eds) (2014) Hyvän metsänhoidon suositukset—METSÄNHOITO. Kehittämiskeskus Tapion julkaisuja, Metsäkustannus Oy, Helsinki

Briceno-Elizondo E, Garcia-Gonzalo J, Peltola H, Kellomäki S (2006) Carbon stocks and timber yield in two boreal forest ecosystems under current and changing climatic conditions subjected to varying management regimes. Environ Sci Policy 9(3):237–252

Chen L, Xiang Y, He Z, Du J, Lin P, Zhu Z (2020) A meta-analysis of the impacts of forest logging on soil $CO_2$ efflux. Sci Cold Arid Reg 12(3):165–179

Garcia-Gonzalo J (2007) Effects of management on timber production and carbon stocks in a boreal forest ecosystem under changing climate: a model-based approach. Dissertations Forestales 42: 1–47 plus four published papers. Faculty of Forest Sciences of the University of Joensuu/the Finnish Society of Forest Sciences

Garcia-Gonzalo J, Peltola H, Zubizarreta-Gerendiain A, Kellomäki S (2007a) Impacts of forest landscape structure and management on timber production and carbon stocks in the boreal forest ecosystem under changing climate. For Ecol Manag 241(1–3):243–257

Garcia-Gonzalo J, Peltola H, Briceño-Elizondo E, Kellomäki S (2007b) Changed thinning regimes may increase carbon stock under climate change: a case study from a Finnish boreal forest. Clim Chang 81:431–454

Ge ZM, Kellomäki S, Peltola H, Zhou X, Wang KY, Väisänen H (2011) Effects of varying thinning regimes on carbon uptake, total stem wood growth, and timber production in Norway spruce (*Picea abies*) stands in southern Finland under changing climate. Ann For Sci 68:371–383

Hynynen J, Eerikäinen K, Mäkinen H, Valkonen S (2019) Growth response of cuttings in Norway spruce stands under even-aged and uneven-aged management. For Ecol Manag 437:314–323

James J, Harrison R (2016) The effect of harvest on forest soil carbon: a meta-analysis. Forests 7: 308

James J, Page-Dumroese D, Bussed M, Palik B, Zhang J, Eaton B, Slesak R, Tirocke J, Kwon H (2021) Effects of forest harvesting and biomass removal on soil carbon and nitrogen: two complementary meta-analyses. For Ecol Manag 485:118935

Jandl R, Lindner M, Vesterdal L, Bauwens B, Baritz R, Hagedorn F, Johnson DW, Minkkinen K, Byrne KA (2007) How strongly can forest management influence soil carbon sequestration? Geoderma 137:253–268

Jurgensen M, Tarpey R, Pickens J, Kolka R, Palik B (2012) Long-term effect of silvicultural thinnings on soil carbon and nitrogen pools. Soil Sci Soc Am J 76:1418–1425

Kaipainen T, Liski J, Pussinen A, Karjalainen T (2004) Managing carbon sinks by changing rotation length in European forests. Environ Sci Pol 7:205–219

Kellomäki S, Väisänen H (1997) Modelling the dynamics of the boreal forest ecosystems for climate change studies in the boreal conditions. Ecol Model 97(1,2):121–140

Kellomäki S, Strandman, H, Peltola H (2019) Effects of even-aged and uneven-aged management on carbon dynamics and timber yield in boreal Norway spruce stands: a forest ecosystem model approach. Forestry 92:635–647. https://doi.org/10.1093/forestry/cpz040

Koivisto P (1962) Kasvu- ja tuotostaulukoita. Growth and yield tables. Communicationes Instituti Forestalis Fenniae 65(8):1–49

Kumpu A, Mäkelä A, Pumpanen J, Saarinen J, Berninger F (2018) Soil CO2 efflux in even-aged and uneven-aged Norway spruce stands in southern Finland. iForest 11:705–712

Lundmark T, Poudel BC, Stal G, Nordin A, Sonesson J (2018) Carbon balance in production forestry in relation to rotation length. Can J For Res 48:672–678

Mäkinen V (1959) Harvennuksen aiheuttamasta muutoksesta nuoren männikön rakenteessa. Summary: on the structural changes caused by thinning in young scots pine stands. Communicationes Instituti Forestalis Fenniae 51(7):1–31

Mäkipää R, Abramoff R, Adamczyk B, Baldy V, Biryol C, Bosela M, Casals P, Dondini M, Filipek C, Garcia-Pausas J, Gros R, Göröryová E, Hashimoto S, Hassegawa M, Immonen P, Laiho R, Li H, Li Q, Luyssaert S, Menival C, Mori T, Naudts K, Santonja M, Smolander A, Toriyama J, Tupek B, Ubeda X, Verkerk PJ, Lehtonen A (2023) How does management affect soil C sequestration and greenhouse gas fluxes in boreal and temperate forests?—a review. For Ecol Manag 529:120637

Nave LE, Vance ED, Swanston CW, Curtis PC (2010) Harvest impacts on soil carbon storage in temperate forests. For Ecol Manag 259:857–866

Nilsen P, Strand T (2013) Carbon stores and fluxes in even- and uneven-aged Norway spruce stands. Silva Fennica 47(4):1024

Olsson BA, Staaf H, Lundkvist H, Bengtsson J, Rosén K (1996) Carbon and nitrogen in coniferous forest soils after clear-felling and harvest of different intensity. For Ecol Manag 82:19–32

Pukkala T (2016) Plenterwald, Dauerwald, or clearcut. For Policy Econ 62:125–134

Pukkala T, Lähde E, Laiho O (2010) Optimizing the structure and management of uneven-sized stands in Finland. Forestry 83(3):131–142

Pukkala T, Lähde E, Laiho O (2011a) Metsän jatkuva kasvatus. Joen Forest Program Consulting 230.

Pukkala T, Lähde E, Laiho O, Salo K, Hotanen J-P (2011b) A multifunctional comparison of even-aged and uneven-aged forest management in a boreal region. Can J For Res 41:851–862

Saunders M, Tobin B, Black K, Gioria M, Nieuwenhuis M, Osborne BA (2012) Thinning effects on the net ecosystem carbon exchange of a Sitka spruce forest are temperature-dependent. Agric For Meteorol 157:1–10

Schlesinger WH, Andrews JA (2000) Soil respiration and the global carbon cycle. Biogeochemistry 48:7–20

Tahvonen O, Pukkala T, Laiho O, Lähde, E, Niinimäki S (2010) Optimal management of uneven-aged Norway spruce stands. For Ecol Manage 260:106–115

Vesterdal L, Dalsgaard M, Felby C, Raulund-Rasmussen K, Jørgensen B (1995) Effects of thinning and soil properties on accumulation of carbon, nitrogen and phosphorus in the forest floor of Norway spruce stands. For Ecol Manag 77:1–10

Vuokila Y (1970) Harsintaperiaate kasvatushakkuissa. Summary: selection from above in intermediate cuttings. Acta Forestalia Fennica 110:1–45

Vuokila Y (1977) Harsintaharvennus puun tuotantoon vaikuttavana tekijänä. Summary: selective thinning from above as a factor of growth and yield. Folia Forestalia 298:1–17
Vuokila Y (1980) Metsän kasvatuksen perusteet ja menetelmät. WSOY, Porvoo
Zhang X, Guan D, Li W, Sun D, Jin C, Yuan F, Wang A, Wu J (2018) The effects of forest thinning on soil carbon stocks and dynamics: a meta-analysis. For Ecol Manag 429:36–43

# Chapter 11
# Effect of Fertilization on Carbon in Boreal Ecosystems

## Contents




**Abstract** In boreal forest ecosystems, carbon and nitrogen cycles interact with impacts on carbon sink/source processes and balance between above- and belowground carbon stocks. The addition of nitrogen increases primary production in the above- and belowground organs of trees and ground vegetation, with a consequent increase of carbon in the ecosystem. The carbon stocks in soil are also linked to plant growth through the subsequent input of aboveground plant/tree litter (e.g., litter from foliage and branches) onto the soil surface and of root litter (e.g., coarse roots, fine





S. Kellomäki, *Forest Management for Timber Production and Climate Change Mitigation*, Managing Forest Ecosystems 44,
https://doi.org/10.1007/978-3-031-71575-4_11

roots) deeper within the soil profile. On the other hand, nitrogen addition may reduce the decay of organic compounds by reducing microbial activity. This reduces carbon dioxide emissions (efflux) from soil, thus increasing soil carbon stocks.

## 11.1 Response of Carbon in Terrestrial Ecosystems to Nitrogen Addition

### *11.1.1 Carbon Source/Sink Dynamics in Interaction with Nitrogen Addition*

In terrestrial ecosystems, the carbon and nitrogen cycles interact, with impacts on carbon sink/source processes and subsequent carbon balance. This holds for boreal forest ecosystems, in which nitrogen addition influences carbon stocks and fluxes through ecosystems, encompassing the carbon sequestration in both above- and belowground processes and pools (Fig. 11.1). The addition of nitrogen, such as through fertilization, tends to increase primary production in the above- and belowground organs of trees and ground vegetation. The carbon stocks in soil are also linked to the balance between plant growth and the subsequent input of aboveground plant litter (e.g., litter from foliage and branches) onto the soil surface and of root litter (e.g., coarse roots, fine roots) deeper within the soil profile (Yue et al. 2016).

Carbon in soil is lost through the decay of soil organic matter and through the transfer of carbon in organic matter dissolved in water (Fig. 11.1). In fertilization, nitrogen addition can reduce microbial biomass, which reduces decay and subsequent carbon dioxide fluxes from soil, thus increasing the carbon stocks in soil. However, nitrogen addition tends to affect the quality of plant litter by decreasing the lignin/nitrogen ratio, with an increase in the labile carbon substrate (Yue et al. 2016), resulting in a higher carbon release through heterotrophic respiration. Thus, nitrogen fertilization increases both carbon input (carbon bound in aboveground plant biomass) and carbon output (through carbon emission). However, these may balance each other, resulting in no change of ecosystem carbon balance. Yue et al. (2016) emphasized that nitrogen addition may even reduce soil carbon stocks because of increased decay related to the changes in properties of soil detritus.

### *11.1.2 Effect of Nitrogen Addition on Carbon Sequestration in Terrestrial Ecosystems*

Yue et al. (2016) further conducted a meta-analysis based on experimental studies of how nitrogen fertilization affects the dynamics of terrestrial carbon fluxes and stocks and subsequent carbon sequestration and stocks in ecosystems representing terrestrial plant species including forest tree species across different biomes. The main findings showed that:

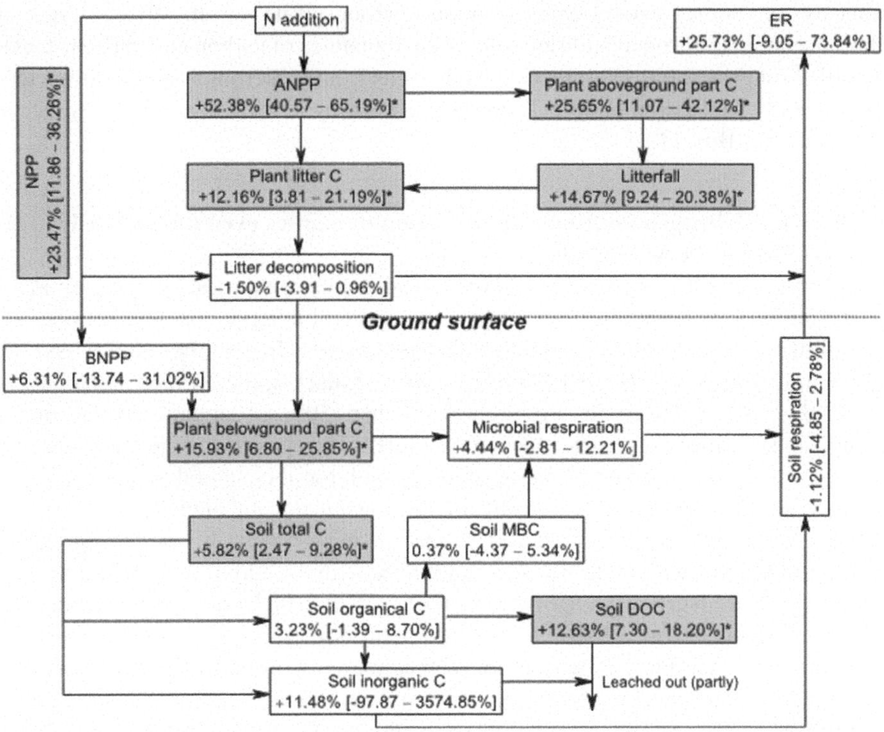


**Fig. 11.1** Responses of the terrestrial ecosystem carbon cycle to experimental nitrogen addition. Numbers indicate mean changes in the carbon pool or flux, with a 95% confidence interval [CI]. Asterisks indicate statistical significance ($P < 0.05$) (Yue et al. 2016). Open Access. Courtesy of Nature Portfolio. Legend: *NPP*, net primary production; *ANPP*, aboveground net primary production; *BNPP*, belowground net primary production; *MBC*, microbial biomass carbon; *TOC*, total organic carbon; *DOC*, dissolved organic carbon; *ER*, ecosystem respiration

- Nitrogen fertilization increased the total soil carbon stock up to 5.82%, with a 95% confidence interval [2.5%, 9.3%].
- Nitrogen fertilization increased the carbon contents of the above- and belowground parts of plants up to 25.7% [11.1%, 42.1%] and up to 15.9% [6.8%, 25.9%].
- Nitrogen fertilization increased aboveground net primary production up to 52.38% [40.58%, 65.19%] and litterfall up to 14.7% [9.2%, 20.4%].
- Nitrogen fertilization had no effect on the carbon influx from the plant litter to the soil through litter decomposition.
- Nitrogen fertilization had no effect on the efflux of carbon from the soil related to microbial respiration and soil respiration.

The meta-analysis indicated that nitrogen fertilization increases soil carbon stocks and carbon in both the above- and belowground parts of plants, increasing the capacity of boreal ecosystems to retain carbon (Yue et al. 2016). Over a chronosequence, nitrogen addition affects the dynamics of carbon and carbon stocks in soil through several ecological, physiological, and microbiological parameters even in early succession in boreal forests as shown by Allison et al. (2010) to occur after wildfire (Box 11.1).

**Box 11.1 Nitrogen addition alters carbon dynamics even during early succession in boreal forests**

Allison et al. (2010) investigated how nitrogen addition may affect the carbon dynamics in soil carbon stock, and several microbiological parameters controlling carbon dynamics, through a chronosequence. In a northern-boreal site (central Alaska, 64° N) in 1999, the canopy of black spruce (*Picea mariana*) stands was destroyed in wildfire. More than 50% of organic matter on the soil surface was lost due to wildfire. The postfire vegetation was dominated by grasses and herbs but allowed for the natural regeneration of trees. This produced a mixture of black spruce with birch (*Betula* sp.) and poplar (*Populus* sp.) in a fertilization experiment. It was based on replicated plots with no fertilization and those with the addition of ammonium nitrogen ($NH_4NO_3$) at a rate of 200 kg ha$^{-1}$ yr.$^{-1}$ and 100 kg ha$^{-1}$ yr.$^{-1}$ in several years. Allison et al. (2010) found nitrogen addition to increase the mean nitrogen availability by two orders of magnitude. Consequently, the aboveground carbon uptake in aboveground net primary production (*ANPP*) increased from 98 ± 25 g m$^{-2}$ yr.$^{-1}$ in the unfertilized plots to 255 ± 57 g m$^{-2}$ yr.$^{-1}$ in the fertilized plots. At the same time, the aboveground biomass increased by 64%. On the other hand, nitrogen addition increased the loss of litter mass by 3–9%. However, the $CO_2$ efflux in the fertilized plots declined from 78.3 ± 6.5 mg $CO_2$-C m$^{-2}$ h$^{-1}$ to 54.4 ± 4.1 mg $CO_2$-C m$^{-2}$ h$^{-1}$ compared to that from the plot with no nitrogen addition. In this experiment, the early-succession carbon cycle responded to nitrogen addition, and declining microbial activity increased the ecosystem carbon stocks. Under excess increase of nitrogen supply, the annual volume growth of trees became saturated or even declined.

## 11.2 Fertilization Effects on Carbon in Boreal Forest Ecosystems

### *11.2.1 Impacts of Fertilization on Growth in Boreal Forests*

In boreal forests, the low temperatures and short nitrogen supply result in low productivity. In these conditions, climatic warming is likely to enhance the nitrogen cycle, including (i) the uptake and binding of nitrogen in growth, (ii) the return of nitrogen to the soil in litter, (iii) and finally the release of nitrogen through the decay of soil organic matter for reuse of nitrogen in growth. The carbon balance in boreal forests is tightly controlled by the availability of nitrogen from natural and anthropogenic sources. In sparsely populated boreal area like in Finland, the nitrogen mainly originates from natural sources, e.g., biological nitrogen fixation and lightning, with a deposition of <4 kg N $ha^{-1}$ $yr.^{-1}$ for boreal forest (http://www.ymparisto.fi). In the most urbanized, industrialized, and agricultural areas, natural nitrogen deposition is overlapped by anthropogenic sources (4–10 or even more kg N $ha^{-1}$ $yr.^{-1}$), including ammonium ($NH_4$) and nitrate ($NO_3$) compounds.

In forest management, fertilization is widely used to maintain and enhance the productivity of forest sites. Fertilization refers to the addition of nutrients (nitrogen, mineral nutrients) on a site to increase forest growth and/or balance the supply of nutrients in the biogeochemical cycle. This especially holds for nitrogen in boreal upland sites, where the short supply of nutrients like phosphorus, potassium, and calcium only seldom limits forest growth (e.g., Nilsen 2001). This contrasts with deep drained peatland soils, where the nitrogen supply is generally adequate, but the short supply of phosphorus, potassium, and calcium frequently limits growth. In very fertile grove-type upland sites (site fertility ≥ *Oxalis-Myrtillus* site type, OMT), a high nitrogen supply may unbalance the nutrient cycle and reduce the otherwise high growth response in relation to the supply of phosphorus, potassium, and calcium.

### 11.2.2 Impacts of Fertilization on Functioning and Structure of Boreal Ecosystems

In an ecosystem context, the fertilization tends to enhance the growth in delay to increasing availability of nutrients (Fig. 11.2). The growth response is dependent on the species, site fertility (i.e., edaphic and climatic conditions), tree maturity, canopy position, seasonal phase, and so forth (e.g., Linder 1995). They enhance the nutrient availability and uptake, which drive ecophysiological processes and carbon uptake and growth in the scale of ecosystem and trees/plant communities. In this context, nitrogen fertilization increases available nitrogen directly in soil, as well as indirectly through nitrogen cycle with the increased production of litter of higher nitrogen content and through the reduced decay of litter. This holds for boreal Scots pine (*Pinus sylvestris*), Norway spruce (*Picea abies*), and birch (*Betula pendula*, *B. pubescens*), as shown in fertilization experiments (e.g., Olsson et al. 2005).

The effect of nitrogen supply on growth is mediated through the biochemical processes that control the photosynthetic rate. Carbon dioxide- ($V_{cmax}$) and radiation-saturated ($J_{max}$) photosynthetic rates depend linearly on the nitrogen content of foliage (e.g., Kellomäki and Wang 1997a, b; Ainsworth and Long 2005; Hyvönen et al. 2008). In coniferous trees, nitrogen supply affects the nitrogen content of needles, the mass of needles through the increasing longevity, mass per needle, and area of needles, with effects on the photosynthetic capacity of shoots and the whole crown (e.g., Jokela et al. 1995; Balster and Marshall 2000; Amponsah et al. 2005).

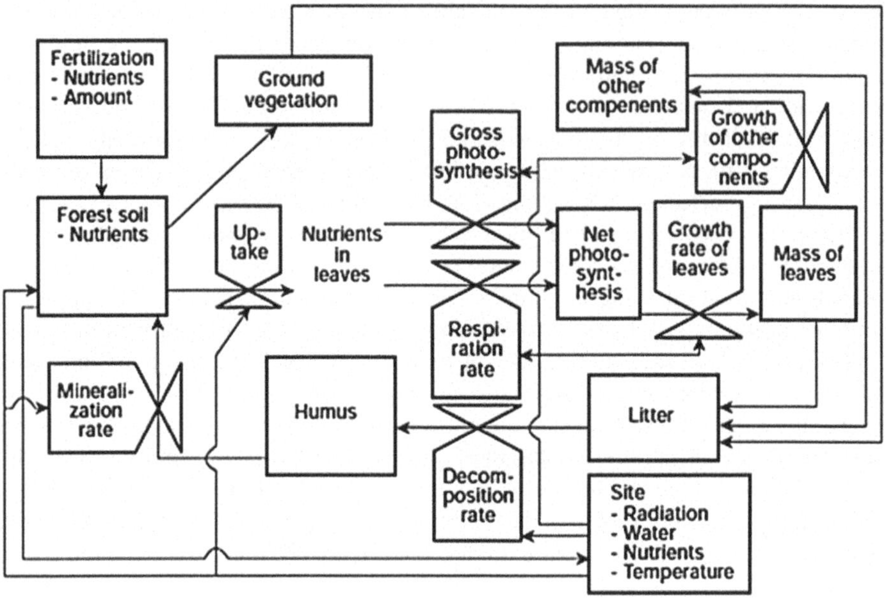


**Fig. 11.2** Schematic depicting how nitrogen fertilization affects the carbon dynamics in boreal upland forest ecosystems, with impacts on stem wood volume growth (Kellomäki 2022, p. 439). Permission of Springer Nature

The increasing nitrogen content in current-year needles tends to enhance the volume growth of Scots pine and Norway spruce in the range of a nitrogen content of 5–10 mg per gram in current-year needles as found by Nilsen and Abrahamsen (2003). In this range, annual stem wood growth more than doubled linearly as a function of the increased nitrogen content in current-year needles.

### *11.2.3 Impacts of Nitrogen Fertilization on Soil Carbon and Nitrogen Mineralization*

Nave et al. (2009) emphasized that nitrogen fertilization affects soil carbon stocks through plant/tree growth, litter production and decay, and stabilization of soil organic matter, as was found for northern temperate forests. In a meta-analysis, Nave et al. (2009) summarized how nitrogen addition affects soil carbon stock, the ratio between carbon/nitrogen (*C/*N), and nitrogen mineralization. Nitrogen addition comprised N-fixing vegetation, nitrogen fertilization, and chronic nitrogen deposition simulated in the experiments. The dataset represented 700 experiments in North America and Europe, where the mean air temperature exceeds 10 °C through 4–8 months per year (Fig. 11.3).

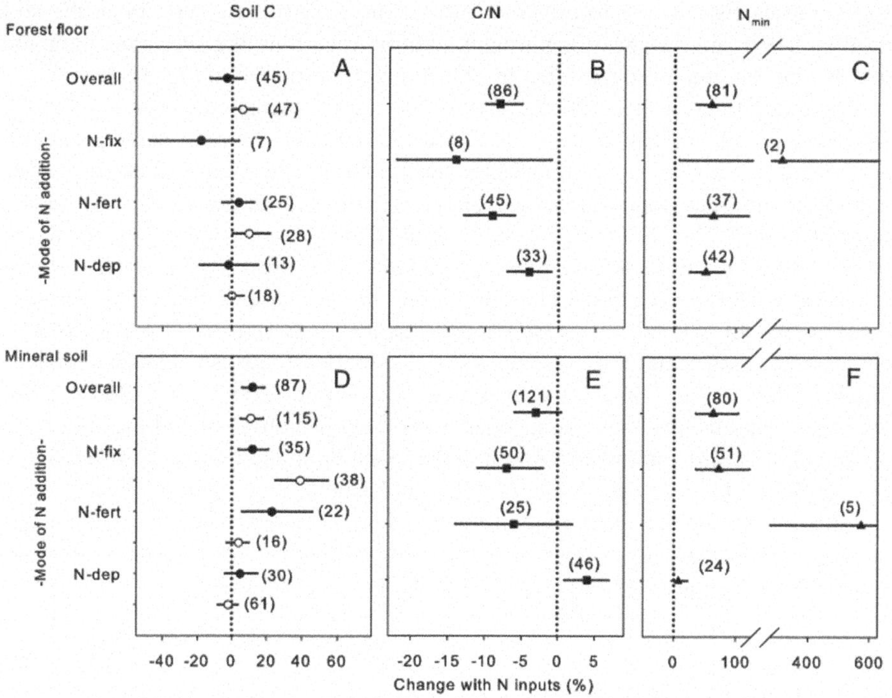


**Fig. 11.3** Effect of nitrogen input on the soil carbon and nitrogen in northern temperate forests (Nave et al. 2009) based on a meta-analysis estimating the magnitude of the change of response to nitrogen input. Permission of Elsevier. The response change ($R$) is $ln(R) = ln\left(\overline{X}^E/\overline{X}^C\right)$, where $\overline{X}^E$ is the mean value of treatment (elevated nitrogen) in observations and $\overline{X}^C$ is the mean value of control observations for a given set of experimental conditions indicated by soil carbon (C), the ratio of carbon/nitrogen (C/N), or nitrogen mineralization ($N_{min}$). Upper: the effects of N inputs on soil *C*, *C/N*, and $N_{min}$ in *forest floors* (panels A–C). Points are means ± bootstrapped 95% confidence intervals, with the number of studies ($k$) in parentheses. Groups with confidence intervals overlapping the dotted reference line (0% change) show no significant effects of N addition. Soil C responses to N inputs are shown separately for pool sizes (filled symbols) and concentrations (open symbols). Lower: the effects of N inputs on *mineral soil* and mineral soil (panels D–F) parameters by biogeographic group. Points are means ± bootstrapped 95% confidence intervals, with the number of studies ($k$) in parentheses. Soil C storage response ratios include pool sizes only. Biogeographic groups are for western conifers (WC), northeast conifers (NEC), and northeast hardwoods (NEH), as indicated in the text

The experimental sites were divided in terms of biogeographical groups (Nave et al. 2009): (i) western conifers (WC), mainly in the western and northwestern USA; (ii) northeastern deciduous trees (NEH) in the northern and eastern USA; and (iii) northeastern conifers, comprising those in the northeastern USA and northern Europe. In the analysis, *soil carbon storage* referred to the soil carbon pool only. Soil carbon included two types of carbon in the experiments. In this regard, soil samples were divided into layers, allowing for the separation of organic and mineral soil components into the forest floor and the mineral soil (5–100 cm in the soil profile). The main findings for nitrogen addition regarding the site, biogeographic tree groups, and the nitrogen input on mineral soil are shown in Fig. 11.3.

According to Nave et al. (2009), any nitrogen inputs increased soil carbon (7.7%) and nitrogen mineralization ($N_{min}$) (62%) while reducing the ratio between carbon and nitrogen (*C/N*, 4.9%). Soil carbon concentrations and carbon stock responded equally to nitrogen inputs, but the storage of soil carbon increased only in the mineral soil (12.2%). The reducing shifts of the *C/N* ratio occurred only in the forest floor (7.8%). However, nitrogen mineralization in the forest floor and mineral soil increased equally in relation to nitrogen input, the increases being 61% and 64%, respectively. In general, the significant differences between the responses of soil parameters to nitrogen input were related to the mode of nitrogen addition, biogeographical factors, and time since treatment. Nave et al. (2009) concluded that the temporal relationships between nitrogen inputs and responses of soil parameters to the increase of nitrogen mineralization in the forest floor and mineral soil were large but transient. The increase followed the decadal-scale changes in the *C/N* ratio in the forest floor and the carbon storage in mineral soil. Temporal patterns in the dataset could have been due to changes in the decomposition of soil organic matter and stabilization or to increased plant productivity and litter inputs.

## 11.3 Effects of Fertilization on Foliage and Stem Wood Growth

### *11.3.1 Foliage Properties Related to Fertilization*

In Finland, the productivity of forest sites is indicated by the site types, which reflect the availability of nitrogen and subsequent effects on photosynthetic capacity. High nitrogen supply, as in fertile sites, also tends to affect the nitrogen content of needles and the structure of shoots of Scots pine (*Pinus sylvestris*), such as the arrangement of needles around shoots and changes in the mass and morphology of needles, as well as their longevity (e.g., Jokela et al. 1995; Pensa and Sellin 2002). Similarly, Brix and Ebel (1969) and Brix (1972) found that the number of needles per shoot in Douglas fir (*Pseudotsuga menziesii*) increases in nitrogen-fertilized trees, with an increase in needle length, width, and needle area per shoot (Table 11.1). Thus, fertilization impacts on total photosynthesis tend to be multiplicative due to the increase of photosynthetic rate per unit foliage area and of foliage area from the increasing uptake and cycling of nitrogen within an ecosystem (e.g., Fagerström and Lohm 1977). This further implies an increase of litterfall, as found by Miller et al. (1976) and Miller and Miller (1976) for Corsican pine (*Pinus nigra*).

**Table 11.1** Effect of nitrogen fertilization on the properties of shoots and needles of Douglas fir (*Pseudotsuga menziesii*) based on Brix and Ebel (1969) and Brix (1972). The numbers in parentheses indicate the response to nitrogen addition relative to that without nitrogen addition

| Structure of crown and properties of shoots and needles | Experiment 1 | | Experiment 2 | |
|---|---|---|---|---|
| | Addition of nitrogen fertilizer, kg ha$^{-1}$ | | | |
| | 0 | 336 | 0 | 250 |
| Needle number per shoot | 210 (100) | 305 (145) | 219 (100) | 250 (114) |
| Needle length, mm | 20.8 (100) | 25.6 (123) | 22.5 (100) | 32.5 (144) |
| Needle width, mm | 1.53 (100) | 1.69 (110) | 1.43 (100) | 1.49 (104) |
| Needle area per shoot, cm$^2$ | 140 (100) | 275 (197) | 63.0 (100) | 79.6 (126) |

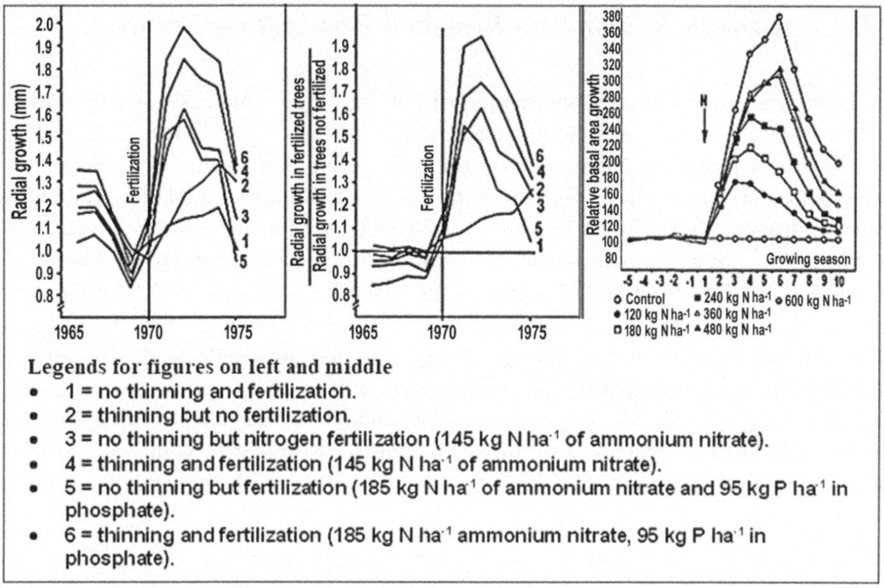


**Fig. 11.4** Effects of varying nitrogen addition on the growth of Scots pine (*Pinus sylvestris*) in boreal forests. Left: absolute and Middle: relative effects of thinning and fertilization on the diameter growth of boreal Scots pine based on Haapanen et al. (1977). Courtesy of Finnish Society of Forest Science. Right: relative growth of basal area in boreal Scots pine stands as a function of nitrogen dose and time since fertilization (Jonsson 1978). Courtesy of Föreningen Skogsädling, Institut för Skogsförbättring

## *11.3.2 Growth Response to Single Fertilization Events*

In general, a single addition of nitrogen increases photosynthesis in the short term (<10 years), or with a delay, as a result of the increase of nitrogen content in foliage, as shown in Fig. 11.4. The response is related to the dose of fertilizer, with a declining response related to increasing nitrogen addition. However, fertilization via increasing addition of nitrogen may also have long-term impacts (>10 years).

This is related to the enhancement of the within-stand nutrient cycle. In this regard, an increasing share of nutrient addition is bound in tree biomass, ground vegetation, and surface soil (e.g., Haapanen et al. 1977; Jonsson 1978). This stored nitrogen gradually becomes available through the decay of litter and soil organic matter originating from trees and ground vegetation. Nitrogen fertilization also increases litterfall, thus increasing the long-term nitrogen supply. Such impacts are affected by site fertility and the properties of trees (species, age, dimensions) and forest structure (spacing), with impacts on total carbon mass and the cycling of carbon in the ecosystem (e.g., Reich et al. 1998; Valinger et al. 2000; Hyvönen et al. 2008).

### *11.3.3 Growth Response to Repeated Fertilization Events*

In general, a single nitrogen application of 150 kg N $ha^{-1}$ increases the growth of boreal Norway spruce and Scots pine by 1–2 $m^3$ $ha^{-1}$ $yr.^{-1}$ over a period of 6–8 years, as found by Nilsen (2001) for both young and mature tree stands. On the other hand, Nohrstedt (2001) and Pettersson and Högbom (2004) found that the single nitrogen dose of 150 kg $ha^{-1}$ was effective with the growth increase by 4–5 $ha^{-1}$ $yr.^{-1}$ over a 5-year period. The similar response was found by Kukkola and Saramäki (1983), who detected a growth increase of 1.3 $m^3$ $ha^{-1}$ $yr.^{-1}$ for Norway spruce and 1.4 $m^3$ $ha^{-1}$ $yr.^{-1}$ for Scots pine in southern Finland over 10–12 years (Fig. 11.5). In northern Finland, the mean response was 63% and 69% of the response in southern Finland for Scots pine and Norway spruce, respectively. Regardless of region, the response was dependent on the site fertility and the requirements of tree species. The growth increase was higher in young stands than in mature stands, where the response decreased along with the aging of trees (Box 11.2).

**Box 11.2 Growth Response to Repeated Fertilization Events**
According to Kukkola and Saramäki (1983), the volume growth of stem wood increase ($\Delta I_t$, $m^3$ $ha^{-1}$ $yr.^{-1}$) is related to (i) the time ($t$) since nitrogen fertilization; (ii) the nitrogen added in fertilization ($N$, kg N $ha^{-1}$) and phosphorus ($P$, kg P $ha^{-1}$); (iii) the type of nitrogen included in fertilization; (iv) the spacing of trees; (v) the dominant height of trees ($H_{\text{dom}}$, m); and (vi) the site fertility as indicated by the site index ($SI$, dominant height in meters at the age of 100 years). Consequently, $\Delta I_t = A_t \times B$, where $A_t = f(t, N, P, T)$ and $B = f(Fre, H_{\text{dom}})$, including the number of trees per hectare ($Fre$) and dominant height ($H_{dom}$) at the time of fertilization. Kukkola and Saramäki (1983) further found that under repeated fertilization, a suitable interval between application would be 4–8 years with a single dose of 100–150 kg N $ha^{-1}$. For this range, the largest response would be obtained when

(continued)

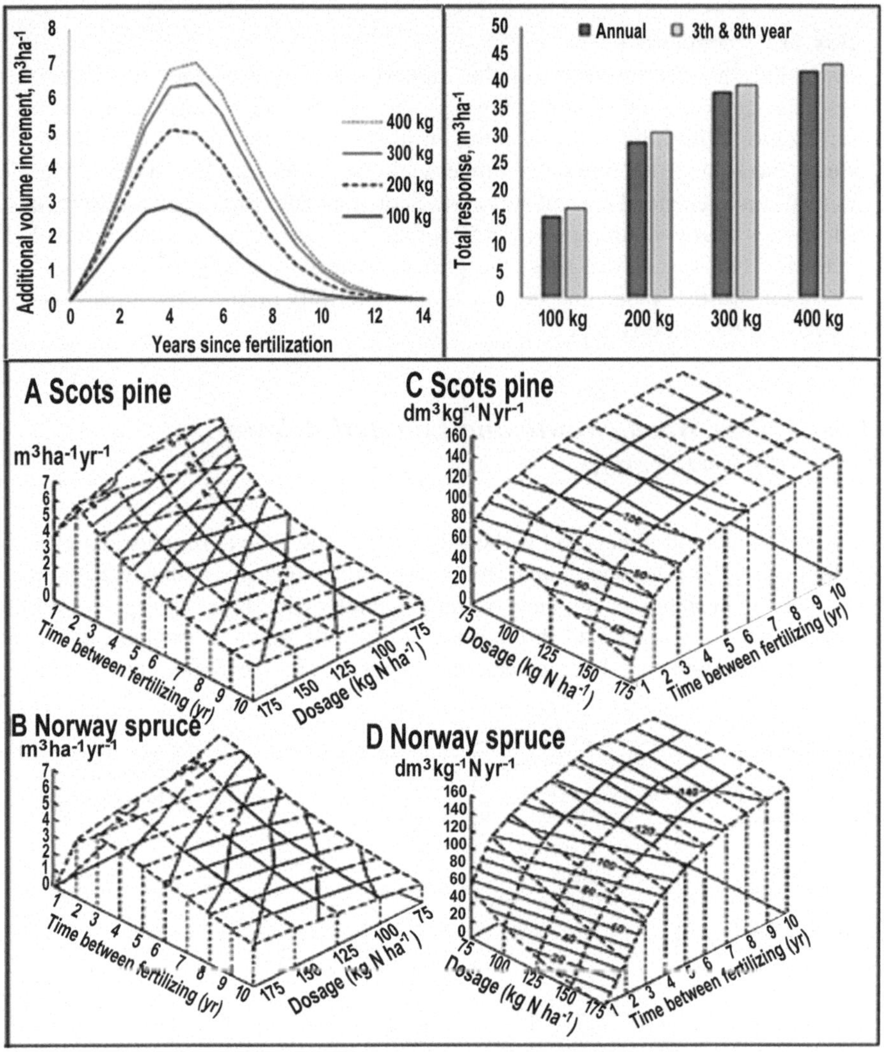


**Fig. 11.5** Growth response of Scots pine (*Pinus sylvestris*) and Norway spruce (*Picea abies*) as a function of nitrogen addition and time between fertilization events. Upper/left: growth response in Norway spruce when nitrogen addition is applied via a single dose of 100, 200, 300, or 400 kg ha$^{-1}$ (Kellomäki et al. 1982; Routa et al. 2011). Courtesy of Oxford University Press. Upper/right: total annual stem wood growth increase and the annual increase in the third and eighth years after fertilization (Pukkala 2017). Courtesy of Springer Open. Lower: effects of the dose of nitrogen fertilizer on the annual growth increase per hectare (A, B) and the annual growth increase per kilogram of nitrogen (C, D) in Scots pine and Norway spruce, respectively (Kukkola and Saramäki 1983). Courtesy of Natural Resources Institute Finland

**Box 11.2** (continued)
using the shortest interval and the highest dose of nitrogen. The largest response per kilogram of nitrogen was obtained using a long interval between fertilization with a low dose of nitrogen in each application. In a simulation study, Pukkala (2017) found that timber yield increased 1 $m^3$ $ha^{-1}$ $yr.^{-1}$ (15%) in medium-aged stand sites in sub-xeric Scots pine and mesic Norway spruce sites when nitrogen addition was 150 kg $ha^{-1}$. In mesic Scots pine sites and herb-rich Norway spruce sites, the growth increase was 0.5 $m^3$ $ha^{-1}$ $yr.^{-1}$ (5%), thus further indicating the effect of site fertility on the growth response to nitrogen addition (Fig. 11.5).

### *11.3.4 Stem Wood Growth and Litterfall Related to Fertilization*

Nitrogen fertilization affects soil carbon through (i) the growth of trees and ground vegetation and the consequent litter production and (ii) changes in the decay of litter and humus, with stabilized amounts of organic matter (SOM, or soil organic carbon stock) on the soil surface and through the soil profile down to groundwater level (Fig. 11.6). Growth and litter production occur in all tree organs (e.g., foliage,

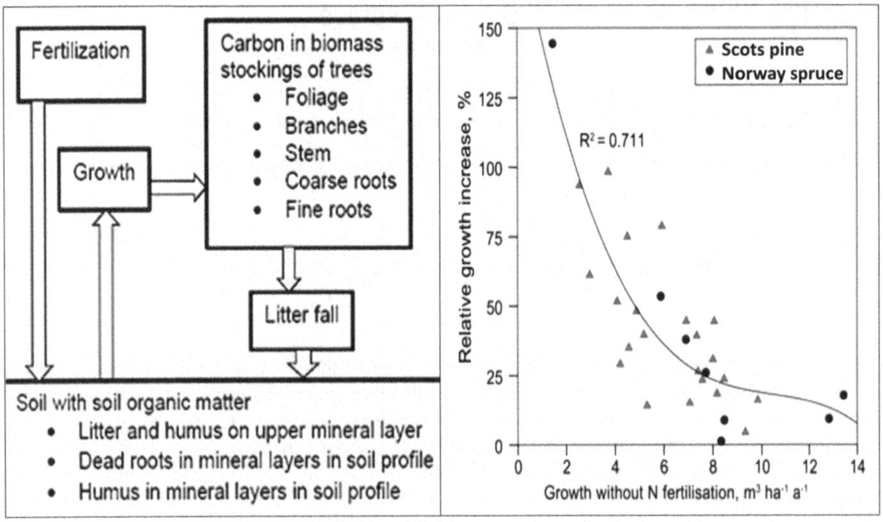


**Fig. 11.6** Litterfall related to stem wood growth under fertilization. Left: overview of effects of nitrogen fertilization on carbon in forest ecosystems, with a focus on trees. Compiled by the author of this book. Right: relative annual increase of stem wood growth induced by nitrogen fertilization as a function of the annual stem wood growth under no nitrogen fertilization, based on experimental findings across (60–70° N) the boreal forest zone in Finland (Saarsalmi et al. 2014). Courtesy of Finnish Society of Forest Science

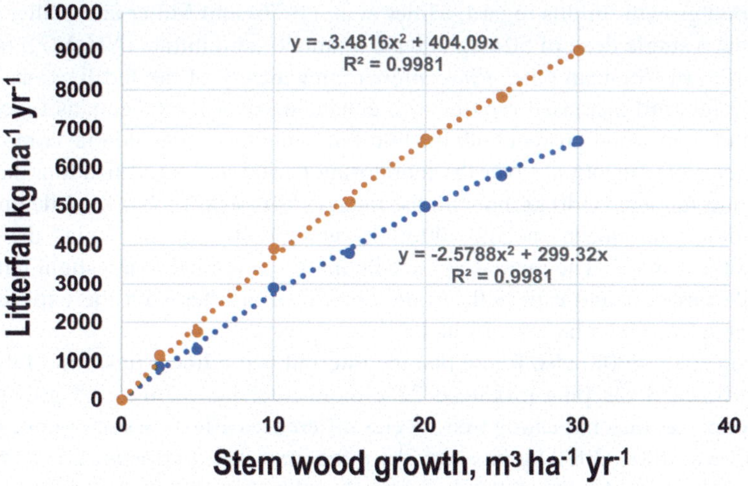


**Fig. 11.7** Effect of fertilization on litterfall, through the enhancement of stem wood growth. Litterfall as a function of volume growth and nitrogen fertilization for boreal Scots pine stands, based on the findings of Matala et al. (2008), assuming no fertilization (blue dots) and nitrogen fertilization (red dots). Compiled by the author of this book

branches, stem, coarse roots, fine roots), thus affecting the amount and properties of soil organic matter (litter on the soil and humus below the litter and in soil profile). In boreal upland forests, tree growth is linked with nitrogen supply in natural conditions, which makes growth and subsequent litter production sensitive to nitrogen fertilization. Furthermore, there are close links between the biomass growth and the mortality of different organs and the whole tree, as demonstrated in Fig. 11.6. Both in Scots pine and Norway spruce, the annual death of needles is directly related to the growth of stem wood. Furthermore, the death of branches is initiated with a few years' delay compared to that of needles and fine roots.

The increasing effect of nitrogen on volume growth also increases growth of other organs of consequent litterfall from trees and ground vegetation. Saarsalmi et al. (2014) found that the relative growth response to nitrogen addition is inversely related to natural nitrogen supply, such as substantial natural nitrogen deposition or nitrogen-rich soil in sites. They also found that the mean annual growth *under no nitrogen fertilization* in boreal forests was in the average 6.2 (2.6–9.8) $m^3$ $ha^{-1}$ $yr.^{-1}$ in Scots pine and 8.1 (1.5–13.4) $m^3$ $ha^{-1}$ $yr.^{-1}$ in Norway spruce. *Under nitrogen fertilization*, the growth increased by 2.2 (0.4–4.7) $m^3$ $ha^{-1}$ $yr.^{-1}$ (35% of that under no fertilization) for Scots pine and by 1.8 (0.1–3.2) $m^3$ $ha^{-1}$ $yr.^{-1}$ (22%) for Norway spruce. In this context, the foliage growth of Scots pine and Norway spruce correlated with the stem wood growth, as shown in Fig. 11.7 for Scots pine, with a 35% increase of litterfall compared to that under no fertilization.

Figure 11.7 further summarizes how litterfall is linked to the growth of stem wood of individual trees based on stand measurements for several boreal and temperate tree species. These experimental studies show a linear dependence of litter on the stem wood growth. In this regard, Miller et al. (1976) and Miller and Miller (1976) found that a single dose of 504 kg N ha$^{-1}$ in ammonium nitrate ($NH_4NO_3$) reduced the litterfall of Corsican pine (*Pinus nigra*) immediately of the fertilization. Thereafter, the litterfall increased rapidly with enhancing over the 6 years as function of increasing stem wood growth with the litterfall being more than double compared to that under no fertilization. Over the monitoring period under fertilization, the mean needle litterfall was 3140 kg ha$^{-1}$ in the range 1780–4900 kg ha$^{-1}$, while under no fertilization, the mean needle litterfall was 1740 kg ha$^{-1}$ in the range 1350–2660 kg ha$^{-1}$. The increase of needle litter was related to the slight reduction of needle turnover and especially to the growth of needles with the expanding of needle area from 8 ha ha$^{-1}$ to 14 ha ha$^{-1}$.

Matala et al. (2008) also found that the litterfall was a linear function of the stem wood growth. Scots pine produced 22% more litter per stem wood growth than Norway spruce, thus indicating tree-species differences affecting soil organic matter. Litterfall was 2000–3000 kg ha$^{-1}$ yr.$^{-1}$ for the stem wood growth of 5–10 m$^3$ ha$^{-1}$ yr.$^{-1}$. However, the litterfall was inversely related to the northward increase of latitude, with 2000–3000 kg ha$^{-1}$ yr.$^{-1}$ of litterfall in southern and up to 1000 kg ha$^{-1}$ yr.$^{-1}$ in the northern Finland. The linear dependence of litterfall on the stem wood growth implies that the increase of stem wood growth due to nitrogen fertilization likely increase the litterfall as a function of stem wood growth up to 20% as estimated based on the findings of Matala et al. (2008).

### *11.3.5 Long-Term Growth and Carbon Sequestration in Response to Fertilization*

Regarding growth and carbon sequestration in response to nitrogen fertilization, Hyvönen et al. (2008) studied how Scots pine and Norway spruce bind carbon in trees (stems, stumps, branches, needles, coarse roots) and in the soil organic layer and upper mineral soil (0–10 cm). The long-term addition of nitrogen over 14–30 years at a rate of 30–50 kg N ha$^{-1}$ yr.$^{-1}$ produced a stronger response than fertilization at a rate of 50–200 kg N ha$^{-1}$ yr.$^{-1}$. A cumulative amount of nitrogen of 600–1800 kg ha$^{-1}$ yielded 25 kg of carbon in trees and 11 kg in soil per kilogram of nitrogen. Hyvönen et al. (2008) also found that fertilizing with a mixture of nitrogen, phosphorus, and potassium (NPK) increased the carbon sequestration compared to that under pure nitrogen fertilization.

Högberg et al. (2006) also investigated the effects of long-term (30 years) nitrogen fertilization on tree growth in a boreal forest. In the experiment, ammonium nitrate ($NH_4NO_3$) was used at two doses of 34 and 68 kg $NH_4NO_3$ ha$^{-1}$ yr.$^{-1}$. A third treatment with 108 kg N ha$^{-1}$ yr.$^{-1}$ was ceased after 20 years to assess recovery of stem growth over the following 10 years. Tree growth initially responded positively to all nitrogen treatments, but the long-term response depended on the rate of nitrogen addition related to the availability of mineral nutrients. However, no growth increase was obtained under the highest nitrogen addition. Under the lowest nitrogen addition, 100 $m^3$ $ha^{-1}$ more stem wood was yielded than in the control. Högberg et al. (2006) assumed that the stem wood density was 500 kg $m^{-3}$, with a carbon content of 50%. This yielded a net carbon gain of 25,000 kg C $ha^{-1}$ under the cumulative nitrogen input of 1020 kg N $ha^{-1}$ (30 years × 34 kg N $ha^{-1}$ $yr.^{-1}$). Consequently, the net gain of carbon in stem wood was 25 kg of carbon per kilogram of nitrogen.

Based on the findings of Hyvönen et al. (2008) and Högberg et al. (2006) and findings from other boreal forest studies, de Vries et al. (2009) concluded that long-term (10–30 years) nitrogen fertilization increases growth in nitrogen-limited forests, but the responses are limited to an average of 20–25 kg of carbon in the trees and 10–25 kg of carbon in the soil per kilogram of nitrogen. This addition of carbon to the ecosystem due to nitrogen addition is comparable to (high) nitrogen deposition (below 50 kg N $ha^{-1}$ $yr.^{-1}$). However, Högberg et al. (2006) found that the higher doses of nitrogen reduced growth.

## 11.4 Impact of Fertilization on Growth of Ground Vegetation and Soil

### *11.4.1 Carbon Uptake in Ground Vegetation Under Fertilization*

Mäkipää (1994) investigated how nitrogen fertilization affects the growth of biomass and carbon in ground vegetation in sites of varying fertility. The experiments were established during 1958–1962 in 36–63-year-old Scots pine and Norway spruce stands in sites of different fertility. The experimental plots were fertilized with nitrogen five to seven times over a 30-year period, the total nitrogen addition being 600–900 kg $ha^{-1}$. The study sites were mainly dominated by Scots pine, except for sites of *Oxalis-Myrtillus* site type (OMT), which were occupied by Norway spruce. At the beginning of the experiment, the tree stands were young in the phases of pre-commercial and first commercial thinning.

Under no fertilization, the mass of mosses (e.g., *Hylocomium splendens*, *Pleurozium schreberi*) and lichens represented more than half of the total mass of the ground vegetation regardless of site fertility (Mäkipää 1994) (Fig. 11.8). The dominance of mosses and lichens declined substantially in the fertilized plots, as also found by Strengbom and Nordin (2008) for boreal forests. However, the effect of nitrogen fertilization on the mass of the field layer dominated by inherently slow-growing dwarf shrubs (e.g., *Vaccinium myrtillus*, *Vaccinium vitis-idaea*, *Calluna vulgaris*, *Linnea borealis*) was nonexistent on fertile sites. However, nitrogen addition increased the mass of *Vaccinium vitis-idaea* but reduced that of *Calluna vulgaris* in poor sites. Concurrently, the biomass was low in grasses and marginal in herbs. Regardless of site fertility, nitrogen addition substantially reduced the mass of

**Fig. 11.8** Effects of nitrogen fertilization on the distribution of biomass in ground vegetation in boreal upland sites of varying fertility (Mäkipää 1994, 1995). Courtesy of Finnish Society of Forest Sciences. Legend for symbols below the columns: 0 = not fertilized; N = fertilized; 113 = fertile site (*Oxalis-Myrtillus* site type, OMT); 26 = moderately fertile site (*Myrtillus* site type, MT); 52 = moderately poor site (*Vaccinium* site type, VT); and 157 and 106 = poor sites (*Calluna* site type, CT)

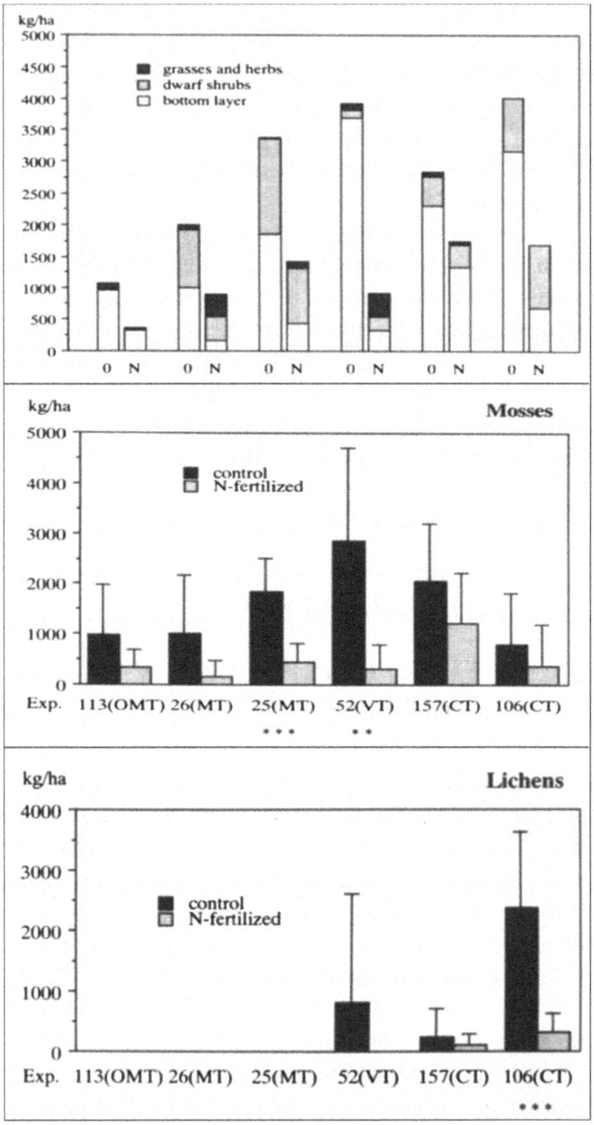


mosses and lichens relative to the total mass of ground vegetation. This held also for dwarf shrubs in fertile sites (e.g., *Vaccinium myrtillus* and *Vaccinium vitis-idaea*).

The fertilization distributes ecosystem carbon from ground vegetation based on the litter and humus layers on soil surface and further contributes carbon in the soil profile down to the depth of 30 cm included in soil samples (Mäkipää 1995). She also found that nitrogen addition increased the amount of carbon in the humus layer (including litter and the humus layer above the mineral soil) and mineral soil based on samples representing these layers from 0 to 10 cm and 10 to 30 cm, respectively.

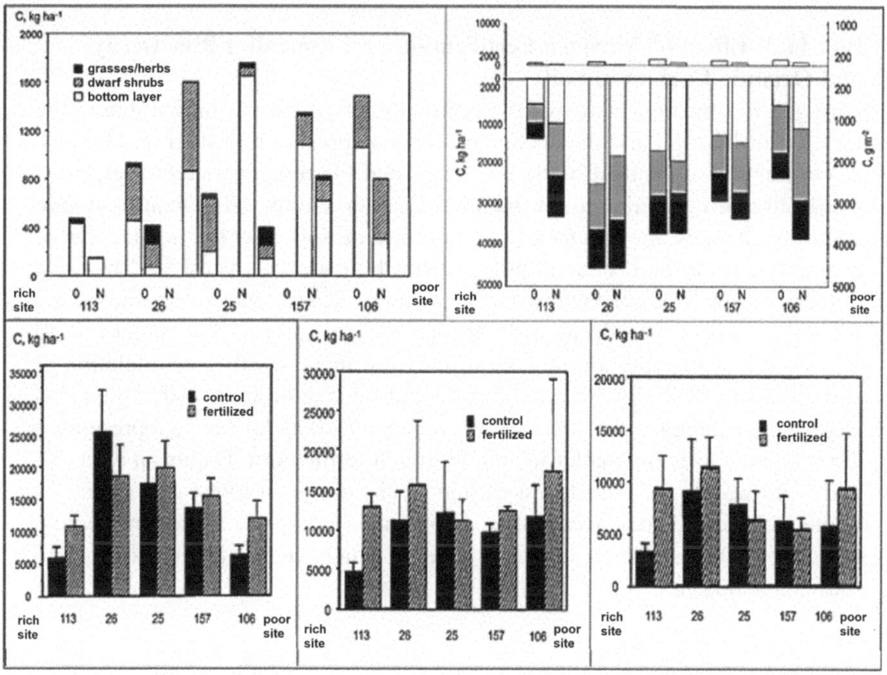

**Fig. 11.9** Carbon uptake in ground vegetation and soil under fertilization (Mäkipää 1994,1995). Permission of Elsevier. Legend for symbols below the columns: 0 = not fertilized; N = fertilized; 113 = fertile site (*Oxalis-Myrtillus* site type, OMT); 25 and 26 = moderately fertile sites (*Myrtillus* site type, MT); 52 = moderately poor site (*Vaccinium* site type, VT); and 157 and 106 = poor sites (*Calluna* site type, CT). Left/upper: effects of nitrogen fertilization on the carbon in ground vegetation. Right/upper: effect of nitrogen fertilization on carbon in the ground vegetation, humus (0–10 cm), and soil profile (10–30 cm), with carbon in mineral soil excluding that in fine roots. Left/lower: effect of fertilization on the amount of carbon in humus layer. Middle/lower: effects of fertilization on carbon in the upper soil layer (0–10 cm). Right/lower: effect of fertilization on humus in the deep soil profile (10–30 cm)

Excluding the site 26, the carbon stock increased by 14–87% in the humus layer and by 15–167% in the mineral soil (Fig. 11.9). The amount of carbon in the humus layer on the fertilized plots was 1900–5530 kg ha$^{-1}$ higher than in the unfertilized plots, depending on site fertility. Similarly, nitrogen addition increased the carbon stock in mineral soil from 8400–20,300 kg ha$^{-1}$ up to 17,400–27,000 kg ha$^{-1}$. Combining the carbon in humus and mineral soil down to 30 cm yielded the total soil carbon, which was 19,000–35,530 kg C ha$^{-1}$ affected by main mechanisms referred in Box 11.3.

**Box 11.3 Effect of Nitrogen Fertilization on Litterfall, Litter Decay, and Organic Carbon in Soil**

Nitrogen fertilization increases the accumulation of carbon in soil due to the increase of litterfall and the decline of the decomposition of litter and humus. According to Smolander (2018), the increase of soil organic matter is related to the decline of lignin due to nitrogen fertilization. On the other hand, nitrogen addition changes the microbe communities in soil due to the decline of available carbon and, concurrently, of the biomass of microbes. This also reduces the biomass of ectomycorrhizal fungi. Consequently, the increasing litterfall and declining decay of litter increase soil carbon. Smolander (2018) estimated that the addition of 1 kg of nitrogen increases the accumulation of carbon in soil by 10 kg. This implies that nitrogen addition by 100 kg $ha^{-1}$ is expected to increase the amount of carbon by 1000 kg $ha^{-1}$, representing 2000 kg $ha^{-1}$ organic matter in soil. Nitrogen fertilization is thus an effective and rapid way to increase forest carbon stocks in boreal forest ecosystems in moderately fertile sites occupied by Scots pine or Norway spruce in the pole and early mature phases of development, which are highly responsive to nitrogen addition.

### *11.4.2 Accumulation of Organic Matter and Humus in Soil Under Fertilization*

Nitrogen fertilization tends to increase tree growth in boreal and temperate forests. Concurrently, the carbon in soil increases. This is the case especially in the upper soil, but even further down through the soil profile, where the nitrogen stock is likely low (Johnson and Curtis 2001) (Fig. 11.10). Nitrogen fertilization increases whole-tree growth, with a consequent increase in litterfall, thus increasing carbon storage in the forest ecosystem. Mäkipää (1994, 1995) and Mäkipää et al. (1998) found that nitrogen fertilization increased the amount of organic matter in humus, except in sites of *Myrtillus* type (MT) (Fig. 11.10). The amount of organic matter increased from 3200 to 10,550 kg $ha^{-1}$, representing a 14–115% increase compared to unfertilized control sites. On the other hand, nitrogen fertilization tends to reduce the $CO_2$ efflux from soil, likely due to the declining decay rate of organic matter (Mäkipää 1994; Bowden et al. 2019).

Previously, Martikainen et al. (1989) found that nitrogen fertilization reduced the rate of $CO_2$ emission from soil in a poor site (*Calluna* site type, CT) occupied by Scots pine. In this case, the nitrogen fertilizers contained urea or ammonium nitrate with lime. A similar reduction of soil $CO_2$ efflux also held for more fertile sites (*Empetrum-Calluna* site type, ECT). However, fertilizing with rapidly soluble nitrogen fertilizers increased the $CO_2$ production in moderately fertile sites (*Myrtillus* site type, MT), as one may expect based on the increased nitrogen content in soil organic matter (Jandl et al. 2007). Similarly, Nohrstedt (1990) studied the

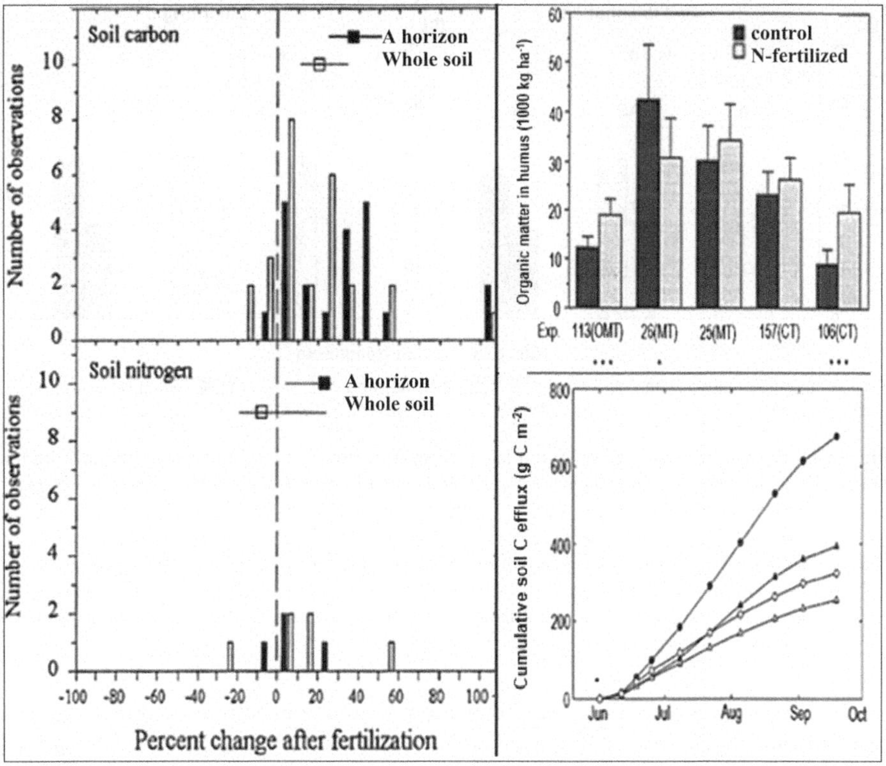


**Fig. 11.10** Effect of nitrogen fertilization on soil carbon and nitrogen, accumulation of carbon, and carbon efflux from soil. Left: distribution of the number of studies showing the effect of fertilization on soil carbon (top panel) and nitrogen (lower panel) in A horizons and whole soils (Johnson and Curtis 2001). Permission of Elsevier. Right/upper: amount of organic matter in the humus layer as a function of site fertility and nitrogen fertilization (Mäkipää 1994). Courtesy of Finnish Society of Forest Science. The numbers linked to site type (*Myrtillus* site type, MT, and *Calluna* site type, CT) refer to the number of experimental sites. Right/lower: cumulative soil carbon efflux [g C $m^{-2}$] in an experiment that combined nitrogen fertilization with girdling of trees (Olsson et al. 2005). Permission of Wiley

properties of soil and soil water 6 years after the fertilization of mature Scots pine (*Pinus sylvestris*) in an old field experiment in boreal conditions. Fertilizer containing ammonium nitrate was applied three times over a 20-year period. In the nitrogen application, the highest dose was 600 kg N $ha^{-1}$, i.e., a total of 1800 kg N $ha^{-1}$ over the period. Nohrstedt (1990) found a high accumulation of organic matter in the forest floor related to fertilization, i.e., a doubling of carbon stock in the forest floor from 5 to 9 Mg C $ha^{-1}$. The response was related to the increased growth, with a consequent increase of litterfall. The declining decay of soil organic matter further increased the accumulation of organic matter in soil.

Soil surface effluxes are related to autotrophic (e.g., root growth and maintenance respiration) and heterotrophic (e.g., decay of organic matter) respiration, both

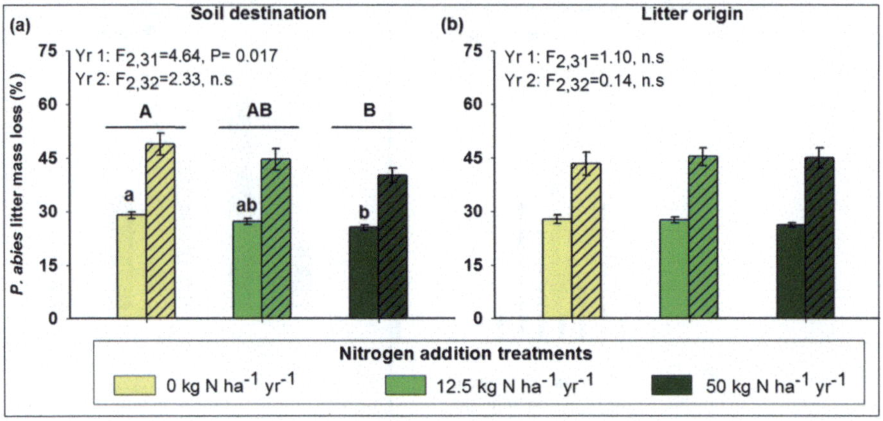


**Fig. 11.11** The mean (±SE) mass loss of Norway spruce (*Picea abies*) litter after decomposing for 1 (open bars) or 2 (hatched bars) years in three soil destinations (a) or three litter origins (b). The soil destinations and litter origins consisted of replicated plots ($n = 5$) treated with three different N addition levels (0, 12.5, and 50 kg N ha$^{-1}$ yr.$^{-1}$) (Maaroufi et al. 2017). Open Access. Courtesy of Nature Portfolio. Capital letters (A or B) on top of each panel group of two bars indicate significant differences between treatments regardless of time, while different lowercase letters (a or b) on top of each group of bars indicate significant differences between treatments determined using Student-Newman-Keuls post hoc tests

affected by nitrogen fertilization. To make a distinction between autotrophic and heterotrophic respiration, Olsson et al. (2005) fertilized and girdled 40-year-old Norway spruces in a moderately fertile site alone or in combination, making a comparison to untreated trees (Fig. 11.10). They found that the total soil $CO_2$ efflux in non-girdled plots in late summer was 40% lower in the fertilized than in non-fertilized plots. At the same time, "the efflux in girdled fertilized and non-fertilized plots was 50% and 60% lower, respectively, than in the corresponding non-girdled controls." Olsson et al. (2005) showed that both heterotrophic and autotrophic soil respiration per unit area were significantly lower in fertilized plots, even though the aboveground production was three times higher in fertilized than in non-fertilized plots. Long-term nitrogen fertilization seemed to reduce decay, nitrogen mineralization, and carbon dioxide emission from soil, thus prolonging carbon retention in soil (Titus and Malcolm 1987). Nitrogen addition likely increases the soil sequestration of carbon by 15–40 kg per kg of nitrogen when the effect of increasing litterfall on carbon sequestration is excluded (Framstadt et al. 2013).

Maaroufi et al. (2017) also investigated how nitrogen fertilization affects the decay rate of litter from boreal Norway spruce (*Picea abies*) (Fig. 11.11) in a 2-year reciprocal transplant experiment. The needle litter originated from a fertilization experiment established 17 years earlier. The measurements dealt with the needles from unfertilized control plots (ambient) and the needles from fertilized plots with low and high nitrogen addition, i.e., 12.5 and 50 kg N ha$^{-1}$ yr.$^{-1}$. According to Maaroufi et al. (2017),

- Litter quality had no clear effect on decay rate, while changes in soil microbiota had a strong effect on decay rate under nitrogen addition regardless of litter quality.
- Decomposition declined in response to nitrogen addition at a rate of 2.14% and 4.34% per year under low and high nitrogen addition, respectively.
- Changes in soil factors were the main pathway through which nitrogen affected litter decay. On the plots with high nitrogen addition, the decay rate decreased about 15% compared to that in the control plots, regardless of litter origin.
- Litter decay rate was correlated with the soil microbiota. Litter decay was positively correlated with *Gram-negative bacteria* (bacteria that do not retain the crystal violet stain used in the *Gram* staining method of bacterial differentiation) and fungal functional groups.
- Increasing accumulation of soil carbon in response to nitrogen addition (fertilization, atmospheric deposition) was induced by the changes in soil microbiota rather than changes in litter quality. This holds for previous studies that show nitrogen fertilization to increase the accumulation of soil carbon in boreal upland sites.

### 11.4.3 Response of Carbon in the Soil Profile to Nitrogen Addition

Carbon in soil profile includes carbon in the roots and in the humus particles mixed with mineral soil particles below the litter and in the humus layers above the mineral soil. The major flux of carbon in mineral soil originates from fine roots distributed through the humus layer (organic layer) and the upper parts of soil. Experimental findings for boreal Norway spruce (*Picea abies*) show that the living fine root density in the organic layer is more than three times higher than in the upper parts of mineral soil of moderate fertility (Leppälammi-Kujansuu et al. 2013a, b) (Fig. 11.12). The density of dead roots follows a similar distribution to that of living fine roots. Nitrogen fertilization further increased the density of living and dead fine roots compared to that in unfertilized plots. Furthermore, soil warming and irrigation increase root density, especially in the organic layer but also in deeper layers of the mineral soil.

The density of fine roots is related to root longevity, which may vary according to the moisture and thermal conditions in soil. Leppälammi-Kujansuu (2014) used the Kaplan-Meier survival curve to define the fraction of fine roots living for a certain amount of time under treatment. The survival curve is defined as the probability of surviving roots for Norway spruce in each time interval when considering time in many small intervals, as shown in Fig. 11.13. The median and mean longevity of fine roots in weeks were estimated by using a regression model (Weibull distribution).

According to Leppälammi-Kujansuu et al. (2014), the median survival of fine roots was under warming with fertilization and irrigation (WFI) was 48 weeks and under warming and irrigation (WI) 58 weeks, respectively (Fig. 11.13). When excluding warming, the survival fine root was 78 weeks under fertilization and

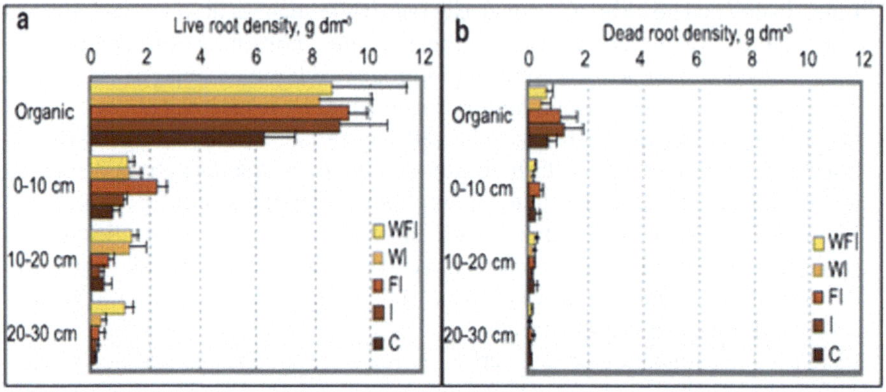


**Fig. 11.12** Distribution of living and dead roots of Norway spruce (*Picea abies*) throughout the soil profile under varying treatments at the Flakaliden experimental site in northern-boreal Sweden (64° N) (Leppälammi-Kujansuu et al. 2013a, b). Open Access. Courtesy of Springer Nature. Legend: WFI, warmed-fertilized-irrigated; WI, warmed-irrigated; FI, fertilized-irrigated; I, irrigated; C, control. Error bars represent the standard error of the treatment mean. *Left*: density of living roots (**a**). *Right*: density of dead roots (**b**) (<2 mm in diameter)

irrigation (FI) and 110 weeks under irrigation (I). Under irrigation (I), the litter production above and below the ground was close to each other: 1341 and 1019 kg ha$^{-1}$ yr.$^{-1}$ (Fig. 11.13) (Leppälammi-Kujansuu et al. 2014). Compared to that under irrigation, the belowground litter increased substantially under fertilization and irrigation (FI) or under warming and irrigation (WI), with the values of 2952 and 2900 kg ha$^{-1}$ yr.$^{-1}$. Under warming, fertilization, and irrigation (WFI), the belowground litter yield increased further up to 4492 kg ha$^{-1}$ yr.$^{-1}$. Regarding treatment, the concurrent changes in aboveground litter production were smaller than those belowground. The aboveground litter fall was the largest when the treatment combined warming, fertilization, and irrigation (WFI) or fertilization was combined with irrigation (FI), with litter yield of 2624 and 2448 kg ha$^{-1}$ yr.$^{-1}$. Fertilization alone or combined with other treatments affected aboveground litter production more than other treatments. Under warming and irrigation, the litter yield was 1338 kg ha$^{-1}$ yr.$^{-1}$, which was a half of that if nitrogen addition was combined with other treatments. Regardless of treatment, the needle litter represented 86–92% of the aboveground litter (Leppilammi-Kujansuu et al. 2014).

## 11.5 Growth Response to Fertilization Combined with Thinning and Genotype Improvement

### *11.5.1 Response to Repeated Fertilization*

Growth in boreal and cold temperate forests is, with few exceptions, limited by scarce availability of nitrogen. One way to satisfy increased growth demands in the

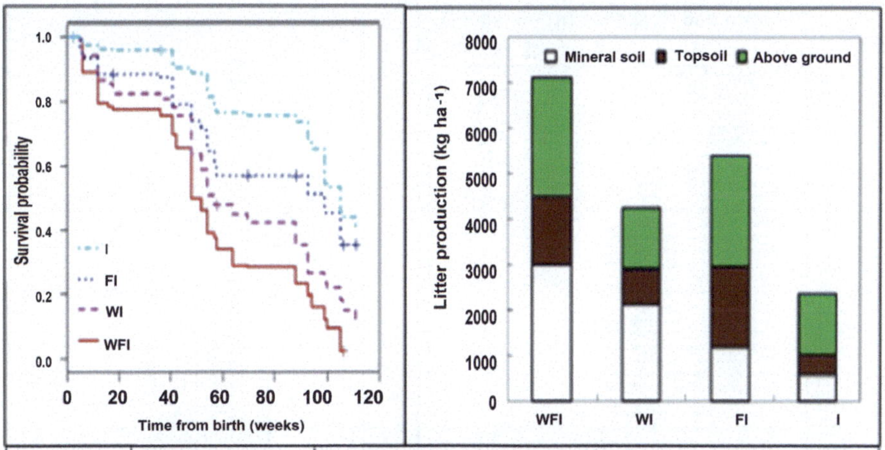


| Treatment | Root longevity, median in weeks ± se | Litter production, g carbon $m^{-2}$ $yr^{-1}$ | | | | |
|---|---|---|---|---|---|---|
| | | Above ground | Layers, below ground | | | Total (above + below ground) |
| | | | 0 - 5 cm | 5 - 25 cm | Sum | |
| WFI | 48±2 | 131 | 75 | 150 | 225 | 356 |
| WI | 58±2 | 67 | 39 | 106 | 145 | 212 |
| FI | 78±8 | 122 | 89 | 59 | 148 | 270 |
| I | 110±7 | 67 | 22 | 29 | 51 | 118 |

**Fig. 11.13** Survival and litter production of fine roots as a function of warming and irrigation at the Flakaliden experimental site in northern-boreal Sweden (64° N) (Leppälammi-Kujansuu et al. 2014). Permission of Springer Nature. Treatments included (i) irrigation (I); (ii) nitrogen fertilization and irrigation (FI); (iii) soil warming and irrigation (WI); (iv) and soil warming, nitrogen fertilization, and irrigation (WFI). Left/upper: survival probabilities of fine roots of Norway spruce as a function of the age of roots from June 11 to October 8, 2010, at 0–25-cm depth in the soil profile. Right/upper: annual belowground (mineral soil: 5–25 cm in the soil profile; topsoil: 0–5 cm upper soil layer) and aboveground litter production. Below: values of longevity of fine roots and annual yield of litter production above- and belowground

short term is nitrogen fertilization. Large-scale experiments on the impacts of nitrogen fertilization alone and in combination with thinning have been conducted in Sweden dating back to the early 1960s (Fig. 11.14). Bergh et al. (2014) analyzed the measurements of these experiments, which included 34 Scots pine sites and 13 Norway spruce sites. The experimental plots included four treatment plots: (i) no thinning, (ii) repeated light thinning, (iii) repeated light thinning with repeated N fertilization, and (iv) repeated light thinning with repeated N + P fertilization. In northern Sweden, 100 kg N $ha^{-1}$ and 150 kg N $ha^{-1}$ were added per each fertilization event for Scots pine and Norway spruce stands, respectively. In southern Sweden, 150 kg N $ha^{-1}$ nitrogen was added in Scots pine stands and 200 kg N $ha^{-1}$ nitrogen in Norway spruce stands. Phosphorus was added at a rate of 100 kg $ha^{-1}$ in fertilization event. Furthermore, several sites also included non-thinned fertilized plots.

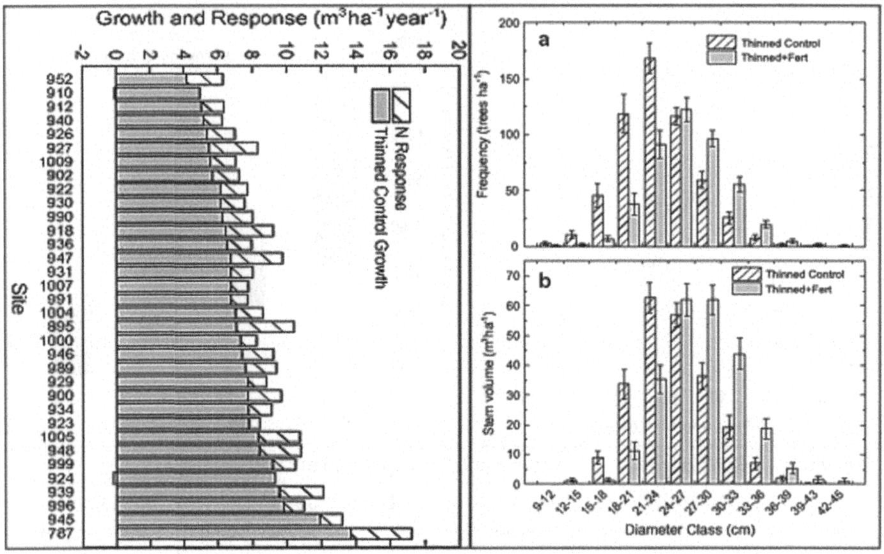


**Fig. 11.14** Response of volume growth to nitrogen fertilization and thinning (Bergh et al. 2014). Permission of Elsevier. Left: response of annual growth to nitrogen fertilization in 34 thinned Scots pine stands. Right: effects of thinning and nitrogen fertilization alone in Scots pine stands. Figure a: Distribution of frequency of trees in different diameter classes of Scots pine. Figure b: Distribution of volume of trees in different diameter classes of Scots pine

According to Bergh et al. (2014), Scots pine unlike Norway spruce was responsive (up to 25% more growth depending on the attribute assessed) to repeated fertilization. On the other hand, the response in non-thinned Scots pines stands continued vigorously to fertilization throughout the 30-year observation period, yielding a higher cumulative volume response than in the thinned stands. The volume growth in thinned stands also continued in response to fertilization, but the growth slowly declined over time. This showed that fertilization and thinning effects were not additive on volume growth (Bergh et al. 2014). Fertilization accelerated stand development in such a way that diameter distributions shifted to larger and potentially more valuable trees. Bergh et al. (2014) concluded that repeated nitrogen fertilization will produce significant and sustained increases in the growth of Scots pine and Norway spruce stands. The findings of Jacobson and Pettersson (2001) support this conclusion. No serious nutrient deficiencies were found under long-term and repeated nitrogen fertilization.

### *11.5.2 Response to Combined Thinning and Fertilization*

In even-aged management, fertilization is widely used together with thinning because increased spacing yields a larger growth response per individual tree than that under fertilization excluding thinning. The appropriate combination of thinning and fertilization increases carbon sequestration and biomass. Jörgensen et al. (2021) investigated how to increase carbon sequestration under regular even-aged management of Scots pine (*Pinus sylvestris*) involving thinning and nitrogen fertilization in field experiments representing a 1300-km-long south-north temperature gradient across Sweden. In 29 forest sites, Jörgensen et al. (2021) studied the interactive effects of fertilization and thinning over a 40-year period on carbon accumulation in standing biomass and the organic horizon (Fig. 11.15).

Jörgensen et al. (2021) found that fertilization without thinning increased the total carbon stock by 50%. The increase was up to 14%, on average, when carbon in the harvested timber was included in the total ecosystem carbon pool. Similarly, fertilization combined with thinning increased carbon stocks regardless of the inclusion (11%) or exclusion (12%) of the biomass removed in thinning. Under no thinning, fertilization increased the total carbon stock by 79%, while the increase was 35% when the removed timber was included in the carbon stock. According to Jörgensen et al. (2021), fertilization increased carbon stocks along a temperature gradient. The relative increase was most evident at high latitudes, representing temperature-limited growth. At the same time, fertilization reduced the soil respiration rate. These findings show that nitrogen fertilization enhances the belowground carbon sequestration by increasing the carbon input into the soil profile and by slowing the decay rate of soil organic matter, thus increasing the total carbon sequestration under management combining timber production and climatic-warming mitigation (Fig. 11.15).

### *11.5.3 Genotype Improvement Combined with Fertilization*

The effects of single management operations, such as forest reforestation with improved seedlings with higher growth or fertilization, on carbon sequestration are evident. Their combined impacts are still poorly understood. Routa et al. (2019) used model simulations to identify how combined genotypic improvement and nitrogen fertilization may affect the growth of Scots pine and Norway spruce and the economic profitability of timber production under the current and warming climate (climate forcing scenario RCP 2.5). The simulations were for moderately fertile sites (*Myrtillus* site type, MT) in mid-boreal conditions in Finland (62° N). The rotation lengths were 50, 60, or 70 years. Nitrogen fertilization with 150 kg N $ha^{-1}$ was conducted once or twice during a rotation. The growth was assumed to be 10% and 20% higher for improved seedlings than for not improved seedlings (Fig. 11.16).

Under the current climate, the baseline management with unimproved seedlings and/or nitrogen fertilization yielded mean growth of 5.6, 5.7, and 6.3 $m^3$ $ha^{-1}$ $yr.^{-1}$ for rotations of 50, 60, and 70 years, respectively (Fig. 11.16). When using improved seedlings alone without fertilization, the growth increased by 10% and 20% (BG 10% and 20%), increasing the timber yield up to 12% and 22% over the rotation relative to baseline management. The use of nitrogen fertilization alone, once, or

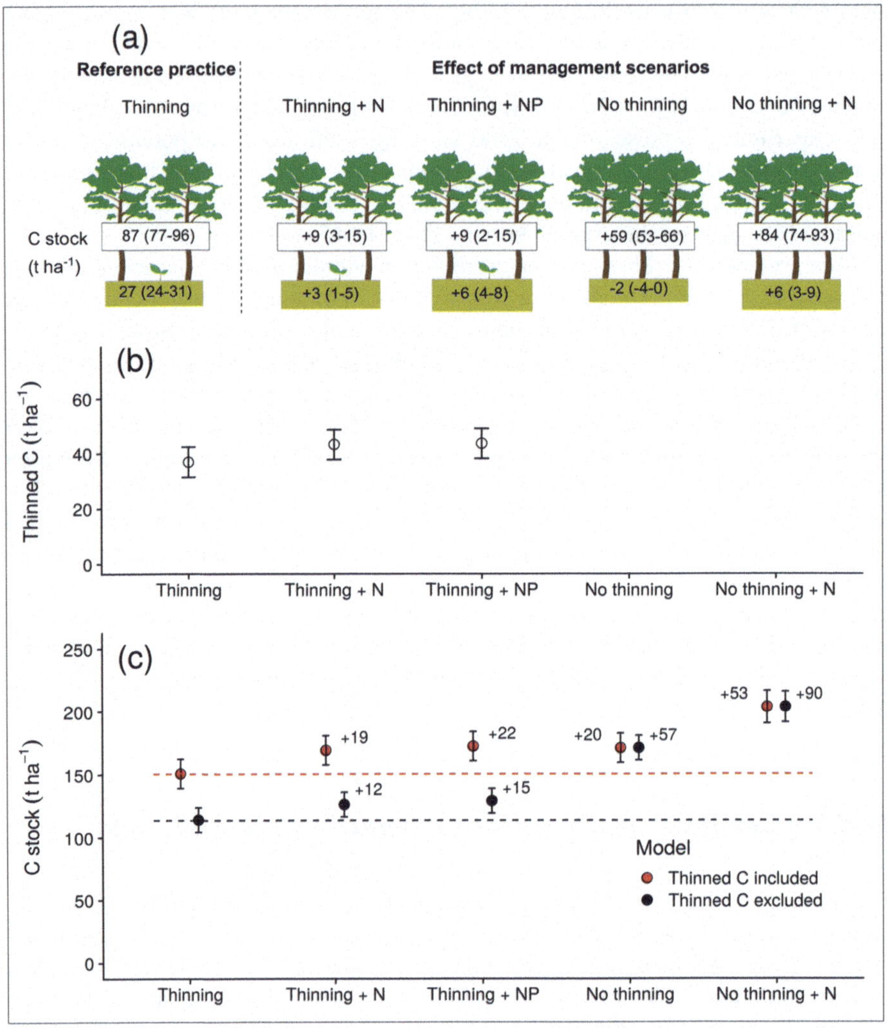


**Fig. 11.15** Effects of nitrogen fertilization and thinning on growth of Scots pine (*Pinus sylvestris*), with the average total nitrogen addition corresponding to 881 kg N $ha^{-1}$ over 40 years (Jörgensen et al. 2021). Open Access. Courtesy of Springer Nature. Legend: red symbols represent total carbon stocks that include removed timber, and black symbols standing carbon stocks without removed timber. Dashed lines show thinning excluding fertilization. Points represent adjusted means, and numbers in parentheses (**a**) and error bars (**b**, **c**) indicate confidence intervals of 95%. Upper (**a**): carbon stocks in trees and soil subjected to thinning including/excluding fertilization. Middle (**b**): carbon removed in thinning. Lower (**c**): ecosystem carbon stocks under different management, with the effects on net carbon sequestration indicated by the numbers beside the points

during the rotation increased the timber yield up to 7% compared to baseline management. The combined use of improved seedlings (BG 20%) and nitrogen fertilization once or twice during the rotation yielded up to a 26–28% increase of

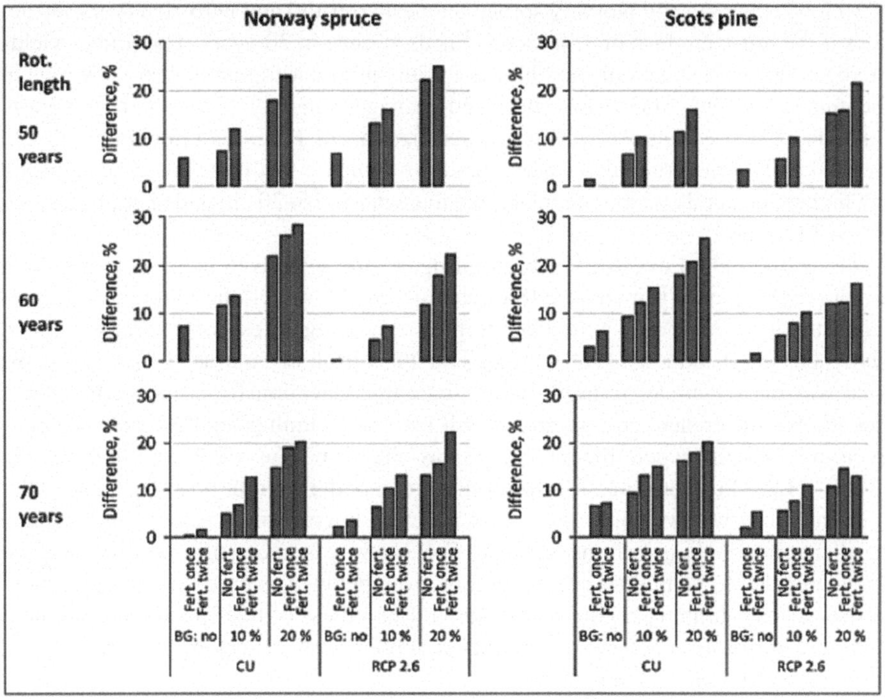


**Fig. 11.16** Effect of improved seedlings of Norway spruce and Scots pine in forest reforestation and nitrogen fertilization on timber production. The rotation length was 50, 60, or 70 years under the current climate (CU) and warming climate (RCP 2.6) (Routa et al. 2019). Permission of Oxford University Press. The growth increase (BG, breeding gain) was assumed to be 10, and 20% higher for improved seedlings higher than for unimproved seedlings

timber yield. Concurrently, the net present value (NPV using the discount rates of 3%) increased up to 32–60% over the 60- and 70-year rotations regardless of tree species or climate. The use of improved seedlings affected timber yield and NPV more than nitrogen fertilization. Furthermore, minor climate change increased timber yield and NPV in Scots pine but not in Norway spruce (Routa et al. 2019).

## 11.6 Fertilization in Management

Nitrogen fertilization is an effective way to increase forest growth under boreal conditions, where a short supply of nitrogen substantially limits forest growth in upland sites. According to Pukkala (2017), the increase of volume growth increases the timber value two to three times higher than the fertilization cost. In model simulations, Pukkala (2017) found that timber yield likely increases 1 $m^3$ $ha^{-1}$ $yr.^{-1}$ (15%) for Scots pine on sub-xeric (poor) sites, versus 0.5 $m^3$ $ha^{-1}$ $yr.^{-1}$

(5%) for Scots pine and Norway spruce on mesic (fertile) and herb-rich (very fertile) sites if the nitrogen dose of 150 kg $ha^{-1}$ is used once in 30 years. The timber yields increased most in stands of medium age, with an increasing amount of saw logs in the future cuttings. Based on optimized management with nitrogen fertilization, Pukkala (2017) concluded: "Nitrogen fertilization of boreal conifer forest should be used mainly in spruce-dominated stands growing on medium sites. The gains are the highest in stands where the mean tree diameter is 16–20 cm and stand basal area is 14–20 $m^2$ $ha^{-1}$."

On an individual tree scale, a high nitrogen supply tends to increase tree growth when growing space is increased through thinning. However, the stand-scale growth tends to decrease along with declining stocking regardless of thinning regime. Hynynen and Kukkola (1989) found that the absolute volume growth of Scots pine was highest under thinning with single-tree selection from above. However, the relative differences in volume growth between thinning regimes were small or even nonexistent when the stocking was the same for each thinning regime (Table 11.2). Furthermore, fertilization increased the volume growth most under thinning from below, while there was no clear interaction between thinning and fertilization affecting volume growth. For Norway spruce, the absolute volume growth tended to be highest when thinning from below was used but the differences between other thinning regimes were small. Regardless of tree species and thinning regimes, fertilization seemed to produce a similar growth response when the post-thinning basal area was the same.

**Table 11.2** Volume growth of Scots pine in a poor site (*Vaccinium* site type, VT) and Norway spruce in a very fertile site (*Oxalis-Myrtillus* site type, OMT) in thinned stands without and with nitrogen fertilization in southern boreal conditions modified from Hynynen and Kukkola (1989). Thinning with single trees refers to the removal of the largest trees. In the experiments 541–543, the fertilization was 180 kg N $ha^{-1}$, while for experiment 544, fertilization was 232 kg N $ha^{-1}$ plus 62 kg P $ha^{-1}$

| Tree species and site fertility | Thinning or selection in spacing | Growth after thinning under no fertilization, % of that under thinning from below | Growth after thinning under fertilization, % of that under thinning from below |
|---|---|---|---|
| Scots pine<br>• Exp. 541 in site of VT type | Below<br>Above<br>Selection | 100<br>96<br>102 | 100<br>76<br>97 |
| • Exp. 542 in site of VT type | Below<br>Above<br>Selection | 100<br>108<br>112 | 100<br>110<br>94 |
| Norway spruce<br>• Exp. 543 in site of OMaT type | Below<br>Above<br>Selection | 100<br>83<br>85 | 100<br>104<br>90 |
| • Exp. 544 in site of OMaT type | Below<br>Above<br>Selection | 100<br>99<br>96 | 100<br>86<br>80 |

## 11.7 Concluding Remarks

In boreal forests, nitrogen fertilization enhances the growth and development of trees, with decelerating growth toward full maturity over the tree life cycle. However, this may reduce the lifespan of trees inducing gradual reduction of growth due to full maturity reached earlier than under no fertilization. Fertilization with only nitrogen may also unbalance the nutrient cycle, especially in fertile sites, where the growth response is sensitive to the addition of phosphorus. Fertilization may further change the allometric mass relations between different organs, likely making trees susceptible to damage by excess wind and/or snow load. This is especially a case if thinning and fertilization are used together. In this respect, the unbalanced mass relations between different organs refer, for example, to the excess increase of foliage in Scots pine in relation to stem growth and strength. Nitrogen fertilization also tends to increase herbivore-induced damage due to the reduction of secondary compounds (e.g., Kytö et al. 1999). This makes plant material more palatable, thus increasing the damage caused by herbivores (e.g., moose).

## References


Ainsworth EA, Long SP (2005) What have we learned from 15 years of tree-air $CO_2$ enrichment (FACE)? A meta-analytic review of the responses of photosynthesis, canopy properties and plant properties to rising $CO_2$. New Phytol 165:351–372

Allison SD, Gartner TB, Marck MC, McGuire K, Treseder K (2010) Nitrogen alters carbon dynamics during early succession in boreal forests. Soil Biol Chem 42:1157–1164

Amponsah IG, Cameau PG, Brockley RP, Lieffers VJ (2005) Effects of repeated fertilization on needle longevity, foliar nutrition, effective leaf area index, and growth characteristics of lodgepole pine in interior British Columbia, Canada. Can J For Res 35:440–451

Balster NJ, Marshall JD (2000) Decreased needle longevity of fertilized Douglas-fir and grand fir in the northern Rockies. Tree Physiol 20:1191–1197

Bergh J, Nilsson U, Allen HL, Johansson U, Fahlvik N (2014) Long-term response of scots pine and Norway spruce stands in Sweden to repeated fertilization and thinning. For Ecol Manag 320: 118–128

Bowden RD, Rice A, Wang J-J, Wurzbacher SJ, Coble AE, Simpson M, Washko SE, Baldauf N, Lijtha K, Wind L, Johnson B (2019) Long-term nitrogen addition decreases organic matter decomposition and increases forest soil carbon. Soil Sci Soc Am J 83(S1):S82–S95

Brix H (1972) Nitrogen fertilization and water effects on photosynthesis and early-late-wood production in Douglas fir. Can J For Res 2:467–478

Brix H, Ebell LF (1969) Effect of nitrogen fertilization on growth, leaf area, and photosynthesis rate in Douglas fir. For Sci 15:189–196

de Vries W, Solberg S, Dobbertin M, Sterba H, Laubhann D, van Oijen M, Evans C, Gundersen P, Kros J, Wamelink GWW, Reinds GJ, Sutton MA (2009) The impacts of nitrogen deposition on carbon sequestration by European forests and heathlands. For Ecol Manag 258:181–1823

Fagerström T, Lohm U (1977) Growth of scots pine (*Pinus sylvestris* L.): mechanism of response to nitrogen. Oecologia 26(4):305–315

Framstadt E, de Wit H, Mäkipää R, Larjovaara M, Vesterdal L, Karltun E (2013) Biodiversity, carbon storage and dynamics of northern forests. TemaNord 507:1–130

Haapanen T, Hari P, Kellomäki S (1977) Effect of fertilization and thinning on radial growth of scots pine. Silva Fennica 13(2):184–189

Högberg P, Fan H, Quist M, Binkley D, Tamm CO (2006) Tree growth and soil acidification in response to 30 years of experimental nitrogen loading on boreal forest. Glob Chang Biol 12: 489–499

Hynynen J, Kukkola M (1989) Harvennustavan ja lannoituksen vaikutus männikön ja kuusikon kasvuun. Summary: effect of thinning method and nitrogen fertilization on the growth of Scots pine and Norway spruce stands. Folia Forestalia 731. 20 p.

Hyvönen R, Persson T, Andersson S, Olsson B, Ågren GI, Linder S (2008) Impact of long-term nitrogen addition on carbon stocks in trees and soils in northern Europe. Biogeochemistry 89: 121–137

Jacobson S, Pettersson F (2001) Growth responses following nitrogen and N-P-K-mg additions to previously N-fertilized scots pine and Norway spruce stands on mineral soils in Sweden. Can J For Res 31:899–909

Jandl R, Lindner M, Vesterdal L, Bauwens B, Baritz R, Hagedorn F, Johnson DW, Minkkinen K, Byrne KA (2007) How strongly can forest management influence soil carbon sequestration? Geoderma 137:253–268

Johnson DW, Curtis PS (2001) Effects of forest management on soil C and N storage: meta-analysis. For Ecol Manag 140:227–238

Jonsson S (1978) Resultat från en tioårig försökserie med hög kvävegivor. Föreningen skogsädling. Inst. Skogsförbättring. Årsbok 1977

Jokela A, Back J, Huttunen S, Jalkanen R (1995) Excess nitrogen fertilization and the structure of scots pine needles. Eur J For Pathol 25:109–124

Jörgensen K, Granath G, Björn D, Lindahl BD, Strengbom J (2021) Forest management to increase carbon sequestration in boreal Pinus sylvestris forests. Plant Soil 466:165–178

Kellomäki S (2022) Management of boreal forests—theories and applications for ecosystem services. Springer Nature

Kellomäki S, Wang KY (1997a) Effects of long-term CO2 and temperature elevation on crown nitrogen distribution and daily photosynthetic performance of scots pine. For Ecol Manag 99: 309–326

Kellomäki S, Wang KY (1997b) Effects of elevated O3 and CO2 on chlorophyll fluorescence and gas exchange in scots pine during the third growing season. Environ Pollut 97(1–2):17–27

Kellomäki S, Puttonen P, Tamminen H, Westman CJ (1982) Effect of nitrogen fertilization on photosynthesis and growth in young scots pines: preliminary results. Silva Fennica 16(4):5169

Kukkola M, Saramäki J (1983) Growth response in repeatedly fertilized pine and spruce stands on mineral soils. Communicationes Instituti Forestalis Fenniae 114:1–155

Kytö M, Niemelä P, Annila E, Varama M (1999) Effects of forest fertilization on the radial growth and resin exudation of insect-defoliated scots pines. J Appl Ecol 36:763–769

Leppälammi-Kujansuu J, Aro L, Salemaa M, Hansson K, Berggren Kleja D, Helmisaari H-S (2013a) Fine root longevity and carbon input into soil from below- and aboveground litter in climatically contrasting forests. For Ecol Manag 326:79–90

Leppälammi-Kujansuu J, Ostonen I, Strömgren M, Nilsson LO, Kleja DB, Sah SP, Helmisaari H-S (2013b) Effects of long-term temperature and nutrient manipulation on Norway fine roots and mycelia production. Plant Soil 366:287–303

Leppälammi-Kujansuu J, Salemaa M, Berggren Kleja D, Linder S, Helmisaari H-S (2014) Fine root manipulation experiment. Plant Soil 374:73–88

Linder S (1995) Foliar analysis for detecting and correcting nutrient imbalance in Norway spruce. Ecol Bull 44:178–190

Maaroufi NI, Nordin A, Palmqvist K, Gundale MJ (2017) Nitrogen enrichment impacts on boreal litter decomposition are driven by changes in soil microbiota rather than litter quality. Sci Rep 7: 4083

Mäkipää R (1994) Effects of nitrogen fertilization on the humus layer and ground vegetation under closed canopy in boreal coniferous stands. Silva Fennica 28(2):81–94

Mäkipää R (1995) Effect of nitrogen input on carbon accumulation of boreal forest soils and ground vegetation. For Ecol Manag 79:217–226
Mäkipää R, Karjalainen T, Pussinen A, Kukkola M (1998) Effects of nitrogen fertilization on carbon accumulation in boreal forests: model computations compared with the results of long-term fertilization experiments. Chemosphere 36(4–5):1155–1160
Martikainen PJ, Aarnio T, Taavitsainen V-M, Päivinen L, Salonen K (1989) Mineralization of carbon and nitrogen in soil samples taken from three fertilized pine stands: long-term effects. Plant Soil 114:99–106
Matala J, Kellomäki S, Nuutinen T (2008) Litterfall in relation to volume growth of trees: analysis based on literature. Scand J For Res 23:194–202
Miller HG, Miller JD (1976) Effect of nitrogen supply on net primary production in Corsican pine. J Appl Ecol 13:233–248
Miller HG, Cooper JM, Miller JD (1976) Effect of nitrogen supply on nutrients in litterfall and crown leaching in a stand of Corsican pine. J Appl Ecol 13:249–256
Nave LE, Vance ED, Swanston CW, Curtis PS (2009) Impacts of elevated N inputs on north temperate forest soil C storage, C/N, and net N-mineralization. Geoderma 173:231–240
Nilsen P (2001) Fertilization experiments on forest mineral soils: a review of Norwegian results. Scand J For Res 16:541–554
Nilsen P, Abrahamsen G (2003) Scots pine and Norway spruce responses to annual N, P and mg fertilization. For Ecol Manag 174:221–232
Nohrstedt H-Ö (1990) Effects of repeated nitrogen fertilization with different doses on soil properties in a Pinus sylvestris stand. Scand J For Res 5:3–15
Nohrstedt H-Ö (2001) Response of coniferous forest ecosystems on mineral soils to nutrient additions: a review of Swedish experiences. Scand J For Res 6:555–573
Olsson P, Linder S, Giesler R, Högberg P (2005) Fertilization of boreal forest reduces both autotrophic and heterotrophic soil respiration. Glob Chang Biol 11:1745–1753
Pensa M, Sellin A (2002) Needle longevity of scots pine in relation to foliar nitrogen content, specific leaf area, and shoot growth in different forest types. Can J For Res 32:1225–1231
Pettersson F, Högbom L (2004) Long-term growth effects following forest nitrogen fertilization in Pinus sylvestris and Picea abies stands in Sweden. Scand J For Res 19:339–347
Pukkala T (2017) Optimal nitrogen fertilization of boreal conifer forest. Forest Ecosystems 4(3): 1–10
Reich PB, Walters MB, Tjoelker MG, Van Der Klein D, Buschena C (1998) Photosynthesis and respiration rates depend on leaf and root morphology and nitrogen concentration in nine boreal species differing in relative growth rate. Funct Ecol 12:395–405
Routa J, Kellomäki S, Peltola H, Asikainen A (2011) Impacts of thinning and fertilization on timber and energy wood production in Norway spruce and scots pine: scenario analyses based on ecosystem model simulations. Forestry 84(2):159–175
Routa J, Kilpeläinen A, Ikonen V-P, Asikainen A, Venäläinen A, Peltola H (2019) Effects of intensified silviculture on timber production and its economic profitability in boreal Norway spruce and scots pine stands under changing climatic conditions. Forestry 92:648–658
Saarsalmi A, Tamminen P, Kukkola M (2014) Effects of long-term fertilisation on soil properties in scots pine and Norway spruce stands. Silva Fennica 48(1):989
Smolander A (2018) Typpilannoitus metsämaan viljavuuden parantajana—kestävyysnäkökohtia maan ja ympäristön kannalta. Metsätieteellinen aikakauskirja 2018:10080
Strengbom J, Nordin A (2008) Commercial forest fertilization causes long-term residual effects in ground vegetation of boreal forests. For Ecol Manag 256:2175–2181
Titus BD, Malcolm DC (1987) The effect of fertilization on litter decomposition in clearfelled spruce stands. Plant Soil 100:297–322
Valinger E, Elfving B, Mörling T (2000) Twelve-year growth response of scots pine to thinning and nitrogen fertilization. For Ecol Manag 134:45–53
Yue K, Peng Y, Peng C, Yang W, Peng X, Wu F (2016) Stimulation of terrestrial ecosystem carbon storage by nitrogen addition: a meta-analysis. Sci Rep 6:19895

# Part IV
# Management of Boreal Forests for Timber Production and Climate Change Mitigation

# Chapter 12
# Carbon Management for Mitigating Climatic Warming

## Contents




**Abstract** Under even-aged management, varying management measures and harvest intensities are used for maintaining/enhancing the potentials of carbon sequestration in boreal forest ecosystems. At the stand/patch scale, carbon sequestration under even-aged management varies substantially depending on the species and age structure between tree stands. Regarding management and harvest, growth and development of tree stands in separate sites with varying site fertility involve the gradual replacement of one population/community of trees by other ones through a chronosequence. Similar replacement further occurs in the ground vegetation, with the replacement of grass- and herb-rich vegetation by the dominance of dwarf shrubs and mosses through the maturation of dominating trees. Combined with management and harvest, this successional development increases the complexity of habitats and diversity of organisms in forest ecosystems. This contributes to carbon sequestration from the scale of individual stands to the whole forest landscape. Under uneven-aged management, the production involves the selection interval and lifespan of different tree species at the individual-tree scale, contributing to the





S. Kellomäki, *Forest Management for Timber Production and Climate Change Mitigation*, Managing Forest Ecosystems 44,
https://doi.org/10.1007/978-3-031-71575-4_12

structural complexity of density and size of trees from the dominance of seedlings and young trees to limited density of mature trees.

## 12.1 Management of Forest Carbon at Tree, Stand, and Landscape Scale

### 12.1.1 Primary and Secondary Ecosystem Production for Ecosystem Services

In forest ecosystem, the stocks of living and nonliving assets in the ecosystem components form the natural capital providing ecosystem services and goods. This implies the benefits obtainable from natural ecosystems or managed ecosystems functioning properly under selected management (Bateman et al. 2011; Sing et al. 2015). The ecosystem goods and services refer to the use and nonuse of the output from ecosystem and services to the processes producing goods or processes usable as such. Over the forest succession, the flow of ecosystem goods and services changes related to the changes in the natural capital provided by the ecosystem. Management is used to control the structure and functioning of ecosystem to produce different goods and services aimed in management representing primary and secondary production.

The primary production (*GPP*, carbon fixation in populations/communities of trees and other plants) produces oxygen and converts photosynthates into organic matter in growth processes. Secondary production ($R_{eco}$, decay of organic matter in population/communities of heterotrophic species) controls the cycles of nutrients, carbon, and water locally in sites and globally in biosphere (Bateman et al. 2011; Sing et al. 2015). Sites with trees and other organisms are spatial units, the functioning of which produces provisioning services, such as wood and non-wood products. The structure of ecosystem further provides regulating and cultural services, such as the climate and flood regulation, esthetic values, and recreation environment. Primary and secondary production drive the sink/source dynamics of the carbon exchange between atmosphere and biosphere. Carbon sink/source dynamics maintains carbon sequestration and stocking in forest ecosystems, thus providing climatic benefits in cooling/warming of climate.

### 12.1.2 Effects on Carbon Sequestration and Stocks Through Production Cycle

Table 12.1 summarizes qualitative effects of selected management operations on the mean carbon stocks under even-aged management addressed previously. From the management point view, the most important are the links to carbon dynamics in a

**Table 12.1** Qualitative effects on mean carbon stocks due to selected management operations, with links to carbon dynamics in a forest ecosystem throughout a rotation compared with the rotation prior to management. The outlined model is based on the management recommendations of Hyvönen et al. (2007) regarding ecosystem carbon dynamics

| | Outline effects of management measures on carbon in forest ecosystem | | |
|---|---|---|---|
| Management measure | Soil carbon stock | Biomass carbon stock | Ecosystem carbon stock |
| **Site preparation for reforestation of clear-cut** | Decreasing | Decreasing, neutral, or increasing | Decreasing, neutral, or increasing |
| **Reforestation**<br>• Tree species choice<br>• Management of seedling stands<br>• Pre-commercial removal/cutting | • Increasing, neutral<br>• Neutral, decreasing<br>• Decreasing, neutral | • Increasing<br>• Increasing<br>• Increasing | • Increasing<br>• Increasing, neutral<br>• Increasing, neutral |
| **Commercial harvest**<br>• Thinning/even-aged<br>• Selection/uneven-aged | • Decreasing, neutral with stem-only harvest<br>• Decreasing with whole-tree harvest | • Decreasing, neutral with stem-only harvest<br>• Decreasing, neutral with whole-tree harvest | • Decreasing, neutral<br>• Decreasing |
| **Protection for pests**<br>• Sanitary cuts<br>• Protection for avoiding fire, wind, snow damages | • Neutral, increasing<br>• Neutral, increasing | • Increasing<br>• Increasing | • Increasing<br>• Increasing |
| **Soil management**<br>• Prescribed burning<br>• Mechanical site preparation<br>• Fertilization | • Neutral with low intensity, decreasing<br>• Decreasing with high intensity<br>• Increasing | • Increasing with low intensity, decreasing<br>• Increasing with high intensity<br>• Increasing | • Increasing with low intensity<br>• Decreasing, neutral with high intensity<br>• Increasing |

forest ecosystem throughout a rotation compared with the rotation prior to management (Hyvönen et al. 2007). In this context, trees and tree stands are carbon sinks when taking up carbon in net primary production more than losing carbon in ecosystem respiration (Hyvönen et al. 2007; Routa et al. 2012). Carbon uptake and loss are linked through litter and dead biomass, depending on the lifespan of biomass produced in net primary production. Litter and dead biomass are decomposed in soil, emitting carbon to the atmosphere at the ecosystem scale as part of the carbon cycle in trees and tree stands in varying phases of growth and development. There are several management options for enhancing the sequestration and storing of carbon in forest ecosystems, such as the use of nitrogen fertilizers, improved regeneration material (seeds/seedlings), high stocking density, and increased rotation length (Table 12.1). The potential of management to increase carbon sequestration through stand-based chronosequences depends on the age and size distribution of trees and tree stands and variability of site fertility in a forest landscape. This affects the temporal and spatial distribution of the future climate benefits related to the long-term sink/source dynamics of carbon in the ecosystem along with the successional changes in forest ecosystems and forest landscapes.

## 12.2 Carbon in Managed Forests at the Landscape Scale

### *12.2.1 Carbon Balance Across Climatic Gradients in Managed Forest Landscapes*

In Nordic countries, even-aged management, or rotation forestry, is the mainstream way to implement the sustainable management and the use of forest resources. Even-aged management combines the high structural variability in the stand-based stocking of biomass throughout rotation.

Concurrently, this implies a variability in sink/source dynamics, indicated by the variability in the carbon balance. This extends through stand-based chronosequences from final felling and regeneration toward the full maturity of trees and final cutting for the next rotation. Carbon balance varies over the landscape, determined by the temporal and spatial variability of young and old tree stands exposed to management and harvesting, controlling the dynamics of stand-based as well as landscape-based variability in carbon balance (Peichl et al. 2022).

The impacts of management and harvest on carbon sink/source dynamics have been widely investigated at the scale of individual stands, including both trees and soil separately in analyses. This emphasizes the effects of regenerative cutting (clear-cutting for planting/seeding or cutting for natural regeneration) on the net carbon losses based on individual sites. Peichl et al. (2022) concluded that "C sequestration occurs in both soil and biomass at higher rates in actively managed forests compared with low-intensive and unmanaged forests in the boreal region" (Fig. 12.1). These findings emphasize the importance of managed boreal forests in mitigating climatic warming, with their concurrent importance in producing timber and biomass for the manufacturing and energy industries.

Peichl et al. (2022) measured the carbon flux in 50 forest stands, mainly of Scots pine (*Pinus sylvestris*) and Norway spruce (*Picea abies*), ranging from 5 to 211 years in age. The measurements were conducted over 3 years (2016–2018) to identify the variability in the annual net ecosystem production (NEP, i.e., the ecosystem carbon balance) across a 68 $km^2$ of managed boreal forest landscape in northern Sweden (64° N) at an altitude range of 138–339 m a.s.l. The annual mean temperature in the area was 2.4 °C and the precipitation was 638 mm. The measurements of carbon flux were conducted with biometric- and chamber-based techniques (Fig. 12.1).

In the ecosystem scale, net primary production rather than heterotrophic respiration regulated the spatial and temporal variations of net ecosystem production (*NEP*) across the heterogeneous mosaic of the managed boreal forest landscape. Successional patterns of net ecosystem production (*NEP*) are relative to those of naturally

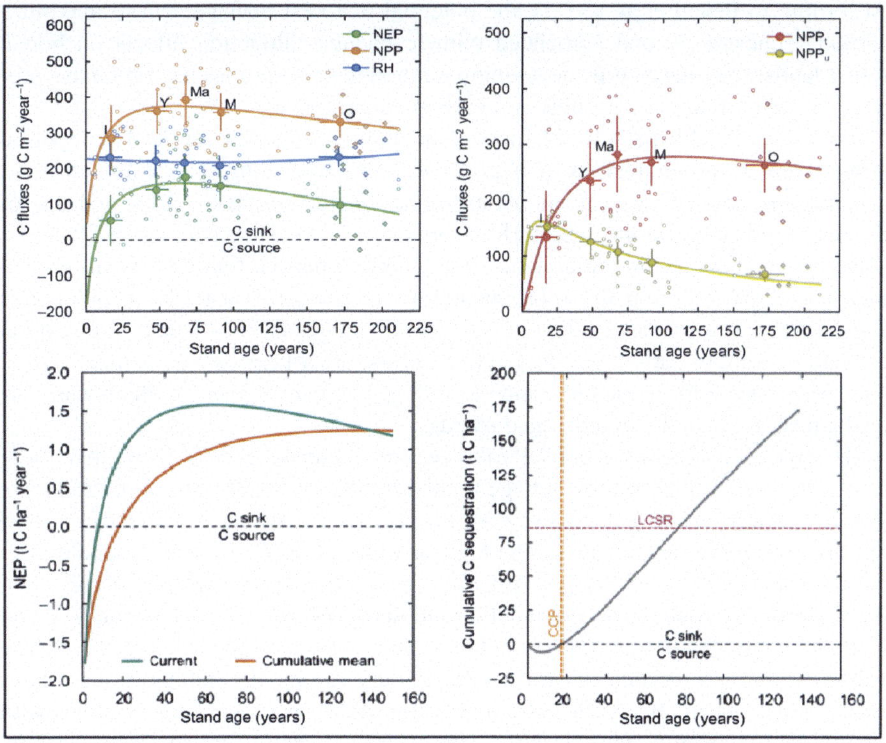


**Fig. 12.1** Main findings on the landscape variability of the carbon balance across a managed boreal forest landscape (Peichl et al. 2022). Open Access. Courtesy of Wiley. Left/upper: annual net ecosystem production (*NEP*) and its components in net primary production (*NPP*) and total heterotrophic respiration ($R_H$) over stand age. Circular open symbols indicate 3-year means (2016–2018) at each of the 50 main forest stands. Circular closed symbols are the means for each of the age classes "initiation (I)," "young (Y)," "middle-aged (Ma)," "mature (M)," and "old (O)." Filled diamonds indicate data for the seven additional old stands. Bars represent 95% confidence intervals. Lines for *NPP* and $R_H$ represent nonlinear best fits for the stand-level data. The fit line for *NEP* was computed as the difference between the lines of *NPP* and $R_{\mathrm{H}}$. Right/upper: annual net primary production of trees ($NPP_{\mathrm{t}}$) and understory ($NPP_{\mathrm{u}}$). Left/lower: current and cumulative mean net ecosystem production (*NEP*, Mg C ha$^{-1}$ yr.$^{-1}$) over stand age, indicating a period of optimal carbon sequestration lasting until the age of 138 years. Right/lower: cumulative net ecosystem production (*NEP*, Mg C ha$^{-1}$) of a boreal forest stand under rotation forestry over an optimal rotation period of 138 years. Vertical yellow and horizontal red dotted lines show the C compensation point (*CCP*, i.e., the age at which the cumulative carbon balance turns from a source to a sink) and the long-term mean C sequestration rate (*LCSR*, i.e., the mean carbon accumulation over multiple 138-year rotation periods)

regenerating boreal forests. This includes the recovery of a carbon sink in the first decade after harvesting due to establishment of a productive understory layer and a sustained carbon sink in old stands (131–211 years). The rotation period for optimum carbon sequestration extended to 138 years. This provided multiple rotations with a long-term carbon sequestration rate of 86.5 Mg C ha$^{-1}$ per rotation.

According to Peichl et al. (2022), the potential of forest management to maximize carbon sequestration and associated climate change mitigation effects on boreal forest landscapes may lie in developing strategies that optimize tree biomass production rather than heterotrophic soil carbon emissions.

Following Peichl et al. (2022), most of the monitoring sites over the 3-year period represented a carbon sink with a positive *NEP*, excluding a recently clear-cut (7-year-old) stand, a 26-year-old stand thinned in the spring of 2018, and one of the seven old stands at the age of 185 years (Fig. 12.1). The pattern of *NEP* across stand age included (i) an initial net carbon emission period, followed by (ii) a rapid increase of *NEP* during the subsequent few decades; (iii) a peak in middle-age stands; and (iv) a gradual decline of *NEP* in the old stands. The length of the initial emission period of carbon was less than 10 years, other periods having a net carbon sink. The mean *NEP* in the old stands was $98 \pm 50$ g C $m^{-2}$ $year^{-1}$, which was 56% of the peak *NEP* in the middle-aged stands.

The age-related changes in *NEP* followed the dynamics of *NPP* rather than those of $R_H$, representing no significant age-related trend, with the mean for different age classes being from $206 \pm 29$ to $232 \pm 52$ g C $m^{-2}$ $yr.^{-1}$. The values of *NPP* increased rapidly during the first decades. The highest value of *NPP* was in the middle-aged class (mean age class of $391 \pm 68$ g C $m^{-2}$ $yr.^{-1}$), thereafter gradually declining in mature and old stands, as occurred in the pattern of *NEP*. On the other hand, *NPP* can be divided into that of trees ($NPP_t$) and of ground vegetation ($NPP_u$) (Fig. 12.1). The values of $NPP_u$ exceeded those of $NPP_t$ rapidly during the initiation phase after clear-cutting. Thereafter, $NPP_u$ declined while $NPP_t$ increased rapidly along with stand age. Opposite patterns in $NPP_t$ and $NPP_u$ induced a limited change in total NPP in the sites with tree stands from middle- to old-aged phases.

### 12.2.2 Cumulative Carbon Loss from Initial Years Compensated for by Net Carbon Uptake

Figure 12.1 further shows the current annual *NEP* and the cumulative mean annual *NEP* trajectories over stand age. According to Peichl et al. (2022), the period of optimal carbon sequestration lasted to the age of 138 years. This point represents the intersection between the current *NEP* and the cumulative mean *NEP*. The cumulative *NEP* over 138 years was biologically optimal, providing a total carbon sequestration of 173 Mg C $ha^{-1}$. Peichl et al. (2022) assumed an annual harvesting rate of 0.78% to maintain stable tree biomass over a 138-year rotation, thus giving a long-term carbon sequestration rate (*LCSR*) of 86.5 Mg C $ha^{-1}$ over multiple 138-year rotations. For the final cutting (clear-cutting or cutting for natural regeneration via seed-tree or shelterwood methods), the results showed that the carbon budget in single forest stands switched dramatically from uptake to loss. According to Peichl et al. (2022), a period of up to 15–20 years in northern boreal forests (64° N) is needed to compensate for the cumulative net carbon loss in the initial years after regenerative cutting. This implies that the carbon compensation point (CCP) is followed by continuous net carbon gain until the end of the rotation period.

## 12.3 Carbon Balance in Forest Landscapes, with Responses to Management and Harvesting

### *12.3.1 Effect of Stand Age Distribution in the Landscape on Growth and Tree Stocking*

The management regime needed to achieve the goals set for timber production and carbon sequestration stocks is dependent on the forest structure. In this regard, Fig. 12.2 shows some basic patterns of how the age classes of trees or tree populations (stands) may be distributed over a forest landscape, with impacts of growth and biomass stocking (Kellomäki et al. 2009).

An *even distribution* with an equal area of each age class of stands, such as in a forest plantation, is suitable for managing forest for any function, but it might not be optimal for any of them (Fig. 12.2). For timber production, a *normal distribution*, with the dominance of trees of intermediate age, is preferable. This holds especially for carbon uptake in the ecosystem but is unlikely to be optimal for ecological and social functions. This is the case also for the *left-skewed distribution*, with the dominance of young populations. A forested area with a *right-skewed distribution* of age class, with the dominance of mature tree populations, will be preferable to the other distributions for ecological and social functions. Consequently, a forested area initially dominated by mature stands provides the maximum potential for timber, whereas areas with a normal age distribution of stands, or those dominated by young stands, provide the highest carbon sequestration in the short term.

Figure 12.2 demonstrates how the structure of a forest affects forest growth. Six Scots pine stands representing age classes of 2, 10, 22, 38, 62, and 82 years were used in the calculations. The volume in the age classes varied from a few $m^3$ $ha^{-1}$ in seedling stands up to 350 $m^3$ $ha^{-1}$ in mature stands. The initial growth and the growth during the tenth year were highest for the stand with an initial age of 38 years. This was fivefold higher compared to the stand with an initial age of 85 years. A forest area of 100 ha was divided into six compartments allocated into separate initial age classes.

The distribution skewed to the dominance of old stands contained the highest stocking, but growth was approximately 15% lower than that for the normal distribution. The total tree growth in the forested area dominated by young stands was lower (one-third of that for the normal distribution) than in the other age distributions. For the even distribution, the total growth was 70–80% of that for the normal distribution. Maximizing the timber production in a forested area would require an initial age structure resembling a normal distribution. Over a rotation, this age distribution would gradually lead to the dominance of old stands, with increasing

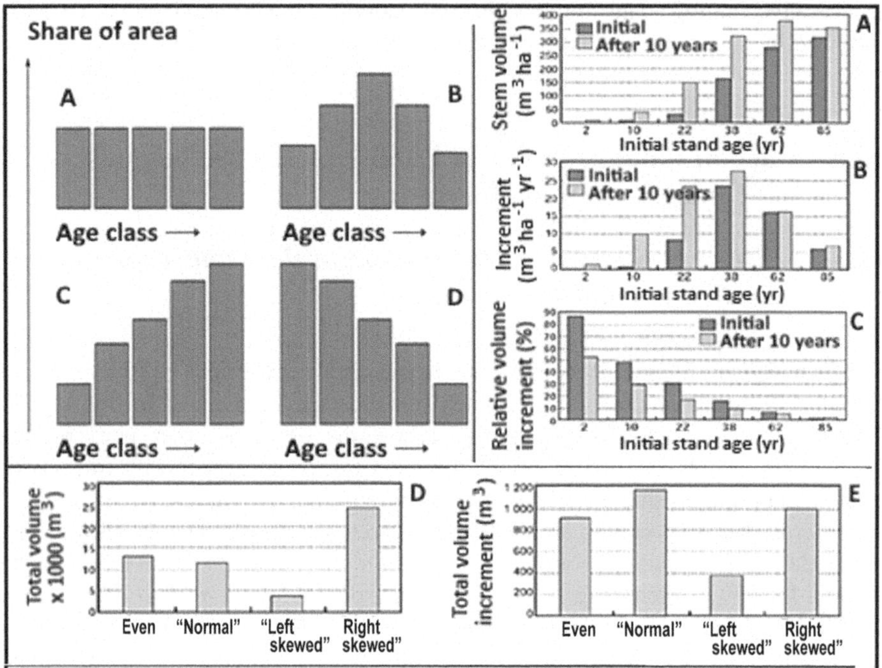


| Parameter | Age of trees at the beginning, yr. | | | | | |
|---|---|---|---|---|---|---|
| | 2 | 10 | 22 | 38 | 62 | 85 |
| Volume at the beginning, $m^3$ $ha^{-1}$ | < 0.1 | 1.3 | 29 | 160 | 278 | 313 |
| Growth at the beginning, $m^3$ $ha^{-1}yr.^{-1}$ | < 0.1 | 0.6 | 8 | 23 | 16 | 6 |
| Volume after 10 years, $m^3$ $ha^{-1}$ | 2.8 | 35 | 148 | 321 | 279 | 353 |
| Growth after 10 years, $m^3$ $ha^{-1}yr.^{-1}$ | 1.5 | 10 | 23 | 27 | 16 | 7 |
| Relative growth at the beginning, % | 86 | 48 | 30 | 15 | 6 | 2 |
| Relative growth after 10 years, % | 52 | 29 | 16 | 9 | 4 | 2 |

**Fig. 12.2** Overview of simulation of how the age distribution of stands in forest landscapes affects growth through a chronosequence. Compiled by the author of this book. Left: some basic patterns of how the age classes of trees/tree populations may be distributed. (**a**) all age classes have the same share of the total population (**b**) the share of populations of intermediate age dominates (**c**) the share of mature populations dominates (**d**) the share of young populations dominates Right: example of how age distribution affects growth and stocking of a tree stand (**a**) stocking at different ages (**b**) growth over 10 years, assuming initial stocking as in panel A (**c**) relative growth over 10 years, assuming initial stocking and growth as in panels A and B (**d**) stocking (**e**) growth over the forested area for the various age distributions. Below: statistics for growth and volume in the initial phase and after 10 years, related to the age of trees at the beginning of calculations

stocking and decreasing growth. Later, young stands would become dominant as a large area of stands previously of intermediate age matured. These would be cut at the end of the rotation, with a consequent shift of the age distribution to the dominance of young stands. The forested area would subsequently have low stocking and low absolute growth but high relative growth. Maximizing timber production over several rotations requires dividing the forest area into subareas representing age distributions dominated by young, intermediate, and old stands so that the total age distribution over the whole forest area is dominated by intermediate stands, with a moderately high total stocking and high growth.

### *12.3.2 Management at Forest Landscape Scale, with Links to Forest Succession*

At the landscape scale, management and harvesting operations represent the suitable subparts, or a set of sites and stands, for facilitating operations that are ecologically, economically, and technically feasible for multiple purposes (e.g., McKinley et al. 2011; Eliasson et al. 2013). The stands exposed to management and harvesting are distributed among stands in different developmental phases. Thus, the forest landscape is composed of individual stands differentiated in relation to the variability in site fertility, tree species composition, and maturity of trees. The structure and functioning in stand-based chronosequences proceed from seedling-phase stands through the pole-phase stands and to fully mature stands suitable for final cutting and reforestation for the next production period or rotation.

The sites and stands in the forest landscape form a parallel set of stand-based chronosequences with varying age and size distribution, depending on the variability in site fertility, tree species composition, maturity of trees, current management and harvesting operations, and natural disturbances affecting the regeneration, growth,

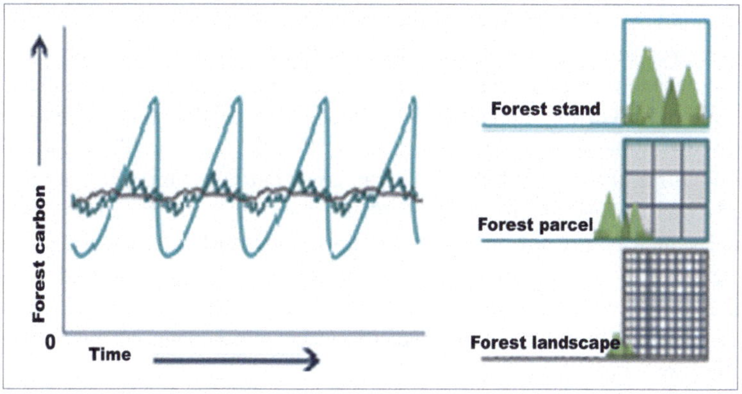


**Fig. 12.3** Overview of forest carbon fluctuation through several rotations over time at varying spatial scales under even-aged management in several rotations: (i) for a single stand (Upper) with wide temporal variability of carbon over time; (ii) for a forest parcel containing separate stands with one at once under management with temporal variability of carbon over time than in single stand (middle); and (iii) for a forest landscape containing several stands with one at once under management with lowest variability of carbon over time (lower) (McKinley et al. 2011). Permission of Wiley

and development of tree stands (Eliasson et al. 2013) (Fig. 12.3). Under even-aged management, the final cutting occurs only once for each rotation period. Over the same time span, final cutting frequently occurs on a fraction of the area in a landscape, defined by the area of mature tree stands ready for regeneration in each chronosequence of stands.

According to Eliasson et al. (2013), the frequency, location, and fraction of land area affected each year by final cutting depend on the age distribution and rotation age of all the stands in the landscape. The subsequent patterns of forest structure and composition are nonrandom related to properties of soil, tree species composition, and successional phase. The impacts of disturbances and management on boreal forest ecosystems are related to the temporal and spatial variability of structure and dynamics through the landscape. Figure 12.3 demonstrates the temporal and spatial variability of carbon storage at different scales in a forest landscape. Under even-aged management, the stand-scale forest carbon varies widely from final cutting to the maturation of trees before the final cutting for the next production cycle. The variability in carbon stocks is lower if the carbon through separate holdings (forest parcels) is considered, and the variability is still lower over the whole forest landscape (McKinley et al. 2011).

### 12.3.3 *Response of Forest Carbon Balance at Landscape Scale to Harvesting*

At the scale of forest landscape, many individual stands are assumed to be harvested at different occasions. Concurrently, losses of carbon from soil and trees in harvesting and gains in growth in the remaining trees in stands become evenly distributed over time. In this regard, Eliasson et al. (2013) investigated the carbon budgets under stem-only harvesting at a single-stand scale and at a landscape scale under different harvesting intensities. The landscape was assumed to consist of many stands of different ages, each representing 1 year of a given rotation period (Fig. 12.4). *First*, the carbon balance during consecutive rotation periods in individual stands was addressed. *Second*, the differences between the carbon budgets in individual stands compared with those of an integrated landscape were examined. *Third*, the consequences of calculating carbon balances based on individual stands or partial landscapes were examined, assuming that aggregation of fluxes in individual forest stands is a minimum requirement for accurate estimates of the actual carbon flux in the forest landscape.

The model-based analyses utilized the Q model (Ågren and Hyvönen 2003) for hemiboreal (56° N) conditions and the Coup model (Ågren et al. 2007) for northern boreal (64° N) conditions in the comparison of carbon budgets between these regions. In both cases, conventional harvesting of the stem wood only was used both at the single-stand scale and at the landscape scale under different harvesting intensities. The landscape was composed of many stands of different ages, each

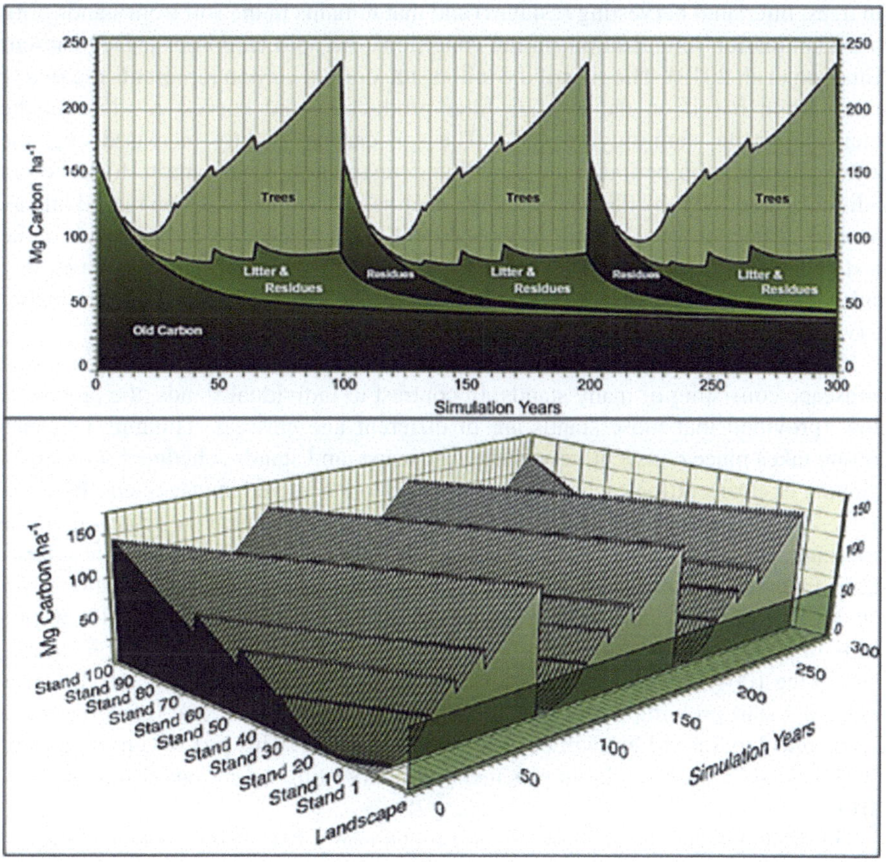


**Fig. 12.4** Forest carbon balance in response to harvesting for Norway spruce (*Picea abies*) in hemiboreal conditions (56° N) under an annual mean temperature of 6.8 °C (Eliasson et al. 2013). Permission of Elsevier. The productivity of the site was 6.3–7.9 $m^3$ $ha^{-1}$ $yr.^{-1}$. The first thinning was conducted in the 33rd year from the initialization of simulation at the rate of 21% of basal area (year/harvest, 33/24). The second thinning was performed in the 48th year at a rate of 27% (48/27) and the third thinning in the 65th year at a rate of 26% (65/26). The final cutting occurred in the 100th year from the initialization of the simulations Upper: stem wood-only scenario for simulated ecosystem carbon over 300 years in an individual stand of Norway spruce. Only stem wood is removed after thinning and final cutting, and logging residues (mainly tops and branches) are left in the stand after each harvesting event. Litter from living trees accumulates in the soil during each rotation. Lower: tree biomass in a chronosequence of 100 identical simulations. The stands are arranged in consecutive 1-year intervals, starting from the year 0. The average carbon content of the trees over all the stands is shown in the front bar (landscape), which is parallel to the *x*-axis

representing 1 year of a given rotation. The simulations dealt with Norway spruce in both southern and northern sites. The carbon dynamics in both sites resembled each other. In Fig. 12.4, the findings for the southern site were addressed to demonstrate how carbon is distributed over the boreal landscape under regular management.

Figure 12.4 shows an individual stand with the ecosystem carbon stock, which is separated into carbon that has been fixed since the beginning of the simulation (in trees, litter, and harvesting residues) and that remains in the soil from stands prior to the simulation initialization (initial conditions, i.e., old carbon or legacy carbon) (Eliasson et al. 2013). The size of the initial stock of old carbon represents the steady state carbon based on the average litter production and harvest residues in the reference or the stem-only scenario. The soil carbon slightly increased over the 300-year simulation period, with a small underestimate for old carbon stock. For an individual stand, it took 25 years until the decline in ecosystem carbon turned into an increase. After the final cutting, it took 40–60 years before the carbon stock in the next generation had reestablished to the level of the previous generation. In cutting and harvesting, the soil carbon declined to 50 Mg $ha^{-1}$ compared to that before harvesting (Eliasson et al. 2013).

Figure 12.4 further shows the introduction of a new management regime in a landscape consisting of many stands. In contrast to individual stands, the process is slow (provided that those stands are of different age classes). Thinning or clear-cutting takes place consecutively in the landscape, and stands scheduled for a given thinning or clear-cutting each year are subjected to the altered management (Eliasson et al. 2013). Sustainable yield under even-aged management is achieved if the landscape comprises a mosaic of stands of different age. This implies that a landscape includes several stands that meet each other in the length of rotation. Figure 12.4 shows that the landscape consists of 100 identical forest stands, despite being of different ages (from 1 to 100 years old). Each of the 100 identical simulations in the 100-year rotation scenario starts consecutively, implying that (i) the trees in Stand 1 start growing in the year after the final cutting in the year 0 and (ii) Stand 2 is identical to Stand 1 but will be clear-cut in the year 1. This process proceeds until all 100 stands are treated in management and harvesting operations (Eliasson et al. 2013).

The findings in Fig. 12.4 are based on stem-only harvesting allowing logging residues to remain on site, thus increasing soil carbon. In the case of whole-tree harvesting, the soil carbon would be lower (Eliasson et al. 2013). On the other hand, the soil carbon lost in the landscape was lower than the carbon harvested in increasing cutting removals. The loss of soil carbon in the first 20 years was 58%, versus 43–78% for the carbon in accumulated harvests, depending on the site productivity. From a landscape perspective, there are many individual stands assumed to be harvested on separate occasions. Subsequently, soil carbon losses from harvests and gains from growth in the remaining stands are evenly distributed over time (Eliasson et al. 2013).

## 12.4 Concluding Remarks

Under even-aged management, forest landscapes are composed of parallel sets of stand-based chronosequences forming varying age and size distributions. The variability in forest landscapes is further affected by natural disturbances, with impacts on the regeneration, growth, and development of tree stands. In management, final cutting frequently occurs on a fraction of the area in a landscape defined by the area of mature tree stands ready for regeneration in a site-specific chronosequence. The frequency and ratio of land area affected in each year by final cutting depend on the age distribution and rotation age of all the stands in the landscape. At this scale, management and harvesting operations represent the suitable subparts, or a set of sites and stands. Each site/stand goes through development from regenerative cutting through the seedling and pole phases to full maturity representing the progress from carbon source to carbon sink, with varying intensity depending on the structure and functioning of the tree stand. The sink capacity of the forest landscape is stable even though the sink/source capacity varies over time in each site/stand.

## References

Ågren GI, Hyvönen R (2003) Changes in carbon stores in Swedish forest soils due to increased biomass harvest and increased temperatures analyzed with a semiempirical model. For Ecol Manag 174:25–37

Ågren GI, Hyvönen R, Nilsson T (2007) Are Swedish forest soils sinks or sources for $CO_2$—model analyses based on forest inventory data. Biochemistry 82:217–227

Bateman IJ, Mace GM, Fezzi C, Atkinson G, Turner K (2011) Economic analysis for ecosystem assessment. Environ Res Econ 48:177–217

Eliasson P, Svensson M, Olsson M, Ågren GI (2013) Forest carbon balances at the landscape scale investigated with the Q model and the CoupModel—responses to intensified harvests. For Ecol Manag 290:67–78

Hyvönen R, Ågren GI, Linder S, Persson T, Cotrufo MF, Ekblad A, Freeman M, Grelle A, Janssens IA, Jarvis PG, Kellomäki S, Lindroth A, Loustau D, Lundmark T, Norby RJ, Oren R, Pilegaard K, Ryan MG, Sigurdsson BD, Strömgren M, van Oijen M, Wallin G (2007) The likely impact of elevated [$CO_2$], nitrogen deposition, increased temperature, and management on carbon sequestration in temperate and boreal forest ecosystems: a literature review. New Phytol 173:463–480

Kellomäki S, Koski V, Niemelä P, Peltola H, Pulkkinen P (2009) Management of forest ecosystem. In: Kellomäki S (ed) Forest resources and sustainable management. Paper Engineers' Association/Paperi ja Puu Oy, Jyväskylä, pp 252–373

McKinley DC, Ryan MG, Birdsey RA, Giardina CP, Harmon ME, Heath LS, Houghton RA, Jackson RB, Morrison JF, Murray BC (2011) A synthesis of current knowledge on forests and carbon storage in the United States. Ecol Appl 21(6):1902–1924

Peichl M, Martinez-Garcia E, Fransson JES, Wallerman J, Laudon H, Lundmark T, Nilsson MB (2022) Landscape-variability of the carbon balance across managed boreal forests. Glob Chang Biol 29(4):1119–1132

Routa J, Kellomäki S, Peltola H (2012) Impacts of intensive management and landscape structure on timber and energy wood production and net $CO_2$ emissions from energy wood use of Norway spruce. Bioenergy Res 5:106–123

Sing L, Ray D, Watts K (2015) Ecosystem services and forest management. Forestry Commission Research Note 20:1–10

# Chapter 13
# Mitigation of Climatic Warming in Forestry

## Contents



**Abstract** In natural conditions, boreal forests retain substantial amounts of carbon in living and dead biomass. Natural disturbances such as wildfires release forest-based carbon, with losses of carbon into the atmosphere. This further holds for the harvesting of timber and biomass for substituting fossil-based materials. The concurrent increase of atmospheric carbon does not disrupt the entry of shortwave solar radiation through the atmosphere, but it reduces the outgoing long-wave radiation into space. Under increasing atmospheric carbon dioxide concentration, more heat is curbed in the lower atmosphere, thus warming the climate. Concurrently, carbon dioxide is taken up in photosynthesis and bound in biomass in plant growth. This carbon retains in the forest-based carbon cycle depending on the amount and properties of biomass and dead organic matter in the ecosystems. The carbon stocks and concurrent reflectance of solar radiation (or albedo) affect the radiative forcing. Radiative forcing refers to changes in the radiative balance and consequent climate cooling/warming in response to the content of carbon dioxide and other greenhouse

 
S. Kellomäki, *Forest Management for Timber Production and Climate Change Mitigation*, Managing Forest Ecosystems 44,
https://doi.org/10.1007/978-3-031-71575-4_13

gases in the atmosphere, with links to the carbon dynamics and management of forest ecosystems.

## 13.1 Mitigating Climate Change Through Management of Boreal Forests

### *13.1.1 Carbon Dynamics in Forest Ecosystems, with Effects on Climatic Warming*

In forest ecosystems, carbon is cycled through net primary production (growth) to decomposing litter and soil organic matter, with the emission of carbon into the atmosphere through autotrophic and heterotrophic respiration (Fig. 13.1). In the atmosphere, carbon dioxide ($CO_2$) does not disrupt the entry of shortwave solar radiation (wavelength $< 900$ nm) into the atmosphere. However, it reduces the egress of outgoing long-wave radiation (wavelength $> 900$ nm) into space. Under

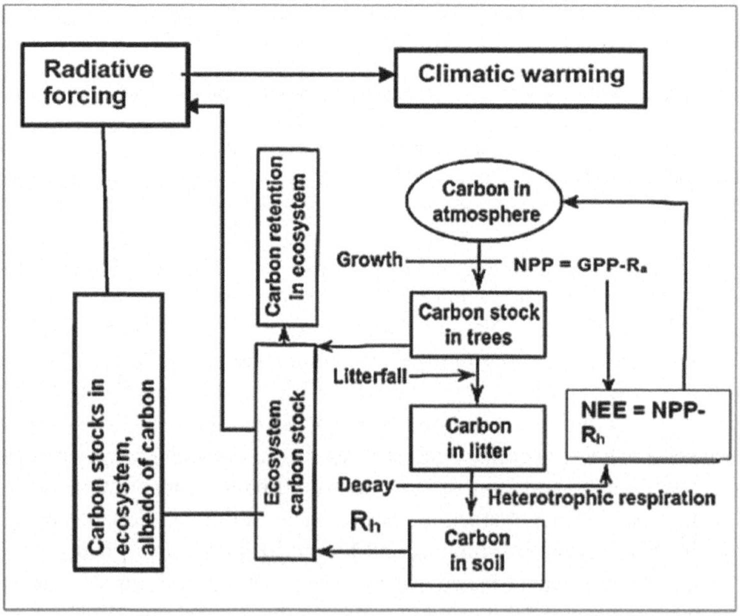


**Fig. 13.1** Overview of flows and stocks of carbon in a forest ecosystem, indicating how the growth and development of plants (trees and other forest vegetation) affect carbon uptake and storage in the ecosystem and carbon emissions from the ecosystem, as well as the carbon in the atmosphere and radiative forcing with the consequent cooling/warming of climate. The difference between the energy absorbed by a given surface and energy radiated indicates the radiative forcing. Compiled by the author of this book

increasing atmospheric carbon dioxide concentration, more heat is curbed in the lower atmosphere, thus warming the climate.

Carbon in trees/plants and soil separately and together indicates the stock of carbon in the ecosystem, where carbon is retained for some time (Fig. 13.1). Concurrently, a part of solar radiation reflects into the atmosphere on canopy and soil surfaces called albedo. This refers to the radiation reflected into the atmosphere in terms of fraction or percentage of incoming radiation on tree canopies and soil surface. The carbon stocks and concurrent reflectance of solar radiation (or albedo) affect the radiative forcing. Radiative forcing refers to changes in the radiative balance and consequent climate cooling/warming in response to the content of greenhouse gases in the atmosphere (IPCC 2007), with links to the carbon dynamics and management of forest ecosystems.

### 13.1.2 Carbon Retention in Forest Ecosystems

In forests, carbon is taken up in gross primary production, cycled through net primary production (growth), and lost in autotrophic respiration in the growth and maintenance of biomass. Carbon bound in net primary production is further lost in heterotrophic respiration through decaying soil organic matter (litter and humus) (Figs. 13.1, 13.2). The balance between net primary production and heterotrophic respiration of carbon indicates the net ecosystem production or net ecosystem exchange, $NEE = NPP—R_h$, where $NPP$ is the net primary growth and $R_h$ the heterotrophic respiration in the ecosystem. Carbon stocks in trees/plants and soil separately and together retain in the ecosystem and form temporal carbon stocks in ecosystem carbon dynamic above and below soil surface. This retention with varying length is indicated by the carbon retention time ($\tau$, years), which is initiated when carbon enters the system and ends when carbon leaves the system (Govers et al. 2013):

$$\tau = \frac{\textit{Capacity of a system to hold carbon}}{\textit{Rate of carbon flow through a system}} \tag{13.1}$$

In forestry and forest-based industry, carbon is also emitted in forest management/harvest/transportation operations and further processing of timber and biomass for materials and energy. Manufacturing processes may also use the carbon in abandoned products, thus recycling carbon in these products. Finally, carbon is emitted into the atmosphere in the combustion of biomass-based fuels and materials burned or left to decay in landfills. Hence, forests and forestry contribute to mitigating climate change by retaining carbon in forest ecosystems and by replacing fossil fuels and materials with biofuels and long-living wood-based materials. The mitigation of climate change also involves the albedo of forest cover, which is dependent on the management/harvest used to sustain/enhance the growth and development of forests.

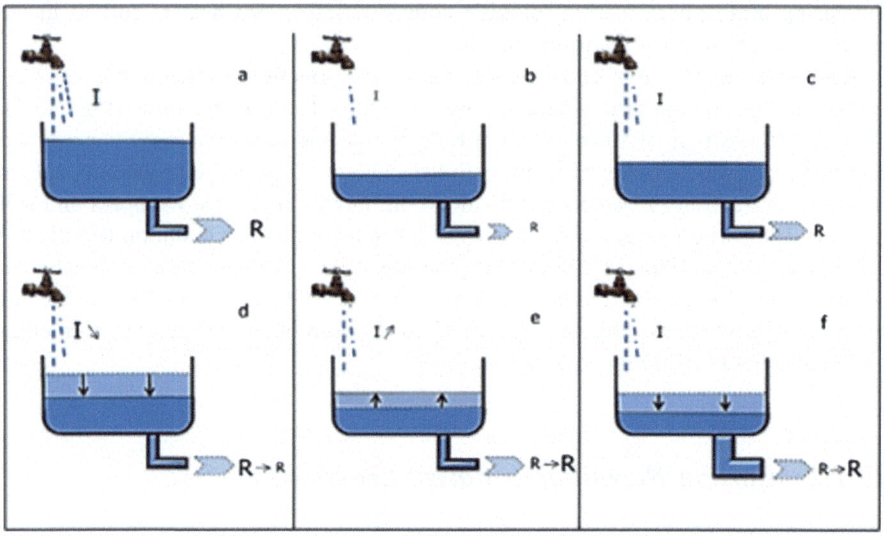


**Fig. 13.2** Overview of the performance of soil organic carbon (SOC) stock using a kitchen sink as an analogy for SOC storage as a function of carbon input in tree growth and litter to the soil and carbon discharged in respiration (Govers et al. 2013). Open Access. Courtesy of Global Environmental Facility. Panels a–c illustrate that various levels of equilibrium in SOC stocks (sink levels) are possible, depending on the amount of carbon input and carbon output (the drain discharge, *R*). Panels d–f illustrate the effect of changes in input or decomposition: if the input from trees (tap discharge) decreases (e.g., due to crop residue removal), the level of the sink (carbon stock) will lower until the drain (respiration) rate is again equal to the new input rate (d). Panels e–f illustrate how the carbon input rate (tap discharge, *I*) increases (e.g., due to enhanced vegetation growth caused by a $CO_2$ increase) with the rise in sink level (SOC stock) (e). Respiration rates may also change; that is, climatic warming and/or soil disturbance may lead to an increase of respiration, leading to a decrease in the SOC stock (f)

Govers et al. (2013) compared the performance of carbon in trees and soil organic matter to that of a kitchen sink in which the water level varies as a function of carbon input to the soil and carbon discharged in respiration. In Fig. 13.2, water input and output vary, with consequent variability in water level. The water level in the sink is analogous to the total carbon in trees and soil organic carbon (SOC), while the tap discharge represents carbon input (*I*) in tree growth and litter and the discharge (*R*) of carbon lost through the decay of soil organic matter and respiration processes of trees.

Figure 13.2 demonstrates through the flow of water in and out of a kitchen sink how the total carbon stock in an ecosystem performs. This also holds for individual stocks such as that of soil organic carbon (SOC). The amount of carbon in soil represents a balance between the carbon input and output. This is independent of the SOC stock in the soil and the respiration, which is proportional to the magnitude of the SOC stock. According to Govers et al. (2013), "the level of water (the SOC stock) is determined both by the rate at which water is added (i.e., the discharge from the tap, independent of the water level in the sink, the organic carbon input) and the rate at which it flows away through the drain (controlled by the water level in the sink, the carbon respiration rate). The equilibrium stock of water results from both the magnitude of the input (tap discharge) as well as that of the output (controlled by the diameter of the drain and the water level). As the output is dependent on the level in the sink (the SOC stock) equilibrium will be reached at a certain water level (or SOC stock)."

The ratio of outflow over total water volume in the sink determines the mean residence time of the water in the sink (Govers et al. 2013):

$$MRT = {}^{SOC}\!/_{R} \tag{13.2}$$

where *MRT* is the mean residence time (yr), *SOC* the total carbon stock in the soil through a given depth (kg C $m^{-2}$), and *R* the respiration rate (kg C $m^{-2}$ $yr.^{-1}$). Respiration is proportional to the total SOC stock. Consequently,

$$MRT = {}^{SOC}\!/_{k \times SOC} \text{ or } MRT = {}^{1}\!/_{k} \tag{13.3}$$

where *k* is the decomposition rate ($yr^{-1}$).

## 13.2 Radiative Forcing Related to Forest Management

### *13.2.1 Albedo and Radiative Forcing*

Radiative forcing can be used to assess how carbon in the forest ecosystem or forest-based production system may affect climate/climatic warming related to (i) the reflective properties of forest cover, i.e., albedo, and (ii) the carbon stocks in the forest ecosystem. Under management, the change of albedo is related to factors including tree species choice, thinning intensity, clear cutting, etc. The albedo depends on the properties and structure of the tree canopy (e.g., density of foliage cover, tree species) as controlled by forest management. Following Kirschbaum et al. (2011), the change in daily radiative forcing ($\Delta R_d$, J $m^{-2}$ $d^{-1}$) related to management depends on the change in albedo:

$$\Delta R_d = Q_s \downarrow \times \Delta a \times (1 - \alpha_{atm}) \tag{13.4}$$

where $Q_s\downarrow$ is the total daily downward radiation, $\Delta a$ is the difference in the albedo between two different management regimes, and $\alpha_{atm}$ is the share of shortwave radiation intercepted by the atmosphere, with a mean value of 20%. The annual change in radiative forcing on a hectare basis ($\Delta R_a$, J $ha^{-1}$ $yr.^{-1}$) related to the change in albedo is (Kirschbaum et al. 2011):

$$\Delta R_a = \sum_{day=1}^{365} 10{,}000 \times \Delta R_d. \tag{13.5}$$

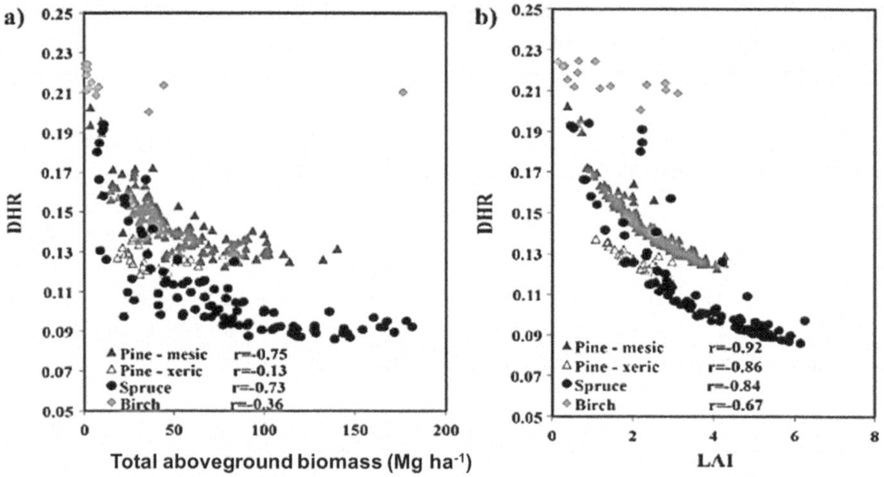


The best fit for the dependence of albedo (a, 0...1) on the carbon stock above ground (m, Mg ha$^{-1}$ in stem, branches, and foliage) was: a $= A \times \exp(-B \times m^C) + D$, with A, B, C, and D being the parameters

| Species | Parameter | | | |
|---|---|---|---|---|
| | A | B | C | D |
| Scots pine | 0.117 | 0.125 | 0.661 | 0.119 |
| Norway spruce | 0.131 | 0.053 | 0.842 | 0.099 |
| Birch | 0.117 | 0.125 | 0.661 | 0.119 |

**Fig. 13.3** Effect of trees species on the albedo related to the properties of canopy of tree stand of different tree species in middle-boreal conditions (Lukeš et al. 2013). Permission of Elsevier. The simulated values of albedo represent black-sky albedo for the zenith angles of 40° and 70° for a latitude of 61° N in midsummer. The shortwave (400–2400 nm) albedo approximates the sum of the spectral albedos weighted by incoming solar irradiances at the given wavelengths. The albedo covers 98% of incoming irradiance, which gives an estimate of the total broadband (300–4000 nm) albedo. Left: pure Scots pine (*Pinus sylvestris*), with albedo related to total aboveground biomass. Right: pure Norway spruce (*Picea abies*), with albedo related to lead area index. Lower: regression model with parameter for estimating the albedo as a function of aboveground carbon mass in trees (stem, branches, foliage) of boreal Scots pine, Norway spruce, and birches

Over a forested area, albedo varies in response to the tree species composition, stocking, and age class distribution of tree stands, which affect the canopy cover. For example, Rautiainen et al. (2011) showed that the albedo in boreal coniferous forest decreased when the tree stand became older and the stocking increased. Similarly, Lukeš et al. (2013) found that the species-specific albedo in middle-boreal forests (61° N) was correlated with the total aboveground biomass, leaf area index (LAI), and canopy cover (Fig. 13.3).

In comparing the effects of different land use options (i.e., management options) on albedo, the incoming radiation needs to be the same for the options to be compared. This is especially the case in climate simulation studies in which stochastic processes affect the comparisons. Therefore, the effects of a climate simulator on the shortwave radiation and albedo due to rounding error is excluded:

$$Q_s\downarrow_1 = \frac{R_{s1}}{1 - a_1} \tag{13.6}$$

$$Q_s\downarrow_2 = \frac{R_{s2}}{1 - a_2} \tag{13.7}$$

where $Q_s\downarrow_1$ and $Q_s\downarrow_2$ are the total daily downward radiation and $R_{s1}$ and $R_{s2}$ are the shortwave radiation for the optional management scenarios 1 and 2, respectively, based on the climate simulator (Strandman et al. 1993). The total daily downward radiation $Q_s\downarrow$ for the management scenarios 1 and 2 is:

$$Q_s \downarrow = \frac{(Q_s\downarrow_1 + Q_s\downarrow_2)}{2}. \tag{13.8}$$

Consequently, the albedo change due to the management change difference is:

$$\Delta a = a_2 - a_1. \tag{13.9}$$

The annual change of radiative forcing per hectare [$R_y$, J ha$^{-1}$ yr.$^{-1}$] is:

$$\Delta R_y = 10{,}000 \times \frac{Q_s\downarrow_1 + Q_s\downarrow_2}{2} \times (a_2 - a_1) \times (1 - \alpha_{atm}) \tag{13.10}$$

where $Q_s\downarrow_1$ and $Q_s\downarrow_2$ are the shortwave radiation [J ha$^{-1}$ yr.$^{-1}$] and $a_1$ and $a_2$ are albedo for the management scenarios 1 and 2.

### 13.2.2 *Carbon Stocks and Radiative Forcing*

When the focus is on a given site with and without tree stands, the change in radiative forcing [$\Delta R_m$, J m$^{-2}$ d$^{-1}$] in relation to the changes in carbon stocks is (Kirschbaum et al. 2011):

$$\Delta R_m = 86{,}400 \times 5.35 \times \ln\left[1 + {}^{\Delta[C]}/_{[C]}\right] \tag{13.11}$$

where [$C$] [ppm] is the background atmospheric $CO_2$ concentration, $\Delta$[$C$] [ppm] is the change in the atmospheric $CO_2$ concentration related to the given management regime, 86,400 is the length of a day in seconds, and 5.35 [W m$^{-2}$] converts $CO_2$ units to radiative forcing. The change in carbon stock $\Delta C$ [Mg of carbon] is converted to the change in atmospheric $CO_2$ under the assumption that 1 ppm of atmospheric $CO_2$ is equal to 2123 Gt of carbon:

$$\Delta[C] = {}^{\Delta C}\!/_{\left(2{,}123 \times 10^9\right)}. \tag{13.12}$$

Consequently, the change in total radiative forcing related to carbon stocks over a year and for the whole Earth [$\Delta R_E$] is:

$$\Delta R_E = \Delta R_m \times 510 \times 10^{12} \times 365 \tag{13.13}$$

where $510 \times 10^{12}$ [$m^2$] is the surface area of the Earth (510 million $km^2$) and 365 is the number of days per year. When assuming an atmospheric $CO_2$ concentration of 390 ppm, the radiative forcing [$\Delta R_E$] for the whole Earth per removal of Mg of carbon is:

$$\Delta R_E = 86{,}400 \times 5.36 \times ln\left[1 + \frac{{}^{\Delta C}\!/_{\left(2.123 \times 10^9\right)}}{390}\right] \times 510 \times 10^{12} \times 365 =$$

$$-104 \text{ GJ } (\text{Mg C})^{-1} \text{ yr}^{-1}. \tag{13.14}$$

## 13.3 Management of Carbon in Ecosystems for Timber and Radiative Forcing

### *13.3.1 Carbon Stocks Related to Forest Management Strategy*

Kellomäki et al. (2021, 2023) investigated how forest management may influence carbon stocks (trees, soil) in boreal Norway spruce (*Picea abies*) stands, with potential impacts on climatic warming. Using an ecophysiological forest ecosystem model, the impact potentials were studied using several management regimes with impacts on the carbon sequestration in trees, soil and the ecosystem, and timber yield obtainable in harvest (Box 13.1) The long-term simulations combined the regeneration, growth, and mortality of trees under the current climate representative of the middle-boreal zone of central Finland (62° N). The climate was defined by a mean annual temperature sum of 1100 d.d., precipitation of 540 mm $yr.^{-1}$, and an atmospheric $CO_2$ concentration of 390 ppm. The simulations were performed for moderately fertile sites (*Myrtillus* site type, MT). At the initialization of the simulations, the carbon content of soil organic matter (SOC down to 30 cm, including carbon in litter, humus, and coarse and fine roots) was 34 Mg carbon $ha^{-1}$. Saw logs and pulp wood were harvested in thinning and final cutting under even-aged management. Under uneven-aged management, selection cuttings were used. The recommended thinning/selection thresholds were employed in Finnish private forestry (e.g., Äijälä et al. 2014).

**Box 13.1 Simulations for the Analysis**

The simulations were divided into two subperiods. In the *first subperiod*, the simulations represented even-aged management following the recommended rotation length and thinning intensity for the use in private forestry (e.g., Äijälä et al. 2014). The simulations were initiated through planting 1800 seedlings per hectare at the year 1 excluding natural regeneration. The initial planting density represented five size classes with 360 seedlings in each (height and diameter in a range of 0.1–0.5 m and 1.0–1.4 cm).

In the *second subperiod*, starting at the year 100, the simulations proceeded using the tree stand and soil with existing properties at the end of the first subperiod to initialize the management for the second period. Over the second period, three optional management strategies were used: (i) even-aged management with three different thinning (from below) intensities, (ii) uneven-aged management with selection cuttings from above, and (iii) no management excluding any cuttings. Under no management and even-aged management, the period 101—400 represented a transition phase (Kellomäki et al. 2021).

In the second subperiod, even-aged management was initiated by clear-cut at the year 100 followed by planting, as in the beginning of first subperiod. The simulation periods extended over 80-, 100- and 120-year rotations (Kellomäki et al. 2023). In the basic management (BT), the thinning intensity followed current recommendations. On the other hand, the thinning intensity was reduced (BT + 20) or increased (BT − 20) to represent a 20% higher or lower basal area thresholds before and after thinning compared to the BT. The timber (saw logs and pulpwood with minimum top diameters ≥16 cm and 6 cm) was harvested in clear-cut at the year 100 and in following thinning over a rotation. Under uneven-aged management, selection cuttings (BCS) were taken at ca. 15-year intervals from above, following recommendations for selection cuttings (Äijälä et al. 2014). The selection cuttings created canopy gaps that allowed the establishment and ingrowth of new seedlings. Timber (saw logs, pulpwood) was only harvested in selection cuttings like in even-aged management. The simulation periods extended for 80, 100, and 120 years, representing the same periods as the rotations under even-aged management.

Under no management (NoMan), trees were allowed to grow and develop without any cuttings. Natural mortality of trees created canopy gaps, which allowed the establishment and ingrowth of new seedlings. In simulations under uneven-aged management and no management, full seed crop potential was assumed. The same simulation periods were used for no management as for the even- and uneven-aged management.

Figure 13.4 shows how different management regimes affect the growth and further carbon stocks in trees, soil, and ecosystem through the first and second subperiods over the 1000-year simulation (Kellomäki et al. 2021). The effects of different management regimes are most comparable in the period 401–1000 years in

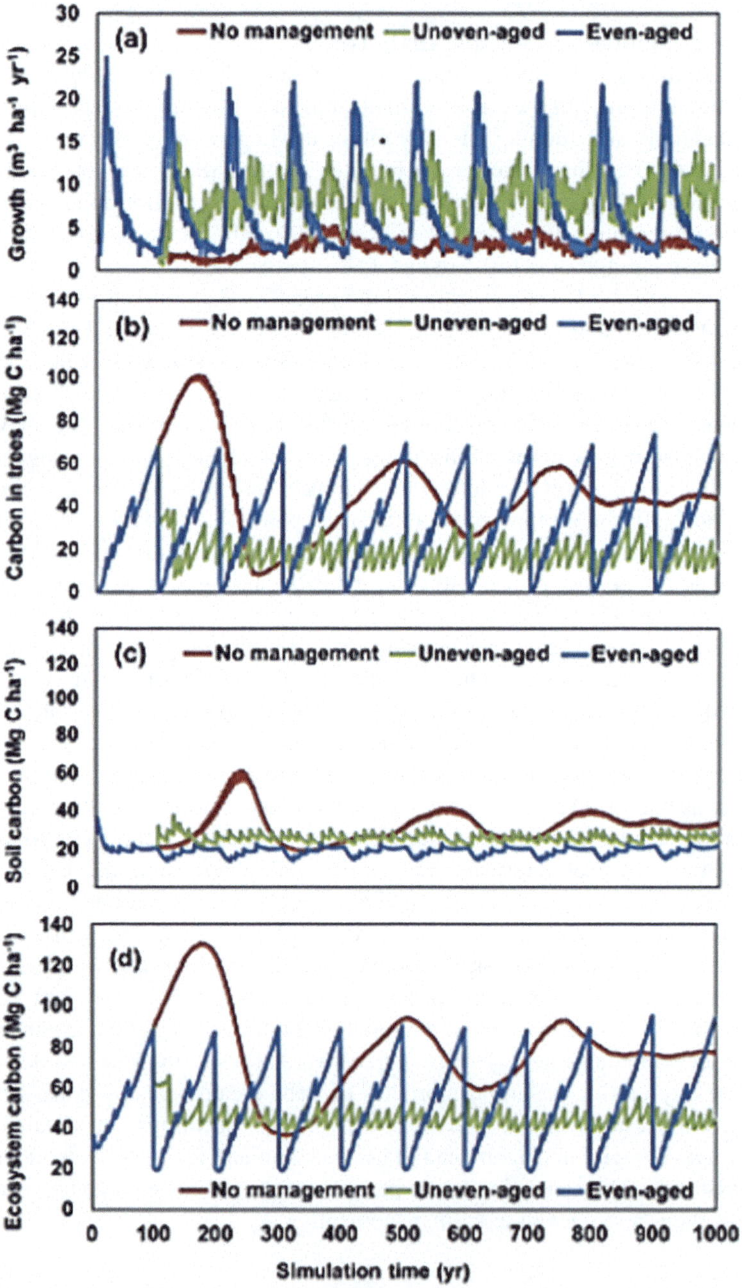


**Fig. 13.4** Simulations over two periods (0–100 and 101–1000 years) on the impacts of even (BT)- and uneven (BSC)-aged management and no management (NoMan) on the carbon sequestration and stock in a boreal forest ecosystem covered 1000 years (Kellomäki et al. 2021). Open Access. Permission of Oxford University Press. Upper/(a): mean growth of stem wood. Upper/lower: mean carbon in trees. Lower/upper: mean carbon in soil. Lower: mean ecosystem carbon stock

**Table 13.1** Long-term mean values for basal area, growth of stem wood, timber yield (saw logs, pulpwood), net present value (NPV, 2% interest rate), and carbon stocks (trees, soil, and ecosystem) for different management regimes over simulation periods of 101–400 and 401–1000 years (Kellomäki et al. 2021)

| | Simulation years 101–400 | | | Simulation years 401–1000 | | |
|---|---|---|---|---|---|---|
| Variables | No management | Even-aged | Uneven-aged | No management | Even-aged | Uneven-aged |
| Mean basal area, $m^2$ $ha^{-1}$ | 17 | 23 | 18 | 18 | 24 | 19 |
| Mean growth, $m^3$ $ha^{-1}$ $yr.^{-1}$ | 2.5 | 6.7 | 8.3 | 3.1 | 6.8 | 8.9 |
| Timber yield, $m^3$ $ha^{-1}$ $yr.^{-1}$ | – | 6.4 | 7.2 | – | 6.5 | 7.1 |
| • Saw logs | – | 3.8 | 3.8 | – | 3.8 | 3.4 |
| • Pulpwood | – | 2.6 | 3.4 | – | 2.7 | 3.7 |
| Carbon stocks, Mg C $ha^{-1}$ | | | | | | |
| • Trees | 47 | 33 | 20 | 45 | 34 | 18 |
| • Soil | 31 | 19 | 26 | 33 | 19 | 26 |
| • Ecosystem (trees and soil) | 78 | 57 | 46 | 78 | 53 | 44 |
| % ecosystem carbon | | | | | | |
| • Trees | 60 | 63 | 43 | 58 | 64 | 41 |
| • Soil | 40 | 37 | 57 | 42 | 36 | 59 |

the transition period past the long-term influence of management in first subperiod on the growth and carbon dynamics. Regardless of management regime (BT, BSC, NoMan), the mean annual growth of stem wood ranged from 3.1 to 8.9 $m^3$ $ha^{-1}$ $yr.^{-1}$. Under no management (old-growth forest), the mean growth was 54% smaller than that under even-aged management. In the latter case, the mean growth was 31% higher than under uneven-aged. However, the total timber yield was 10% larger under uneven-aged management than under even-aged management.

Mean carbon stocks in the trees ranged from 18 to 45 Mg C $ha^{-1}$, in the soil from 19 to 33 Mg C $ha^{-1}$, and in the ecosystem (trees and soil) from 44 to 78 Mg C $ha^{-1}$ for all three management regimes over the years 401–1000. Mean carbon stocks in trees, soil, and the ecosystem were 34, 70, and 47% larger under no management than even-aged management. Under uneven-aged management, mean carbon stocks in trees and the whole ecosystem were 46 and 16% lower, but in the soil, they were 37% higher than under even-aged management. The carbon in trees contributed 41, 58, and 64% to the total ecosystem carbon stocks under uneven-aged, no management, and even-aged management, respectively.

Table 13.1 further summarizes the growth and carbon dynamics in a forest ecosystem under different management regimes for the periods of 101–400 and 401–1000 years. Under no management, the mean basal area was 17–18 $m^2$ $ha^{-1}$ for both simulation periods. For even-aged management, the mean basal area was 23–24 $m^2$ $ha^{-1}$ with four thinning over a 100-year rotation. Under uneven-aged management, selection cuttings were performed at an interval of 15–20 years. The mean basal area was 18–19 $m^2$ $ha^{-1}$ (20% lower than that under even-aged management). The differences between different management regimes were repeated in the growth and carbon stocks. Under no management, mean annual stem wood growth was 2.5–3.1 $m^3$ $ha^{-1}$ $yr.^{-1}$, while under even-aged management, the mean stem wood growth was 6.7–6.8 $m^3$ $ha^{-1}$ $yr.^{-1}$. This was less than under uneven-aged management, with 8.3–8.9 $m^3$ $ha^{-1}$ $yr.^{-1}$ of stem wood. However, the timber yield (saw logs, pulpwood) was equal for both management regimes. This holds also for net present value (NPV, 2% interest rate) of timber per hectare and year. Concurrently, the total amount of carbon in the ecosystem was 52–53 Mg $ha^{-1}$ for even-aged management and 46 Mg $ha^{-1}$ for uneven-aged management (Kellomäki et al. 2021).

Carbon stock was higher in trees under even-aged management than under uneven-aged management, whereas the opposite situation occurred for the carbon stock in soil. In both simulation periods, the total carbon stock remained lower than that under no management, which yielded a total mean carbon stock of 78 Mg $ha^{-1}$. In this case, the carbon stock was 45–47 Mg $ha^{-1}$ in trees and 31–33 Mg $ha^{-1}$ in soil. Under no management, the share of carbon was 58–60% in trees and about 40% in soil. Under even-aged management, the contribution was 63–64% for trees and 36–37% for soil and, under uneven-aged management, 41–43% for trees and 57–59% for soil regardless of simulation period. Thus, the carbon contributions of trees and soil to the total ecosystem carbon under no management and even-aged management resembled each other, while uneven-aged management increased the carbon contribution of soil and reduced that of trees (Kellomäki et al. 2021).

### *13.3.2 Carbon Stocks and Timber Yield Related to Management Intensity*

Based on model simulations, Kellomäki et al. (2023) further analyzed how thinning intensity and rotation length may affect radiative forcing in managed Norway spruce (*Picea abies*) forests under the current climate (Fig. 13.5). Even-aged management using the baseline thinning (BT) from below over an 80-year rotation length provided a reference for other management regimes. The optional management regimes included BT thinning with ±20% changes in basal area thinning thresholds over rotations of 80, 100, and 120 years. The optional management regimes also included uneven-aged management (BSC), with selection cuttings (from above) at 15-year intervals, and no management (NoMan), over the simulation periods of 80–120 years.

Table 13.2 further summarizes the mean growth and carbon stock in trees, soil and ecosystem, timber yield, and net present value (NPV) of timber yield in relation to management (Kellomäki et al. 2023) with the following main findings:

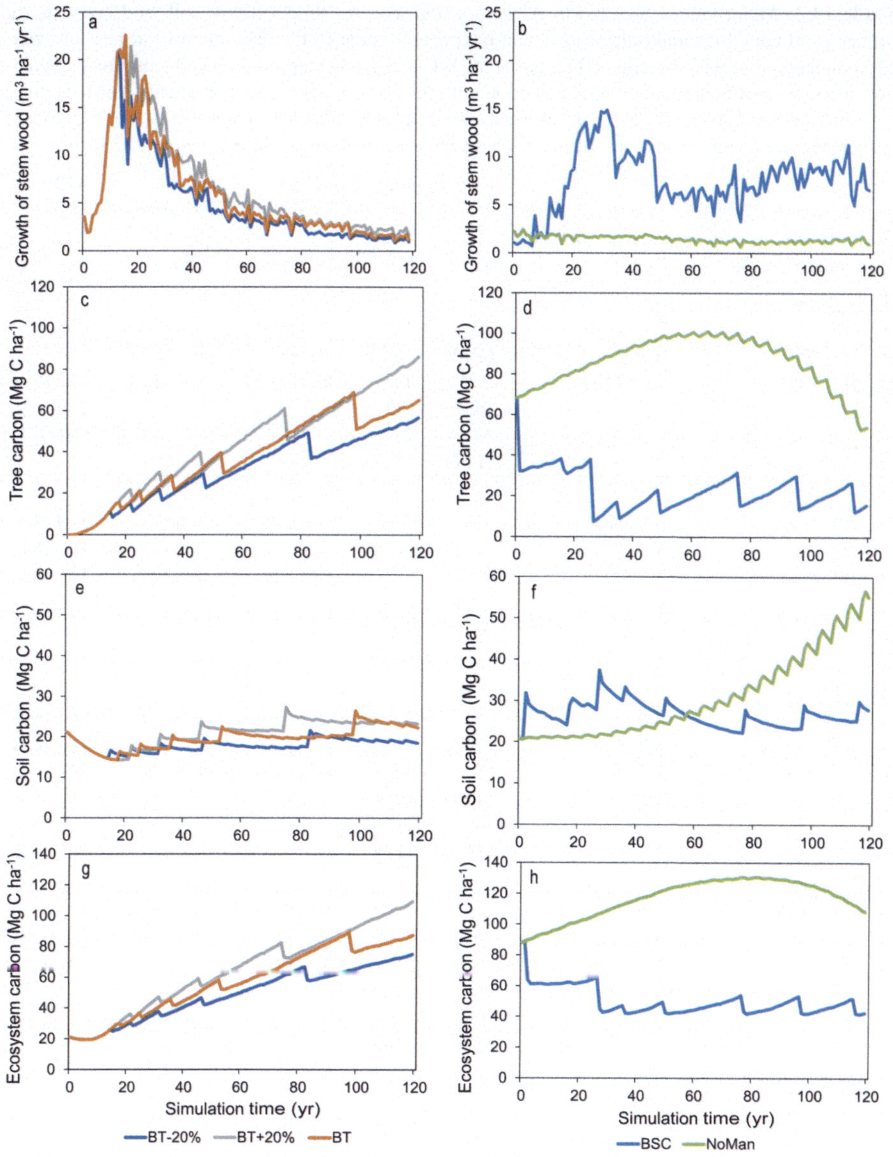


**Fig. 13.5** Stem wood growth and carbon stock in trees, soil, and the ecosystem over time in relation to management strategy (Kellomäki et al. 2023). Open Access. Courtesy of Oxford University Press. Left: mean annual growth of stem wood and carbon stock in trees, soil, and the ecosystem for even-aged management with baseline (BT), reduced (BT + 20), and increased (BT-20) thinning intensities (a, c, e, g). Right: mean annual growth of stem wood and carbon stock in trees, soil, and the ecosystem and for uneven-aged management (BSC) and no management (NoMan) (b, d, f, h). Under both management strategies, soil carbon includes carbon in humus and litter and in coarse woody litter. The simulated values extended over a 120-year rotation length or simulation period

**Table 13.2** Mean values for growth of stem wood, carbon stocks (trees, soil, and ecosystem), timber yield (saw logs and pulp wood), and net present value (NPV, 2% interest rate) under even-aged management with baseline (BT), reduced (BT + 20) and increased (BT-20) thinning intensities, uneven-aged management with selection cuttings (BSC), and no management (NoMan), over rotations of 80, 100, and 120 years or corresponding simulation. In parentheses, the values are percentage values of those from the BT with an 80-year rotation (Kellomäki et al. 2023)

| Regime and simulation period | Growth, $m^3$ $ha^{-1}$ $yr.^{-1}$ | Carbon stock, Mg C $ha^{-1}$ | | | Timber, $m^3$ $ha^{-1}$ $yr.^{-1}$ | NPV (2%), € $ha^{-1}$ $yr.^{-1}$ |
|---|---|---|---|---|---|---|
| | | Trees | Soil | Ecosystem | | |
| Rot. 80 yr | | | | | | |
| • BT | 7.9 (100) | 26 (100) | 19 (100) | 45 (100) | 7.7 (100) | 235 (100) |
| • BT-20 | 6.7 (84) | 22 (85) | 17 (92) | 40 (88) | 6.6 (85) | 229 (98) |
| • BT + 20 | 9.0 (113) | 31 (117) | 21 (114) | 52 (116) | 8.7 (114) | 245 (104) |
| Rot. 100 yr. | | | | | | |
| • BT | 6.8 (86) | 33 (125) | 20 (105) | 53 (117) | 6.6 (85) | 194 (82) |
| • BT-20 | 5.7 (72) | 27 (101) | 17 (93) | 44 (98) | 5.5 (71) | 190 (81) |
| • BT + 20 | 7.7 (97) | 37 (139) | 22 (116) | 58 (129) | 7.4 (96) | 196 (83) |
| Rot. 120 yr | | | | | | |
| • BT | 6.0 (76) | 39 (147) | 21 (113) | 60 (133) | 5.7 (74) | 161 (69) |
| • BT-20 | 5.0 (63) | 31 (119) | 18 (95) | 49 (109) | 4.7 (61) | 158 (67) |
| • BT + 20 | 6.8 (86) | 44 (165) | 22 (118) | 66 (146) | 6.4 (83) | 163 (69) |
| BSC period | | | | | | |
| • 100–180 yr | 7.2 (91) | 24 (92) | 27 (147) | 52 (115) | 8.2 (106) | 173 (74) |
| • 100–200 yr | 7.3 (93) | 24 (90) | 27 (144) | 51 (112) | 7.8 (101) | 145 (62) |
| • 100–220 yr | 7.5 (95) | 23 (87) | 27 (144) | 50 (110) | 7.4 (97) | 124 (53) |
| NoMan period | | | | | | |
| • 100–180 yr | 1.6 (20) | 89 (337) | 24 (131) | 114 (252) | | |
| • 100–200 yr | 1.5 (19) | 90 (339) | 27 (146) | 117 (259) | | |
| • 100–220 yr | 1.5 (19) | 86 (326) | 31 (165) | 117 (259) | | |

*Mean annual growth rate* of stem wood ranged from 5.0 to 9.0 $m^3$ $ha^{-1}$ $yr.^{-1}$, depending on thinning intensity and rotation length under even-aged management (Fig. 13.5):

- Under baseline thinning (BT), increased rotation length from 80 to 120 years reduced the mean growth by 24%.
- Under reduced thinning intensity (BT + 20), mean growth was 13% higher for the 80-year rotation reduced by 3–14% under the 100- and 120-year rotations compared to an 80-year rotation of BT.
- Under increased thinning intensity (BT-20), the mean growth was 16–37% lower depending on the rotation length than under an 80-year rotation of BT.
- Mean growth was 5–9% lower under uneven-aged management, and 79–80% lower under no management compared to BT, depending on the length of the simulation period.

*Mean annual timber yield* under even-aged management was close to that of mean annual growth of stem wood regardless of thinning intensity and rotation (Fig. 13.5):

- Extension of the rotation length from 80 to 120 years reduced the NPV (with a 2% interest rate) by 18–31% under BT.
- Under reduced thinning intensity (BT + 20), the NPV was 4% higher for the 80-year rotation, but 17–31% lower for the 100- and 120-year rotations, than that under an 80-year rotation with BT with 80-year rotation.
- Under increased thinning intensity (BT-20), the NPV was 2–33% lower depending on rotation length than under an 80-year rotation of BT.
- Under uneven-aged management, the NPV was 26–47% lower depending on the length of simulation period than under an 80-year rotation of BT.

*Mean carbon stocks* in the ecosystem, under even-aged management, ranged from 40 to 66 C Mg $ha^{-1}$, depending on thinning intensity and rotation length (22–44 Mg C $ha^{-1}$ in trees and 17–22 Mg C $ha^{-1}$ in soil) (Fig. 13.5):

- Extension of rotation from 80 to 120 years increased mean ecosystem carbon stocks by 17–33% under BT.
- Under reduced thinning intensity (BT + 20), mean carbon stocks in the ecosystem were 16–46% higher depending on rotation length than that under an 80-year rotation of BT (7–65% and 14–18%) higher in trees and soil, respectively.
- Under increased thinning intensity (BT-20), the mean carbon stocks in the ecosystem were – 12 to +9% (–5 to +19% in trees and – 8 to –5% in the soil).

*Under uneven-aged and no management,* mean carbon stocks in the ecosystem were up to 15 and 159% higher than under baseline thinning (BT) with 80-year rotations. High carbon stocks in soil under uneven-aged (up to 47% higher in soil, but 13% lower in trees) and no management (up to 65% higher in soil and up to 239% higher in trees) were due to natural regeneration and high mortality of trees (Fig. 13.5).

## 13.4 Radiative Forcing Related to Albedo and Carbon Stocks in Forestry Perspective

### *13.4.1 Albedo- and Carbon-Based Radiative Forcing in Boreal Forests*

The calculated mean albedo for years 401–1000 ranged from 0.12 to 0.15 depending on the management regime in a simulation study by Kellomäki et al. (2021) (Fig. 13.6; Table 13.3). Compared to the albedo under even-aged management, the albedo was 15% lower under no management and 10% higher under uneven-aged management. The differences were related to the differences in aboveground biomass over the simulation period. In clear-cut areas, the albedo was 0.23, thus being substantially higher than that under even-aged management (Kellomäki et al. 2021).

Based on albedo, the mean radiative forcing varied in the range of 5.0–7.2 GJ $ha^{-1}$ $d^{-1}$ over a 600-year period (Fig. 13.7; Table 13.3). Under no management, the forcing was 22% higher than that under even-aged management, while the forcing under uneven-aged management was 15% higher than that under the reference management scenario. For ecosystem carbon stocks (in trees and soil), the mean radiative forcing (cooling effect) for years 401–1000 was −4.3, −6.6, and − 13.7 GJ $ha^{-1}$ $d^{-1}$ under uneven-aged, even-aged, and no management, respectively. The mean cooling effect of carbon in trees was highest under no management (−13.8 GJ $ha^{-1}$ $d^{-1}$), followed by even-aged (−9.6 GJ $ha^{-1}$ $d^{-1}$) and uneven-aged management (−5.2 GJ $ha^{-1}$ $d^{-1}$). Correspondingly, the mean warming effect (increase in radiative forcing) of carbon in the soil was higher under even-aged management (2.9 GJ $ha^{-1}$ $d^{-1}$) than under no and uneven-aged management (0.9 GJ $ha^{-1}$ $d^{-1}$). No management reduced the total radiative forcing (−6.5 GJ $ha^{-1}$ $d^{-1}$) through the combined effect of changes in albedo and ecosystem carbon stocks over years 401–1000. In contrast, even-aged (−0.8 GJ $ha^{-1}$ $d^{-1}$) and uneven-aged (0.7 GJ $ha^{-1}$ $d^{-1}$) management induced minor changes in radiative forcing compared to the reference scenario. The beneficial effect of increasing carbon stocks in trees was negated by adverse changes in albedo and soil carbon. However, the no management option resulted in a significant net cooling effect, whereas for even-aged and uneven-aged management, the opposing effects of changes in albedo and carbon stocks canceled each other out, with a small net effect (Kellomäki et al. 2021).

### *13.4.2 Radiative Forcing Related to Management for Timber*

Radiative forcing refers to changes in the radiative balance and consequent climate cooling/warming in response to the content of greenhouse gases in the atmosphere (IPCC 2007). The total radiative forcing ($\Delta RF_{tot}$) is the sum of radiative forcing based on albedo ($\Delta R_{yr}$) and on ecosystem carbon ($\Delta RF_E$) over a given period. The total forcing (or combined forcing) ($\Delta \mathrm{RF}_{tot}$) includes the reflecting changes in the

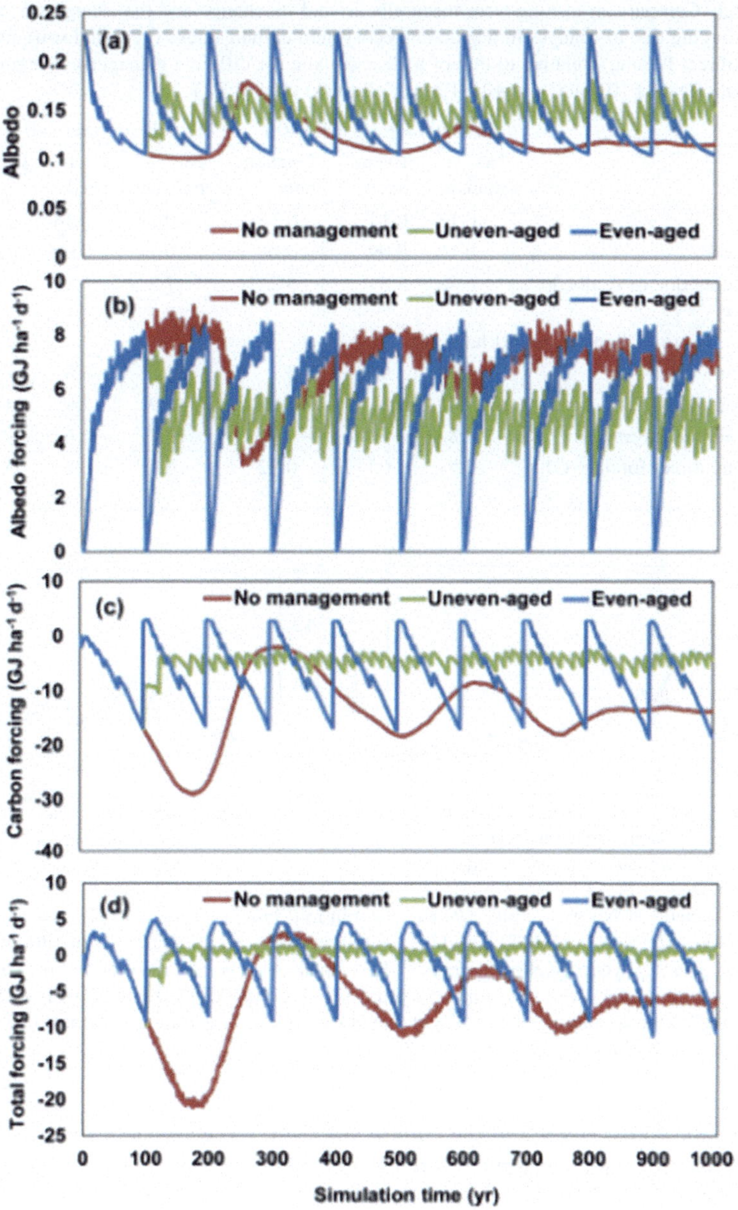


**Fig. 13.6** Albedo- and carbon-based radiative forcing in forests (Kellomäki et al. 2021). Open Access. Courtesy of Oxford University of Press. Figure includes (**a**) albedo and radiative forcing due to (**b**) changes in albedo (albedo forcing), (**c**) ecosystem carbon stocks (trees and soil) alone (carbon forcing), and (**d**) together (total forcing). Simulations were done under three different management regimes over a 1000 period. Albedo for a clear-cut site is shown as a gray line in (**a**). The clear-cut area is assumed to have the albedo of 0.23

**Table 13.3** Comparison of long-term mean albedo and its change and the changes in the mean radiative forcing due to changes in albedo and ecosystem carbon stocks (trees and soil), alone and together (direct forcing), relative to that of a clear-cut site for different management regimes over the simulation years 101–400 and 401–1000 (Kellomäki et al. 2021)

| | Simulation years 101–400 | | | Simulation years 401–1000 | | |
|---|---|---|---|---|---|---|
| Variables | No manage | Even-aged | Uneven-aged | No manage | Even-aged | Uneven-aged |
| Albedo | 0.13 | 0.14 | 0.15 | 0.12 | 0.14 | 0.15 |
| • Δ albedo | 0.10 | 0.09 | 0.08 | 0.11 | 0.09 | 0.08 |
| • Δ forcing due to Δ albedo, GJ ha$^{-1}$ d$^{-1}$ | 6.3 | 5.8 | 5.2 | 7.2 | 5.9 | 5.0 |
| Δ forcing due to carbon stocks, GJ ha$^{-1}$ d$^{-1}$ | | | | | | |
| • Carbon in trees | −13.4 | −9.3 | −5.7 | 12.8 | −9.8 | −5.2 |
| • Carbon in soil | −0.4 | 3.0 | 0.8 | −0.9 | 2.9 | 0.9 |
| • Carbon in ecosystem | −13.8 | −6.3 | −4.9 | −13.7 | −6.6 | −4-3 |
| • Δ Total direct forcing, GJ ha$^{-1}$ d$^{-1}$ | −7.5 | −05 | 0.32 | −6.5 | −0.8 | 0.7 |

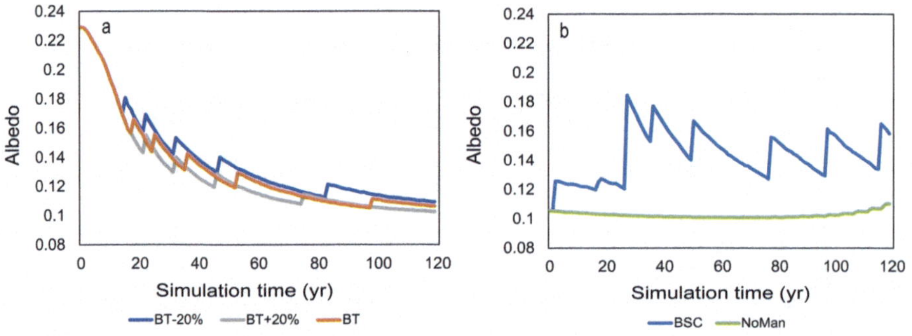


**Fig. 13.7** Albedo of Norway spruce (*Picea abies*) in mid-boreal forests (62° N) on a moderately fertile site (*Myrtillus* site type, MT) based on simulations using an initial density of 1800 seedlings per hectare over 120 years (Kellomäki et al. 2023). Open Access. Courtesy of Oxford University Press. Left: albedo for even-aged management with baseline (BT), reduced (BT + 20), and increased (BT-20) thinning intensities. Right: albedo for uneven-aged management (BSC) and for no management (NoMan)

albedo of the forest cover and the carbon stocks in the ecosystem (trees and soil). One obtains the total mean annual forcing if dividing the total radiative forcing over the simulation period by the length of the period ($t = n$, years):

$$\Delta RF_{tot} = \left( \sum_{t=1}^{t=n} (\Delta R_{yr}) + \sum_{t=1}^{t=n} (\Delta RF_E) \right). \tag{13.15}$$

Management affects the reflectance of radiation through changes in biomass (e.g., stocking of trees and soil), and the properties of biomass in forest ecosystem.

Kellomäki et al. (2021, 2023) investigated how forest management may affect radiative forcing and concurrent timber yield in a managed chronosequence of Norway spruce (*Picea abies*) in middle-boreal conditions (62° N) as detailed in Box 13.1.

The focus of the simulations was on (i) the mean annual growth of stem wood; (ii) timber yield (harvested stem wood for saw logs and pulp wood (diameter for pulp wood >6 cm) and for saw logs >16 cm), (iii) carbon stocks in the forest ecosystem (trees and soil), (iv) radiative forcing based on the changes in ecosystem carbon stocks, (v) radiative forcing based on changes in albedo, and (vi) combined radiative forcing based on changes in albedo and ecosystem carbon stocks. In the simulations, the baseline management for even-aged (BT) and uneven-aged (BSC) were used following the current recommendations for the intensity of thinning/selective cutting. In both management strategies, the harvest intensity was further increased/reduced by 20% of the basal area remaining in harvesting. Simulations excluding management were also done for evaluating how management affects radiative forcing compared to that in naturally grown trees through chronosequences.

### *13.4.3 Radiative Forcing Based on Albedo and Ecosystem Carbon Stocks Related to Management*

Under even-aged management, mean albedo was in the range of 0.129–0.150, depending on thinning intensity and rotation length (Figs. 13.7 and 13.8):

- Mean albedo decreased with extended rotation length from 80 to 100 or 120 years by 5–9% under the baseline thinning (BT). Under reduced thinning intensity (BT + 20), mean albedo was 3–11% lower for the rotation from 80 to 100 or 120 years, compared to BT with the 80-year rotation.
- Under increased thinning intensity (BT-20), mean albedo was 4% higher for the 80-year rotation, but 1–5% lower at longer rotations of 100 or 120 years, compared to BT with the 80-year rotation.
- Under uneven-aged and no management, mean albedo was up to 2% and 29% lower than that under an 80-year rotation of BT.

Under even-aged management, albedo-based radiative forcing was in the range of 5.0–6.4 GJ $ha^{-1}$ $d^{-1}$ relative to a clear-cut site prior to reestablishment of forest cover, depending on thinning intensity and rotation length (Figs. 13.7 and 13.8). Both reduced thinning intensity and extended rotation length from 80 to 100 or 120 years increased albedo-based radiative forcing. Under uneven-aged and no management, the albedo-based radiative forcing was 5.5 and 8.0 GJ $ha^{-1}$ $d^{-1}$.

The ecosystem carbon stock affecting radiative forcing (relative to a clear-cut site before reestablishment of forest cover) ranged from −13.1 to −18.8 GJ $ha^{-1}$ $d^{-1}$ under even-aged management (Fig. 13.8; Table 13.3). Reduced thinning intensity and extended rotation length from 80 to 100 or to 120 years decreased forcing due to

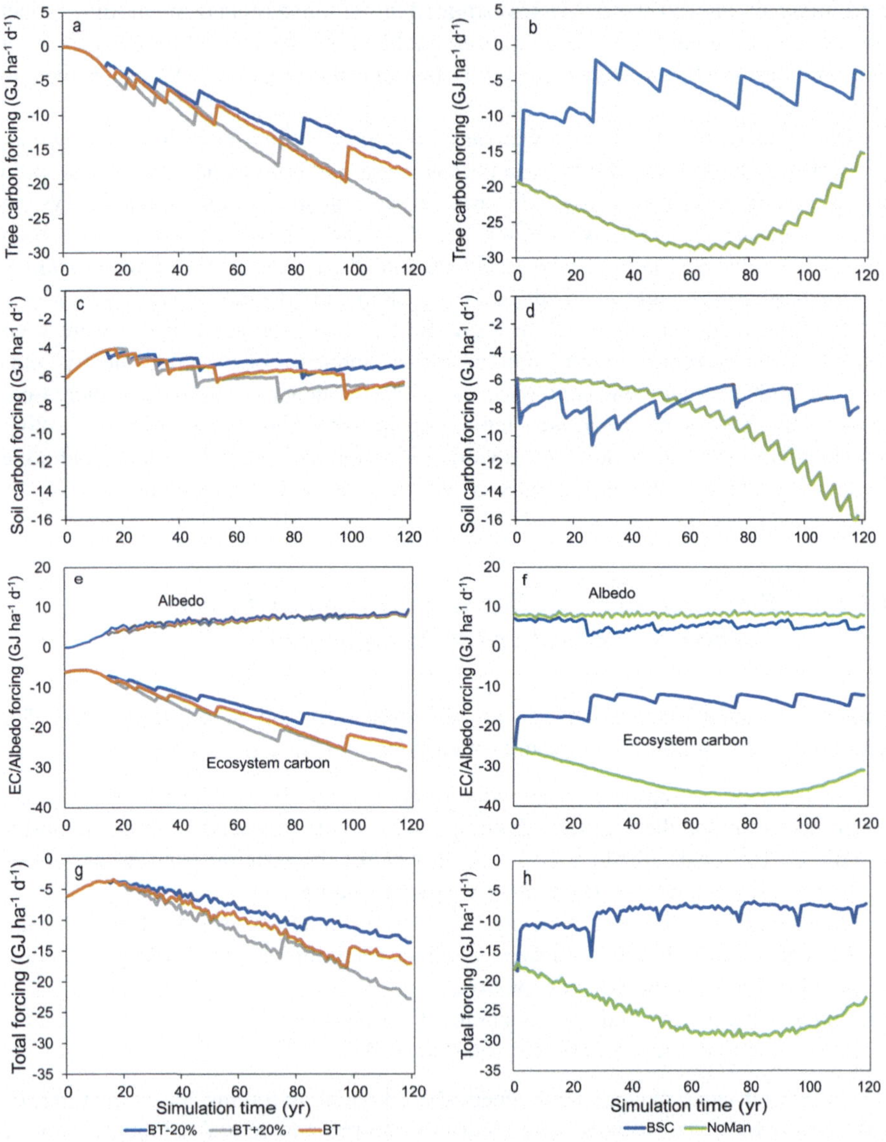


**Fig. 13.8** Albedo- and carbon-based forcing related to management (Kellomäki et al. 2023). Open Access. Courtesy of Oxford University Press. Radiative forcing is based on changes in tree, soil, and ecosystem carbon stocks, albedo, and total radiative forcing for even-aged management with baseline (BT), reduced (BT + 20), and increased (BT-20) thinning intensities (a, c, e, g), and for uneven-aged management (BSC) and no management (NoMan) (b, d, f, h), over a 120-year rotation length or simulation period

the increasing ecosystem carbon stock. Under uneven-aged and no management, the ecosystem carbon stock-based forcing was up to −13.9 and − 33.4 GJ $ha^{-1}$ $d^{-1}$. The maintenance of higher carbon stocks (e.g., reduced thinning intensity), as well as extended rotation length or no management, increased the relative contribution of tree carbon stocks to radiative forcing compared to that of soil carbon stock. Changes in ecosystem carbon stock affected the total radiative forcing more than that of albedo, which was not very sensitive to changes in tree carbon stocks (e.g., to thinning intensity).

### *13.4.4 Radiative Forcing, Timber Yield, and Profitability Related to Management*

Table 13.4 further summarizes how management affects the mean albedo and radiative forcing based on changes in albedo and carbon stock alone and together over time (Kellomäki et al. 2023). The simulations represent even-aged management with baseline (BT), reduced (BT + 20), and increased (BT-20) thinning intensities. Under uneven-aged management, only the basic selective cutting (BSC) was used over the simulation periods corresponding to the rotation length of 80 and 120 years. Under no management (NoMan), the corresponding simulation periods were used. Albedo differences and radiative forcing-based changes in albedo and ecosystem carbon stocks were calculated relative to a clear-cut site prior to reestablishment of forest cover. The numbers in parentheses are percentage values relative to those from an 80-year rotation of BT.

Table 13.4 shows that the total radiative forcing ranged from −7.1 to −12.4 GJ $ha^{-1}$ $d^{-1}$ under even-aged management and up to −8.4 and − 25.4 GJ $ha^{-1}$ $d^{-1}$ under uneven-aged and no management, respectively. Concurrently, reduced thinning intensity and extended rotation lengths from 80 to 100 or to 120 years decreased the total radiative forcing compared to 80-year rotations of baseline management (BT) and enhanced cooling effect as shown in Fig. 13.9. Under even-aged management with BT and/or increased thinning intensity with rotation lengths from 80 to 100 years, the opposing effects of changes in albedo and ecosystem carbon stocks largely canceled each other out, i.e., there was only a small net climatic effect. This was also the case under uneven-aged management. No management resulted in the highest net cooling effect (Kellomäki et al. 2023).

Total radiative forcing was dominated by changes in the total ecosystem carbon stocks (in trees and soil), while radiative forcing was related to albedo changes with decreased climate benefits (Table 14.3). The simulations showed that both thinning and selective cutting increased albedo slightly due to the decrease of aboveground forest biomass. However, thinning and selective cutting affected forest carbon stocks relatively more than albedo and, consequently, also radiative forcing based on changes in carbon stocks. This retained the dominant role of forest carbon stocks in mitigation of climate change. These stand-based findings suggest that the time (period) chosen for assessment of findings has a significant effect on the

**Table 13.4** Mean albedo and radiative forcing based on changes in albedo and carbon stock alone and together under even-aged management with baseline (BT), reduced (BT + 20), and increased (BT-20) thinning intensities, uneven-aged management with selective cutting (BSC), and no management (NoMan), over rotations of 80, 100, and 120 years or corresponding simulation periods. Albedo differences and radiative forcing are calculated relative to a clear-cut field before reestablishing a forest cover. In parentheses, the values are percentage values of those from the BT with an 80-year rotation length (Kellomäki et al. 2023)

| Regime and simulation period | Albedo, fraction of total (0–1) | Δ Albedo | Δ Forcing albedo, GJ ha$^{-1}$ d$^{-1}$ | Δ Forcing carbon stock, GJ ha$^{-1}$ d$^{-1}$ | | | Δ Total forcing, GJ ha$^{-1}$ d$^{-1}$ |
|---|---|---|---|---|---|---|---|
| | | | | Trees | Soil | Ecosyst. | |
| Rot. 80 yr | | | | | | | |
| BT | 0.145 (100) | 0.085 (100) | 5.4 | −7.5 | −5.4 | −12.9 | −7.5 (100) |
| BT-20 | 0.150 (104) | 0.079 (94) | 5.0 | −6.4 | −5.7 | −12.1 | −7.1 (95) |
| BT + 20 | 0.140 (97) | 0.089 (105) | 5.6 | −8.8 | −6.0 | −14.8 | −9.2 (123) |
| Rot. 100 yr | | | | | | | |
| BT | 0.138 (95) | 0.092 (108) | 5.8 | −9.4 | −4.9 | −14.3 | −8.5 (113) |
| BT-20 | 0.143 (99) | 0.086 (102) | 5.5 | −7.6 | −4.9 | −12.5 | −7.0 (93) |
| BT + 20 | 0.134 (93) | 0.095 (112) | 6.0 | −10.5 | −5.1 | −15.6 | −9.6 (128) |
| Rot. 120 yr | | | | | | | |
| BT | 0.132 (91) | 0.097 (115) | 6.2 | −11.1 | −6.0 | −17.1 | −10.9 (145) |
| BT-20 | 0.138 (95) | 0.092 (108) | 5.8 | −9.0 | −6.3 | −15.3 | −9.5 (127) |
| BT + 20 | 0.129 (89) | 0.100 (119) | 6.4 | −12.5 | −6.3 | −18.8 | −12.4 (165) |
| BSC period | | | | | | | |
| 100–180 yr. | 0.142 (98) | 0.086 (102) | 5.5 | −6.9 | −7.0 | −13.9 | −8.4 (112) |
| 100–200 yr. | 0.143 (99) | 0.086 (102) | 5.5 | −6.8 | −7.0 | −13.8 | −8.3 (111) |
| 100–220 yr | 0.144 (99) | 0.085 (101) | 5.5 | −6.6 | −7.0 | −13.7 | −8.2 (109) |
| NoMan period | | | | | | | |
| 100–180 yr. | 0.102 (71) | 0.127 (150) | 8.0 | −25.4 | −6.8 | −32.2 | −24.2 (327) |
| 100–200 yr. | 0.102 (71) | 0.127 (150) | 8.0 | −25.5 | −7.0 | −32.5 | −24.5 (327) |
| 100–220 yr | 0.103 (71) | 0.127 (150) | 8.0 | −24.6 | −8.8 | −33.4 | −25.4 (339) |

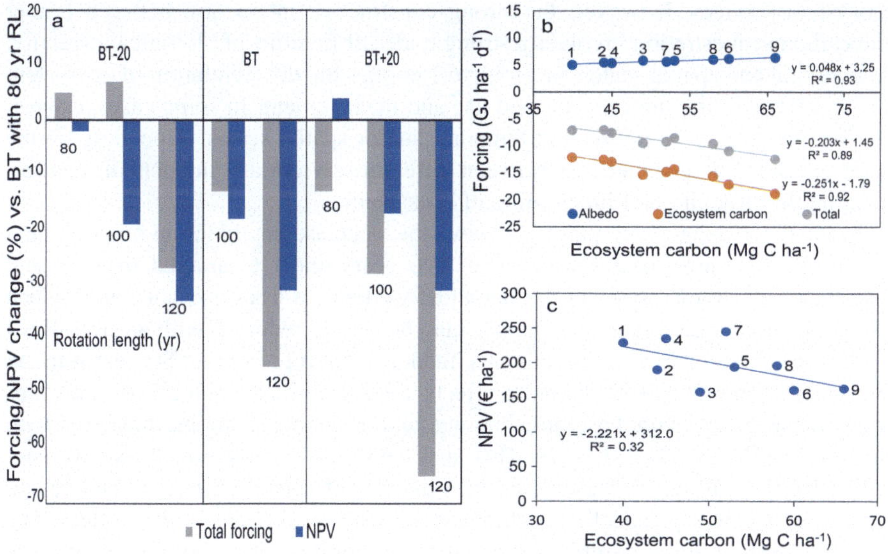


**Fig. 13.9** Radiative forcing and net present value (NPV) of timber under management using varying rotation length and thinning intensity in even-aged management as a function of total ecosystem carbon (Kellomäki et al. 2023). Open Access. Courtesy of Oxford University Press. Left: percentage changes in total radiative forcing and net present value (NPV, interest rate 2%) of timber for different even-aged management regimes. They are indicated with rotation length (number below columns) and thinning intensity (BT-20, BT, and BT + 20 above columns) relative to the baseline even-aged management (BT) with a rotation length of 80 years. Right/upper: relationship between the mean ecosystem carbon stocks (Mg C $ha^{-1}$) and radiative forcing (GJ $ha^{-1}$ $d^{-1}$) changes related to the ecosystem carbon stocks (red line below) and to albedo (blue line upper) and the combination of these two (violet line in the middle). Right/lower: net present value (NPV, interest rate 2%) for even-aged management regimes, with different thinning intensities (BT, BT-20, BT + 20) and rotation lengths from 80 to 120 years, as a function of the ecosystem carbon stock. Numbers 1–9 in the subfigures on the right indicate the simulations used in compiling the figures

interpretation of results. Under even-aged management, the most profitable options were clear-cutting with planting if using 80-year rotation. The extension of rotation with delay of the final cutting reduced the profitability. Under no management, the net cooling was highest, but no timber or harvesting income was obtained. Suitable incentives are likely needed to promote management regimes that increase the climatic benefits of forestry (Kellomäki et al. 2023).

## 13.5 Concluding Remarks

Management of varying intensity directly and indirectly controls the structure and functioning of forest ecosystems and affects carbon sequestration and carbon stocks. They have strong synergetic effects on the supply of timber and biomass and other ecosystem services. However, the strong constructive interaction between timber and carbon sequestration involves a strong trade-off in terms of biological diversity. At temporal and spatial scales, this trade-off implies the differentiation of ecosystem services through the intensity of land use and management. In some cases, ecosystem services might be optional at temporal and/or spatial scales, depending on the trade-off and synergistic effects between different services and impacts of management on the structure and functioning of ecosystems.

From the management point of view, the decrease in thinning intensity and extension of rotation length from 80 to 120 years under even-aged management may increase carbon stocks in the boreal ecosystem (in trees and soil) and reduce albedo. Consequently, radiative forcing may be clearly reduced, with an overall net cooling effect compared to the reference management representing 80-year rotation and baseline thinning (BT). Cooling effects are also quite the same between even-aged management with baseline thinning and/or higher thinning intensity with rotation lengths of 80–100 years. This also held under uneven-aged management with selective cuttings. In these cases, the opposing effects of changes in albedo and carbon stocks largely cancelled each other out, with a small net cooling effect. The no management regime resulted in the highest net cooling effect, but with no income from timber harvesting.

## References

Äijälä O, Koistinen A, Sved J, Vanhatalo K, Väisänen P (eds) (2014) Hyvän metsänhoidon suositukset—METSÄNHOITO. Kehittämiskeskus Tapion julkaisuja, Metsäkustannus Oy, Helsinki

Govers G, Merck R, van Oost K, van Wesemael B (2013) Soil organic carbon management for global benefits: a discussion paper. Paper presented at the workshop "Soil Organic Carbon Benefits": a scope study, 10th–12th September 2012, Nairobi. Workshop organized by the Scientific and Technical Advisory Panel of the Global Environmental Facility

IPCC (2007) Climate change 2007: impacts, adaptation, and vulnerability. In: Working group II contribution to the fourth assessment report of the intergovernmental panel on climate change. Cambridge University Press, Cambridge

Kellomäki S, Väisänen H, Kirschbaum MUF, Kirsikka-aho S, Peltola H (2021) Effects of different management options of Norway spruce on radiative forcing through changes in carbon stocks and albedo. Forestry 94:588–597

Kellomäki S, Strandman H, Kirsikka-Aho S, Kirschbaum MUF, Peltola H (2023) Effects of thinning intensity and rotation length on albedo and carbon stock based on radiative forcing in boreal Norway spruce stands. Forestry 96(4):518–529

Kirschbaum MUF, Whitehead D, Dean S, Beets PN, Shepard JD, Ausseil A-GE (2011) Implications of albedo changes following afforestation on the benefits of forests as carbon sinks. Biogeosciences 8:3687–3696

Lukeš P, Stenberg P, Rautiainen M (2013) Relationship between forest density and albedo in the boreal zone. Ecol Model 261–262:74–79

Rautiainen M, Stenberg P, Mõttus M, Manninen T (2011) Radiative transfer simulations link boreal forest structure and shortwave albedo. Boreal Environ Res 16:91–100

Strandman H, Väisänen H, Kellomäki S (1993) A procedure for generating synthetic weather records in conjunction of climatic scenario for modelling of ecological impacts of changing climate in boreal conditions. Ecol Model 70(3–4):195–220

# Part V
# Management of Carbon in Boreal Forests for Timber and Climatic Benefits: Overall Review

# Chapter 14
# Management of Carbon in Boreal Forests for Climatic Benefits: Overall Review

## Contents




**Abstract** On a global scale, the carbon in boreal forest ecosystems represents a large proportion of the land-based carbon stored in aboveground biomass (trees, ground vegetation) and soil organic matter (litter, humus) and living/dead roots of trees and ground vegetation. Intensified forest management increases ecosystem carbon stocks by increasing primary production in different phases of the growth and development of trees and ground vegetation through succession of forest chronosequences. Intensive forest management includes the use of fast-growing genotypes of tree species, mechanical soil preparation of low to moderate intensity, appropriate thinning, and fertilization. Similar management measures can also be used in extensive management. However, the intensity, frequency, and management investment per unit land area are lower than in intensive management. Carbon sequestration and carbon stocks have strong relationships and synergetic effects with the supply of timber and biomass as management intensity increases. However, the strong relationships and constructive interaction between timber and carbon sequestration involve a strong trade-off in relation to biological diversity.





S. Kellomäki, *Forest Management for Timber Production and Climate Change Mitigation*, Managing Forest Ecosystems 44,
https://doi.org/10.1007/978-3-031-71575-4_14

## 14.1 Links of Carbon in Forest Ecosystems with Management and Climatic Warming

### *14.1.1 Carbon in Forests Related to Ecosystem Dynamics and Management and Harvest*

Currently, international forest policy emphasizes the increasing role of forests in responding to the production of timber and biomass and maintenance of biodiversity while reducing net carbon dioxide emissions for mitigating climatic warming. In this regard, forestry refers to the use of forested land to producing different ecosystem services, which have ecological, economic, and social value locally and globally. In this regard, various management strategies and measures are available for increasing carbon sinks and retaining carbon in ecosystems, with the concurrent focus on the sustainable production of timber and biomass and mitigation of climatic change.

In forest ecosystems, a large proportion of the carbon is stored in aboveground biomass (trees, ground vegetation) and soil organic matter (litter, humus) and living/dead roots of trees and ground vegetation through different phase of chronosequences in the scale of forest landscape. This is summarized in Table 14.1 depicting how the ecosystem structure and functioning prior to regenerative cutting change in relation to different phases of chronosequences under even-aged management.

According to Table 14.1, carbon sequestration and stocking occur in (i) vegetation (trees, ground vegetation), (ii) organic matter on soil (litter) and soil organic matter in upper soil profile, and (iii) the profile of mineral soil down to the parent material. Under even-aged management, the carbon sequestrations and stocks oscillate substantially over time depending on the developmental phase of tree stands and management/harvest regime. Carbon sequestration and stocks recover more rapidly in trees and ground vegetation than those in soil, with a delayed response to carbon sequestrated in trees and ground vegetation. Following, overall comments are evident concerning, how management effects on carbon dynamics through forest chronosequences (Table 14.1):

- Regenerative cutting, as clear-cutting, reduces most of the carbon stocks in removed trees and reduces substantially litterfall and enhances decay of soil carbon inherent from the previous production cycle. Decay of soil carbon may further be enhanced by harvest and soil management in mixing with mineral soil below organic matter on the surface of mineral soil.
- Following clear-cut, the mass and species composition of ground vegetation shift from the dominance of dwarf shrubs and mosses to the dominance of grasses and herb with smaller carbon stock than in vegetation before cutting and soil management.
- During reforestation and management of seedling stands, soil carbon further reduces, but the reduction rate is declined by the increasing growth of seedlings

**Table 14.1** Structure of boreal forest ecosystem in terms of tree cover, ground vegetation, organic soil, and inorganic soil prior to management and harvest, with the impacts on the dynamics of carbon linked to even-aged management related to the successional structure of chronosequences through the phase-by-phase growth and development over a production cycle (rotation) on moderately fertile sites (*Myrtillus* site type, MT). Compiled by the author of this book

| Ecosystem structure (vegetation and soil) | Ecosystem structure with changes related to management and harvesting over a chronosequence | | | | |
|---|---|---|---|---|---|
| | Ecosystem structure prior to regenerative cutting | Regenerative cutting and soil management reforestation | Management in seedling phase after reforestation | Precommercial management in seedling and pole phase for biomass and pulp wood | Commercial management maturing phases for pulpwood, saw logs, and biomass |
| **Tree cover** | Ready for regeneration, large carbon stocks in soil and trees, with slight decline of growth and carbon sequestration | Clear-cut area or shelterwood area for regeneration, removal of most legacy carbon, i.e., half, or even more of carbon prior to deforestation | Planting/seeding or natural regeneration initiating uptake of de novo carbon for recovery of ecosystem carbon balance in enhancing growth and litterfall | Spacing and cleaning of seedling stands, harvesting for energy biomass, reduction of carbon in trees and carbon increase in ground vegetation | Spacing and cutting of timber (pulpwood, saw logs) and biomass, Fertilization combined with thinning to increase carbon uptake and stocks |
| **Ground vegetation** | Climax type, with dominance of mosses and dwarf shrubs, scattered herbs, and grasses, with carbon stock on soil surface dominated by litter, mosses, and dwarf shrubs | Declining mosses and dwarf shrubs, increasing growth of herbs and grasses, scattered natural tree seedlings established after reforestation, increasing carbon sequestration | Dominance of herbs and grasses, with recovery of dwarf shrubs and mosses, planted/seeded and natural seedlings with enhancing growth, carbon sequestration and carbon stocking in trees, ground vegetation and soil | Reducing dominance of herbs and grasses, recovery of dwarf shrubs and mosses, reducing carbon sequestration and stocks below shading trees, increasing carbon stocks | Dominance of mosses and dwarf shrubs, scattered herbs and grasses, stable carbon sequestration and stocks under shading trees, declining carbon uptake in trees due to gradually declining growth in maturing trees |
| **Organic soil** | Litterfall from trees dominating litter and soil organic matter, representing large carbon stocks under stable or declining litterfall related | Increasing litter from harvesting, decay and decline of legacy carbon due to increasing soil temperature and moisture, with carbon loss over 10– | Increasing litter with de novo carbon, further decrease of legacy carbon, recovery of carbon in humus and increasing carbon in soil due to | Increasing litterfall and increasing humus layer, mainly due to de novo carbon from the increasing growth of trees and litterfall, increasing | Increasing litterfall due to growth of trees and ground vegetation, increasing humus layer, and increasing soil carbon due to logging |

(continued)

**Table 14.1** (continued)

| Ecosystem structure (vegetation and soil) | Ecosystem structure with changes related to management and harvesting over a chronosequence | | | | |
|---|---|---|---|---|---|
| | Ecosystem structure prior to regenerative cutting | Regenerative cutting and soil management reforestation | Management in seedling phase after reforestation | Precommercial management in seedling and pole phase for biomass and pulp wood | Commercial management maturing phases for pulpwood, saw logs, and biomass |
| | to declining growth of trees, and ground vegetation | 20 years after cutting before recovery of carbon accumulation | increasing growth and above soil litterfall and belowground root litter through soil profile | carbon stocks below soil surface due to root litter | residues, increasing carbon sequestration and retention of carbon in soil stocks due to fertilization |
| **Inorganic soil** | Humus particles mixed with particles of mineral soil | Declining/increase of organic matter in soil, with mixing of humus particles in mineral soil | Increased mixing of humus particles in mineral soil, carbon accumulation in soil | Increased mixing of humus particles in mineral soil, carbon accumulation through soil profile | Increased mixing of humus particles in mineral soil, carbon accumulation through soil profile |

and litterfall and the recovery of the dominance of dwarf shrubs and mosses and growth and mass of herbs and grasses.

- During pre-commercial phase, the enhanced growth of seedlings, with increasing litterfall compensate previous losses of soil carbon due to in decay and reduced litterfall. Recovery of growth and mass of dwarf shrubs and mosses further increase the soil carbon, which rapidly converges the level prior to regenerative cutting.
- During commercial management, regular harvest reduces the carbon bound in trees while increasing the spacing and carbon uptake and litterfall in remaining trees, with balancing of the carbon losses in harvest. Logging residues further compensates for the loss of soil carbon as does the recovered growth and mass of dwarf shrubs and mosses.

Management makes it possible to control the carbon dynamics in forest ecosystems as timber and biomass and retain carbon for mitigating climate change. Concurrently, management affects the uptake (sequestration) of carbon and the accumulation of carbon (stocks), with impacts on the combined forcing related to the ecosystem carbon stocks and the reflectance of solar radiation (albedo), with the combined impacts on cooling/warming (Table 14.1):

### *14.1.2 Carbon Dynamics in Forest Ecosystems with Links to Management and Harvest*

Carbon dynamics in forest ecosystems mediate the interaction between atmospheric carbon and primary production and carbon lost in atmosphere through autotrophic and heterotrophic respiration and disturbances (Fig. 14.1). These processes control the sequestration and accumulation of carbon in ecosystem including in living organic matter (biomass) in trees and ground vegetation and dead organic matter in litter and humus in soil. Carbon stocking and the consequent reflectance of solar radiation alone/combined effect on radiative forcing with impacts on climatic cooling/warming. In managed forests, carbon stocks provide timber and biomass for wood-based manufacturing, thus forming temporary carbon stocks in forest-based products outside the ecosystem. Harvesting of wood (pulpwood, saw logs) provides raw materials for short-living pulp and paper products and long-living wood products, such as those used in the construction industry.

Wood and biomass may also be harvested for energy production, which may also be based on abandoned wood-based materials (Fig. 14.1). Carbon in wood-based products reverts immediately to the atmosphere through combustion. This hold for the decomposition and return of carbon to the atmosphere. This takes from a few years up to many decades depending on the resistance of materials in use and/or in decomposing processes. Carbon bound in wood-based materials is also emitted through burning in energy production and/or through the decomposition of materials abandoned in landfills. This yields uncertainty in estimating the time for the forest-

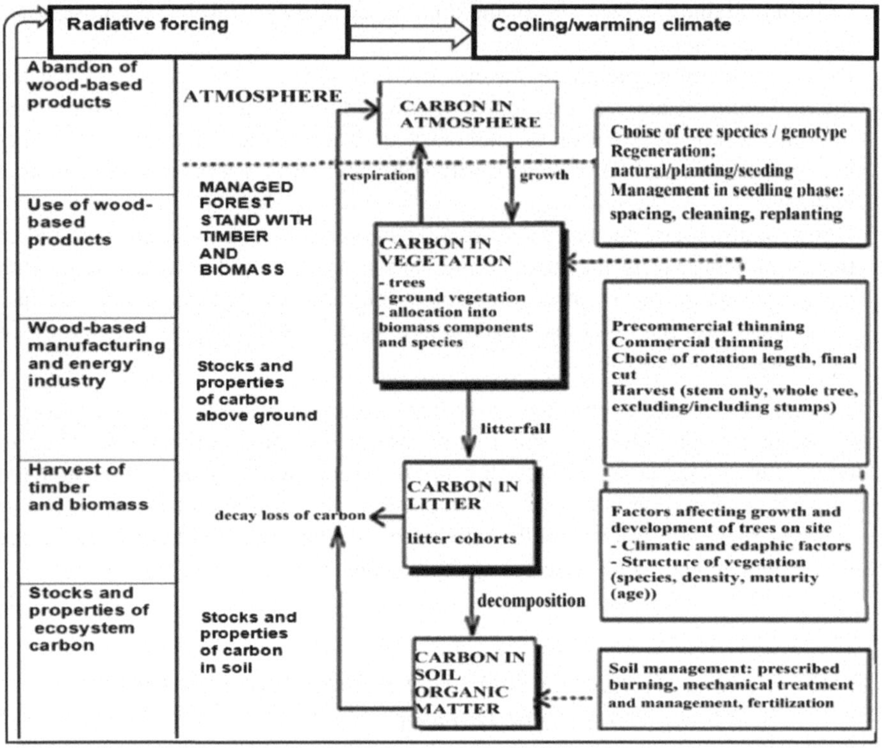


**Fig. 14.1** Outlines of structure and functioning of forest ecosystems that affect the dynamics of major carbon stocks and fluxes. These are linked to the management and production of timber and biomass for the manufacturing and energy industries, with impacts on radiative forcing and climatic warming/cooling. Compiled by the author of this book

based carbon being lost to the atmosphere. Concurrently, it is still poorly known how long forest-based carbon stocked outside forest ecosystems is capable to mitigate climatic warming (Fig. 14.1) (e.g., Hurmekoski et al. 2020).

### 14.1.3 Carbon Dynamics Related to Management and Harvest Through Chronosequences

Figure 14.2 further summarizes the main findings addressed previously in the context of successional dynamics under even-aged management, with the main examples based on the experiences for northern Europe. The critical phase of carbon loss and uptake is harvesting, especially final cutting for reforestation. The balance between carbon loss and uptake is also lost in soil management for improving regeneration success. The dual carbon recovery is gradually regained through

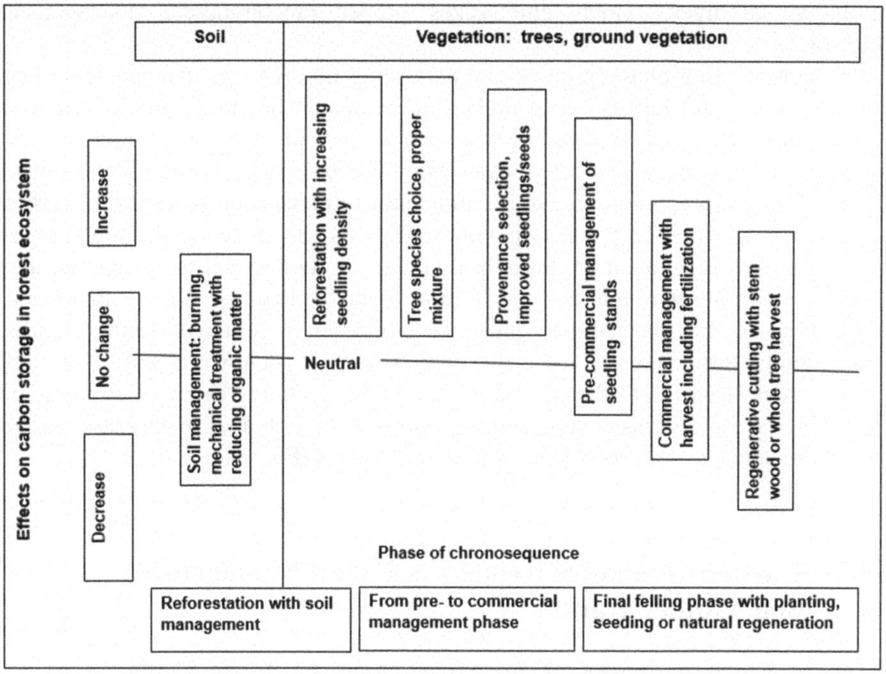


**Fig. 14.2** Phase-by-phase summary of how forest management and harvesting likely affect carbon in soil and vegetation through a chronosequence in managed boreal forest under even-aged management. Compiled by the author of this book

regeneration. This recovery is dependent on tree species choice or species genotype, soil management, and the density of established seedlings in the clear-cut area. On the other hand, natural regeneration with non-intensive soil management reduces the loss of soil carbon. This especially holds for the shelterwood method of reforestation due light soil management required to create the necessary seedbeds, compared to more intensive soil disturbance in clear-cutting for establishing forest plantations.

In even-aged management, Vuokila (1980) divided the management following reforestation and growth and development of tree stands into a phase of increasing stocking and a phase of harvesting. In forest plantations, the carbon taken up in the phase of increasing stocking rises with increased seedling density, especially when using improved seedlings/seeds (Fig. 14.2). This substantially accelerates the recovery of the carbon balance to meet the potential that the site fertility makes possible for a given tree species. In this regard, the suitable selection of tree species is also important in the long term for wood production and further litter production to maintain/increase soil carbon. On poor sites, single-species selection may be the only feasible option, such as of Scots pine in large areas in northern Europe. The use of a mixture of several species may be beneficial but is limited to fertile sites that allow the successful growth of several tree species concurrently. High seedling density increases litterfall and total growth of tree stand. This may reduce the growth of individual trees, which emphasizes proper precommercial management (Fig. 14.2).

In the harvesting phase, commercial harvesting involves the management operations necessary for harvesting of timber and biomass prior to the final cutting for regeneration. The loss of ecosystem carbon is dependent on the frequency and intensity of thinning, which affect stem wood carbon stocks and the carbon input in soil. Furthermore, whole-tree harvesting tends to reduce soil (ecosystem) carbon more than stem wood-only harvesting does. The loss may be reduced if high carbon storage in trees is maintained through the phase of commercial management, thus delaying a proportion of the economic returns to the final cutting. On the other hand, fertilization in the management phase is profitable in terms of future returns. Concurrently, fertilization increases the carbon stocks in stem wood and in litterfall. Carbon stock in soil is further increased due to the declining decay of soil organic matter. These dual impacts substantially improve the potential to increase carbon stocks in managed forest ecosystems in boreal forests (Fig. 14.2).

## 14.2 Carbon Dynamics Related to Forest Management and Timber Harvest

### *14.2.1 Forest Carbon Under Short-Term vs. Long-Term Management*

Management and harvesting cycles contain operative and strategic dimensions. In this context, sustainable management involves the long-term goal of obtaining timber and biomass while maintaining an effective carbon sink. Management and harvesting regimes can be divided into operations, or measures, for implementing the production cycle at the scale of the forest landscape or only at the scale of individual sites and tree stands. Regarding the dynamics of forest carbon, Hyvönen et al. (2007) divided the effects of management and harvesting into short- and long-term ones for identifying how timber and biomass production should be combined for mitigating climate change through one management cycle to the next one.

*In the short term*, human-induced disturbance events such as harvesting via clear-cutting and thinning and natural disturbances damage through fire, insect outbreaks, and windstorms, have a large effect on the carbon cycle of forest ecosystems (over years to decades) (Figs. 14.1 and 14.2). Such disturbance reduce the foliage of tree cover and kills trees and ground vegetation. This tends to turn a site into a temporary source of carbon dioxide due to the reduction of carbon uptake and increase of carbon loss through the decay of dead material. After any stand-replacing disturbance, such as clear-cutting, the consequent regrowth after regeneration and regaining of carbon takes several years, as found through eddy covariance measurements (e.g., Rannik et al. 2002). In boreal Scots pine (*Pinus sylvestris*) seedling stands, for example, about 20 years are needed to counterbalance ecosystem respiration and thus return a site with vegetation from a net source to a net carbon sink. The length of the regrowth phase is dependent on the reforestation and management of seedling stands prior to the thinning phase, with effects on the amount of carbon bound in soil and vegetation.

*In the long term*, the carbon lost in regenerative cutting (clear-cutting, shelterwood cutting, seed tree cutting) is replaced by the subsequent regrowth and development of established seedlings and recovery of ground vegetation (Figs. 14.1 and 14.2). However, the decay of harvesting residues tends to dominate the net ecosystem exchange during the early recovery phase. When the established seedlings are still small relative to ground cover species, the ecosystem respiration is dominated by the decomposition of harvesting residues. Later, heterotrophic respiration becomes more closely related to current production of trees and litterfall as a tree stand matures under even-aged management. Concurrently, the increasing interval between disturbance events may allow the progressive accumulation of carbon to continue. Management controls productivity by changing the competition through the seedling and thinning phases, thus modifying ecosystem carbon exchange (Hyvönen et al. 2007). As the vegetation recovers after a disturbance and the canopy closes, a transition from carbon source to carbon sink occurs at a rate determined by the productivity of trees and ground vegetation.

### *14.2.2 Forest Carbon Under Intensive vs. Extensive Management*

Ameray et al. (2021) analyzed how management practices and strategies affect carbon dynamics in boreal, temperate, and tropical forests in relation to the intensity of management. They divided the management into three classes based on intensity: (i) no management (old-growth forest), (ii) extensive management, and (iii) intensive management. In boreal forestry, no management refers to forests managed only for amenity, recreation, and conservation values. The difference between extensive and intensive management lies in the frequency and intensity of their management/harvesting operations in sites and tree stands through the production cycle.

In extensive management, the intensity, frequency, and management investment per unit land area are lower than in intensive management. Extensive forest management is an intermediate compromise between conservation and intensive forest management strategies. Extensive management includes thinning, selective cutting, and natural regeneration based on moderate removal intensity. On the other hand, intensive management includes a larger number of management operations: planting, use of improved seedlings, site preparation, fertilization, increased protection, and silvicultural treatments, with varying frequency and intensity over a relatively short rotation (Ameray et al. 2021). Intensive forest management enhances forest carbon sequestration capacity by using fast-growing species, mechanical soil preparation of low to moderate intensity, and fertilization. Prolonged rotation is also used

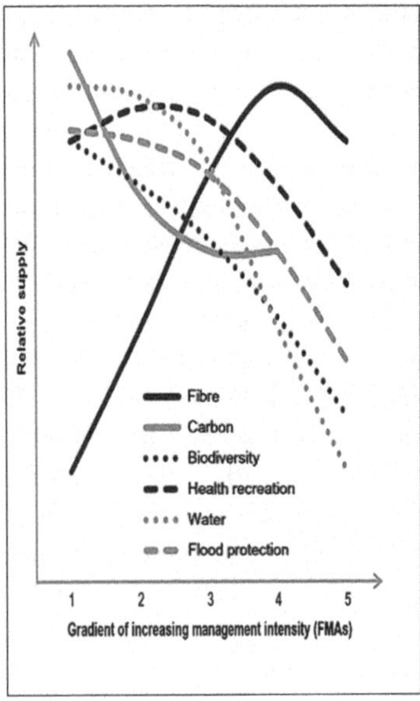


| Management intensity | Examples of management patterns |
|---|---|
| **FMA1: No management**<br>Unmanaged forest nature reserve allowing necessary infrastructure | Interventions limited to infrastructure for conservation and recreation |
| **FMA2: Low-intensity management**<br>Close-to-nature forestry, e.g., uneven-aged management using long selection intervals. Low thinning intensity in even-aged management and natural regeneration | Single tree and group selection, retaining older trees that partly compensate the low carbon stocks in seedlings and young trees dominating tree layers |
| **FMA3: Moderate-intensity management**<br>Management for joint production of several services, e.g., in even-aged forestry | Integrative production of timber and some other services, long rotation increases the carbon storage in forests |
| **FMA4: High-intensity management**<br>Management for production of single services in even-aged forestry | Management of even-aged cultures of selected tree species, small mixes of other species, stem-wood harvesting, reduced carbon storage |
| **FMA5: Highest-intensity management**<br>Short-rotation forestry for production of single services, increased carbon uptake | Management of even-aged cultures of selected tree species only for biomass, whole-tree harvesting, reduced carbon storage |

**Fig. 14.3** Management of forest carbon under intensive vs. extensive operations and strategies. Left: outlines of how management intensity affects the supply potential of selected services, as adapted from Sing et al. (2015, 2018). Permission of Oxford University Press. Right: management categories from no management to intensive management

to increase carbon stocks in ecosystems while still providing services like timber in a profitable way.

Similarly, Duncker et al. (2012) proposed five categories of management intensity, with a focus on the production of ecosystem services (Fig. 14.3): (i) no management (FMA1, natural forest), (ii) low-intensity management (FMA2, close-to-nature management), (iii) moderate-intensity management (FMA3, multiservice management), (iv) high-intensity management (FMA4, single-service management), and (v) highest-intensity management (FMA5, intensive single-service management using short rotations and high-productivity species). These classes align with those of Ameray et al. (2021): (i) *the class of old-growth forest* is equal to no management (FMA1); (ii) *the class of extensive management* is equal to low- and moderate-intensity management (FMA2 and FMA3); (iii) *the class of intensive management* is equal to high-intensity; (iv) and highest-intensity management (FMA4 and FMA5). Intensive management refers to practices that increase forest productivity and timber quality in the shortest period to meet industrial needs in a profitable way, as are widely used in commercial forestry.

The production/supply of different forest-based ecosystem services is related to the management intensity, ranging from no management to very intensive management (Fig. 14.3). In this context, no management (FMA1) refers to naturally growing and developing forests, which provide conservation and recreation services. Concurrently, intensive management (FMA4, FMA5) refers to forest plantations producing fiber. For services under the management of low/moderate intensity (FMA2–FMA3), carbon sequestration has a strong synergetic effect with the supply of timber and biomass. However, the strong constructive interaction between timber and carbon sequestration involves a strong trade-off in relation to biological diversity. At temporal and spatial scales, this trade-off implies the differentiation of ecosystem services in separate stands/forest landscapes through land use and management. Regarding land use, ecosystem services might be spatially optional, depending on the trade-off and synergistic effects between different services and the impacts of management on the structure and functioning of ecosystems.

## 14.3 Radiative Forcing Through Management Cycle

Regarding even-aged management, Table 14.2 shows the overall effects of management operations on radiative forcing through a chronosequence based on changes in carbon- and albedo-based forcing. They have opposite effects on the forest-based climatic benefits: an increasing ecosystem carbon increases cooling, while an increasing albedo reduces warming. The same management operations are applicable in even-aged and uneven-aged management. Under even-aged and uneven-aged management, carbon- and albedo-based forcing are influenced by the changes in the amount/properties of carbon in trees and ground vegetation and soil, as affected by thinning/selective cutting. The main difference between even-aged and uneven-aged management is regenerative cutting with soil management (prescribed burning, mechanical treatment), which is not used under uneven-aged management.

Concurrently, the increase in forest cover decreases albedo and increases forcing with warming under even-aged management, while forcing reduces related to cutting and harvesting (Table 14.2). On the other hand, reforestation and seedling establishment increase the carbon-based forcing with reducing warming while reducing the albedo-based forcing with increasing warming. This tendency increases or reduces with the growth and development of seedlings until the spacing and cleaning of the seedling stand and subsequent precommercial management. The dominance of biomass-based forcing with cooling is evident during the phase of commercial thinning. In this developmental phase, fertilization in combination with thinning phase increases further cooling effect due to increased tree growth and biomass and litterfall induced by nitrogen fertilization. Cooling effect is further increased due to reduced decay of soil organic matter under nitrogen fertilization, with a subsequent increase of carbon in soil. Throughout the management cycle, stem-only harvesting maintains/increases ecosystem carbon and forcing, while whole-tree harvesting tends to reduce ecosystem carbon and the subsequent reduction of cooling compared to the stem-only harvest.

Effects of individual management operations on ecosystem carbon have been comprehensively addressed in several reviews (e.g., DeLuca and Boisvenue 2012; Ameray et al. 2021; Mäkipää et al. 2023). Such studies help to experimentally identify the multiple effects of individual management measures on the structure

**Table 14.2** Effects of varying forest management operations on ecosystem carbon, with effects on changes in radiative forcing with impacts on cooling/warming based on carbon stocks and albedo through a rotation under an even-aged management cycle over a chronosequence

| Measures throughout production cycle | Effect on forcing based on ecosystem carbon | Effect on forcing based on albedo | Effect on combined radiative forcing |
|---|---|---|---|
| **Regenerative cutting**<br>• Clear-cutting<br>• Shelterwood cutting<br>• Seed-tree cutting | Reduces carbon in trees with increase/decline of carbon in ground vegetation and the soil surface, with reduction of carbon-based cooling | Increases albedo with decreasing warming due to increase of reflectance from the changing forest cover and ground vegetation cover | Reduces carbon and increases albedo with reducing combined forcing with warming due to reduction of carbon stocks in trees and ground vegetation and carbon loss from the soil surface |
| **Soil management**<br>• Prescribed burning<br>• Mechanical treatment | Burning reduces carbon in ground vegetation and soil surface, with further decay of carbon in disturbed soil with decreasing ground vegetation, with reduced carbon-based forcing | Reduces albedo, especially in burning, with increasing warming due to reducing ground vegetation and reducing reflectance with increasing albedo-based forcing | Reduces carbon, especially in burning and soil preparation, with effects on forcing linked with biomass and albedo related to reduction of ground vegetation and soil organic matter with reducing cooling |
| **Reforestation**<br>• Shelterwood cutting<br>• Seed-tree cutting<br>• Planting<br>• Seeding | Reduces carbon in mature trees due to regenerative cutting, initiating carbon sequestration for the next production cycle, effects lasting for decades, with increasing carbon-based forcing | Increases albedo following removal of trees due to the changing ground cover. Gradually reduced albedo due to the establishment and growth of seedlings, with gradual increase of albedo-based warming | Reduces forcing with the removal of trees and increases reflectance with changing ground cover. Gradual increase of forcing due to the establishment and growth of seedlings, gradual increase of combined forcing with cooling |
| **Management in seedling phase**<br>• Cleaning<br>• Spacing | Reduces carbon in seedling stands but increases carbon uptake in the remaining seedlings and in ground vegetation, with gradual recovery of carbon stocks with increasing cooling | Reduces albedo-based forcing with gradual increase of forcing for increasing growth of trees and ground vegetation and reducing reflectance in soil and trees | Reduction of carbon-based forcing due to decline of biomass but enhances growth of trees and ground vegetation, with gradual increase of biomass and decrease of albedo, with combined reducing of forcing with increasing cooling |
| **Management in precommercial phase** | Reduces carbon in seedling stands but increases carbon uptake in the remaining trees | Reduces albedo and increases forcing with the increase of growth in the remaining trees | Reduces carbon with cooling compensated for by increasing growth in the |

(continued)

**Table 14.2** (continued)

| Measures throughout production cycle | Effect on forcing based on ecosystem carbon | Effect on forcing based on albedo | Effect on combined radiative forcing |
|---|---|---|---|
| • Spacing<br>• Harvesting | and ground vegetation, with gradual decrease of carbon-based forcing with cooling | and ground vegetation and increase of albedo-based forcing | remaining trees and ground vegetation, with gradual increase of cooling and reducing albedo, with combined increase of cooling |
| **Management in commercial phase**<br>• Commercial thinning<br>• Fertilization | Reduces carbon in trees with increasing growth of remaining trees, thus increasing litterfall for fertilization and increasing soil organic matter, with increase of carbon-based cooling | Reduces albedo and increases warming due to increased growth of the remaining trees and ground vegetation, with increasing albedo-based warming | Reduces carbon and forcing but later increases forcing due to increased growth and litter from remaining trees and reduced decay of soil carbon with increasing cooling combined along with increase of albedo-based warming |
| **Rotation**<br>• Reduced<br>• Elongated | Reduction of rotation reduces the mean ecosystem carbon, whereas elongation of rotation reduces the growth of mature trees, with reduction/increase of carbon-based forcing with cooling/warming depending on rotation length | Reduction of rotation increases the reflectance of radiation with reduction of warming, whereas elongation of rotation has the opposite effects, with reducing/increasing albedo-based forcing with warming depending on rotation length | Reduction of rotation reduces the ecosystem carbon with reduction of cooling, whereas elongation increases total ecosystem carbon stock and reduces albedo with reduction/increase of combined forcing depending on the rotation length |

of forest ecosystems in the short term. However, there are still many open questions regarding the long-term effects of individual management measures, with cascading impacts on the long-term dynamics of forest ecosystem structure and functioning controlling ecosystem carbon stocks. This is among the main questions regarding how carbon in managed boreal forests is lost/emitted and taken up/sequestered, with impacts on carbon stocks and albedo in different developmental phases through management and harvesting cycles under even-aged and uneven-aged management. Moreover, this is especially important issue regarding even-aged management through forest stands forming chronosequences compiling any forest landscape, with effects of carbon- and albedo-based forcing on cooling/warming with links to management and harvest of forest resources. Under uneven-aged management, the interaction between forest and atmosphere is more stable, with small fluctuating in biomass between selective cuttings in removing trees and increasing regeneration and ingrowth of seedlings and young trees through dominating forest cover. In uneven-aged management, there seem to be less management options in controlling climatic warming than under even-aged management.

## 14.4 Concluding Remarks

The effects of management operations on radiative forcing through a chronosequence are based on the changes in carbon-based and albedo-based forcing. Under even-aged and uneven-aged management, carbon- and albedo-based forcing are influenced by changes in the amount/properties of carbon in vegetation and soil, as affected by cutting and harvesting in thinning/selection. The selective removal of trees and regeneration based on the remaining trees are combined under uneven-aged management, with a shorter treatment interval than under even-aged management. Under uneven-aged management, this does not facilitate high carbon accumulation in the remaining trees because large stocking of trees reduces the establishment and the growth of seedlings under the shading of overstory trees. In both management strategies, the long-term carbon stock depends on the frequency and intensity of thinning/selective cutting. The litterfall and soil carbon stocks under uneven-aged management are stable as affected by litterfall compared to that under even-aged management. However, carbon uptake and retention are dependent on how harvesting is implemented in terms of tree species preference, site fertility, and the management objectives emphasized in forestry with single or multiple dimensions.

## References

Ameray A, Bergeron Y, Valeria O, Girona MM, Cavard X (2021) Forest carbon management: a reports. Curr Forestry Rep 7:245–266

DeLuca TH, Boisvenue C (2012) Boreal forest soil carbon: distribution, function, and modelling. Forestry 85(2):162–184

Duncker PS, Barreiro SM, Hengeveld GM, Lind T, Mason WL, Ambrozy S, Spiecker H (2012) Classification of forest management approaches: a new conceptual framework and its applicability to European forestry. Ecol Soc 174:51

Hurmekoski E, Myllyviita T, Seppälä J, Heinonen T, Kilpeläinen A, Pukkala T, Mattila T, Hetemäki L, Asikainen A, Peltola H (2020) Impact of structural changes in wood-using industries on net carbon emissions in Finland. J Ind Ecol 24(4):899–912

Hyvönen R, Ågren GI, Linder S, Persson T, Cotrufo MF, Ekblad A, Freeman M, Grelle A, Janssens IA, Jarvis PG, Kellomäki S, Lindroth A, Loustau D, Lundmark T, Norby RJ, Oren R, Pilegaard K, Ryan MG, Sigurdsson BD, Strömgren M, van Oijen M, Wallin G (2007) The likely impact of elevated [$CO_2$], nitrogen deposition, increased temperature, and management on carbon sequestration in temperate and boreal forest ecosystems: a literature review. New Phytol 173:463–480

Mäkipää R, Abramoff R, Adamczyk B, Baldy V, Biryol C, Bosela M, Casals P, Dondini M, Filipek C, Garcia-Pausas J, Gros R, Göröryová E, Hashimoto S, Hassegawa M, Immonen P, Laiho R, Li H, Li Q, Luyssaert S, Menival C, Mori T, Naudts K, Santonja M, Smolander A, Toriyama J, Tupek B, Ubeda X, Verkerk PJ, Lehtonen A (2023) How does management affect soil C sequestration and greenhouse gas fluxes in boreal and temperate forests?—a review. For Ecol Manag 529:120637

Rannik Ü, Altimir N, Raittila J, Suni T, Gaman A, Hussein T, Hölttä T, Lassila H, Latokartano M, Lauri A, Natsheh A, Petäjä T, Sorjamaa R, Ylä-Mella H, Keronen P, Berninger F, Vesala T, Hari P, Kulmala M (2002) Fluxes of carbon dioxide and water vapour over scots pine forest and clearing. Agric For Meteorol 111:187–202

Sing L, Ray D, Watts K (2015) Ecosystem services and forest management. In: Research note FCRN020. Forestry Commission, Bristol

Sing L, Metzger MJ, Paterson JS, Ray D (2018) A review of the effects of forest management intensity on ecosystem services for northern European forests with a focus on the UK. Forestry 91:151–164

Vuokila Y (1980) Metsän kasvatuksen perusteet ja menetelmät. WSOY, Porvoo

# Appendix

Units and conversions commonly used in analyzing the carbon dynamics in forest ecosystems with producing forest biomass for timber and energy and mitigating climatic warming.

**Prefixes and units**

| Prefix in SI system | Explanation | Value |
|---|---|---|
| T | tera- | $10^{12}$ |
| G | giga- | $10^{9}$ |
| M | mega- | $10^{6}$ |
| k | kilo- | $10^{3}$ |
| h | hecto- | $10^{2}$ |
| d | deci- | $10^{-1}$ |
| c | centi- | $10^{-2}$ |
| m | milli- | $10^{-3}$ |
| μ | micro- | $10^{-6}$ |
| n | nano- | $10^{-9}$ |
| p | pico- | $10^{-12}$ |

**Energy**

| Unit | Transformations |
|---|---|
| J | $1\ J = 1\ N{\cdot}m = 1\ kg \cdot m^2 \cdot s^{-2} = 1\ W{\cdot}s = 0.239\ cal = 10^7\ erg$ |
| W·h | $1 W{\cdot}h = 3.6\ kW \cdot s = 3.6\ kJ = 0.86\ kcal$ |
| MJ | 1 MJ = 0.278 kWh |
| cal | 1 cal = 4.1868 J |
| kcal | $1\ kcal = 1.163\ W \cdot h$ |

 
S. Kellomäki, *Forest Management for Timber Production and Climate Change Mitigation*, Managing Forest Ecosystems 44,
https://doi.org/10.1007/978-3-031-71575-4

## Pressure

| Unit | Transformations |
|---|---|
| MPa | 1 MPa = $10^6$ Pa = 10 bar |
| bar | 1 bar = $10^5$ N · $m^{-2}$ = $10^5$ Pa = 100 J · $kg^{-1}$ = $10^6$ erg · $cm^{-3}$ |
| bar | 1 bar = 750 Torr = 0.9869 atm |
| atm | 1 atm = 1.0132 bar = 760 Torr |

## Amount and concentration

| Unit | Transformations |
|---|---|
| Molarity | mol · $kg^{-1}$ of liquid |
| Ppm | 1 ppm = $10^{-6}$ mol · $mol^{-1}$; 1 μg · $g^{-1}$; 1 μl · $l^{-1}$ |
| Ppb | 1 ppb = $10^{-9}$ mol · $mol^{-1}$; 1 ng · $g^{-1}$; 1 nl · $l^{-1}$ |
| Ppm | 1 ppm $CO_2$ = 1.82 mg · $m^{-3}$ = 41.6 μmol · $m^{-3}$ = 0.101 Pa (at the temperature of 20 °C and pressure 101.3 kPa) |

## Radiation and energy

| Transformations |
|---|
| 1 W · $m^{-2}$ = 1 J · $m^{-2}$ $s^{-1}$ = 31.53 MJ · $m^{-2}$ · $a^{-1}$ |
| 1 mol photon = 1.8 · $10^5$ J (when λ 650 nm) . . . 2.7 · 105 J (when λ 450 nm) |
| 1 cal · $cm^{-2}$ · $min^{-1}$ = 6.98 · $10^2$ W · $m^{-2}$ = 6.98 · $10^5$ erg · $cm^{-2}$ · $s^{-1}$ |
| 1 erg · $cm^{-2}$ · $s^{-1}$ = 1.43 · $10^{-6}$ cal · $cm^{-2}$ · $min^{-1}$ = $10^{-3}$ W · $m^{-2}$ |

## Gas exchange

| Transformations |
|---|
| 1 g $CO_2$ (exchange) ≈ 0.73 g $O_2$ (exchange) |
| 1 g $O_2$ (exchange) ≈ 1.38 g $CO_2$ (exchange) |
| Diffusion $D_{CO2}$ = 0.64 $D_{H2O}$ |
| Diffusion $D_{H2O}$ = 1.56 $D_{CO2}$ |
| 0.03 $\%_{vol}$ $CO_2$ = 300 μ · $l^{-1}$ = 282 μbar = 28 Pa $CO_2$ (partial pressure) |
| 1 μl · $l^{-1}$ = 1.963 μg $CO_2$ · $l^{-1}$ (at a pressure of 1013 mbar and temperature 0 °C) |
| 1 mg $CO_2$ · $dm^{-2}$ · $h^{-1}$ = 0.028 mg $CO_2$ · $m^{-2}$ · $s^{-1}$ = 0.63 μmol $CO_2$ · $m^{-2}$ $s^{-1}$ |
| 1 mg $CO_2$ · $m^{-2}$ · $s^{-1}$ = 36 mg $CO_2$ · $dm^{-2}$ · $h^{-1}$ = 22.7 μmol $CO_2$ · $m^{-2}$ · $s^{-1}$ |
| 1 μmol $CO_2$ · $m^{-2}$ · $s^{-1}$ = 0.044 mg $CO_2$ · $m^{-2}$ · $s^{-1}$ = 1.58 mg $CO_2$ · $dm^{-2}$ · $h^{-1}$ |
| 1 mg $H_2O$ · $dm^{-2}$ · $h^{-1}$ = 1.54 μmol $H_2O$ · $m^{-2}$ · $s^{-1}$ |
| Conductance (at temperature of 20 °C and pressure 101.3 kPa) |
| 1 cm · $s^{-1}$ ≈ 0.416 mol · $m^{-2}$ · $s^{-1}$ |
| 1 mol · $m^{-2}$ · $s^{-1}$ ≈ 0.024 mm · $s^{-1}$ |

## Biomass

| Transformations |
| --- |
| 1 g DM · $m^{-2}$ = $10^{-2}$ Mg · $ha^{-1}$ |
| 1 g DM ≈ 0.42 - 0.51 g C ≈ 1.5 – 1.7 g $CO_2$ |
| 1 g C ≈ 2 – 2.22 g DM ≈ 3.1 – 3.4 g $CO_2$ |
| 1 g $CO_2$ ≈ 0.59 – 0.66 g DM ≈ 0.27 – 0.30 g C |
| 1 g $CO_2$ = 3.67 g [=44/12] C |
| Biomass = Volume [$m^3$] x Density of mass [kg $m^{-3}$] |
| Carbon content in stem wood [kg C $m^{-3}$]:<br>Scots pine 0.3091; Norway spruce 0.3715; birch 0.4152 |

*DM* Dry Mass

**SPRINGER NATURE**

# GPSR Compliance

*The European Union's (EU) General Product Safety Regulation (GPSR) is a set of rules that requires consumer products to be safe and our obligations to ensure this.*

*If you have any concerns about our products, you can contact us on ProductSafety@springernature.com*

In case Publisher is established outside the EU, the EU authorized representative is:

Springer Nature Customer Service Center GmbH
Europaplatz 3
69115 Heidelberg, Germany

The manufacturer's authorised representative in the EU is Springer Nature Customer Service Centre GmbH, Europaplatz 3, 69115 Heidelberg, Germany. If you have any concerns regarding our products, please contact ProductSafety@springernature.com

Printed and bound by CPI Group (UK) Ltd, Croydon, CR0 4YY
30/09/2026
02225598-0011